Science Education

NEW DIRECTIONS IN MATHEMATICS AND SCIENCE EDUCATION

Volume 31

Scope

Mathematics and science education are in a state of change. Received models of teaching, curriculum, and researching in the two fields are adopting and developing new ways of thinking about how people of all ages know, learn, and develop. The recent literature in both fields includes contributions focusing on issues and using theoretical frames that were unthinkable a decade ago. For example, we see an increase in the use of conceptual and methodological tools from anthropology and semiotics to understand how different forms of knowledge are interconnected, how students learn, how textbooks are written, etcetera. Science and mathematics educators also have turned to issues such as identity and emotion as salient to the way in which people of all ages display and develop knowledge and skills. And they use dialectical or phenomenological approaches to answer ever arising questions about learning and development in science and mathematics.

The purpose of this series is to encourage the publication of books that are close to the cutting edge of both fields. The series aims at becoming a leader in providing refreshing and bold new work—rather than out-of-date reproductions of past states of the art—shaping both fields more than reproducing them, thereby closing the traditional gap that exists between journal articles and books in terms of their salience about what is new. The series is intended not only to foster books concerned with knowing, learning, and teaching in school but also with doing and learning mathematics and science across the whole lifespan (e.g., science in kindergarten; mathematics at work); and it is to be a vehicle for publishing books that fall between the two domains—such as when scientists learn about graphs and graphing as part of their work.

Science Education

An International Course Companion

Edited by

Keith S. Taber
University of Cambridge, UK

and

Ben Akpan
Science Teachers Association of Nigeria, Nigeria

SENSE PUBLISHERS
ROTTERDAM/BOSTON/TAIPEI

A C.I.P. record for this book is available from the Library of Congress.

ISBN: 978-94-6300-747-4 (paperback)
ISBN: 978-94-6300-748-1 (hardback)
ISBN: 978-94-6300-749-8 (e-book)

Published by: Sense Publishers,
P.O. Box 21858,
3001 AW Rotterdam,
The Netherlands
https://www.sensepublishers.com/

Printed on acid-free paper

TABLE OF CONTENTS

PREFACE

Science education is a relatively broad and dynamic area. The premium which is being placed by humanity on science, technology, and engineering as pivots for future growth and development means that science education will, well into the future, continue to play a significant role in advancing the frontiers of development efforts. Incidentally, while various books abound in support of science education programmes, there is a dearth of books that are concise and at the same time written to cover much of the contents of science education programmes at undergraduate and postgraduate levels. This book, *Science Education: An International Course Companion*, is intended to fill this gap providing introductory readings on topics relevant to both undergraduate and post-graduate courses in science education. The chapters cover the major course offerings in science education globally. The content is therefore suitable for supporting the implementation of various national curricula such as the Next Generation Science Standards (NGSS) in the USA. The authors are drawn from various countries, and indeed continents, thus making the book uniquely international both in content and authorship. The conciseness of the readings in the book with each chapter being limited to about 5,000 words in length but at the same time containing sufficient material for undergraduate and post-graduate programmes makes it a truly comprehensive companion for science education. In terms of pedagogy, chapters in the book are arranged in sequence in line with science education programmes globally. There is an introductory chapter followed by 40 other chapters which are arranged in groups of 3–8 around nine themes or sections. Within each theme, there is generally a sequencing that ensures a succeeding chapter is built on the previous one.

The book opens with an introduction to science education as a scholarly field. This is followed by a group of five chapters in Section I which focuses on the nature of science and science education and provides a grand reflection of the nature of science including beliefs, epistemology, and interdisciplinarity. Section II explores thinking and learning in science education through a discussion of the learning theories, scientific reasoning, nature of student conceptions and the role of intuition and insight in science education. The science curriculum – its development and implementation, worldwide initiatives, and integration with mathematics and engineering – is the key idea coming through in Section III. Section IV examines science teaching – instructional practices, inquiry teaching, models and modelling, context-based teaching, assessment, and challenges faced in teaching biology, chemistry, and physics. Resources for science teaching are the focus of Section V which discusses role of laboratory, practical work, emerging technologies, and 21st century skills. The importance of informal science education is highlighted in Section VI through a discussion of educational visits and public understanding of science. Section VII explores equity, indigenous and gifted learners, and sustainable development within the thematic framework of inclusive science education. The theme of the last section of the book is science teacher education. This section highlights pedagogical content knowledge, teacher preparation, research perspectives and skills, further professional development, and role of science teacher associations in science education.

We commend this book for use by undergraduate and post-graduate students in science education and their teachers as well.

Ben Akpan

INTRODUCTION

KEITH S. TABER

1. SCIENCE EDUCATION AS A FIELD OF SCHOLARSHIP

SCIENCE EDUCATION PRACTICE AND RESEARCH

Science education is a key area of activity internationally. Science education is a major field of practice, with science (and individual science disciplines) being taught and learnt at various levels, both formally (for example in schools) and through more informal approaches (such as the learning that takes place when people visit science museums) all around the world. In most countries, science is seen as a key component of schooling, and higher education in science subjects is usually considered of major importance for meeting societal needs such as ensuring the 'supply' of scientists, engineers and other professionals working in scientific fields and for ensuring sustainable economic development.

This major field of practice is supported and explored through the academic study of science education. Science education research (SER) is a well established, major area of research (Fensham, 2004) that can inform the practice of science education. This present chapter considers the nature of science education as a scholarly field: one that is both about, and looks to inform, the practice of science education. The subsequent chapters in this international companion to science education offer introductions to some of the key research areas in science education that can inform practitioners (such as classroom teachers and college lecturers, education officers of scientific societies, outreach officers in university science departments, educators working in museums and science centres, subject officers in examination and curriculum authorities, and so forth) in their work. In many of the chapters authors draw upon research to make specific suggestions for effective teaching practice.

The Focus of Science Education as a Field of Scholarship

Research is the process of developing new public knowledge: knowledge that helps us understand phenomena better, and – hopefully – informs actions in the world that are more successful in achieving our goals. *Science education*, as a research field is therefore concerned with developing knowledge about the learning of science, and the teaching of science. This knowledge helps us better understand phenomena such as:

K. S. Taber & B. Akpan (Eds.), Science Education, 3–19.

- why some science topics are usually considered more difficult than others...
- ...and why some students do not experience the same learning difficulties as their classmates;
- why students often misunderstand some science concepts;
- why some classroom activities undertaken in science lessons are more motivating than others;
- the different ways teachers can organise practical work;
- how science textbooks influence teaching approaches in some countries;
- how teaching models and analogies influence students' developing understanding of science concepts.

Science teachers should not only know about the findings of pedagogical research, but have some insight into the processes by which such knowledge is constructed. This chapter explains how research in science education is similar to, and yet different from, research in science, and outlines some key ideas used to discuss educational research and the common approaches and tools used.

EDUCATIONAL RESEARCH AND SER

Educational research is primarily concerned with the processes of teaching and learning (Pring, 2000), and SER is a subset of educational research where the focus of what is being learnt and/or taught is science curriculum content. That may seem to suggest that science education as a field is little more than the application of more general research in education to science learning contexts. However, that would be too simplistic an assumption. Teaching and learning are very complicated processes, and because they are complex there are many aspects of teaching and learning that we cannot simply treat generically. That is, we cannot simply say that (for example) learning happens 'this' way and teaching is best done 'that' way. As researchers are dealing with complex systems, context is often very important. That context can involve many components.

So it may make an important difference whether the class consists of 6–7 year olds or 13–14 year olds; it may matter whether the class is mixed-ability or a 'set' or 'stream' of similar ability students; it might make a difference whether the language of instruction is English, or French, or Japanese or some other language (as different languages offer different resources for communication and learning); and it often makes a substantive difference whether the subject of a class is romantic poetry; the causes of the industrial revolution; trigonometry; redox reactions; reproduction in flowering plants; electromagnetic induction; or the merits of nuclear power supply. It may also make a difference whether the class is taking place in a well designed and well resourced classroom in a peaceful and wealthy society or in a hastily put together shack, with its corrugated iron roof noisily vibrating in the wind, in a poor, war-ravaged country. It makes a difference whether the teacher has minimal qualifications, or has earned a subject specialist degree and post-graduate teaching qualification. The reader can readily see that these examples can be multiplied many times.

Research can inform all such contexts – but application will not be uniform in every context, and research carried out in one place can not always be assumed to tell us much about a very different context. Some principles are generally applicable – but how they can be applied may still vary according to teaching and learning context. What works with a class of 15 high-aspiring, self-assured, fee paying students in an elite selective school may not work with a class of 40 students from an area of poverty where the children are under-fed, and need to fit school around working to help their families. It cannot be assumed that a pedagogy that has been shown to be effective when employed by a highly qualified, highly skilled, experienced teacher who has been trained in the particular techniques will work when implemented by all other teachers. For that matter, an innovative approach may sometimes be more effective when used by an enthusiastic novice teacher committed to the pedagogy than when employed by a much more experienced teacher who already 'knows' it is not going to be successful.

Levels of SER

Curriculum area is then one aspect of teaching and learning context that can make a difference. Because of this, much general educational research that has been carried out in other subject teaching contexts *may be* relevant to science teaching. Yet this cannot be assumed to be so: it is often necessary to test how theories, principles and recommendations for practice deriving from other areas of research actually apply in science teaching and learning contexts. It is also the case that the unique features of science as an area of human activity lead to particular pedagogic issues that may not arise in other curriculum areas – so science education has its own specific foci and emphases. An obvious example is that of practical work in science and the science teaching laboratory which is a particular focus that only arises in science teaching. A less obvious example perhaps, but an important one nonetheless, is the extent to which learners who come to science classes often have existing alternative conceptions of science topics which are often inconsistent with the ideas they are expected to learn.

It has been suggested (Taber, 2013) that educational research carried out in science teaching and learning contexts can be considered to fit one of three levels of SER (see Table 1). This typology is of course just a model, and it is not suggested that all studies obviously and unambiguously fit into one of the categories. However, it may be a helpful way of thinking about SER.

In this scheme some research carried out in science classrooms (labelled collateral SER) is about general educational issues, and the choice of a science teaching and learning context is often little more that a convenience. Imagine a researcher interested in a general educational question who asks teachers at a school to volunteer to take part in a study. Perhaps the science teacher is interested in taking part, and

Table 1. Research carried out in science learning contexts varies in the extent to which it is specific to science education

Level of SER	*Focus*	*Examples*	*Comment*
Collateral	General educational issue	Does how long a teacher waits after asking a question, before selecting a student to respond, influence the quality of response? Can peer tutoring be used as a means of challenging the most able learners in a class?	These are relevant to science teaching, but findings are likely to apply to other curriculum subjects just as much.
Embedded	Wider educational issue to be understood within the context of science teaching and learning	How can dual-encoding theory (about the cognitive processing of verbal and visual information) support learning about the circulatory system? How can tasks with higher order cognitive demand be incorporated into studying the classification of living things?	These are principles relevant to teaching across the curiculum, but where application needs to be related to the specifics of disciplinary subject matter.
Inherent	Issue arising from the specific of science teaching and learning.	Do students appreciate the affordances of chemical equations in linking between laboratory phenomena and submicroscopic models of matter? What are pupils' moral and aesthetic responses to dissection in school science?	These are questions that only arise in science education, as they relate to something specifically about subject content.

the history teacher is not. In another school it may have been the geography teacher or maths teacher who offered to help. Surveys that are carried out in large numbers of classrooms may include some science classes as part of a more diverse sample. It may be then that there can then be some analysis by subject – but only when the sample size and method used to build the sample allow meaningful comparisons of that type.

Embedded SER

Other research may explore more general educational issues, but in such a way that the location of the research in the science class is a principled choice. So there are many educational theories, principles and ideas which are quite general, but which could be tested in regard to their value in specific (science) teaching and learning contexts. So the notion of 'scaffolding' learning is a general idea about how to support learners in gaining competence in new areas – but the idea needs to be operationalised in particular contexts. We might ask how to effectively scaffold

- learning about the theory of natural selection;
- learning to undertake the calculations needed when using titrations in quantitative chemical analysis;
- learning how to test the principle of conservation of momentum in the school laboratory.

and so forth.

Another example is the general principle that teachers should include tasks requiring higher order cognitive skills (such as synthesis and evaluation) in their lessons. SER might look at what that means when teaching 7 year olds about types of animals, or 13 years olds about the periodic table, or 18 year olds about nuclear reactions, or setting up a museum exhibit about the extinction of the dinosaurs.

There are many such areas of general educational theory that are believed to apply widely in teaching but where more specific research (embedded SER) is needed to see how they might be best applied in particular science teaching contexts. However there are also issues that arise specifically from the teaching and learning of science that would not be directly relevant elsewhere in the curriculum. Thus some studies in science teaching and learning contexts can be considered inherently *science* education studies (inherent SER).

Inherent SER

For example, there are many common alternative conceptions about science topics that learners bring to science classes. Most people have an intuitive notion of how forces are related to the motion of objects. These ideas are often very similar, and are at odds with the formalism of physics taught in school science and college physics courses. Research suggests that not only do pupils often enter school with these ideas already established, but they very often retain the ideas despite teaching. That presents a very specific (but rather important) science teaching challenge, that is reflected in many science topics in the curriculum. This has motivated a great deal of research about the nature of learners' science ideas (see Chapter 9: '*The Nature of*

Student Conceptions in Science'), how these might change over time, and the kinds of teaching able to bring about desired conceptual change.

Another major issue in science education concerns how to organise science practical work such that it is a 'minds-on' and not just a 'hands-on' activity (see Chapter 29: *'Minds-on Practical Work for Effective Science Learning'*) as often during school laboratory work students are too busy thinking about the organisation and safe manipulation of materials to actually reflect very much on the purpose of the activity and the theoretical significance of their observations. This has become a major focus of concern in science education.

Other examples of foci of inherent SER would be

- how to teach learners about the nature of scientific ideas such as theories, principles and laws – for example, that theories can be well-evidenced, robust, explanatory schemes, and yet should be considered as provisional, theoretical knowledge;
- how to teach students about the affordance of particular representations such as chemical equations, and circuit diagrams, or the kinds of 'tree' diagrams used in cladistics;
- how to best teach evolutionary ideas to students from communities which reject evolution on religious and cultural grounds;
- how can teaching models of energy be developed which offer authentic reflections of the scientific concept, but are not too abstract for lower secondary school students to engage with…

Issues such as these arise from the specifics of teaching and learning science, and so are intrinsic to the professional concerns of science teachers, whilst being of little direct relevance to teachers in other curriculum areas.

SCIENCE TEACHER CLASSROOM RESEARCH

It is the existence of topics motivating inherent SER that supports the existence of a distinct scholarly field of science education. Such a field is both dynamic and permeable. It is dynamic because active topics of interest change over time (Fensham, 2004; Gilbert, 1995), and it is permeable both in terms of ideas and scholars. Ideas from many academic fields become adopted and adapted in SER, and ideas first developed in the field may come to be used more widely. Some people who contribute to SER spend their entire careers doing so. Others may shift into or out of the field to and from cognate areas (such as educational psychology, sociology of education, mathematics education). Some researchers continue to work in the field in parallel with working on projects in other areas. There are also subfields, such as physics education research, chemistry education research, biology education research, astronomy education research, etc., within SER, and a wider field of 'STEM educational research' which encompasses issues of interest across teaching subjects such as science, technology, engineering and mathematics (often collectively known as STEM).

A very important, if less obvious, part of the field comprises of classroom teachers who carry out research motivated by issues arising in their own science teaching practice. There are barriers to teachers and other professionals making major contributions to the field. An obvious one is time, as busy teachers often have very little capacity to undertake research on top of their teaching and pastoral duties. Teachers may also find it difficult to access research literature, much of which is behind paywalls: although increasingly material is available through open or free access. Another potential barrier is lack of knowledge and skills in research methods – although this is now often recognised as something that should be included in science teacher education.

Furthermore much research carried out by teachers is context-directed. That means that rather than seeking to explore a major theoretical issue in the field that might apply widely, the teacher is primarily concerned to address an immediate problem or issue in practice: why do *these* students not 'get' conservation of energy? Why is *this* class not motivated to study molluscs? How can I teach *my* students about climate change in a way that is informative, but allows them to make up their own minds about what needs to be done?

Often the teacher's issue or problem would be shared by many others around the world, and perhaps their solution may work elsewhere – but their motivation and focus is appropriately on changing their own practice rather than claiming a new generalisable theory or approach. Often very useful teacher research is of the form (Whitehead, 1989):

a. recognising I am not happy about some aspect X of the science teaching and learning here;
b. finding out what existing research suggests might work;
c. trying out some promising ideas;
d. finding something that seems to work better to incorporate in classroom practice from now on.

It may be that much context-directed research of this kind offers little that is new to inform other teachers, but it can still make a real difference to students in the specific research context. Some research undertaken by science teachers in their classroom however does offer new ideas and deserves to at least be reported to colleagues at science teacher association meetings and in practitioner journals (such as *School Science Review*) or more specific disciplinary teaching magazines (such as *Education in Chemistry*).

WHAT MAKES A FIELD?

A scientific or academic field is an area of activity having a sufficient level of organisation and coherence to have become widely recognised as an entity. So the field of science education is a social construct – it exists to the extent that enough people considered to be authoritative about such matters (which is itself a matter of social agreement) consider it exists. If a scholarly field is a social construct, it does not exist in the same way that perhaps a tree, or a mountain or the planet jupiter exist. These things exist as natural objects independently of whether people know of them and how they think about them. Social constructs depend on a kind of social consensus that could disappear.

Moreover many of the *objects* of research in education are also social constructs that do not have an independent physical existence. Think about such notions as exemplary teaching, an orderly classroom, a productive group discussion, effective learning, a good classroom environment or a naughty child. Such things 'exist' but how they are understood depends upon the perceptions of those who construe them in particular ways. The child who interrupts learning activities to ask awkward questions may be seen as disruptive and naughty, or as gifted and inquisitive, perhaps depending on cultural or institutional norms. That is, the same child, doing the same things, could be construed very differently. A school science lesson where all the children sit quietly in rows and write down what the teacher says would in some cultural contexts (in particular places and at particular times) be seen as a good lesson, when at other times and in other places this would be seen as unsuitable for promoting effective learning. In some contexts such a classroom would be assumed to be demotivating for the children – but that need not necessarily be the case, as this depends upon children's expectations and the norms they have assimilated.

All research has to engage with issues of ontology (the nature of things that we might investigate), epistemology (how we come to knowledge about those things) and axiology (the values that inform our choices of action in the world). This applies to research in natural science as much as in science education (see Table 2), although because science disciplines are often organised into well established research traditions (Kuhn, 1996) the novice researcher gets inducted into the shared assumptions of those already working in the field and comes to take much for granted. In educational research these issues tend to be more often explicitly discussed when writing about research.

As research is progressive (that is as what is found out in research today suggests questions for further enquiry), questions that may be considered things to think about when designing research today, may have themselves been the subject of enquiry previously. The ontological and epistemological questions in Table 2 are matters that can be informed by empirical research (and the axiological questions may be explored through philosophical enquiry). To adopt these technical terms, a reader might pose the question that: whilst it might seem a good idea that there should be a field of research to support science teaching (an axiological consideration), if research fields have the rather tenuous status of being socially constructed (an ontological claim), how can we know it is reasonable to talk of there really being a field of science education (an epistemological matter)? This is a reasonable

Table 2. Research (in science, and in science education) raises questions about (a) the nature of things we might enquire into; (b) how robust knowledge can be constructed; and (c) how we should act in the world

Type of issue:	*Relates to:*	*Examples from science*	*Examples from education*
Ontological	The nature of things	How should an acid be understood (e.g. Brønsted-Lowry, Lewis model)? Are atomic orbitals real or just useful fictions? What is dark energy?	What is learning? How is knowledge represented and structured in a person's mind? What is effective teaching?
Epistemological	How we know things	Does preparing samples for observation under the electron microscope modify the sample and change what is there to be seen? How can we deduce the existence of very short-lived particles from patterns seen in high energy collider detectors? How can we identify functional groups in organic chemicals from absorption patterns in spectra?	What counts as evidence of learning? When do researcher beliefs and expectations get communicated to teachers and learners and influence research outcomes? Are classrooms that are observed by researchers changed by the presence of the observer?
Axiological	Acting in ways informed by our values	Is it right to induce or breed disease in animals for purposes of medical research? Should scientists undertake research intended to develop lethal weapons? When is it fair to study participants to run clinical trials on unproven drugs to test their efficacy?	When is it fair to teach children using unproven techniques/resources to test their efficacy? How much do we need to tell people about a study to consider they have given us informed consent? Is it fair to use classes as controls in research when they cannot benefit from the intervention being tested?

question, but there is a good deal of evidence that the SER field is a meaningful and useful construct for making sense of much scholarly activity (Fensham, 2004). For example, we might take the following as useful indicators:

- There are national and international associations concerned with science education;
- People can take degrees in science education – including both taught and research—based postgraduate ('higher') degrees;
- There are specified posts in the field (such as lecturer in science education; professor of science education) and sometimes university departments of science education;
- There are national and regional research associations specifically concerned with SER (for example the *European Science Education Research Association*; the *Australasian Science Education Research Association*; and the *National Association for Research in Science Teaching* based in the United States);
- There are quite a number of research journals in science education, and a major journal devoted to publishing reviews of research in the field (see the further reading at the end of this chapter);
- Major publishers produce book series in science education;
- There are regular national and international conferences in science education;
- There are established 'handbooks' as key reference works in the field (see the further reading at the end of this chapter).

Such indicators show that much scholarly activity is commonly recognised as an entity, SER. Moreover, just as in the natural sciences (Lakatos, 1970), it is possible to find evidence of specific research programmes within science education (Erickson, 2000; Gilbert, 1995; Solomon, 1993; Taber, 2006) where research topics are addressed over series of studies that adopt the same set of starting points, and build iteratively on each other.

HOW IS RESEARCH CARRIED OUT IN SCIENCE EDUCATION?

According to the historian of science Thomas Kuhn, science is characterised by communities which share a 'disciplinary matrix' – which includes such features as common assumptions, key discipline-specific exemplars, theoretical, methodological and analytical tools. Scientists working in the same field and within related research programmes are likely therefore to agree on core concepts for making sense of the field, general experimental or observational approaches, the kind of instrumentation that is useful and how to analyse data sensibly. So particular techniques (e.g., electron microscopy, high energy particle colliders; genomic sequencing; PET scans, etc) may tend to become recognised as standard approaches in a field. Knorr Cetina (1999) has characterised the very different ways that scientists work in the two different fields of molecular biology and high energy physics – where the different nature of the subject matter and the consequent epistemological challenges lead to different ways of organising laboratories (and even understanding what a laboratory is), as well as distinct sets of core concepts, core assumptions, and core techniques.

Science education is not as finely structured as many areas of the natural sciences. Often problems are not as well defined, and researchers may work across a range of problems and kinds of research questions. A wide range of methods are used in science education (National Research Council Committee on Scientific Principles for Educational Research, 2002), and because the phenomena being studied are often complex, it is sometimes considered that methodological pluralism and/or analytical pluralism is needed to address problems – so complementary methods of data collection (National Research Council Committee on Scientific Principles for Educational Research, 2002) and analysis (Taber, 2008) may be used in the same study.

Experimental Research in Education

Some SER uses experimental methods. A common (but often quite weak) form of educational experiment involves teaching two parallel classes by different methods to see which approach produces the better educational outcomes. There are a number of serious problems with this kind of research design. For one thing there are so many uncontrolled and often unknown variables. Any experienced teacher knows that parallel classes can be very different to teach as every student is unique and classes behave in ways that depend upon the complex interactions between the individual students. Moreover teachers with similar qualifications and levels of teaching experience cannot be assumed to be 'equivalent'. Having the same teacher teach both classes does not solve this challenge as few teachers are equally as skilled at (or committed to) several different teaching approaches. Even such factors as the time of day when classes are scheduled (which may be different for the classes) or the teaching rooms used (different levels of light, noise, arrangement of resources, etc) can make a difference – although such details are seldom reported in studies.

To some extent these problems can be reduced if research is carried out on a large scale, with many classes randomly assigned to the two different conditions, as most of the incidental factors will (probably) largely cancel out if the sample of classes is big enough. But such studies are difficult to organise (and expensive to resource). They still can not allow for such factors as teacher and student beliefs about which approach is better (expectancy effects that can influence outcomes) or problems of allowing for the novelty of an innovation – which might be motivating for students, or may sometimes seem to threaten familiar routines.

Another problem with many studies evaluating new teaching approaches, curriculum or resources is the tendency of teachers to need to have taught something new several times through with different classes before they become proficient and outcomes reflect the full potential of the approach. Often studies are comparing teachers trying something new with teachers carrying out their normal practice – carefully honed teaching routines. Teaching and learning are complex, and teachers refine their skills over time based on – sometimes subtle – feedback on how students respond to different approaches, sequences, models, activities etc. Teachers also develop a better understanding of pedagogies as they reflect on trying things out with a variety of classes, and gradually come to optimise a fit between teaching style; personal repertoire of anecdotes, examples, and teaching skills; pedagogy; local resources; and student characteristics. Evaluations of innovations then often suffer from either being enhanced by the enthusiasm of pioneers, or being handicapped by the unease of seasoned professionals moving out of their comfort zone to work in ways they have not yet been able to practise and mould.

There are also ethical issues to be considered. Children (and teachers) cannot be assumed to be available as research subjects. Rather, participants must offer informed consent to be part of a research project, especially where their talk or written work may be published. Usually when children are involved parental permission is needed before children can be closely observed or interviewed or tested for research purposes.

A particular problem occurs when researchers claim to be comparing some innovative approach (constructivist/progressive/reform-based teaching) with 'traditional' teaching – if traditional teaching involves asking teachers to teach in ways research already suggests are less effective. This may sound fanciful, but it is not unusual to see studies where it is reported that the control class did not involve discussion work, or access to multimedia resources, or hands-on practical work, etc., but was restricted to listening to a teacher, reading a text book, and making notes or answering written questions. Perhaps sometimes that genuinely is the typical practice in the research context and so provides a fair comparison condition – but it is not acceptable to set up such a situation in the hope of showing that an alternative approach is more successful, when there is already a great deal of evidence suggesting that this will be the case. This can be avoided by testing innovative approaches against pedagogy that is already recognised to be effective. Rather than ask 'is this approach better than a stereotype of unsatisfactory outdated pedagogy?' the question becomes 'how well does this innovative approach compare with other pedagogies known to be effective?'

Teachers testing out new ideas in their own classrooms often adopt an 'action research' approach where new ideas are tried out, and carefully evaluated, then modified if need be, then evaluated again (as many times as seems appropriate) to inform future practice. The lack of a control or comparison condition limits the inferences that can be drawn from this kind of research, but the teacher-researcher is aware of the provisional nature of their findings, and will not assume that what has worked well with one class will always work well in the future. The attitude here is to adopt evidence-based practice, but to always be open to collecting further evidence. As in science itself, we should always be open to new information that may change our minds.

There are approaches, such as design research and lesson study, that allow teachers to work together, sometimes with specialist researchers, to test out teaching approaches and resources that may be generalisable across classes – at least within certain bounds (in teaching a certain topic; in working with pupils of a certain age;

etc.). These approaches also tend to use iterative cycles of research, where repeated modified applications help optimise instructional design or resources. However, the science teacher is in clinical practice in the sense that, like a doctor treating patients, every case is somewhat idiosyncratic. The teacher needs to be alert to the sense in which every new class presents a new test of teaching practice and a potential reason for seeking to develop that practice further.

Alternatives to Experimental Research

Experimental research is a major tool in the natural sciences where it is often possible to identify, measure and control potentially confounding variables. Educational phenomena are seldom easily subject to laboratory-type testing and researchers in science education have other kinds of research tools more suited to exploring complex phenomena. Researchers can use methodologies such as surveys, case studies, ethnography and grounded theory to develop new knowledge (Taber, 2014).

Surveys sample a population to answer a research question of some generality. For example surveys might be developed to answer questions such as:

- how many years of classroom teaching do science teachers typically have before being made heads of school science departments – and does this vary by gender?
- what proportion of 16 year old students understand the nature of ionic bonding?
- which science topics do primary school children most enjoy?
- how much time is typically given over to group discussion in secondary school biology lessons?
- to what extent do school textbooks offer historical context to scientific discoveries?
- to what extent do students see college teaching about atomic structure (or photosynthesis, or genetics, or...) as building upon, rather than contradicting, school science?

Questions such as these require data to be collected from either an entire population (which is seldom possible) or a sample considered representative of that population. This type of ('nomothetic') research is focused on what is usual, normal or typical. Sometimes it makes sense to look for the average or typical in this way, but sometimes research is focused on the complex nature of teaching and learning – with many interacting factors at work – which cannot be meaningfully characterised through averages. Here 'idiographic' approaches such as case studies may be used to explore individual cases in great detail, and report them with 'thick description' so the reader appreciates the context of the case. The case is one unique example among many:

- the science curriculum in one country or region;
- the marking scheme for a particular physics paper;
- the presentation of atomic structure in one textbook;
- the teaching of acids by one teacher to one class;
- one student's ideas about the nature of the world at the micro-scale.

Findings may not be generalisable in the sense that they can be assumed to apply in other contexts, but may be informative because they highlight nuances and subtleties that can only be identified through detailed study of individual cases. Sometimes cases are deliberately chosen NOT to be typical, as when an exemplary teacher might be studied to find out what make them such a good teacher. For example, when Karina Adbo interviewed Swedish upper secondary students about their thinking about the chemistry topics they studied in school she chose to report a case study of a student with an atypically rich conceptualisation because this provided particular insight into how a students' initial ideas channelled the way his thinking developed in response to teaching (Adbo & Taber, 2014).

Researchers may adopt ideas and approaches from anthropology if they are exploring a cultural issue. One example would be a book by David Long (Long, 2011) reporting an ethnographic account of how students in the USA responded to scientific ideas about evolution when these were widely rejected in their local communities. Another strategy, grounded theory is a coherent research approach that tests and retests ideas and theory developed in a particular context until there is strong evidence that key issues are well understood. It requires extensive engagement in the research context and an open-ended evolving research design. Unlike in experimental research where samples and instruments and endpoints are determined before starting data collection, grounded theory research lets ongoing analysis of data inform decisions about what new data to collect, and the research continues until this process no longer provides substantially new insights.

These research methodologies then offer diverse types of research strategy. There is also a wide range of specific research methods or techniques adopted in SER, and some of these are used (but in somewhat different ways) across methodologies. For example, surveys may use observation or written instruments. The latter may be questionnaires to survey opinions, perceptions and the like, or assessments (tests) to explore levels of scientific knowledge.

Observation in a survey is likely to use a structured format – with fixed categories and indicators for coding what is seen. Because a case study would normally have a more open-ended research question, observations of the case are more likely to allow the observer to decide what it is pertinent to record. In both approaches there is inevitably a level of interpretation in making observations based on what is seen: but in the former case it has been previously decided what indicators are likely to reflect features of interest. In grounded theory studies, research observation may shift from being open-ended to more structured during the research as initial analysis suggests conjectures to test out.

Surveys may use structured interviews, but more often educational research uses less structured interviews where the researcher probes the participant using a more conversational approach. A conversation is less reproducible between participants, but may be needed to find out about the complexity of someone's ideas, especially when they may not have mastered a canonical use of scientific terminology. Much of the work into learners' ideas in science has used more open ended interview approaches. There have new techniques developed for this kind of work, such as the interview-about-instances (Gilbert, Watts, & Osborne, 1985) which used a pack of visual images as foci for prompting thinking about situations ("is there any force here…?") This is an example of a research technique developed in science education which can be used much more widely.

CONCLUSIONS

Science education comprises a major area of professional activity concerned with teaching, supported and informed by an active field of scholarship and research. SER is well established as a field, employing a diverse range of research methodologies, and has led to a great deal of knowledge about student thinking, common learning difficulties, effective teaching approaches and so forth – albeit in some science topics more than others. Effective science teachers seek to make sure their work is informed by this research. However, every institutional context has its own quirks, every class is different, and every student is unique. Each teacher has their own strengths, weaknesses and style. So even when relevant research is available to inform our teaching, there is a sense that every lesson is potentially a further field trial of the research results we seek to apply. That is something science teachers should appreciate – if they adopt evidence-based teaching, then their professional work takes the form of a personal research programme within the scholarly field of science education.

FURTHER READING

There are quite a number of research journals concerned with science education. Among the most highly thought of are:

International Journal of Science Education
Journal of Research in Science Teaching
Research in Science Education
Science Education

There is also a journal devoted to in-depth reviews of areas of research in science education

Studies in Science Education

There are also more specialist journals such as

Chemistry Education Research and Practice
Journal of Biological Education
Journal of Geoscience Education
Physics Education

There is also a number of standard reference works in science education that include scholarly articles on a range of topics:

Fraser, B. J., Tobin, K. G., & McRobbie, C. J. (Eds.). (2012). *Second international handbook of science education*. Dordrecht: Springer.
Gunstone, R. (Ed.). (2015). *Encyclopedia of science education*. Dordrecht: Springer Reference.
Lederman, N., & Abell, S. K. (Eds.). (2014). *Handbook of research in science education* (Vol. 2, pp. 457–480). New York, NY: Routledge.
Matthews, M. R. (Ed.). (2014). *International handbook of research in history, philosophy and science teaching*. Dordrecht: Springer.

An introduction to educational research written with classroom teachers and preparing teachers in mind is:

Taber, K. S. (2013). *Classroom-based research and evidence-based practice: An introduction* (2nd ed.). London: Sage.

REFERENCES

Adbo, K., & Taber, K. S. (2014). Developing an understanding of chemistry: A case study of one Swedish student's rich conceptualisation for making sense of upper secondary school chemistry. *International Journal of Science Education, 36*(7), 1107–1136. doi:10.1080/09500693.2013.844869
Erickson, G. (2000). Research programmes and the student science learning literature. In R. Millar, J. Leach, & J. Osborne (Eds.), *Improving science education: The contribution of research* (pp. 271–292). Buckingham: Open University Press.
Fensham, P. J. (2004). *Defining an identity: The evolution of science education as a field of research*. Dordrecht: Kluwer Academic Publishers.
Gilbert, J. K. (1995). Studies and fields: Directions of research in science education. *Studies in Science Education, 25*, 173–197. doi:10.1080/03057269508560053
Gilbert, J. K., Watts, D. M., & Osborne, R. J. (1985). Eliciting student views using an interview-about-instances technique. In L. H. T. West & A. L. Pines (Eds.), *Cognitive structure and conceptual change* (pp. 11–27). London: Academic Press.
Knorr Cetina, K. (1999). *Epistemic cultures: How the sciences make knowledge*. Cambridge, MA: Harvard University Press.
Kuhn, T. S. (1996). *The structure of scientific revolutions* (3rd ed.). Chicago, IL: University of Chicago.
Lakatos, I. (1970). Falsification and the methodology of scientific research programmes. In I. Lakatos & A. Musgrove (Eds.), *Criticism and the growth of knowledge* (pp. 91–196). Cambridge: Cambridge University Press.
Long, D. E. (2011). *Evolution and religion in American education: An ethnography*. Dordrecht: Springer.
National Research Council Committee on Scientific Principles for Educational Research. (2002). *Scientific research in education*. Washington, DC: National Academies Press.
Pring, R. (2000). *Philosophy of educational research*. London: Continuum.
Solomon, J. (1993). Four frames for a field. In P. J. Black & A. M. Lucas (Eds.), *Children's informal ideas in science* (pp. 1–19). London: Routledge.
Taber, K. S. (2006). Beyond constructivism: The progressive research programme into learning science. *Studies in Science Education, 42*, 125–184.
Taber, K. S. (2008). Of models, mermaids and methods: The role of analytical pluralism in understanding student learning in science. In I. V. Eriksson (Ed.), *Science education in the 21st Century* (pp. 69–106). Hauppauge, NY: Nova Science Publishers.
Taber, K. S. (2013). Three levels of chemistry educational research. *Chemistry Education Research and Practice, 14*(2), 151–155. doi:10.1039/C3RP90003G

Taber, K. S. (2014). Methodological issues in science education research: A perspective from the philosophy of science. In M. R. Matthews (Ed.), *International handbook of research in history, philosophy and science teaching* (Vol. 3, pp. 1839–1893). Dordrecht: Springer Netherlands.

Whitehead, J. (1989). Creating a living educational theory from questions of the kind, 'How do I Improve my Practice?'. *Cambridge Journal of Education, 19*(1), 41–52. doi:10.1080/0305764890190106

SECTION I

NATURE OF SCIENCE AND SCIENCE EDUCATION

KEITH S. TABER

2. REFLECTING THE NATURE OF SCIENCE IN SCIENCE EDUCATION

WHY TEACH SCIENCE?

Science is now an accepted, indeed often a core, part of the school curriculum around the world. However, no matter how much time is put aside for teaching science, there always has to be a severe selection of material as there is much more potential science content than could realistically be fitted within a pupil's school career. In selecting curriculum, we should always keep in minds our purposes for teaching science. There are a number of good reasons that might be suggested for teaching science. In particular it is worth considering the following arguments:

- It is important to teach science because of the need for future scientists, engineers, technologists, and others who will need a strong science background for their work.
- It is important to teach science as it is an important aspect of modern culture and everyone should appreciate this aspect of culture.
- It is important to teach science because a knowledge of science is needed for citizenship in modern technological societies.

The first argument has two aspects. Societies need a supply of suitably qualified people to work as scientists, doctors, engineers and so forth, and that requires sufficient pupils completing school who are qualified and motivated to enter science and related areas in further and higher education. The other aspect of this is that many young children do aspire to be scientists, or to work in areas applying science such as medicine and engineering. Perhaps not all have the potential to fulfil their aspirations, but schools should give pupils suitable opportunities (through suitable science and mathematics teaching, for example) such that those with the desire and aptitude are able to progress to scientific careers.

The second argument is related to the importance of a liberal education. School should introduce young people to all of the important elements of their culture, so they are in a position to engage with that culture through their lives. This would include such areas as music and fine art (which in some educational contexts might include both indigenous traditions as well as those of more Western 'classical' traditions), but would also include such areas as politics and science. The idea here is that schooling should enable anyone to feel enabled to visit an art gallery, or attend a concert, or to visit a natural history museum or read an article in a popular science

K. S. Taber & B. Akpan (Eds.), Science Education, 23–37.

magazine, and to have sufficient background to appreciate and not feel alienated by that aspect of culture. It could be suggested that a 'liberal' education enables a person to feel they can join in with an intelligent conversation about different aspects of a society's culture. In modern societies, that would include aspects related to science and technology.

The final argument to be considered here goes beyond feeling able to join in a conversation about science, but rather is based on the assumption that to function effectively in a modern democratic industrially advanced society – or indeed in a society aspiring to be democratic and/or technologically advanced – the citizen needs to have a basic understanding of science. The citizen who is advised by a doctor about treatment options for themselves or a sick relative can only make an informed decision if they understand some basic science. The citizen who wishes to live their life in an environmentally responsible way (without producing undue waste and pollution) needs a basic understanding of some science so they can make choices about their purchases and sensible recycling behaviour. The citizen asked to vote in an election where different options are presented as best meeting future power supply needs (e.g. should the country invest in new nuclear power stations?) can only make an informed choice when they understand some basic science.

These different purposes are not necessarily contrary to each other, but they do bring different emphases. Ideally a good science education is meeting all of these needs by providing a curriculum which allows some students to qualify for higher level study, and leads all students to know enough basic science to make informed choices, and to feel comfortable with engaging with science-based issues when they arise.

BALANCING SCIENCE PROCESS AND SCIENCE PRODUCT IN THE CURRICULUM

Given that the development of a science curriculum necessarily involves a selection of content from the vast amount of science that could be taught, it is important to make principled choices (see Chapter 13: '*Curriculum Development in Science Education*'). Indeed, there is evidence that in some ways 'less is more'. This has been seen for example in England where a prescribed National Curriculum set out a large number of topics from across the sciences that all students should study during their schooling. This was seen to ensure that everyone knew something about what were considered important topics in biology, chemistry, physics, and earth and space sciences. Yet with so many topics to 'cover' teachers had limited time to delve into topics in any depth. Often students who found science difficult tended to feel they were always moving onto new material before they had really got to come to terms with the previous topic. Those students who performed at high levels, the 'gifted' learners, tended to feel that science was a subject where they were constantly being given more material to learn – but with limited opportunity for the kind of in-depth treatments needed to challenge them. Unfortunately the curriculum tended to deter a lot of pupils from wanting to study science further once they reached the end of compulsory schooling. Of course that did not apply to all students: but many of those potentially suitable for scientific careers thought other academic areas offered more opportunity for deep engagement, and many of the rest left school bewildered by science rather than enchanted with it. So a very dense curriculum seems to generally fail in meeting the key purposes for science education discussed above.

The other aspect of this particular curriculum that did not meet its designers' intentions concerned the extent to which it enabled learners to develop a feel for the nature of science. That is, a good science curriculum needs to not only teach *some* science, but also teach *about* science. There needs to be a balance between teaching some of the products or outcomes of science (such as the periodic table; the theory of natural selection; the ideal gas law) and teaching about the processes of science – how science goes about producing new knowledge.

This is important in terms of our reasons for teaching science. A young person who aspires to be a scientist or work in a related field (as a doctor, or an engineer) certainly needs to know some science. Universities and other advanced educational institutes will expect to select students who already have a good background in key basic science, and who have demonstrated they are able to learn and apply scientific ideas. The more science students know at the end of school, the easier it is for those teaching on advanced courses. However, as long as students are carefully selected for the subject they go on to study, it is not actually difficult for the university to teach material not covered in school. A good understanding of fundamental ideas that demonstrates strong interest and aptitude is more useful than a broad, but shallow, knowledge across a wide range of topics.

However, as well as some background knowledge, the future scientist should have a good feel for the nature of the work they will do if they qualify in and enter a scientific field. That is school science should give them a feel for what it is to be a scientist and do science. This consideration also applies in terms of a liberal education. Scientific knowledge moves on very quickly. Some of the science a person learns in school will be discredited or substantially modified during their adult life. During that life, quite a lot of the science learnt in school will be of limited importance to new developments, and whole new areas of science with major applications will open up that were never mentioned in school as they were unanticipated. What will not substantially change is the nature of science as a cultural activity which produces, evaluates, develops and sometimes demotes, scientific knowledge.

This argument becomes even more important in terms of the third purpose of science education discussed above – to prepare young people for citizenship. Inevitably most of the 'products' of science in the school curriculum tend to be pretty secure knowledge claims that are no longer the subject of active disagreement. Yet the science that people are asked to take a view on in the political or civic realm tends to be in areas where there remains controversy. One example, nuclear power, has already been mentioned. Other areas include such important themes as global warming, deforestation, and biodiversity, where even when there is a clear majority of scientists arguing that science suggests urgent and new policies are needed, some other scientists will appear in the media denying that this is the case. If school science is presented as a 'rhetoric of conclusions' (Schwab, 1962) – as a series of accounts of consensual, settled science – then students are not prepared to understand how they should respond to bitter arguments between different scientists who are each claiming the scientific evidence supports their view. Yet, actually, that kind of argumentation is typical of the scientific process – which is quite unlike the straightforward accretion of successive models, theories, laws, etc., that science education can easily portray with the benefit of decades or even centuries of hindsight. In science, the account presented in school can reflect the 'winners' of various scientific debates rather than the argumentation that was at the heart of the scientific process itself.

The argument here is not that school science education should be about the processes of science *instead of* the content – even if that were possible (as some contexts are needed to effectively teach about the processes). The argument is that teaching about the nature of science is essential to a science education that wishes to prepare future scientists, cultured members of society, and informed citizens, and that accordingly great care is needed to balance the teaching about science itself as a cultural and intellectual activity, and teaching about some of the important, fascinating, and highly applicable, scientific knowledge that this cultural activity we call science has produced.

THE NATURE OF 'THE NATURE OF SCIENCE'

Having established that there are good reasons to teach students about the nature of science as a key part of school science, it is important to acknowledge a number of potential problems. These issues may explain why despite many high profile calls for the importance of teaching about the nature of science (e.g., Clough & Olson, 2008; Duschl, 2000; Matthews, 1994), the nature of science is still not well reflected in the school curriculum in some countries. These issues are:

- science is a broad area of activity, so it is not always very obvious what is common to all of science;
- there is not always strong consensus on how to best understand, and so represent in teaching, the nature of science;
- scholarship about the nature of science from areas such as philosophy, history, psychology and sociology can be quite technical and specialised, and is often too sophisticated for most school learners;
- there is less expertise amongst science teachers, curriculum developers and textbook and other resource authors, regarding the nature of science compared with the level of expertise in areas of science themselves.

The last point is something that can be overcome in time, if teaching about the nature of science within the school curriculum is recognised as a priority (as argued here). The other points are important, but need not be major impediments. Indeed the diversity of science may be seen as a positive feature in a sense, as it implies that teaching about the nature of science should focus on features that are common across the wide range of sciences.

The lack of consensus on some aspects of the nature of science (for example, exactly how to distinguish scientific fields from activities we would not consider science) is important, but actually there seems to be a widespread agreement on those key features of science that need to be represented in school science (as discussed below). The issue of the sophistication of the level of professional scholarship in areas related to the nature of science does present a challenge, but in principle this is no different than when teaching about scientific content itself. School science includes many content areas where scientific thinking is nuanced and where the detailed scientific theory or model is too sophisticated for school age students.

In developing curriculum, complex and abstract scientific ideas are represented in curricular models that offer learners the essence of those ideas at a suitable level of complexity to allow them to be grasped as meaningful. Topics such as the theory of natural selection, the nature of chemical bonding in metals, or the formation of heavier elements in stellar nucleosynthesis – to offer just a few examples – are not suitable for teaching in school at the level of current scientific knowledge, but can be taught through appropriate simplifications that are accessible to learners whilst offering an authentic basis for later progression in understanding. Finding the optimal level of simplification (Taber, 2000) for presenting such topics is a key task for science education, and this is true of representing aspects of the nature of science as well as aspects of science content knowledge (Taber, 2008).

KEY ASPECTS OF THE NATURE OF SCIENCE

There is a vast literature on the nature of science or what is sometimes called 'science studies'. The aim in school science is to get across a flavour of some key features of our understanding of the nature of science. This brief introductory account is intended simply to alert readers to some important topics and ideas. There are many good sources for learning more (see the list at the end of the chapter for some examples), and other chapters in this section fill out much more detail on some of these themes. The focus here will be on:

- The nature of scientific knowledge
- The nature of scientific method
- The limits of science
- The cultural embeddedness of science
- Logic and creativity in science
- The human aspect of science

- The institutional aspect of science
- The rhetorical nature of science

These are presented below as short vignettes on distinct themes, but the astute reader will notice many areas of overlap and connection. As a teacher, it is important to remember that in teaching about these areas the aim is to introduce students to perspectives, rather than to seek to teach models and theories from science studies as if they are facts. In effect, the teacher should try to adopt a social sciences or humanities pedagogy where the am is to help students understand the different perspectives, rather than to accept them as 'true' accounts.

THE NATURE OF SCIENTIFIC KNOWLEDGE

If this approach to teaching seems outside the normal way of working for science teachers, then it may be useful to bear in mind that an appreciation of the nature of science suggests that a common problem with school science teaching is that it often presents science content as true accounts of nature, so that students see science as about facts (see Chapter 4: '*Beliefs and Science Education*'). Yet primarily science is not factual in nature, but theoretical. The essence of science is developing explanatory schemes that make sense of extensive volumes of data and that have predictive value. Scientists often talk as if they are describing how nature is, but they are actually presenting theories and models and other kinds of constructs that derive from the human imagination. Scientists invent categories such as acids and stars which helpfully put order into how we can think about a very complex universe. But often these categories only approximately work. Think about a category such as homo sapiens. A little thought suggests that although we have little difficulty telling humans from non-humans today, it is not always so clear cut whether hominid fossil remains belonged to individuals we would consider part of our own species. Chemists have changed their minds over time in how best to characterise acids and oxidising agents. Physicists have changed their minds about the nature of time and space and for many purposes use Newtonian models they now believe to be flawed (but still very useful) representations of reality.

Scientists refer to laws as if they are universally applicable descriptions of aspects of nature – but usually on the basis of data collection that is limited. (The evidence that 'universal gravitation' applies across the universe is necessarily indirect given how little of the universe we have been able to visit.) Students often think that theories are scientists' guesses or hunches that they are waiting to prove by experiments. Yet actually theories are the very basis of scientific knowledge. They are far more than guesses, as they must be based on extensive evidence, but they are always open to being surpassed when new data or a new interpretation of existing data comes along. All scientific knowledge is technically provisional – that is, in principle open to re-examination in the light of new information. This leads to one of the major challenges in teaching about the nature of science – how science offers knowledge that is generally robust and reliable, yet always somewhat tentative in nature.

THE NATURE OF SCIENTIFIC METHOD

A simplistic account of science has scientists testing their ideas by doing experiments that will prove or falsify their ideas. An experiment ideally explores a phenomenon under laboratory conditions, where variables of interest can be manipulated and measured and the potential effects of confounding variables controlled by keeping values constant. This is a problematic simplification in at least two regards. For one thing, not all scientists do experiments as such. In some branches of science it may be impractical or unethical to undertake experiments. It is not possible to manipulate the conditions at the centre of stars, or compare how life develops on a planet under different starting conditions. It is not generally considered acceptable to subject people to potentially dangerous conditions to see how their physiology reacts (although such research has been undertaken in the past).

So often scientists working in some scientific disciplines use observational approaches, looking for 'natural experiments' where features of interest naturally vary and allow conditions to be compared. Scientists also use simulations and models to test their ideas, being aware that the results are only as good as the (inherently uncertain and limited) simulation or model. One well known philosopher of science, Paul Feyerabend (1975/1988) argued that there is no such thing as the scientific method, but rather than scientists have to develop their own customised methods that will work in their own areas of research.

Even where genuine experiments are possible, the simple logic of 'proving' or refuting a hypothesis is over-simplistic. An experimental prediction may be correct for a reason other than the verisimilitude (closeness to the truth) of the hypothesis that led to its prediction. It is always possible to produce alternative theories to explain any set of data (even if sometimes the alternatives seem cumbersome and forced). Any experimental data set intended to test some general hypothesis is necessary sampling a very small proportion of the population of possibly relevant events. (Consider how you would test for certain that adding salt to water *always* lowered its melting temperature; or that the human heart *always* has four chambers; or that the electron *always* has a charge of magnitude 1.6×10^{-19}C.)

The difficulty of proving general statements from a limited sample of instances (known as 'the problem of induction') led the famous philosopher of science Karl Popper to recommend that scientists focus more on refutations which at least seemed to rule out hypotheses where experiments did not agree with theoretical predictions (Popper, 1989). However, this is just as problematic. Experiments can go wrong for all kinds of reasons – impure chemicals, laboratory (e.g. technician) error, instrument error, faulty power supplies, unexpected and unnoticed temperature fluctuations, and so forth. Moreover, most modern science uses complex apparatus of measurement and analysis that relies on its own theory of instrumentation. A hypothesis that is correct may seem to need to be rejected if the theory behind the instrumentation is flawed – so scientists need to be wary of too easily rejecting ideas as well as being careful about when considering them to be supported. Science is a more complex business that many school practical activities would suggest!

THE LIMITS OF SCIENCE

One key question in the philosophy of science is the demarcation of science: how we can distinguish what is and is not science. It is fairly straightforward to list some good candidates: physics, chemistry, biology, astronomy, geology, etc. It may be less obvious if we should include psychology (certainly some parts, but all?). For example there has been discussion over whether Freud's theory and practice of psychoanalysis should count as scientific. Claims that aspects of the social sciences are genuinely scientific also lead to debate. Marx suggested he had a scientific take on history (but many commentators would not consider his research programme as scientific), and there have schools of sociology set up to adopt a model based upon natural science. Given that natural sciences do not seem to have a common characteristic method (see above), it does not seem reasonable to exclude other scholarly areas on grounds of methodology. By its nature, history does not involve the setting up of controlled experiments – but it does present theories which can be tested against new data and cases. That is not so different from some work in astronomy.

One philosopher of science, Imre Lakatos (1970), has suggested that the criterion for scientific work is the existence of what are referred to as 'progressive' research programmes – where there is a programme of activity informed by a set of pre-established tenets (core commitments) and where the interplay between the development of theory and collection of new data continues to be productive. Although applying this criterion requires judgement and is not straightforward, this does offer an inclusive approach that allows areas of work which admit diverse methodologies, such as science education (Taber, 2014), to be considered scientific.

One contentious question is whether aspects of indigenous cultures should be included as scientific. Such cultures often have long-standing traditions of using traditional ecological knowledge to harvest nature in sustainable ways: yet unlike in Western science, such knowledge is not separated out from other aspects of culture. So often this knowledge is learnt through legitimate peripheral participation in cultural activities (such as farming), as knowledge in action, and is commonly integrated with strong spiritual values reflecting the assumption that people, the rest of the biota, and the land (and seas and rivers) are spiritually connected as part of an interdependent creation. The atheoretical nature of this traditional technological knowledge, often learnt through practice and through the use of narrative and ritual, makes it quite distinct from how scientific knowledge is understood in formal scientific traditions (this issue is explored further in Chapter 34: '*Science Education and Indigenous Learners*').

Another, related, issue is the limits of science itself. Some scientists seem to feel that science can (and perhaps will) ultimately explain everything, whilst other scientists see science as an important way of knowing, but one that has a limited range of application (so that there are some aspects of human experience that will always be beyond scientific explanation). There is *a sense* in which anything in the natural world *could* be reduced to a description in terms of particles, forces, energy etc. So – in principle at least – it may be possible one day to explain why a person falls in love with one suitor and not another in terms of physics: however, even if such an account was feasible, it would not be presented in terms that would seem to relate to the human experience of love.

This links to an important issue in the philosophy of science – how sciences 'reduce' to each other. Even in chemistry, a discipline closely linked to physics, there are concepts which *could* (in principle) be redescribed in purely (but in some cases necessarily convoluted) physical terms but which reflect emergent phenomena at the 'level' of chemistry and which are useful as chemical concepts in their own right (acidity, oxidation, resonance, hydrogen bonding, electrophile, halogen, indicator, nucleophilic substitution, covalent bond, etc., etc). A reductionist perspective has historically proved very valuable in science. Yet increasingly scientists are recognising that complex systems often need to be studied at different levels, and that important new phenomena can arise when systems become complex. A particularly important example might be life itself emerging from the evolution of increasingly complex physico-chemical systems and providing the phenomena studied in biology.

Science teachers should be careful not to imply in their teaching that science (the best means we have of developing knowledge of the natural world) is able to tell us everything about everything. There may well be areas that will always be outside the effective remit of science, and features of human cognition may limit how well we can understand even the natural world.

THE CULTURAL EMBEDDEDNESS OF SCIENCE

An important debate about the nature of science is the extent to which scientific discoveries are dependent on the cultures that produce them. Ideally science is independent of culture, as it is intended to be an objective quest for discovering true knowledge of the natural world. However, we have seen above that scientific knowledge is theoretical, and so based on constructs humans have developed to best describe and explain observations and measurements of nature. As there is no foolproof method of developing scientific knowledge that is absolutely certain (again, see above) all scientific knowledge is limited by human understanding and the available data.

Scientists are people who use their imaginations to develop ideas that might represent aspects of nature – ideas that they then test as best they can. Inevitably scientists' thinking is influenced by the widespread ideas in the society where they they live. So, for example, scientists often develop metaphors and analogies as a basis for scientific conjectures – but they are limited to drawing upon sources they are already familiar with. Thomas Kuhn, a physicist who moved into historical studies of science, argued that once a particular way of thinking about the world became familiar, and its affordances had been worked out in detail by scientists, it became much harder to see how some alternative scheme might be at least as useful – even if it dealt better with known flaws in existing theories (Kuhn, 1970). Kuhn suggested that different theoretical frameworks, with their different ways of seeing the world, were incommensurable (could not be measured against each other). He meant it was difficult to evaluate different frameworks objectively, as the evaluator would always be working from within their own existing worldview. Kuhn thought that science could make progress towards knowledge that better represented the true nature of things: but that this process was difficult because scientists can never completely step outside of the assumptions inherent in their habitual ways of making sense of the world.

LOGIC AND CREATIVITY IN SCIENCE

Science is often associated with logical thinking, and this is indeed an important feature of science (see Chapter 8: '*Scientific Reasoning During Inquiry*'). Logic is needed to work out predictions consistent with particular hypotheses or models, and logic is needed to interpret data in terms of different principles, laws and theories, and to construct arguments to persuade other scientists of the validity of conclusions.

Yet science relies on creative thought as well as logic. Logic is needed when testing out ideas, but first scientists have to come up with the ideas to test. It is naive to think that scientists can move directly from data to scientific knowledge, as data always have to be interpreted in terms of some conceptual scheme. That scheme is an imaginative construction of the human mind. Science proceeds though the complementary roles of creative (expansive, imaginative, divergent) and logical (rational, closed, linear) thought (Taber, 2011).

Often the scientists who become most well known do so not because they were more logical than other scientists, but because they were able to use their imaginations to develop possible new ways of thinking which could then be compared to data. For some scientists, such as Einstein, this imaginative process is primarily visual – they are able to imagine pictures that represent novel relationships and concepts. Visualisation is also important in running thought experiments (mental simulations) that may be useful in ruling out some options without needing to run real experiments, and which may help predict the outcomes to be expected in experiments according to particular hypotheses.

Much human knowledge is tacit in nature, and this includes much of the knowledge of professional scientists (Polanyi, 1962). Scientists develop intuition based upon their implicit knowledge (see Chapter 10: '*Tacit Knowledge in Science Education*'). Imaginative processes, such as visualisation, can be very important in providing explicit awareness of a scientist's tacit knowledge (see Chapter 11: '*Developing Visual/Spatial Thinking in Science Education*').

THE HUMAN ASPECT OF SCIENCE

Science is in principle an objective activity. There is a stereotype of the scientist who has put aside personal feelings to focus on scientific work – sometimes to the neglect of such personal needs as sleep and food. Many scientists see their work as in the interest of wider humanity and/or for the joy of better understanding nature – and at times they will become engrossed to the exclusion of distractions.

Realistically, though, scientists are human with all the usual flaws. They may cling to their pet theories in the face of contrary evidence. They often seek professional advancement if not financial rewards. Some covet awards and titles and prestigious honours. Sometimes some scientists may show prejudice – towards their close colleagues, or to their co-nationals, or against those of different faith or ethnicity.

There is a major literature on issues around gender and science – both questions of whether Western formal science is inherently masculine in nature (for example in focusing on controlling nature, rather than relating to it), to the exclusion of women and the detriment of science, and whether female scientists today still regularly face sexism from individual scientists and institutions.

There are many historical cases that can illustrate these themes. An especially potent one concerns the discovery of the double helix structure of DNA. As well as relating to a iconic scientific discovery this work has been much documented. It illustrates the extent of co-operation within science (with and between institutions: Crick, Watson, Wilkins) as well as competition (again within and between institutions: Wilkins with Franklin; Crick and Watson, with Pauling). It reveal how prejudices, friendships, and chance, can play a role in science. It also reveals how science can proceed through examining mistakes (such as Linus Pauling's three strand structure for DNA) and through the interaction between creative exploration and tedious laboratory work (relating results from Franklin's meticulous preparation and analysis of X-ray photographs to Crick's theoretical work on helical diffraction and Watson's exploratory model building).

THE INSTITUTIONAL ASPECT OF SCIENCE

The same case study can illustrate some of the institutional features of modern science, where the work of individual scientists relies upon institutional support in a laboratory, and may be subject to local norms and practice – as when Rosalind Franklin (co-discoverer of the structure of DNA, see above) discovered she was not allowed to take refreshments in the same common room as her male colleagues, and was therefore excluded from the informal professional conversations that inevitably take place in such settings. That particular indignity is less likely today. However, modern scientific research laboratories are places of hierarchy, protocols and procedures, and financial restraints (Knorr Cetina, 1999; Latour & Woolgar, 1986).

From an anthropological perspective, science is a sub-culture with its own rituals and priesthood. The scientific societies, the journals, the research funding councils and the formal conferences, are essential institutions in supporting scientific debate and in ultimately recognising what counts as successful science. Science is a relatively democratic enterprise in the weight given to the peer review process (such that anyone can publish in the top journals if their work is judged as original and rigorous), but inevitably as a human activity can only take place within a supporting structure of formal institutions. The stereotype of the lone scientist making great breakthroughs in their shed or basement is – with the very occasional exception like James Lovelock (who invented the electron capture detector, surveyed the levels of chlorofluorocarbons (CFCs) in the atmosphere, and proposed the Gaia theory of the biosphere) – now a historical anachronism.

THE RHETORICAL NATURE OF SCIENCE

It follows then that success in science does not in practice mean discovering the truths of nature (as we can never be sure how well our theories give an account of nature, and how long they might go unchallenged) but rather persuading the scientific community, or that part of it working in the same field at least, that particular scientific results and ideas are important and progress the field forward. This then depends upon argument: making a case that data can be best interpreted in a certain way, and persuading those who may currently think quite differently about certain natural phenomena (see Chapter 5: '*Epistemic practices and scientific practices in science education*'). In recent years it has been increasingly recognised that authentic science education needs to have a strong focus on engaging students in argumentation (see Chapter 12: '*Language, Discourse, Argumentation, and Science Education*').

Given human nature, once scientists are convinced of some idea or some particular interpretation of data, they will tend to want to persuade others to their way of thinking. The scientific paper is in effect a rhetorical structure for best presenting a particular interpretation of certain data such that it seems to offer evidence for a particular model, theory, principle, or other such construct (Medawar, 1963/1990). In presenting this argument, the author(s) will select and sequence material to make a case, and will necessarily exclude much information (possibly including some collected data) that is considered less relevant to the knowledge claims being made. Even when scientists are scrupulously honest in their attempts to be objective, other scientists approaching the same evidence base from different perspectives might have made different judgements about what was relevant and should be included, and how the presented data should best be interpreted as scientific evidence. Peer reviewers generally do not have access to omitted details that authors feel should be excluded from their papers.

The scientific literature should therefore not be seen as a series of factual and objective accounts of nature, but rather as a cumulative collection of knowledge claims, each based on some limited data, interpreted through particular frameworks of understanding, and evaluated as of merit by referees chosen as suitable experts by journal editors. Scientific knowledge is therefore not only uncertain, but in areas of current research still in flux. Only in retrospect, once research activity in some programme is long exhausted, can observers start to see that area of knowledge as relatively unproblematic.

Teaching science involves helping learners to appreciate the value of the unfamiliar constructs used in science. Just as scientists orchestrate evidence and present carefully structured arguments to persuade their colleagues of claims made in scientific papers, so similar rhetorical moves are made by science teachers in reconstructing scientific concepts with their students (Lemke, 1990; Ogborn, Kress, Martins, & McGillicuddy, 1996). Science teachers can reflect the nature of science in their teaching by giving learners insight into those rhetorical processes.

CONCLUSION

In many countries, school science tends to focus on areas of well-established science, where scientific knowledge appears firm and not currently under debate. Such knowledge is still provisional rather than absolute (as new evidence could be uncovered and presented at any time) but can too easily be presented as factual (rather than theoretical) and obviously following from data (that is the data presented as evidence in the papers now seen, after the event, as most significant) rather than being an interpretation based on human imagination.

Many areas of science that have reached such impasses can contribute to a useful science education, but if they are taught as unproblematic they are stripped of the nature of the very scientific activity which produced them. This can be avoided by careful presentation and phrasing, and the inclusion of some of the debate and uncertainty that led up to their wide acceptance as robust scientific knowledge. Science teaching that meets our key aims needs to give students an authentic feel of scientific processes, whether through historically contextualising established science; through authentic enquiry activity in the classroom; or the inclusion in the curriculum of examples of current scientific controversies where there is not yet any wide consensus, and so where competing knowledge claims, based on incommensurate interpretations of data, invite genuinely open-ended consideration. Ideally school science education will include all three of these elements to allow learners to learn *about* science itself, alongside learning some science. Science teachers need to regularly consider how they will represent the nature of science in their own science teaching – a theme developed in the next chapter (on '*History and Nature of Science in Science Education*').

FURTHER READING

Brown, S., Fauvel, J., & Finnegan, R. (Eds.). (1981). *Conceptions of inquiry*. London: Routledge.
Chalmers, A. F. (1982). *What is this thing called science?* (2nd ed.). Milton Keynes: Open University Press.

Useful Classroom Resources

Allchin, D. (2013). *Teaching the nature of science: Perspectives and resources*. Saint Paul, MN: SHiPS Educational Press. (This book presents a strong argument for teaching case studies about the nature of science, and includes examples that can be used in the classroom.)
Osborne, J., Erduran, S., & Simon, S. (2004). *Ideas, evidence & argument in science: In service training pack*. London: Kings College London.
Taber, K. S. (2007). *Enriching school science for the gifted learner*. London: Gatsby Science Enhancement Programme. (This book and resource pack includes activities around several nature of science themes.)

REFERENCES

Clough, M. P., & Olson, J. K. (2008). Teaching and assessing the nature of science: An introduction. *Science & Education, 17*(2–3), 143–145.
Duschl, R. A. (2000). Making the nature of science explicit. In R. Millar, J. Leach, & J. Osborne (Eds.), *Improving science education: The contribution of research* (pp. 187–206). Buckingham: Open University Press.
Feyerabend, P. (1975/1988). *Against method* (Rev. ed.). London: Verso.
Knorr Cetina, K. (1999). *Epistemic cultures: How the sciences make knowledge*. Cambridge, MA: Harvard University Press.
Kuhn, T. S. (1970). *The structure of scientific revolutions* (2nd ed.). Chicago, IL: University of Chicago.
Lakatos, I. (1970). Falsification and the methodology of scientific research programmes. In I. Lakatos & A. Musgrove (Eds.), *Criticism and the growth of knowledge* (pp. 91–196). Cambridge: Cambridge University Press.
Latour, B., & Woolgar, S. (1986). *Laboratory life: The construction of scientific facts* (2nd ed.). Princeton, NJ: Princeton University Press.
Lemke, J. L. (1990). *Talking science: Language, learning, and values*. Norwood, NJ: Ablex Publishing Corporation.
Matthews, M. R. (1994). *Science teaching: The role of history and philosophy of science*. London: Routledge.
Medawar, P. B. (1963/1990). Is the scientific paper a fraud? In P. B. Medawar (Ed.), *The threat and the glory* (pp. 228–233). New York, NY: Harper Collins. (Reprinted from: The Listener, Volume 70: 12th September, 1963)
Ogborn, J., Kress, G., Martins, I., & McGillicuddy, K. (1996). *Explaining science in the classroom*. Buckingham: Open University Press.
Polanyi, M. (1962). *Personal knowledge: Towards a post-critical philosophy* (Corrected version ed.). Chicago, IL: University of Chicago Press.
Popper, K. R. (1989). *Conjectures and refutations: The growth of scientific knowledge* (5th ed.). London: Routledge.
Schwab, J. J. (1962). The teaching of science as enquiry (The Inglis Lecture, 1961). In J. J. Schwab & P. F. Brandwein (Eds.), *The teaching of science*. Cambridge, MA: Harvard University Press.
Taber, K. S. (2000). Finding the optimum level of simplification: The case of teaching about heat and temperature. *Physics Education, 35*(5), 320–325.
Taber, K. S. (2008). Towards a curricular model of the nature of science. *Science & Education, 17*(2–3), 179–218. doi:10.1007/s11191-006-9056-4

Taber, K. S. (2011). The natures of scientific thinking: Creativity as the handmaiden to logic in the development of public and personal knowledge. In M. S. Khine (Ed.), *Advances in the nature of science research: Concepts and methodologies* (pp. 51–74). Dordrecht: Springer.

Taber, K. S. (2014). Methodological issues in science education research: A perspective from the philosophy of science. In M. R. Matthews (Ed.), *International handbook of research in history, philosophy and science teaching* (Vol. 3, pp. 1839–1893). Dordrecht: Springer Netherlands.

MICHAEL P. CLOUGH

3. HISTORY AND NATURE OF SCIENCE IN SCIENCE EDUCATION

INTRODUCTION

The history and nature of science (HNOS) is a phrase used in science education that encompasses issues such as what science is, how science works, characteristics of scientists, and how scientific knowledge is developed and comes to be accepted by the scientific community. Answers to these questions often seem fairly obvious to most people, particularly teachers and students of science. But an abundance of studies report that the general public, science teachers and their students have significant misconceptions about the HNOS. This chapter addresses why accurately portraying the HNOS is important for both science teachers and students, prevalent HNOS misconceptions, and how to incorporate HNOS instruction in a manner that effectively bolsters understanding of both HNOS and science content.

HNOS IN SCIENCE EDUCATION

Accurately understanding important features of the HNOS is an important aspect of scientific literacy, and a longstanding goal of science education. This is reflected in its being part of most contemporary science education reform documents. But the value of HNOS extends beyond understanding the characteristics of science, scientists and scientific knowledge. When thoughtfully and seriously considered, understanding the HNOS:

- *Helps teachers understand students' difficulties learning science ideas and the tenacity of misconceptions.* The HNOS makes clear that very intelligent scientists struggled to understand the natural world, and how many tenaciously held to ideas that the scientific community has now abandoned. Those scientists had reasons, often good reasons, for committing to those ideas, and even for disagreeing with colleagues who proposed new ways of understanding and explaining phenomena. Teachers are in a better position to understand and assist struggling students if they understand the many historical examples illustrating scientists' struggles to understand phenomena, and how intelligent individuals rejected new ideas or only slowly and with difficulty came to understand the superiority of those new ideas.
- *Assists teachers in understanding why telling and showing do not compel students to change their thinking.* The struggles of scientists noted in the previous bullet occurred despite other scientists explaining the idea they advocated and providing

K. S. Taber & B. Akpan (Eds.), Science Education, 39–51.

evidence they maintained supported their thinking. The HNOS makes clear that the difficulties of understanding and accepting colleagues' ideas occurred in spite of their carefully considering the arguments and evidence of those colleagues. But disagreements continued because data and arguments may be interpreted in a variety of ways. Of course, providing explanations with evidence is important both in science and science teaching, but doing so is often insufficient for bringing about a change in thinking.

- *Assists students in understanding the complexity of learning science and identify with past scientists' struggles, thus increasing students perseverance* (Arya & Maul, 2012; Hong & Lin-Siegler, 2012). Understanding the HNOS can help students better understand their own struggles learning science. As a result, rather than thinking they are incapable of understanding science, they are more likely to persist in their effort to learn.
- *Improve students' attitude toward and interest in science and science education.* Those who have a more accurate view of the HNOS see that science is done by people of all cultures, see science as a creative endeavor that involves interacting with people, and possess improved attitudes toward science, scientists, and science-related careers.
- *Plays a role in socio-scientific decision-making* (Mitchell, 2009; Herman, 2015). For instance, many people deny global climate change, biological evolution, and other important science ideas, in part, because they wrongly think that good science demands control-treatment experiments. However, the HNOS illustrates that for many scientific questions, that approach is either not possible or not appropriate. For much of astronomy, ecology, geology and other fields of study, a control-treatment experimental approach is not possible or appropriate, yet much of the knowledge those disciplines have produced is as well-established as that resulting from control-treatment experiments.
- *Understanding science content.* Many science ideas are counter-intuitive, and are only understood by abandoning our everyday approach to making sense of phenomena. For instance, deeply understanding the law of pendulum motion requires an understanding of the idealized (and impossible) conditions that law is based upon and the value of having such a scientific idea. Understanding biological evolution is, in part, dependent upon understanding methodological naturalism and how well-supported scientific ideas need not always be based on experiments or make specific predictions.

MISCONCEPTIONS REGARDING THE HNOS

HNOS misconceptions like those appearing in Table 1 are widespread, but hardly surprising given the way that science and scientists are portrayed on television, in movies, and in other popular media. However, school science is also to blame. Science textbooks typically ignore information about the work of scientists, how questions and ideas regarding the natural world arise, the disagreements about the meaning of data, and how the scientific community came to eventually reject and accept particular ideas (Leite, 2002). As Postman (1995) noted:

> ...textbooks are concerned with presenting the facts of the case (whatever the case may be) as if there can be no disputing them, as if they are fixed and immutable. And still worse, there is usually no clue given as to who claimed these are the facts of the case, or how "it" discovered these facts (there being no he or she, or I or we). There is no sense of the frailty or ambiguity of human judgment, no hint of the possibilities of error. Knowledge is presented as a commodity to be acquired, never as a human struggle to understand, to overcome falsity, to stumble toward the truth. (p. 116)

Table 1. Common misconceptions regarding the HNOS

- Disagreements regarding competing scientific explanations for natural phenomena are resolved through polling scientists on their view of the best explanation.
- Science and those who do science can and should be free from emotions and bias.
- Scientific ideas arise directly from data.
- Data supporting a contentious scientific idea demands that doubting scientists drop their objections to the idea.
- Data that is at odds with a prevailing science idea should result in the rejection of that idea.
- Science, when well done, produces ideas that are proven to be "true". Scientific knowledge falling short of that status is unreliable.
- While creativity and inventiveness assist scientists in setting up their research, the resulting science ideas are discovered, much like finding something.
- Scientific models are exact copies of reality.
- Science is equated with technology, and all science research is thought or expected to be in some way directed at solving societal problems.
- Science research follows a step-by-step scientific method and carefully adhering to this systematic method accounts for the success of science.
- The status of, and relationship between, scientific laws and theories is misunderstood.
- Methodological naturalism is equated with philosophical materialism.

When textbooks do make an effort to convey characteristics of science and scientists, it is often done in superficial ways that wrongly sanitize the actual workings of science and scientists, thus bolstering many of the misconceptions appearing in Table 1. Moreover, highly directive cookbook activities so ubiquitous in science classes reinforce many of the same misconceptions. Lab reports, while written for the sole purpose of communicating results of investigations and justification for conclusions reached, misportray how science is really done (Medawar, 1963) and promote many of the same misconceptions, including wrongly portraying scientific research as following a step-by-step scientific method. Finally, teachers' language when teaching science often distorts the HNOS (Munby, 1976). For instance, asking students "What does the data tell you?" and inappropriately using words such as "theory", "law", "prove" often distort the HNOS.

Several of these HNOS misconceptions coalesce, forming an overarching image of science and scientists that, while incorrect, makes sense and thus requires considerable effort to change. That said, much is known about teaching the HNOS in a manner that promotes among students more accurate understandings that are held long after a course ends (Clough, 1995; Herman & Clough, 2016).

EFFECTIVELY TEACHING THE HNOS

While promoting an understanding of the HNOS has been a persistent goal of science education, science teachers at all levels have struggled to accurately and effectively promote this goal for a variety of reasons including, but not limited to, their own misconceptions regarding the HNOS, uncertainty regarding how to effectively teach the HNOS, and the paucity of curriculum materials to support HNOS teaching. However, science teachers committed to HNOS instruction have successfully integrated it extensively in their classrooms (Herman, Clough, & Olson, 2013a).

Important HNOS Issues Worth Addressing in Science Education

Many issues regarding the HNOS are complex and contextual, but for the purposes of science teaching and learning, general agreement exists regarding ideas that ought to be addressed. However, even these generally agreed upon ideas have nuances that depend on contextual factors. Eflin, Glennan and Reisch (1999, p. 112) caution that "Just as science educators stress that science is more than a collection of facts, we emphasize that a philosophical position about the nature of science is more than a list of tenets." Rather than listing HNOS ideas that both teachers and students may wrongly interpret as facts to be taught and learned verbatim, HNOS issues should be addressed as questions like those found in Table 2. Addressing HNOS matters as questions rather than tenets encourages both teachers and students to think about the HNOS issue and consider how different contexts may call for more nuanced answers to the questions.

HNOS Instruction should be Deliberately Planned

Effectively promoting a deep and robust understanding of the HNOS first demands that science teachers intentionally plan how they will teach HNOS ideas, just as they overtly plan how to promote understanding of science content objectives. Such effort requires that they genuinely value HNOS learning, not merely in a general sense, but for reasons like those noted earlier in this chapter (Herman, Clough, & Olson, In press). HNOS instruction should be planned to challenge prevalent HNOS misconceptions and encourage student actions like those in Table 3. Effectively

Table 2. Example HNOS questions worth exploring in science education (From Clough, 2011)

- In what sense is scientific knowledge tentative? In what sense is it durable?
- To what extent is scientific knowledge based on and/or derived from observations of the natural world? In what ways is it justified on grounds other than observational evidence?
- To what extent are scientists and scientific knowledge subjective? To what extent can they be made less subjective?
- To what extent is scientific knowledge socially and culturally embedded? In what sense does scientific knowledge transcend particular cultures?
- In what sense is scientific knowledge invented? In what sense is it discovered?
- How does the notion of a scientific method distort how scientists actually work? In what sense are particular aspects of scientists' work guided by protocols?
- In what sense are scientific laws and theories different types of knowledge? How are they related to one another?
- How are observations and inferences different? In what sense is an observation an inference?
- How is the private work of scientists similar to and different from what is publicly shared in scientific papers?

Table 3. Example student actions that convey HNOS understanding (From Clough, 2011)

- Accurately describe the differences and interactions between basic science, applied science and technology.
- Articulate why contemporary science explains natural phenomena in naturalistic terms with no recourse to the supernatural.
- Provide arguments against a universal scientific method.
- Explain how imagination and creativity are crucial in doing science.
- Explain and provide examples illustrating how scientists develop ideas to account for data, and how data does not tell scientists what to think.
- Justify why well-supported science ideas, while durable, may be re-examined, modified, and replaced. Explain why this possibility of change is a strength of science.
- Accurately explain how scientific laws and theories are different types of knowledge, yet how they relate to one another?
- Provide examples illustrating that science has both a collaborative and competitive character.
- Identify inaccurate stereotypes of scientists.
- Provide examples of how science and society impact one another.

planning for HNOS instruction entails several important features that include making overt to students the HNOS issues being addressed, creating successful contexts for addressing HNOS ideas, and asking questions that assist students in developing more accurate HNOS conceptions.

Making HNOS Instruction Overt to Students

Lessons that merely have students take part in activities, complete readings, or watch multimedia that accurately portray the HNOS are not effective at altering their mistaken notions regarding what scientists are like and how science works. This is because learners use what they already know—in this case their existing HNOS misconceptions—to make sense of what they encounter. Consequently, they will miss or unknowingly interpret and modify aspects of accurate HNOS experiences so that they appear to fit what they already think (Abd-El-Khalick & Lederman, 2000;

Table 4. Teacher questions that draw students' attention to the HNOS (From Clough, 2011)

- How does your work in this laboratory activity illustrate that you did not follow a step-by-step scientific method? How is your work similar to the work of scientists?
- How does the work of [insert scientist or scientists] illustrate that data does not tell scientists what to think, but instead that creativity is part of making sense of data?
- The word "theory" in science is often wrongly interpreted by people as meaning "guess", "opinion", or a not well substantiated claim. How does that meaning not capture the confidence we have in kinetic molecular theory? [This question is most effective when asked after students have studied and are coming to understand the power of the theory. The question can be asked in the context of any well-established theory such as atomic theory, the theory of plate tectonics, the theory of evolution, etc.]
- How does the DNA work of James Watson, Francis Crick, Maurice Wilkins, Rosalind Franklin and Linus Pauling illustrate that doing science involves both collaboration and competition?
- Consider the model of the atom and the evidence that supports it. How does this work illustrate that science ideas are developed to account for data (i.e. data do not tell scientists what to think)?
- In what ways does this portion of your textbook distort what real science is like? [This question must wait until students have first developed more accurate views of the HNOS, but then may be asked most anywhere with typical science textbooks.]
- How does the process by which science came to understand the link between asteroids and dinosaurs illustrate that science requires creativity and does not follow a linear process (see http://undsci.berkeley.edu/article/0_0_0/alvarez_01)?
- What prior knowledge did you use in developing your laboratory procedure and analyzing your data? How does this illustrate that scientific theories guide researchers in determining what questions to ask, how to investigate those questions, and how to make sense of data?

Tao, 2003). Teachers must therefore include in HNOS lesson planning how students' attention will be drawn to targeted NOS issues in a manner that encourages students to mentally engage in what they are experiencing and more accurately compare it with the ideas regarding the HNOS that they already hold. This demands that teachers think about the kinds of questions they will ask and have students respond to during discussions, assigned readings, laboratory activities and other activities. Table 4 presents examples of questions that overtly draw students' attention to HNOS ideas in a manner that requires them to think deeply about those ideas in light of commonly held HNOS misconceptions.

Important Contexts for HNOS Instruction

Promoting a deep and robust HNOS understanding also requires that it be addressed throughout the school year in a variety of contexts. Table 5 situates instruction regarding the nature of science (NOS) in three broad categories on a continuum.

Decontextualized NOS instruction. The first category is decontextualized in the sense that NOS instruction experiences (e.g., black box and other types of puzzle-solving activities) draw similarities to how science works, but the context is devoid of science content and the workings and words of actual scientists. Decontextualized NOS instruction is useful for introducing and addressing NOS issues without complicating matters with science content. However, disconnected from science content and the work and words of scientists, students will unlikely alter their misconceptions regarding how authentic science really works and what scientists are actually like.

Moderately contextualized NOS instruction. Moderately contextualized NOS instruction is associated with science content, but links to the authentic words or work of scientists are absent or superficial. Using students' experiences in inquiry activities to illustrate how their varied approaches illustrates that scientists do not follow a step-by-step method or how their struggles to make sense of data reflects that data do not tell scientists what to think is an example of moderately contextualized NOS instruction. Teaching science through inquiry is instrumental in effective HNOS instruction because it raises opportunities – planned and unplanned – for HNOS instruction (Herman, Clough, & Olson, 2013b). However, unless these experiences and important NOS ideas are overtly connected to the genuine work of scientists (e.g., scientists using varied investigative methods and their difficulties and disagreements interpreting data), students can easily maintain that they and their situation are not the same as scientists who are more intelligent and have access to better equipment. Thus, moderately decontextualized NOS instruction, like decontextualized NOS instruction, is important, but insufficient, for promoting a genuine and long-lasting accurate view of the NOS.

Table 5. Contexts for teaching about the nature of science (After Clough, 2006)

	Decontextualized NOS Instruction	*Moderately Contextualized NOS Instruction*		*Highly Contextualized NOS Instruction*
Connection to science content	None	Decontextualized NOS activities associated with, but still separate from, content	Inquiry science content activities overtly linked to NOS ideas	Seamless
Connection to actual work of scientists	Absent	Superficial	Superficial	Extensive
Example activities	Black box activities Gestalt images Puzzle solving	A black box activity included with, but not meaningfully connected to, instruction regarding atomic structure	Raising NOS issues in a student inquiry activity determining the mass % of water in a hydrate	Authentic historical and contemporary science examples

Highly contextualized NOS instruction. Highly contextualized NOS instruction incorporates historical and contemporary stories of authentic scientists and science research to overtly draw students' attention, using questions like several appearing in Table 4, to important HNOS ideas. These stories provide needed evidence for many students that the NOS ideas addressed in decontextualized and moderately contextualized NOS instruction accurately reflect authentic science. Such accounts need not be lengthy, but they should engage and intrigue students and over time compel them to alter their previous HNOS misconceptions. Highly contextualized NOS instruction alone may appear sufficient for effective NOS instruction, but as noted earlier, the HNOS misconceptions that students bring to science classes interfere in accurately interpreting historical and contemporary science stories.

Scaffolding between contexts. Decontextualized, moderately contextualized, and highly contextualized NOS instruction all play important roles for promoting deep and robust HNOS understanding. To summarize, decontextualized instruction is important for introducing HNOS ideas, moderately contextualized instruction is important for embedding HNOS instruction in everyday science content instruction (crucial so that HNOS will be addressed consistently throughout the school year), and highly contextualized instruction is important for convincing students that what they are learning about the HNOS accurately reflects what scientists and doing science are like. But students often need assistance in making connections between contexts as they wrestle with the HNOS. The role of teachers is always crucial as illustrated by the following questions, modified from Clough (2015), that exemplify how to assist students in making desired meaning while ensuring they are mentally engaged in making sense of their HNOS experiences.

- *Example question scaffolding between a moderately contextualized and decontextualized NOS experience.* How were your efforts to develop a procedure during your inquiry activity similar to your experience with the black box activity you experienced earlier this school year? In what sense did your work in both instances deviate from what is often called "the scientific method"?
- *Example question scaffolding between a highly contextualized and decontextualized NOS experience.* How were scientists' difficulties making sense of the DNA X-ray crystallography data similar to your struggles earlier this year to make sense of the black box activity data? Some people think data tells people what to think. What would you say to a person who thought that?
- *Example questions scaffolding between moderately and highly contextualized NOS experiences.* How was your effort to make sense of data in our conservation of matter inquiry activity similar to and different scientists' work regarding the same question about nature? How do they illustrate the important HNOS idea that data do not tell researchers what to think?
- *Example question scaffolding back and forth between all three broad NOS instruction contexts.* What do both your and scientists' efforts noted in the prior

bullet have in common with your effort to make sense of the data you collected in the black box activity we conducted earlier this school year? How does this illustrate that researchers create ideas that account for/make sense of data?

The wording of these questions also makes apparent important HNOS ideas. As the school year progresses, questions asked should be more open-ended so that students identify relevant HNOS issues. For example, "What about the HNOS did [insert black box activity and classroom inquiry activity] have in common with scientists' efforts to determine the structure of DNA?

HNOS Learning Must be Assessed

Science teachers who plan for and effectively teach the HNOS as described above can be assured that their efforts will improve students' HNOS understanding. Nevertheless accurately identifying how well individual students understand and can apply particular HNOS ideas, and what struggles and misconceptions remain, demands incorporating HNOS assessments throughout the school year. Moreover, students often place more effort on what appears to be of consequence in a course and "assessment gives clear messages to students about what is important in the subject" (Dall' Alba et al., 1993, p. 633). As with HNOS instruction, HNOS assessment should occur throughout the school year in a variety of contexts including but not limited to exams, quizzes, laboratory activities, and readings. Many of the questions appearing in Table 4 could make fine HNOS assessment questions. Table 6 includes examples of assessment questions I incorporated throughout the school year as a high school teacher to assess my students' HNOS understanding.

SUMMARY

Bullet Point List of Main Chapter Ideas

- A deep and robust HNOS understanding has considerable value for science teachers and students, for science literacy, and socio-scientific decision-making.
- Several HNOS ideas are worth teaching, but they should be explored as questions rather than tenets.
- HNOS instruction should be deliberately planned and implemented, taking into account common and tightly held HNOS misconceptions.
- Because HNOS misconceptions are often tied together and make sense, truly changing those mistaken notions requires that HNOS instruction be incorporated throughout the school year.
- Effective HNOS instruction overtly draws students' attention to targeted HNOS ideas in a manner that requires students to mentally engage and wrestle with those ideas.

Table 6. Example HNOS Assessment questions (From Clough, 2011)

- How is an understanding of the nature of science important when looking at the biological evolution/creation/intelligent design public education controversy? [Question on an exam addressing biological evolution]
- How does the "Plant and Animal Cells" lab demonstrate that theory must precede observation? [Question to be answered in a cell biology laboratory report]
- In our genetics unit, you learned that at one time scientists, looking at the same data, disagreed whether DNA or protein was the genetic material. What does this and similar kinds of disagreements about the meaning of data illustrate about how science works? [Question on a biology exam addressing genetics]
- Science textbooks often claim that scientific laws are discovered. Using the conservation of mass law as an example, critique this claim. [Question on a chemistry exam addressing conservation of mass and balancing chemical equations.]
- People often wrongly think that scientific laws are superior to scientific theories. Use what you have learned about gas laws and kinetic molecular theory to correct this misconception. [Question on a chemistry exam addressing gases]
- List and defend at least three ways that your laboratory work to determine the products of the following chemical reaction

  $$NaHCO_{3\,(aq)} + CaCl_{2\,(aq)} \rightarrow \text{? products}$$

  accurately portrayed the NOS. List three ways it did not accurately portray the NOS. [Question on an exam addressing stoichiometry]
- Reflect on all the thinking you did in this inquiry laboratory activity. What scientific theories were guiding your thinking and explain how they guided your thinking. [Question that can be asked in most any science content inquiry laboratory activity where students have to make decisions such as how to set up their investigation, assess what data are relevant and irrelevant, how to account for their data, and what conclusion(s) are possible and probable]
- Compare and contrast how your science textbook presented the structure of the atom with the historical account presented in class. List at least five ways how your textbook's presentation of this content misportrayed the HNOS.

- Promoting a deep and robust HNOS understanding demands that the HNOS be taught in a variety of contexts along the decontextualized to highly contextualized continuum, with extensive scaffolding that assists students in drawing appropriate meaning from instruction.
- As with all cognitive objectives, students' understanding of HNOS should be assessed, but in a way that requires justification for positions rather than mere recall of NOS ideas.
- When HNOS is effectively taught, students learn how science is done along with the evidence and reasoning in support of science ideas, thus developing a deeper understanding of both HNOS and science content.

RECOMMENDED RESOURCES

Clough, M. P. (2015). *Role of visual data in effectively teaching the nature of science*. In K. D. Finson & J. Pedersen (Eds.), *Application of visual data in K-16 science classrooms*. Charlotte, NC: Information Age Publishing. Retrieved from http://www.infoagepub.com/products/Application-of-Visual-Data-in-K-16-Science-Classrooms

Clough, M. P. (1997). Strategies and activities for initiating and maintaining pressure on students' naïve views concerning the nature of science. *Interchange, 28*(2–3), 191–204.

National Academy of Sciences. (1998). *Teaching about evolution and the nature of science*. Washington, DC: National Academy Press. Retrieved from http://www.nap.edu/catalog/5787/teaching-about-evolution-and-the-nature-of-science

Understanding Science: How Science *Really* Works. http://undsci.berkeley.edu/index.php

Story Behind the Science: Bring Science and Scientists to Life. http://www.storybehindthescience.org/

The Science Teacher, Vol. 71, No. 9, November 2004. Special issue addressing the history and nature of science.

REFERENCES

Abd-El-Khalick, F., & Lederman, N. G. (2000). The influence of history of science courses on students' views of nature of science. *Journal of Research in Science Teaching, 37*(10), 1057–1095.

Akindehin, F. (1988). Effect of an instructional package on preservice science teachers' understanding of the nature of science and acquisition of science-related attitudes. *Science Education, 72*(1), 73–82.

Arya, D. J., & Maul, A. (2012). The role of the scientific discovery narrative in middle school science education: An experimental study. *Journal of Educational Psychology, 104*(4), 1022–1032.

Clough, M. P. (1995). Longitudinal understanding of the nature of science as facilitated by an introductory high school biology course. *Proceedings of the Third International History, Philosophy, and Science Teaching Conference, University of Minnesota, Minneapolis, MN*, 212–221.

Clough, M. P. (2006). Learners' responses to the demands of conceptual change: Considerations for effective nature of science instruction. *Science & Education, 15*(5), 463–494.

Clough, M. P. (2011). Teaching and assessing the nature of science: How to effectively incorporate the nature of science in your classroom. *The Science Teacher, 78*(6), 56–60.

Dall' Alba, G., Walsh, E., Bowden, J., Martin, E., Masters, G., Ransden, P., & Stephanou, A. (1993). Textbook treatments and students' understanding of acceleration. *Journal of Research in Science Teaching, 30*(7), 621–635.

Eflin, J. T., Glennan, S., & Reisch, G. (1999). The nature of science: A perspective from the philosophy of science. *Journal of Research in Science Teaching, 36*(1), 107–117.

Herman, B. C. (2015). The influence of global warming science views and socio-cultural factors on willingness to mitigate global warming. *Science Education, 99*(1), 1–38.

Herman, B. C., & Clough, M. P. (2016). Teachers' longitudinal NOS understanding after having completed a science teacher education program. *International Journal of Science and Mathematics Education, 14*(1), 207–227.

Herman, B. C., Clough, M. P., & Olson, J. K. (2013a). Teachers' NOS implementation practices two to five years after having completed an intensive science education program. *Science Education, 97*(2), 271–309.

Herman, B. C., Clough, M. P., & Olson, J. K. (2013b). Association between experienced teachers' NOS implementation and reform-based science teaching practices. *Journal of Science Teacher Education, 24*(7), 1077–1102.

Herman, B. C., Clough, M. P., & Olson, J. K. (In Press, On-Line First). Pedagogical reflections by secondary science teachers at different NOS implementation levels. *Research in Science Education.* doi:10.1007/s11165-015-9494-6. Retrieved from http://link.springer.com/article/10.1007/s11165-015-9494-6

Hong, H., & Lin-Siegler, X. (2012). How learning about scientists' struggles influences students' interest and learning in physics. *Journal of Educational Psychology, 104*(2), 469–484.

Khishfe, R., & Abd-El-Khalick, F. (2002). Influence of explicit and reflective versus implicit inquiry-oriented instruction on sixth graders' views of nature of science. *Journal of Research in Science Teaching, 39*(7), 551–578.

Lederman, N. G. (1992). Students' and teachers' conceptions of the nature of science: A review of the research. *Journal of Research in Science Teaching, 29*(4), 331–359.

Leite, L. (2002). History of science in science education: Development and validation of a checklist for analysing the historical content of science textbooks. *Science & Education, 11*(4), 333–359.

Medawar, P. B. (1963/1990). Is the scientific paper a fraud? In P. B. Medawar (Ed.), *The threat and the glory: Reflections on science and scientists.* New York, NY: HarperCollins.

Mitchell, S. (2009). *Unsimple truths: Science, complexity and policy.* Chicago, IL: University of Chicago Press.

Postman, N. (1995). *The end of education: Redefining the value of school.* New York, NY: Vintage Books.

Tao, P. K. (2003). Eliciting and developing junior secondary students' understanding of the nature of science through a peer collaboration instruction in science stories. *International Journal of Science Education, 25*(2), 147–172.

KEITH S. TABER

4. BELIEFS AND SCIENCE EDUCATION

INTRODUCTION

Beliefs and Commitments

Science teaching is not about persuading students to believe things. Indeed, it will be suggested in this chapter that it is – usually – inappropriate for science teachers to think about learning objectives in terms of what their students should *believe*.

This chapter will consider the nature of belief, and the nature of scientific knowledge, and explain why the role of the science teacher is *not* to tell the students things we want them to believe, but rather to teach students things we want them to doubt! The chapter will also explore some things that scientists generally need to accept to make it sensible for them to work in science. We might consider these as scientific beliefs, although it will be suggested that the term 'commitments' is preferred as it carries different associations to 'belief'. The chapter also considers the potential for science teaching to be resisted due to students' existing beliefs.

Things that Students Might Believe

There are many things we teach in science that students might go away from science lessons 'believing'. They might believe that:

- the pH of a strong acid is 1
- pure water has a pH of 7
- the methane molecule is a tetrahedral shape
- the Sun gives out light and heat
- ionic bonds form when metal atoms donate electrons to non-metal atoms
- the molecule ATP has energy-rich phosphate bonds
- a child gets their genes from their two parents equally
- the particles in solids are tightly packed with no space between them
- oxidation is the addition of oxygen or the loss of hydrogen
- everything is made of atoms

The reader might wish to pause at this point and consider their own response to each of these statements. You might consider:

K. S. Taber & B. Akpan (Eds.), Science Education, 53–67.

- Which, if any, of these statements are worthy of belief?
- Would you want your students to believe all, or some, of these statements?
- Are there any statements here that you would be concerned about your students believing?

Is Science a Factual Subject?

One issue that has recurred in some studies of science teaching and learning is an association of science with facts. Sometimes some younger children are actually attracted to science because of its high factual content: after all, our students often think, scientists can be certain about the world because they do experiments to prove things and so can know things 'for a fact'. However, whilst a factual association attracts some younger pupils, older students – adolescents – often find that science lessons – in some countries at least – comprise an endless parade of (what they perceive to be) facts to be learnt. This is often considered to make science difficult and even boring. Indeed even some of the students with the greatest potential for achievement in science – those with the most creative imaginations for example – may find classes packed with things to learn (rather than think about) dull.

Scientists certainly do collect a lot of information about the world, and much of this may seem 'factual' – it is certainly not fiction. However, science is centrally about understanding and explaining (and so being able to make predictions with some confidence) and so is inherently about developing *theoretical knowledge*, rather than simply accumulating facts. Theoretical knowledge is never certain, as it coordinates concepts and categories and generalisations, rather than simply cataloguing facts (Taber, Billingsley, Riga, & Newdick, 2015).

To consider an extreme example, to state that scientists at the European Organisation for Nuclear Research (CERN) have observed the Higgs boson (an elementary particle predicted to exist under the conditions of the early universe according to some theories) may seem like a simple statement of fact. However, the process of 'discovering' this particle involved thousands of scientists working for years with especially designed (and extremely expensive) apparatus and looking for very indirect evidence of the Higgs in extremely complex and voluminous data from especially built detectors. The observation depended upon building, testing and calibrating apparatus, and designing analytical protocols able to offer the scientists confidence to 'know' when they found the 'signal' they were seeking amongst the 'noise' (Knorr Cetina, 1999).

The 'discovery' of the Higgs boson at CERN in 2012 is then not a fact in the normal sense of the word, but the outcome of a highly complex process of construction and interpretation – a theoretical deduction more than an empirical observation. It may make sense *to believe* that scientists at CERN have reported their conclusion that they have observed Higgs: that might reasonably seem to be a fact; but it may be less appropriate *to believe* that scientists now know that the Higgs does exist, i.e. 'for a fact'. As scientists we should consider such knowledge as substantially theoretical, and open to possible revision in the future.

Things that are no Longer True

One problem with believing what we are taught in school science is that scientific knowledge moves on. There are a great many instances that could be considered, but here are just a few examples from the main branches of science.

When I was at school I was taught that there were two main kingdoms of living things – the animals and the plants. The biota of planet earth has not changed significantly since I completed school, but science has – and more sophisticated ways of understanding living things has been developed. To the scientist, animals and plants (and fungi) are in important ways quite similar compared to the rather different bacteria and archaea. Of course, plants have not actually become any more like animals (or animals any more like plants). Rather, as scientists have developed their understanding of the nature of living things, new ways of thinking about how to most usefully classify organisms have been proposed. So animals and plants now *seem* much more alike to scientists than they once did.

In particular, during the twentieth century cellular and genetic perspectives became increasingly central to the life sciences. Theoretical perspectives can be considered to be like mental filters that enable us to see the world in particular ways. It is almost impossible for a biologist trained today to appreciate what biology was like in the time of, say, Charles Darwin as – once acquired – the theories, ideas, and conceptual frameworks offered by modern biology inherently influence our very ways of looking at and thinking about the living world (Kuhn, 1970).

Another telling example concerns the noble gases (helium, neon, argon, etc.), which used to be called the inert gases. The name 'inert' gases implied that these elements did not undergo any chemical reactions. The noble gases do tend to be pretty non-reactive, so this was not a completely foolish idea. In part, the notion that the noble gases did not react was associated with a misuse of the octet rule in chemistry. The octet rule is a useful heuristic that can help work out likely charges on simple ions (Mg^{2+} not Mg^{3+} or Mg^{-}) and the stoichiometry of simple compounds (NH_3 not NH_2 or NH_4) on the basis that stable species tend to display particular electronic structures. However, this is only a general indication of stability: there are relatively stable species where the rule is not 'obeyed' and it certainly does not suggest that any species where the rule is followed will be so stable it does not react. Indeed, most chemical reactions occur between reactants that *are* stable according to the octet rule. We now know that some of the noble gases do form compounds, but it seems likely that this discovery was long delayed by the 'fact', a belief in effect, that these were 'inert' gases (Greenwood, 1964).

A similar brake on scientific progress was the idea that DNA codes for RNA, which codes for proteins. This scheme for a one-way flow of information became known as the 'central tenet' of molecular biology, and because it came to be believed (i.e. seen as factual knowledge, not theoretical knowledge) it delayed the recognition of occasions when there are exceptions to the scheme. Such exceptions include retroviruses that are able to use their RNA to code for new DNA within a host cell. These exceptions are not of purely academic interest as for example HIV, the virus associated with AIDS, is a retrovirus of immense significance to human wellbeing.

Another example of note comes from cosmology. The astronomer Fred Hoyle proposed a possible explanation of the origin of life on earth in terms of it being seeded from space in the form of very primitive organisms. The suggestion that living organisms, even simple ones, might survive being in space was generally considered fanciful. However, space scientists now find this idea perfectly reasonable – as we now know of extremophiles that can survive conditions that would not have been thought able to support life a few decades ago. A parallel case might be the geneticist Barbara McClintock who suggested the idea that genes sometimes 'transposed', or 'jumped', around a chromosome at a time when other scientists 'knew' (i.e. believed) this was not possible (Keller, 1983). McClintock's particular scientific heresy eventually became scientific orthodoxy, and she was awarded the Nobel Prize for her pioneering work.

As a final example, consider how it is that a heavy, largely metal, aircraft can fly. (The reader may wish to pause and consider if they know the scientific explanation here.) Physics textbooks and websites often suggest this is because of the Bernoulli Effect. A brief outline of this effect is given in the Textbox, and anyone not familiar with the effect may wish to refer to the Textbox before continuing.

> If one considers a closed system comprising a circuit of pipes of different radii, containing a circulating incompressible fluid, then the fluid must flow around all parts of the circuit at the same rate. If there are 10 ml of fluid passing some point each second then the flow rate must also be 10 ml/s at every other part of the circuit (as the fluid cannot 'back up' anywhere as there is no free space for that). Yet if the pipes are of different radii then in order to maintain a constant rate of flow (i.e. in mass or volume per second) the *speed* at which the fluid flows must vary around the circuit – the fluid must move faster in the narrower pipes so that the same amount of fluid (with its narrower cross-section) passes points there in any given time.
>
> Bernoulli suggested, from considerations of the conservation of energy, that where the pipes were narrower, and the fluid moved faster (so had more associated kinetic energy), the fluid pressure would be less. This can be tested, and shown to be the case.

Textbox: The Bernoulli Effect

The careful reader will have spotted that aircraft were not involved in the account of Bernoulli's principle offered in the Textbox (and indeed were not around when he published his idea in 1738). The ability of planes to stay in the air is often explained in terms of the cross-section of the wing (aerofoil). The modern aircraft wing has a shape such that there is a longer distance from the front to back of the wing across the upper surface than across the lower surface. Therefore, the argument goes, the air must move faster over the top surface of the wing to reach the trailing edge as soon, and so (according to the Bernoulli principle) the pressure is lower than on the bottom surface. The difference in pressure leads to a net upward force, which provides the lift the aircraft needs (i.e. the force that balances its weight when in flight). This may seem a convincing explanation, worthy of being believed. After all, it draws on some basic scientific ideas about force, and pressure, and the conservation of energy. However, many aircraft can fly upside down (as during air displays for example) when the 'lift', as well as the weight, should (according to the Bernoulli based account) act downwards.

Wind tunnel tests show that the airflow over the two surfaces of an aerofoil wing does *not* take the same amount of time (Babinsky, 2003) – and indeed there is no reason why it should, as the atmosphere is not an incompressible fluid trapped in a closed pipe! This should remind us not to believe everything we find in science textbooks. It is then more in keeping with the nature of science for teachers to encourage students to question and critique, than to believe, what they are taught in science lessons.

BELIEF

What is Meant by Belief?

A person's beliefs are those things they take to be true and certain. A person can certainly have beliefs about matters that can be tested empirically. Usually, however, when we think of people's beliefs we are concerned with matters where there is a strong commitment that is unlikely to be readily revised. Such commitments are often about matters that cannot be subjected to straightforward empirical testing.

Consider some examples of things that a person might believe:

- that they live in a material world that can be experienced through their senses;
- that they share the world with others who experience the world in similar ways, through similar perceptions, ideas, and feelings;
- that this world was created and is maintained by a supernatural being, God, who exists beyond the material realm of the creation;
- that illness is a result of the activity of spirits, immaterial beings that are either mischievous or evil, or that are being invoked through magical practices of another person.

People tend to become very committed to these kinds of beliefs, and do not tend to change such beliefs readily.

Metaphysical Commitments

A person's beliefs about the basic nature of the world that are *not* open to empirical testing are sometimes referred to as being metaphysical commitments (after the Aristotelian term 'metaphysics'). Consider, for example, the belief that we live in a material world that we experience through our senses. This might seem a pretty sound belief – after all, it may seem self-evident that this is indeed the case. However, philosophers might suggest that we have no absolute certainty about this. Perhaps we are dreaming that we live in the material world and are only imagining we sense it (sometimes when we dream we seem to be experiencing something real); or perhaps our brain has been carefully removed and is bathed in a sustaining fluid, whilst false information is being fed into its sensory nerve connections; or perhaps we have been involved in an accident and some novel new technique has been used to transfer our consciousness into an artificial neural matrix; or perhaps an omniscient God decided to think of a world He might create with sentient beings to inhabit it – and being all-knowing did not need to actually go beyond imagining it.

Some of these scenarios may seem fanciful, but the point is that we would have no way *to be certain* that the world we think we perceive is real and not an illusion, dream, or distorted perception of some kind. (The reader is challenged to suggest a test of this assumption – bearing in mind that they cannot assume sensory information is valid as that is part of what is being tested.) Even the apparent consistency of the world relies on a commitment to a belief that the memories we have now actually do reflect some previous experiences and are not being formed anew at this moment. It is not suggested here that such metaphysical commitments are unreasonable (although memories are not as fixed and reliable as people tend to think), just that they are just that – commitments (beliefs) that we logically have to adopt prior to examining any empirical evidence.

Our belief in others who experience the world much like us is a theoretically based belief – as again we can never be sure. Leaving aside the idea that other people might be imaginary, or automatons made to look like people, or even that they are evolved beings like us, but lack the conscious experience we uniquely do, we can have no way of knowing directly how other people experience the world. We can empirically test whether other people classify the same objects as 'red' as we do, but have no way of knowing just what their mental experience of seeing red is. Perhaps they experience red as we experience red – but perhaps they experience red as we experience blue, or indeed in a way that is different to anything we experience.

It seems *reasonable* that other people experience emotions as we do (they are genetically similar, and living in the same environment, and have learned to respond to what are usually seen as emotionally charged situations as we do) but it is not possible to be sure what they are feeling. They say they feel happy, or sad, or guilty or elated, and we may think we know what that is like – but we only know for sure what that is like *for us*, and not them. Yet in everyday life, we can only function in the social world by adopting a 'theory [sic] of mind' (Wellman, 2011) and assuming that – in effect *committing to the belief that* – other people's experience of the world is basically similar to our own.

METAPHYSICAL COMMITMENTS OF SCIENCE

It was suggested above that science should not be considered to be about belief, and that science teaching is not about getting students to believe in scientific ideas – which are often theoretical. However, science does involve the adoption of something like a belief system, a set of metaphysical commitments, which are necessary to make science a viable enterprise. So scientists do accept the existence of a material world that exists beyond our imaginations. This means that scientists believe that even though people may have different viewpoints or biases, or may have different evidence available to them, we all share the same material world, and so there is an objective level of reality that is common to everyone. For example, the scientific model of how atoms emit radiation of different frequencies is theoretical knowledge that is in principle open to revision, but whatever the reality actually is, it will be the same for any of us. This can be contrasted to a relativist stance that sees reality as relative to a particular context. Relativism may be sensible in the social world (ideas about what is an ordered and productive classroom may vary across different national contexts), but not in the natural sciences. If a researcher wanted to study the nature of progression from school science to university science then it would be foolish to expect there to necessarily be one answer that applied across different times and cultural contexts. Yet scientists assume that the charge to mass ratio of an electron remained the same in 1950, 1980 or 2010, and regardless of whether the measurement was made in the Netherlands, or in Nigeria, or in New Zealand.

So science makes a commitment to the possibility of objective observations and measurements – the idea that replacing one qualified and careful researcher by another should not in principle make a difference to the results of an observation. Scientists also make a metaphysical commitment to the stability of the material world. That is, despite observing many changes, scientists believe that at some fundamental level there is a stable reality such that foundational principles that apply today will have applied in the past and will continue to apply in the future. If the universe is expanding, if the sun will burn out, if perhaps even the universal gravitational constant [sic] may change during the evolution of the universe, there are – scientists assume – still underlying fixed principles from which these changes logically follow (and which scientists can aspire to investigate).

Scientists also make a metaphysical commitment to the idea that the world is knowable and can be understood. Scientists believe that human senses (often aided by various apparatus constructed by human ingenuity) are sufficient for observing the universe and that human cognition is sufficient to understand it. Scientists will differ in the degree to which they think humans are capable of understanding nature – so some think there are limits, and others feel that one day everything will be revealed through science (a 'scientistic' notion). However, for anyone to sensibly work as a scientist, they need to commit to the belief that human perception and cognition are sufficient and suitable to develop useful (if not absolute and unerring) knowledge of the world.

Scientific Values

These metaphysical commitments are often supported by a value system that most scientists will adopt – consisting of what have been labelled as constitutive values of science because they are "the source of the rules determining what constitutes acceptable scientific practice or scientific method" (Longino, 1983, p. 8). Scientific values include always seeking to avoid bias influencing scientific work, evaluating scientific work in terms of evidence and arguments rather than the authority of particular scientists, always being open to reconsidering a conclusion in the light of new evidence, always exploring alternative explanations before adopting a persuasive explanation – and in general to question everything and never take anything for granted. These are of course ideals, which are lived up to varying degrees by the real humans doing science.

Some scientists may also adopt personal metaphysical commitments that inform their scientific work which are not common commitments across all scientists. One common heuristic that may become a commitment for scientists is that the simplest explanations with the least number of auxiliary assumptions are to be preferred (Occam's razor). Some scientists are also guided by metaphysical commitments that nature is elegant, beautiful and/or highly symmetrical at a fundamental level (Girod, 2007).

Some centuries ago a common stance among scientists in Europe was natural theology – the idea that the world was the work of God, and doing science was a way to come to know God through his works. If, on religious grounds, a scientist believes that people can understand nature (because God has set up the world such that people can come to know His works) then this can motivate the metaphysical commitment that science can provide reliable knowledge of the world – i.e., that human senses and cognition are suitable for making good sense of nature (Yeo, 1979).

Scientism and Materialism

There are various forms of what is known as 'scientism' (Stenmark, 1997). As suggested above, some people take very optimistic views about science, such as that everything can be explained, that science is able to explain it, and even that in time scientists *will* understand and explain everything. That is quite an act of faith, and probably only a minority of scientists adopt such extreme scientistic views.

Some people who adopt scientism consider that science is the *only* source of genuine and reliable knowledge. A complementary position to considering that everything should be open to scientific explanation is to adopt the attitude that only things that *are* open to scientific investigation should be considered as real.

This implies atheism – the denial of *the possibility* of the existence of a God or any other entity considered to be supernatural (i.e., to exist outside or beyond the natural world). This is a form of materialism sometimes called ontological materialism. Some scientists suggest that all their scientist colleagues should adopt this position as a common commitment, but very many scientists would reject such a view (Berry, 2009). In an earlier time it was suggested that the natural stance for a scientist was to remain uncertain about God or other entities which were not open to scientific investigation – something sometimes labelled agnosticism. (It is common for people who are unsure of the existence of God to call themselves agnostic, but the true agnostic holds that it is *not possible* in principle to be certain about the existence or otherwise of God.)

Many scientists around the world have religious beliefs (Coll, Lay, & Taylor, 2008) and so are clearly neither ontological materialists nor agnostics. But there is a sense in which scientists are usually expected to be 'agnostics-when-at-work' or what is termed *methodological* materialists. This does not involve denying the existence of the supernatural, but rather committing to limiting scientific explanations to natural causes. It is widely considered that a 'God of the gaps approach' (where God is invoked as an explanation for the things science cannot yet explain) is both bad science and bad theology. Instead, scientific explanations should be limited to material causes and mechanisms, but are not considered to exclude the possibility of a complementary explanation. For example, many scientists may believe both that the world is as it is because this is God's will and creation, but also that science (cosmology, evolutionary biology, geology, etc.) tells us about the mechanisms by which this creation was brought about. Clearly these are independent beliefs and individuals may commit to both, one, or neither.

A commitment to methodological materialism leads to a rejection of 'intelligent design' ideas that suggest that because science cannot (yet) explain all of the details of evolution, we should assume the hand of some deliberate designer. Whether such a designer (God?) exists or not, scientists are committed to explaining all aspects of the natural world in terms of natural causes and mechanisms. Just because it may not be obvious how – for example – natural selection led to the bacterial flagellum, that is not a good reason to give up looking and simply explain it away as a 'design' feature. If a student asked a science teacher why a microscope was able to magnify an image, it would be a poor teacher whose 'explanation' was limited to replying that it had been designed to do just that! That might be an accurate response, but not a scientific explanation.

STUDENTS' BELIEFS AND LEARNING SCIENCE

Students come to their science lessons with various beliefs about the natural world, and sometimes – quite often actually – these are inconsistent with scientific accounts. For the present discussion, we will distinguish three categories of student beliefs:

- those which concern natural mechanisms and causes for natural phenomena;
- those which concern supernatural causes underlying the natural world;
- those which concern supernatural mechanisms and causes for natural phenomena.

It is important here to distinguish beliefs that relate to values and practices from those that are actually about how the world works. For example, students from some indigenous cultures believe in showing respect for, and recognising dignity in, all living things. This may mean such students are uncomfortable about being taught in a science classroom where living animals are kept in cages. Such beliefs are not inherently anti-scientific, even though they lead to rejecting some common scientific practices. (We might note that some scientific practices in relation to animals that were commonplace at one time are now widely seen as unethical and unacceptable.) Faced with such a situation the sensitive science teacher may decide the inclusive approach is to respect the values of these students and not keep caged animals in the classroom.

Alternative Conceptions of Natural Mechanisms

There is a vast literature revealing that students have their own takes on many natural phenomena (Duit, 2009). Often these ideas are inconsistent with scientific accounts, and sometimes (but by no means always) students are strongly committed to their alternative conceptions. One example would be that most of the material in a tree has come from the ground through the roots. Such conceptions can seriously interfere with science teaching (and they are the topic of Chapter 9: *The Nature of Student Conceptions in Science*).

Religious Beliefs that may be Compatible with Science

Many students come to science lessons with strong religious beliefs. They may for example believe that the world was created by God and is maintained by His grace. As suggested above, such beliefs are neither supported, nor contradicted, by science and such beliefs are shared by many scientists and science teachers in parallel with the conviction that science offers us the best way to understand the natural mechanisms at work in the world.

A problem for the science teacher here is that students may not appreciate that scientific and religious accounts can be understood to be complementary (Taber, Billingsley, Riga, & Newdick, 2011). So, for example, a student may consider that the big bang model of the origin of the universe necessarily competes with the idea of the world being God's creation. Here the science teacher needs to help students appreciate that science is not seeking to refute or compete with religion, but rather offers a different layer of explanation. An analogy might be a scientific analysis of a painting, which can reveal information, for example, about the pigments used, without competing with an artistic evaluation of its aesthetic qualities or a biographical account of why the painter made that particular artwork.

Beliefs in Supernatural Mechanisms that Compete with Science

Sometimes, however, students may have strongly committed beliefs that *are* inconsistent with scientific accounts of the natural world. One example is some of the teachings of 'young earth creationism' in some branches of Christianity (Long, 2011). These may include the belief that the earth is no more than 10 000 years old (rather than more like 4 600 000 000 years suggested by scientific evidence) and that each main type of living thing was created separately with evolution only producing variations on these basic kinds (where the scientific account has all living things on earth descending from one common ancestral type). As another example, students from some communities may hold alternative notions of ill health and disease based on beliefs in such ideas as magical spells and spirit possession (Foster, 1976). These ideas are clearly inconsistent with scientific models of disease based on – for example – infection with microorganisms.

In this situation the science teacher will wish to persuade her students of the merits of the scientific accounts – especially when these are well established and supported by overwhelming evidence. However, it may be counter-productive to proceed by setting out to show students their existing beliefs are false. Often student beliefs of this kind are strongly based on family and community commitments, and reinforced by (and associated with the authority of) parents and elders, so setting up a competition between science and existing beliefs may be counterproductive and damage students' attitudes to science as a school subject and more generally.

TEACHING SCIENCE AS THEORETICAL KNOWLEDGE

Earlier in this chapter it was emphasised that science should not be taught as facts to be believed, but rather as theoretical knowledge. This means we are teaching viable explanations consistent with evidence, but bearing in mind that sometimes there may be other possible explanations that could fit the evidence as well, and that we should always be aware the evidence base could shift. Science presents models that are simplifications, and often have limited application (or at least are broad generalisations that have only been tested under limited ranges of conditions.)

Whilst not a total solution to the challenge of student beliefs being inconsistent with scientific accounts, such an approach to teaching science avoids a head-on confrontation between what we teach in science and what our students may believe. Our job is not to persuade students to believe that natural selection or the germ theory of disease is true – but to get them to see why these are strongly motivated by the extensive evidence available; to understand the scientific account; and to be able to see how to apply the ideas where relevant (Taber, In press).

This also addresses the pedagogic problem that most of what we teach in school science comprises of curriculum and teaching models that simplify the science, and will need to be developed, extended, and complemented if the student continues with their studies. Students should never need to be told that what they have studied before was 'wrong' and must be 'forgotten' – but they will need to progress to more sophisticated models (see Table 1). We should avoid teaching over-simplifications, but as long as we teach scientific ideas as models (and students have been taught to understand what that means, see Chapter 20: *Models and modelling in science and science education*) and not facts, then shifting to more sophisticated accounts will be less problematic.

Table 1. If ideas learnt about science are seen as 'facts' then they may impede the development of more sophisticated scientific accounts

Proposition that might be learnt in science	*Why belief in the proposition as a fact may impede science learning*
The pH of a strong acid is 1	This is what universal indicator paper suggests when testing typical lab samples – but the pH depends upon both strength and concentration. A very dilute solution of a strong acid could have, e.g. pH4, and at the usual concentration 'bench' acids have pH<0.
Pure water has a pH of 7	The degree of ionisation of water is temperature dependent so the pH (and pOH) of pure water is greater than 7 at 0˚C and less than 7 at 100˚C.
The methane molecule is a tetrahedral shape	It is not clear how we best understand the shape of something on the scale of molecules where matter is more 'fuzzy'. An electron density map will not follow a tetrahedral pattern for example. (We get a tetrahedron by treating the hydrogen centres as points, joining them up, and considering the result as a solid shape. That does not fully represent the actual arrangement of matter in the molecule.)
The Sun gives out light and heat	Heat is often defined in science as energy transferred due to a difference in temperature, so then the light (and u.v. and infrared radiation) emitted by the sun and absorbed by the cooler earth is part of the net heat transfer.
Ionic bonds form when metal atoms donate electrons to non-metal atoms	The common ways of forming ionic compounds in the school lab (precipitation reactions, or neutralisation followed by evaporation) involve ionic bonds formed between ions that already existed in the reactants. Even when there is a direct reaction between the elements, the reactants do not comprise discrete atoms.

Table 1. (Continued)

Proposition that might be learnt in science	*Why belief in the proposition as a fact may impede science learning*
The molecule ATP has energy-rich phosphate bonds	Breaking the 'energy rich' bond in ATP (as in breaking any bond) requires an input of energy,
A child gets their genes from their two parents equally	The Y chromosome has less genetic material than the X chromosome. (The Y chromosome is like an incomplete X chromosome that seems to be slowly getting smaller over successive generations.) Mitochondrial genes are only passed down the female line.
The particles in solids are tightly packed with no space between them	Yet we teach that a solid can contract or expand when the temperature changes according to the amount of space the particles have to vibrate around!
Oxidation is the addition of oxygen or the loss of hydrogen	Or it could be the loss of an electron, or an increase in the formal oxidation state. These definitions may apply when there is no oxygen or hydrogen involved.
Everything is made of atoms.	This could mean everything was made *from* atoms – but actually this is a way of conceptualising, not a 'historically' accurate notion – little of the material in the universe from which material on earth has formed is atomic (rather than plasma, or molecules, or ions). This could alternatively mean that everything *contains* atoms – but again this is a way of conceptualising (like thinking of a person as just a torso, head, arms and legs, placed together rather than connected through continuous tissues) as very few common materials actually comprise of atomic matter. E.g. a molecule is a distinct arrangement of subatomic particles, not just atoms placed next to each other.

CONCLUSIONS

Science teachers should not see the science curriculum as comprising of known facts to be believed and taught as truths. Rather, science teaching should be about presenting students with the models, principles, theories etc. developed by scientists and trying to get them to see why, on the basis of observations and other evidence, scientists feel these are the most useful ways currently available of thinking about the world (but not proven, absolute truths). Such a perspective gives a better reflection of the authentic nature of science, helps avoid clashes with some students' religious and cultural beliefs, and provides a more educationally sound way of helping students cope with the progression of increasingly more sophisticated models they will meet as they progress through their science education.

FURTHER READING

Barbour, I. G. (2000). *When science meets religion: Enemies, strangers or partners?* San Francisco, CA: HarperCollins.

Reiss, M. J. (2008). Should science educators deal with the science/religion issue? *Studies in Science Education, 44*(2), 157–186.

Taber, K. S. (2013). Conceptual frameworks, metaphysical commitments and worldviews: The challenge of reflecting the relationships between science and religion in science education. In N. Mansour & R. Wegerif (Eds.), *Science education for diversity: Theory and practice* (pp. 151–177). Dordrecht: Springer.

REFERENCES

Babinsky, H. (2003). How do wings work? *Physics Education, 38*(6), 497.

Berry, R. J. (Ed.). (2009). *Real scientists real faith*. Oxford: Monarch Books.

Coll, R. K., Lay, M. C., & Taylor, N. (2008). Scientists and scientific thinking: Understanding scientific thinking through an investigation of scientists views about superstitions and religious beliefs. *Eurasia Journal of Mathematics, Science & Technology Education, 4*(3), 197–214.

Duit, R. (2009). *Bibliography – Students' and teachers' conceptions and science education*. Kiel, Germany: IPN – Leibniz Institute for Science and Mathematics Education.

Foster, G. M. (1976). Disease etiologies in non-Western medical systems. *American Anthropologist, 78*(4), 773–782.

Girod, M. (2007). A conceptual overview of the role of beauty and aesthetics in science and science education. *Studies in Science Education, 43*(1), 38–61. doi:10.1080/03057260708560226

Greenwood, N. N. (1964). Chemistry of the noble gases. *Education in Chemistry, 1*(1), 176–188.

Keller, E. F. (1983). *A feeling for the organism: The life and work of Barbara McClintock*. New York, NY: W H Freeman and Company.

Knorr Cetina, K. (1999). *Epistemic cultures: How the sciences make knowledge*. Cambridge, MA: Harvard University Press.

Kuhn, T. S. (1970). *The structure of scientific revolutions* (2nd ed.). Chicago, IL: University of Chicago.

Long, D. E. (2011). *Evolution and religion in American education: An ethnography*. Dordrecht: Springer.

Longino, H. (1983). Beyond "Bad Science": Skeptical reflections on the value-freedom of scientific inquiry. *Science, Technology, & Human Values, 8*(1), 7–17. doi: 10.2307/688902

Stenmark, M. (1997). What is scientism? *Religious Studies, 33*(01), 15–32. doi:10.1017/S0034412596003666

Taber, K. S. (In Press). Knowledge, beliefs and pedagogy: How the nature of science should inform the aims of science education (and not just when teaching evolution). *Cultural Studies in Science Education.*

Taber, K. S., Billingsley, B., Riga, F., & Newdick, H. (2011). Secondary students' responses to perceptions of the relationship between science and religion: Stances identified from an interview study. *Science Education, 95*(6), 1000–1025. doi: 10.1002/sce.20459

Taber, K. S., Billingsley, B., Riga, F., & Newdick, H. (2015). English secondary students' thinking about the status of scientific theories: Consistent, comprehensive, coherent and extensively evidenced explanations of aspects of the natural world – or just 'an idea someone has'. *The Curriculum Journal, 26*(3), 370–403. doi:10.1080/09585176.2015.1043926

Wellman, H. M. (2011). Developing a theory of mind. In U. Goswami (Ed.), *The Wiley-Blackwell handbook of childhood cognitive development* (2nd ed., pp. 258–284). Chichester, West Sussex: Wiley-Blackwell.

Yeo, R. (1979). William Whewell, natural theology and the philosophy of science in mid nineteenth century Britain. *Annals of Science, 36*(5), 493–516. doi:10.1080/00033797900200341

MARÍA PILAR JIMÉNEZ-ALEIXANDRE
AND BEATRIZ CRUJEIRAS

5. EPISTEMIC PRACTICES AND SCIENTIFIC PRACTICES IN SCIENCE EDUCATION

INTRODUCTION: STUDENTS' PARTICIPATION IN THE EPISTEMIC GOALS OF SCIENCE

There is a growing consensus in considering that learning science involves students' participation in the epistemic goals of science (Duschl, 2008; Kelly, 2008) or that, as Duschl (2008) proposes, science education should balance conceptual, epistemic and social learning goals. By epistemic goals we mean goals related to how we know what we know, to how scientific knowledge is constructed. Thus for instance understanding the criteria for evaluating explanations, theories or models, or the criteria for choosing one explanation over alternative ones. The main argument of this chapter is that these purposes may be achieved through placing scientific practices at the centre of science teaching and learning, in an approach that pays attention to their epistemic and social dimensions, besides the conceptual ones. This would mean shifting the focus towards the development and modification of epistemic claims (Duschl & Jiménez-Aleixandre, 2012), of claims related to scientific knowledge, in a perspective conceptualizing epistemic cognition as a practice (Kelly, 2016).

The chapter discusses, first, characterizations of epistemic cognition and epistemic practices, as well as the relationships between scientific and epistemic practices; second, characterizations of scientific practices and the translation of these theoretical approaches to policy; third, how to support students' engagement in the practices of modelling, argumentation and planning and carrying out investigations.

EPISTEMIC COGNITION AS A PRACTICE

We may say that the purpose of epistemic practices is to generate knowledge about the world. Epistemic practices (EP) are characterized in a variety of ways. For Kelly (2008) they constitute particular social practices, which are "patterned set of actions, typically performed by members of a group based on common purposes and expectations, with shared cultural values, tools and meanings" (Kelly, 2008, p. 99). He defines epistemic practices as "the specific ways members of a community propose, justify, evaluate, and legitimize knowledge claims within a disciplinary framework" (ibid, p. 99), and distinguishes three types within them, associated with producing, evaluating and communicating knowledge. Drawing from a sociocultural

K. S. Taber & B. Akpan (Eds.), Science Education, 69–80.

perspective and the notion of learning through participation in activities, Kelly (2016) conceptualizes epistemic cognition as a practice, proposing that epistemic practices are constructed in social interaction, and that they include interactionally accomplished understandings of knowing. According to Wickman (2004), who shares this sociocultural, situated cognition approach, epistemic practices reveal students' underlying *practical epistemologies* or epistemologies used in specific practices. This perspective focuses on practical epistemologies as actions, rather than as beliefs, considering that students' and teachers' *actions* are situated in an activity.

A complementary perspective, grounded on philosophy and psychology, is the AIR (Aims, Ideals, Reliable processes) model, developed by Chinn, Buckland and Samarapungavan (2011). Chinn et al. characterize epistemic cognition in terms of aims, standards and criteria (ideals), and reliable processes for attaining epistemic achievements. In this model, epistemic aims are goals related to finding things out, understanding them and forming beliefs. As Chinn et al. note, knowledge is the most discussed epistemic aim. Standards and criteria relate for instance to the specific standards people use to evaluate knowledge claims, or to select evidence. Standards may refer also to the consistency of a belief or knowledge system. The third component concerns the cognitive and social processes by which knowledge and other epistemic aims are achieved.

The appropriation of criteria for justifying knowledge or for revising models is a relevant component of epistemic cognition. Duschl (2008) argues that an understanding of criteria for evaluating knowledge claims, that is, deciding "what counts" (as evidence, as justification, etc.), is as important as an understanding of conceptual frameworks for developing knowledge claims. Duschl's conclusion is that conceptual and epistemic learning should be concurrent in science classrooms. He suggests a need for balancing conceptual, epistemic and social goals in science education, doing so by focusing on three integrated domains: (1) the conceptual structures and cognitive processes used when reasoning scientifically; (2) the epistemic frameworks used when developing and evaluating scientific knowledge; and (3) the social processes and contexts that shape how knowledge is communicated, represented, argued, and debated.

There are studies focusing on epistemic practices (EP), others focusing on scientific practices (SP), and sometimes these terms are used interchangeably. We think that, for analytical research purposes, it may be necessary to treat them as different notions, although there is a degree of overlapping between them, in particular in classroom contexts where they may blend. Students engaged in SP may be at the same time involved in EP, as discussed in the last section. Tentatively, we suggest that we can think of epistemic practice as a broader construct and of scientific practices as epistemic practices in the context of specific learning contexts or content areas. Figure 1 represents this overlapping. There are, however, some scientific practices – for instance measuring – which are not epistemic, and thus the overlapping is not complete.

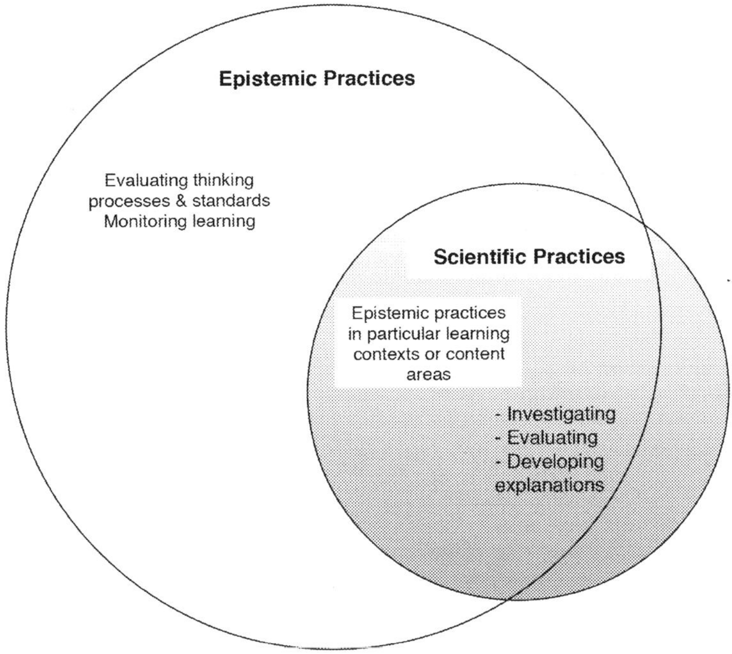

Figure 1. Relationships between epistemic practices and scientific practices

These perspectives represent a shift in focus from examining learners' epistemological beliefs, towards examining their engagement in epistemic practices. This change is translated into policy or evaluation documents. Thus for instance, the Program for International Students Assessment's (PISA) draft framework for 2015 (OECD, 2013) places as one of science education goals helping students to become scientifically literate citizens. However, as this framework acknowledges, scientific literacy requires not only knowledge about scientific concepts and theories but also about scientific practices and how they enable science to advance.

One of the aims of engaging students in scientific practices is to build knowledge about the nature of scientific endeavour, and about how knowledge is constructed; in other words to promote the development by students of an understanding of what scientists do (Osborne, 2011). The notion of teaching and learning sciences in a way consistent with scientific work is not a new one, it had been advanced by Dewey; what is new is the approach framing this participation as part of a coherent whole.

SCIENTIFIC PRACTICES AT THE CENTRE OF SCIENCE TEACHING

How can we promote students' participation in the epistemic goals of science? A way of achieving it is through placing scientific practices at the centre of science teaching and learning. Doing so would mean paying attention to the epistemic and social dimensions of science education, besides the conceptual ones. The notion of practices embodies a move "from viewing science as a set of processes to emphasizing, also, the social interaction and discourse that accompany the building of scientific knowledge in classrooms" (Reiser, Berland, & Kenyon, 2012, p. 8). For these authors, the practices involve doing the work of building knowledge in science and understanding why we build, test, evaluate and refine knowledge as we do. This is coherent with an approach that views science as consisting of a set of scientific practices (Osborne, 2014). Osborne's argument is that science education needs to include "explaining how we know what we know or why we believe what we do" (Osborne, p. 580), for doing so will contribute to a commitment to evidence as the epistemic basis of beliefs. He considers this commitment as one of science major contributions to contemporary culture, one that promotes rationality and critical thinking.

The Role of Activity and Purpose in Practice-based Approaches

It needs to be emphasized that a defining feature of scientific practices is *activity*. Students should be *engaged* in scientific practices, *carrying out* modelling, argumentation or investigation. It is also an engagement in discourse and social interactions, rather than only in experiments or hands-on. In his approach to epistemic cognition in practice Kelly (2016) proposes learning contexts where meanings are defined and socially negotiated around *purposeful activity*. Berland et al. (2016) offer a framework, the Epistemologies in Practice (EIP) for this practice-based approach to science education. The EIP seeks to distinguish students' reflective participation in constructing and evaluating knowledge from mere attainment of skills. For Berland and colleagues, "understanding science as participation in practices offers an explanation for these challenges: this perspective underscores that the work of science is part of an ensemble of activity such that the tasks are part of a coherent network of purposeful action" (Berland et al., p. 2). The EIP approach emphasizes two aspects of students' engagement in scientific practices: their epistemic goals for knowledge construction, and their epistemic understanding of how to engage in it. We agree with these authors in the relevance of purpose and purposeful activity in science education at all age levels. Monteira and Jiménez-Aleixandre (2016) discuss the role of active purposeful observation in kindergarteners' engagement in the scientific practice of using evidence. In the context of kindergarten they define it as prolonged systematic observation that has a clear focus, is guided by the teacher, recorded, explicitly discussed, and used to test claims and to revise initial models.

Scientific Practices in Policy Recommendations

The practices approach is being translated from science education research to policy documents, namely the National Research Council (NRC, 2012) framework and the New Generation Science Standards, NGSS (Achieve, 2013) in the U.S., which propose that science education be built around three dimensions: scientific and engineering practices, crosscutting concepts, and disciplinary core ideas. These suggestions, grounded on the idea that science requires both knowledge and practice, are framed in studies documenting that students' understanding of science and their competency in performing scientific investigations require understanding the specific disciplinary practices of science. This chapter focuses on the dimension of scientific practices. As the NRC framework acknowledges, the term *practices* instead of others such as *skills*, emphasizes that engagement in scientific inquiry requires coordination of skills with knowledge specific to each practice. It also emphasizes that students should *engage* themselves in the practices rather than merely learn about them. According to the NRC, the advantages of a focus on practices are that it avoids first, the tendency to overemphasize experiments over argumentation, critique or modelling; second the tendency to teach procedures in isolation from science content. It also promotes the acknowledgement of the existence of a broad spectrum of methods, rather than one single "scientific method".

The NRC (2012) framework frames scientific and engineering practices in three spheres of activity: Investigating, evaluating and developing explanations and solutions. Osborne (2011; 2014) provides a rationale for that model grounded on psychology and philosophy.

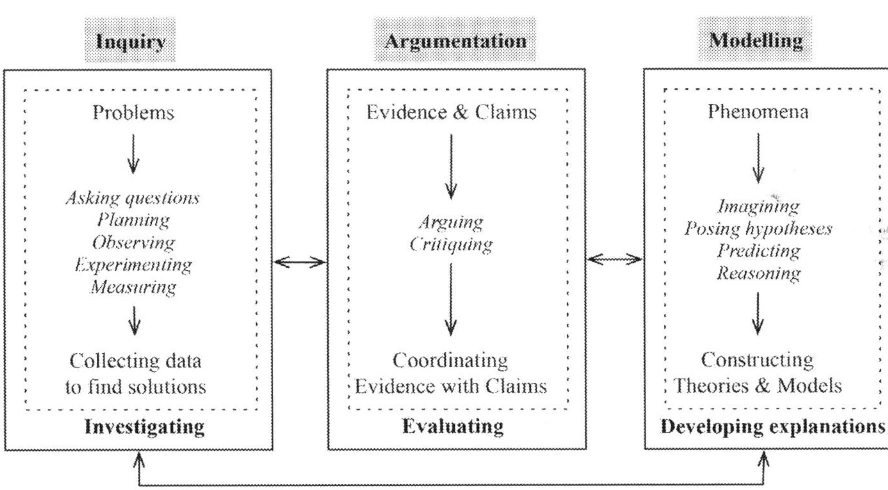

Figure 2. Three broad practices of scientific activity (modified from NRC, 2012)

Figure 2, based on the NRC (2012) framework represents how the three spheres of science are interrelated. In this modification we have associated the three domains or spheres from NRC (in bold at the bottom of each box) to the three scientific competencies (in bold over each box) from the PISA 2015 draft framework

(OECD, 2013): Evaluate and design scientific inquiry (inquiry), interpret data and evidence scientifically (argumentation), and explain phenomena scientifically (modelling). We propose that these three PISA competencies correspond to the three broad practices or spheres of science in the NRC. Operations making part from each practice are summarized in italics, and we have introduced some new ones in order to be coherent with the nature of each practice. For instance "planning" is introduced in "Investigating" to be coherent with the practice of *Planning and carrying out investigations*. The main goal for the operations is listed in each domain: investigating in order to collect data to find solutions to problems; evaluating evidence, data and claims in order to coordinate them; interpreting phenomena and developing explanations in order to produce theories and models.

Thus the activity of science may be synthesized in three spheres or overarching practices:

- *Investigating*: which involves asking questions, identifying problems, planning and carrying out investigations, or analysing and interpreting data.
- *Developing explanations*: which involves posing hypotheses, interpreting phenomena, formulating predictions or constructing and using theories and models.
- *Evaluating*: which involves selecting appropriate evidence, contrasting explanations against available evidence, comparing alternative explanations and critiquing them, or constructing arguments from evidence.

The three spheres contain the eight scientific practices proposed by the NRC (2012) and reproduced in Table 1:

Table 1. The eight scientific practices proposed by the NRC (2012)

Asking questions and defining problems	Using mathematics and computational thinking
Developing and using models	Constructing explanations and designing solutions
Planning and carrying out investigations	Engaging in argument from evidence
Analysing and interpreting data	Obtaining, evaluating, and communicating information

These eight practices may overlap in some cases; they are intertwined (Bell et al., 2012), and are not carried out in isolation. In the next section we discuss examples of how to support student' engagement in them.

ENACTING SCIENTIFIC PRACTICES IN SCIENCE CLASSROOMS

How are scientific practices enacted in the classroom? According to Bybee (2011), when students engage in scientific practices "activities become the basis for learning about experiments, data and evidence, social discourse, models and tools, and mathematics, and for developing the ability to evaluate knowledge claims, conduct empirical investigations, and develop explanations" (Bybee, 2011, p. 38). Therefore school science programs need to actively involve students through investigations and activities, including hands-on, laboratory work and minds-on tasks. In this section we discuss examples of tasks designed in order to engage students in investigating, argumentation and modelling.

Investigating: Which Toothpaste is Ineffective in Preventing Cavities?

This laboratory task, designed for secondary school students, requires them to *plan and carry out an investigation*, in order to compare the effectiveness of two toothpastes for preventing tooth decay. An excerpt of the handout is: "A campaign aimed at preventing tooth decay was conducted in schools, giving students two brands of toothpastes (x and y). Soon after, it was found that students having brushed their teeth with one of the toothpastes had more cavities than students using the other brand. So, we need to find out which toothpaste does not prevent cavities in order to withdraw it from shops. *Design an experiment to find out which toothpaste is less effective*. To do this you can use clamshell pieces to simulate teeth and hydrochloric acid to simulate the environment created in the mouth after eating carbohydrates." The task required two 50-minute sessions, one for planning the experiment, and a second one to implement it. Table 2 unpacks some components of the three broad scientific practices enacted by 9th grade students (Crujeiras & Jiménez-Aleixandre, 2015).

These operations, empirically identified in our study, are aligned with the 5D components proposed by Duschl and Bybee (2014) for the practice of *Planning and Carrying out Investigations* (PCOI): (1) Deciding what and how to measure, observe, and sample; (2) Developing or selecting procedures/tools to measure and collect data; (3) Documenting and systematically recording results and observations; (4) Devising representations for structuring data and patterns of observations; and (5) Determining if (1) the data are good (valid and reliable) and can be used as evidence, (2) additional or new data are needed, or (3) a new investigation design or set of measurements are needed. The focus of the task is on investigating, but the three practices are intertwined.

Evaluating Knowledge: Argumentation in Socio-Scientific Contexts

Evaluating knowledge is a practice that plays a crucial role in building scientific knowledge. In science knowledge claims are contrasted with available evidence in order to be accepted. Argumentation is this process of knowledge evaluation, involving connecting evidence to claims through justifications (Jiménez-Aleixandre & Erduran, 2008), these last called "reasoning" by McNeill and Krajcik (2012).

Table 2. Scientific practices students engage in when performing the task. Legend: SP = scientific practice; I = investigating; M = modelling; A = argumentation

SP	*Operation*	*How is it carried out in the task*
I	Planning the investigation	To decide how to identify the ineffective toothpaste: • *Proposing samples, equipment and procedure*, for instance: a) To use three shells (one for each toothpaste, and one control), weighing them. b) To measure a small volume (e.g. 10 mL) of HCl in a test tube. c) To place each shell inside a balloon, and the balloon at the top of the test tube and dropping the shell into the acid (or other procedures). • *Recording data*, for instance: a) Once the shell contacts the acid and the gas release starts, to measure the time until the balloon stands up. b) To record time values. • *Selecting a criterion in order to identify the ineffective toothpaste,* as: the balloon that stands up sooner will be the one containing the shell washed with the ineffective toothpaste. • *Considering fair testing and reproducibility criteria*: To use equal volumes of HCl, and clamshells of the same weight in all the experiments; to repeat each experiment at least twice.
	Carrying out the investigation	• *Carrying out* the chemical reaction with each shell and measuring the time that takes to each balloon to stand up. • *Collecting experimental data and representing them* in tables and graphs.
M	Using theoretical models	• *Understanding and applying the models* of chemical reaction and inhibition processes.
	Using analogies and simulations	• *Understanding and applying the relationship between the elements used to simulate* tooth decay (e.g. clamshells and HCl) and their targets in the natural world.
A	Interpreting evidence	• *Deciding if the data collected are valid and sufficient* in order to identify the ineffective toothpaste. *Applying the criteria* to the data.
	Linking claims to evidence	• *Concluding* which toothpaste is ineffective in preventing cavities and *justifying* the conclusion in the light of the data.

Engaging in argumentation means not only comparing alternative explanations and selecting the one that best fits with evidence, but also critiquing the ones that are unsatisfactory.

There is a wealth of research papers about argumentation, and resources including learning tasks (e.g. McNeill & Krajcik, 2012; see also the resources section). Two instances, summarized in Figure 3, are part of teaching sequences (Puig, Bravo, & Jiménez-Aleixandre, 2012) about genetics and ecology.

1. How do you explain black sprinters' achievements in athletics?
Since the athletics world championship in Rome in 1987, when there were three white finalists in the 100 m, black sprinters took all the final positions in the Olympic and World Championships. There are different explanations to these achievements:

A) This is due to their genes
B) This is due to the influence of factors such as nourishment, training, etc.
C) This is due to a combination of A and B

Your tasks:
1) From the available pieces of information, choose which ones support A, which ones B, and which ones C.
2) Choose the best explanation and justify your choice based on the different pieces of information available.
3) From the pieces of information provided: Which ones do you think that constitute evidence and why?

(*Eight pieces of information are provided, in the booklet in www.rodausc.eu*)

2. Resources management in a bay
A small town on the seaside was hit by a hurricane. Afterwards, many people were homeless, their harvests destroyed, and most of their cattle was lost. Currently the main resources they have for surviving is a small bay, where several fish populations exist, including sardines, herring and salmon. You are a NGO team, sent in order to help the people in the town to manage the bay, so that it provides them with food for several months while their crops are able to grow again and cattle can be raised. *Your objective in this task is to decide how to manage the bay in order to feed the population for as long as possible.* You will need to arrange the most efficient way of using the fishing resources available, and to elaborate a plan, explaining how you would carry it out.
(*Four data sets are provided, in the booklet in www.rodausc.eu*)

Figure 3. Two argumentation tasks (Puig, Bravo, & Jiménez-Aleixandre, 2012)

Both tasks are authentic, socio-scientific, drawn from real life, and require coordinating scientific explanations, the models of gene-environment interaction in the first example, and energy transfer in ecosystems in the second, with complex data. Results from their implementation are summarized in Puig et al. (2012), showing the difficulties that students (and the public), experience in making sense of science news and in relating pieces of evidence to complex scientific explanations through justifications. Students need to engage in argumentation tasks in order to learn that practice.

Developing Explanations and Models: Epistemic Criteria for Good Models

This practice engages students in developing explanations and models about natural phenomena; about how does the natural world work, and why it works that way (Berland et al., 2016). Scientific models are based on evidence, and thus argumentation and developing explanations are related. Duschl (2008) represents the relationships between evidence and explanation in three steps or transformations: (1) selecting or generating data to become evidence; (2) using evidence to ascertain patterns of evidence and models; and (3) employing the models and patterns to propose explanations. As Duschl points out, in each transformation students need to make epistemic judgments about "what counts" as data, evidence or explanations.

The example about modelling means to illustrate the overlapping of epistemic and scientific practices in classroom settings. It is drawn from the work of Pluta, Chinn and Duncan (2011) in the PRACCIS project, which focuses on model-based reasoning. The tasks had as a goal the development by students of criteria for good scientific models, in other words, epistemic criteria. 7th grade students completed a series of model-evaluation tasks. In the first task they were presented with contrasting cases of models, for instance 12 representations of volcanoes including models, non-models and debatable cases, and they were asked to select the ones they thought were models and discuss their ideas with a partner, in order to think about what distinguishes a model from a non-model. In the second task, students were asked to compare seven pairs of models about familiar phenomena, and to consider several questions, some general as which model was better (or if they were equally good), and some more precise, for instance about purposes: "which of these two models is better if you want to explain…?". Finally, students individually generated and wrote six criteria about good models. The criteria were coded in three levels: (a) primary criteria, central to the practices of science, such as reflecting the explanatory goals of models, or the fit of models with evidence; (b) secondary criteria, which contribute to the epistemic aims of science, such as mentioning data; and (c) criteria that were vague or suggested misconceptions about the practices of science. The authors suggest the need for an emphasis on epistemic criteria in science instruction.

Pluta et al.'s study shows that students engaged in scientific practices, as evaluating models, may be at the same time involved in epistemic practices, in this case in developing epistemic criteria for good models. We consider this an appropriate implication for concluding our discussion of scientific and epistemic practices.

SUMMARY: SUPPORTING ENGAGEMENT IN SCIENTIFIC AND EPISTEMIC PRACTICES

- Learning science involves students' participation in the epistemic goals of science, goals related to how scientific knowledge is constructed.
- Science education's epistemic goals may be achieved through placing scientific practices at the centre of science teaching and learning.

- A defining feature of scientific practices is *activity*. Students should be *engaged* in scientific practices, *carrying out* modelling, argumentation or investigation.
- Science instruction needs to actively involve students in scientific practices through investigations, evaluation activities and development of epistemic criteria.

ACKNOWLEDGEMENTS

Work supported by the Spanish Ministerio de Economía y Competitividad (MINECO); Contract grant number: EDU2012-38022-C02-01. We thank Anat Zohar for her suggestions about the relationships between epistemic practices and scientific practices.

FURTHER READING

ADI (Argument Driven Inquiry) web page. Downloable instructional materials for argument-driven investigations. http://www.argumentdriveninquiry.com

McNeill, K. L., & Krajcik, J. (2012). *Supporting grade 5–8 students in constructing explanations in science. The claim, evidence and reasoning framework for talk and writing.* New York, NY: Pearson Allyn & Bacon.

NGSS (Next Generation Science Standards) Web Seminar Series, by National Science Teachers Association (NSTA) to help implementation in the classroom. http://learningcenter.nsta.org/products/symposia_seminars/NGSS/webseminar.aspx

PRACCIS (Promoting Reasoning and Conceptual Change in Science). Focus on model-based reasoning. https://sites.google.com/a/gse.rutgers.edu/praccis-promoting-reasoning-and-conceptual-change-in-science/

Puig, B., Bravo-Torija, B., & Jiménez-Aleixandre, M. P. (2012). *Argumentation in the classroom: Two teaching sequences.* Santiago de Compostela: Danu. (in English, Spanish & Galician www.rodausc.eu).

Organisation for the Economic Cooperation and Development (OECD). (2013). *PISA 2015 Draft science framework.* Paris: Author.

Zembal-Saul, C., McNeill, K. L., & Hershberger, K. (2013). *What's your evidence? Engaging k-5 students in constructing explanations in science.* New York, NY: Pearson Allyn & Bacon.

REFERENCES

Achieve. (2013). *Next generation science standards: For states, by states.* Washington, DC: National Academies Press.

Bell, P., Bricker, L., Tzou, C., Lee, T., & Van Horne, K. (2012). Exploring the science framework: Engaging learners in scientific practices related to obtaining, evaluating and communicating information. *The Science Teacher, 79*(8), 31–36.

Berland, L. K., Schwarz, C. W., Krist, C., Kenyon, L., Lo, A. S., & Reiser, B. J. (2016). Epistemologies in practice: Making scientific practices meaningful for students. *Journal of Research in Science Teaching, 53*(in press). doi:10.1002/tea.21257

Bybee, R. (2011). Scientific and engineering practices in K-12 classrooms. Understanding a framework for K-12 science education. *The Science Teacher, 78*, 34–40.

Chinn, C. A., Buckland, L. A., & Samarapungavan, A. L. A. (2011). Expanding the dimensions of epistemic cognition: Arguments from philosophy and psychology. *Educational Psychologist, 46*(3), 141–167.

Crujeiras, B., & Jiménez-Aleixandre, M. P. (2015, 31 August–4 September). *Students' engagement in planning an investigation about tooth decay: Epistemic operations in the chemistry laboratory*. Paper presented at the ESERA conference, Helsinki.
Duschl, R. (2008). Science education in three-part harmony: Balancing conceptual, epistemic and social learning goals. *Review of Research in Education, 32*, 268–291.
Duschl, R. A., & Bybee, R. W. (2014). Planning and carrying out investigations: An entry to learning and to teacher professional development around NGSS science and engineering practices. *International Journal of STEM Education*, 1–12. doi:10.1186/s40594-014-0012-6
Duschl, R. A., & Jiménez-Aleixandre, M. P. (2012). Epistemic foundations for conceptual change. In S. M. Carver & J. Shrager (Eds.), *The journey from child to scientist: Integrating cognitive development and the education sciences* (pp. 245–262). Washington, DC: American Psychological Association.
Jiménez-Aleixandre, M. P., & Erduran, S. (2008). Argumentation in science education: An overview. In S. Erduran & M. P. Jiménez-Aleixandre (Eds.), *Argumentation in science education: Perspectives from classroom-based research* (pp. 3–27). Dordrecht: Springer.
Kelly, G. J. (2008). Inquiry, activity and epistemic practice. In R. A. Duschl & R. E. Grandy (Eds.), *Teaching scientific inquiry: Recommendations for research and implementation* (pp. 99–117). Rotterdam: Sense Publishers.
Kelly, G. J. (2016). Methodological considerations for interactional perspectives on epistemic cognition. In J. A. Greene, W. A. Sandoval, & I. Bråten (Eds.), *Handbook of epistemic cognition* (pp. 393–408). New York, NY: Routledge.
Monteira, S. F., & Jiménez-Aleixandre, M. P. (2016). The practice of using evidence in kindergarten: The role of purposeful observation. *Journal of Research in Science Teaching, 53* (in press). doi:10.1002/tea.21259
National Research Council. (2012). *A framework for K–12 science education: Practices, crosscutting concepts, and core ideas*. Washington, DC: National Academies Press.
Osborne, J. (2011). Science teaching methods: A rationale for practices. *School Science Review, 93*(343), 93–103.
Osborne, J. (2014). Scientific practices and inquiry in the science classroom. In N. G. Lederman & S. K. Abell (Eds.), *Handbook of research on science education* (Vol. II, pp. 579–599). New York, NY: Routledge.
Pluta, W. J., Chinn, C. A., & Duncan, R. G. (2011). Learners' epistemic criteria for good scientific models. *Journal of Research in Science Teaching, 48*(5), 486–511.
Reiser, B. J., Berland, L. K., & Kenyon, L. (2012). Engaging students in the scientific practices of explanation and argumentation. Understanding a framework for K-12 Education. *Science and Children, 49*(8), 8–13.
Wickman, P.-O. (2004). The practical epistemologies of the classroom: A study of laboratory work. *Science Education, 88*(3), 325–344.

JOANNE BROGGY, JOHN O'REILLY AND SIBEL ERDURAN

6. INTERDISCIPLINARITY AND SCIENCE EDUCATION

INTRODUCTION

In recent years there has been an increased urgency to better educate young people in the fields of science, technology, engineering and mathematics (STEM) and increase their understanding of the importance of the role STEM education in economic growth, technological innovation, national health and food production. A possible solution to the shortfall in understanding and more frequently, poor attitudes, towards STEM is the introduction of interdisciplinary teaching which highlights how knowledge across disciplines is interrelated in the natural world. This chapter will examine how science can be taught using an interdisciplinary approach by presenting the possible relationships that can be developed with other subjects and disciplines.

DEFINING INTERDISCIPLINARY IN SCIENCE TEACHING

An interdisciplinary approach to science teaching involves the learning that crosses subject boundaries to facilitate a better learning experience for students. This approach encourages students to explore and integrate multiple perspectives from different subject disciplines, sub-disciplines and areas of expertise (Golding, 2009) thereby leading the process of innovation in education (Bauerle, Hatfull, & Hanauer, 2014). An important distinction must be made between "multidisciplinary" and "interdisciplinary" approaches. In the case of the former, different subjects contribute independently to a theme under investigation, maintaining a separate identity. In the latter case, common learning and concepts are identified across multiple subjects to address a theme. In this case subjects overlap with each other and the theme and are truly integrated, employing interdisciplinary skills such as literacy, numeracy and research and seeking to understand how the subjects speak to each other and the theme. In order to teach science efficiently there are many aspects of its interdisciplinarity that require consideration. The integration of physics, chemistry and biology within science is an obvious example of how science lends itself to an interdisciplinary approach, in that all three subjects can be combined to facilitate a more comprehensive understanding of the issue under examination.

Interdisciplinary teaching in science classrooms is represented through a variety of forms including; integrated science courses, coordinated science courses and

K. S. Taber & B. Akpan (Eds.), Science Education, 81–90.

subject-focused course (Nagle, 2013) and multidisciplinary (Klein, 1990). These courses vary in the methodological approach used in integrating the individual disciplines. Integrated science courses provide students the opportunity to explore and study the various concepts from multiple disciplines within individual lessons and across units (Nagle, 2013). In coordinated science courses several disciplines are taught in a year in a logical sequence, with links identified between content-focused units. A subject-focused course, as the name suggests, focuses on one subject (for example biology, physics or chemistry) and includes extensive interdisciplinary opportunities provided by scientific questions.

WHY TEACH WITH AN INTERDISCIPLINARY APPROACH?

The quality of education that teachers provide to the students in a classroom is highly dependent upon what teachers do *in* the classroom. At a time when there is a demand for a skilled workforce in STEM careers who can tackle complex real life problems wisely we must ensure the teaching of science and STEM subjects is effective by providing opportunities for this. There are several arguments put forward in support for an integrated, interdisciplinary approach claiming that it makes sense to integrate the sciences as in real life where knowledge and experience are not separated (Eurydice, 2011). The approach provides opportunities for improved understanding, for example through answering big questions, exploring an issue, solving problems or completing a final project where the learner can explore clear and relevant links across the curriculum. Often approaching complex problems or concepts (for example climate change and world poverty) from single disciplines (i.e. biology), is not enough to provide understanding and interdisciplinary education is required so students can develop knowledge from different disciplines and learn how to purposefully and reflectively integrate and synthesise different perspectives in order to advance understanding and solve problems (Golding, 2009). The approach also provides students with an experience which will help them address misconceptions they may hold about the nature of science and rigid notions about the scientific method (Nagle, 2013).

The use of interdisciplinary teaching not only improves the student experience but can also have a positive impact for teachers as it will encourage teachers of different subjects to collaborate. Teachers together can develop materials or adapt existing teaching materials and approaches for their classrooms. The use of interdisciplinary teaching can also help in developing students' epistemic position. Within interdisciplinary classrooms often students are presented with contrary disciplinary perspectives that may confuse or frustrate them and unless they have developed a clear understanding of the knowledge and conceptions of the nature of interdisciplinary teaching they will be unable to make sense of the information. Whether the student takes a simple or sophisticated epistemic position (Perry, 1981) will determine how they respond to their experience of the multiple credible, contradictory answers and alternatives features of interdisciplinary teaching. The development of a sophisticated epistemic position would allow students to see multiple perspectives they are confronted with and engage in "dialogical reflective thinking to make reasonable judgements" (Golding, 2009, p. 18).

INTEGRATING HISTORY AND PHILOSOPHY OF SCIENCE IN SCIENCE EDUCATION

The integration of history and philosophy of science (HPS) in science education has been advocated as a goal since at least the 1960s. Contemporary applications of HPS in science education (e.g. Matthews, 2014) follow earlier arguments (e.g. Connelley, 1969) which were predominantly justified on the grounds that science needs to be connected to its social and historical roots. Since science teaching has traditionally embraced little or no reference to the cultural, personal and historical contexts in which science occurs, learners of science do not develop an understanding of science as a human endeavour. HPS has thus been advocated as an instrument for improving students' motivation and attitudes towards science.

The recent compilation of a handbook (Matthews, 2014) consisting of 3 volumes is testimony to the significance that the science education community has placed on the role of HPS in science education. The Handbook demonstrates that HPS contributes significantly to the understanding and resolution of the numerous theoretical, curricular and pedagogical questions and problems that arise in science and mathematics education.

In recent years, some work in history and philosophy of science has focused on the exploration of how particular domains of science can contribute to knowledge in philosophy of science. For example, Dagher and Erduran (2014) have reviewed how explanations and laws in biology and chemistry contrast, illustrating their relevance in biology and chemistry textbooks and offering some implications for education. Some practical examples have been proposed to illustrate how the 'big' ideas from philosophy of science can be applied at the level of the classroom. For example, following a review of the literature on how chemical laws relate to laws in physics, Erduran (2007) identifies the following example to promote debate and discussion in chemistry lessons. Students can be presented with two alternative theories about the nature of laws:

- Claim 1 The periodic law and the law of gravitation are similar in nature. The term "law" can be used with the same meaning for both of them.
- Claim 2 The periodic law and the law of gravitation are different in nature. The term "law" cannot be used with the same meaning for both of them.

Erduran (2007) explains that these claims could be presented with evidence that would support either claim, both or neither. For example, the statement "a law is a generalization" could support both claims while "the periodic law cannot be expressed as an algebraic formula while the law of gravitation can be" could support the second claim. The task for the students would be to argue for either claim and justify their reasoning. Further statements can be developed that would act as evidence for either, both or neither claim. The outcome of such activities would be improved understanding of the nature of laws in science in general and in particular branches of science.

INTEGRATING ECONOMICS OF SCIENCE IN SCIENCE EDUCATION

There is now a formalized field of study called "economics of science". Professionals in this field aim to understand the behaviour of scientists, the distribution of resources and the financial operation of scientific institutions. Wibble (1998) has explored how science can be understood as a market. Furthermore, commercialisation and commodification of science have been key areas of research (e.g. Irzik, 2013). Radder (2010) defined commodification as "the pursuit of profit by academic institutions through selling the expertise of their researchers and the results of their inquiries" (p. 4). The commercial nature of science is related to the production of scientific knowledge as private property which can be sold. The "science market" then can create market barriers to hinder free consumption of scientific knowledge. In order to understand the commercial nature of science, researchers resort to making a demand and supply analysis, positioning scientific knowledge as a commodity.

It should be noted that there is policy and curricular relevance for why economics of science should be considered in science education. In many parts of the world, policy documents (e.g. OECD, 2006) have advocated the contextualisation of science in its wider societal, cultural, political as well as economic dimensions. However the precise articulation of what aspects of economics of science are relevant for science education remains underdeveloped in the science education research literature. Erduran and Mugaloglu (2013) have proposed some practical resources to situate some examples from the economics of science literature at the level of the classroom. For instance, by drawing on an example on the discovery/invention dichotomy debate in economics of science, they produced a practical resource that teachers can use with students to promote understanding of some key issues that relate to the financial dimensions of science in relation to genetic cloning and patent rights. Other similar examples relate to the inclusion of entrepreneurship skills in pre-service science teacher education (e.g. Kaya, Erduran, & Birdthistle, 2015).

INTEGRATING MATHEMATICS AND SCIENCE EDUCATION

The argument to integrate mathematics and science education seems a logical one, especially in relation to the physical sciences. Chapter 16 (Curriculum Implications of the Integration of Mathematics into Science) will look at this area in detail; however the discussion of the interdisciplinary nature of science would not be complete without addressing it in relation to its association with mathematics. Famously, Galileo Galilei asserted that mathematics is the language of science. Integration is advocated by international groups such as the International Council of

Associations for Science Education and national agencies in many countries. There is, however, a lack of clarity about what mathematics-science integration actually looks like in practice.

Hurley (2001) reviewed empirical studies on mathematics-science integration and determined that there was a reasonable increase in science achievement resulting from integration. This effect increased significantly with large effects on achievement in science being associated with higher levels of integration. Hurley (2001) notes that integration is difficult to define given the complexities of timetabling, sequencing and the relative emphasis on the subjects integrated. The studies reviewed also lacked a careful conceptualisation of integration itself. Pang and Good (2000) explore this issue in detail, calling for more sophisticated understanding and explicit discussion of the nature of the two disciplines, highlighting that science seeks to understand the world through empirical evidence external to the field itself while mathematics is concerned with internal, logical deduction. They also indicated that, in common with Hurley's findings, mathematical concepts were rarely regarded as of primary importance in integrated approaches, but rather were in the service of science learning.

This has significant implications for those designing an interdisciplinary approach to mathematics and science instruction where clarity of the purpose of integration must be established before practical decisions can be made. Furner and Kumar (2007) provide some suggestions for developing interdisciplinary approaches for mathematics-science integration, emphasising a Problem Based Learning approach and drawing on integration when there is clearly overlapping content in both disciplines, rather than trying to artificially force an interdisciplinary approach.

Resources are available to assist interdisciplinary planning such as "Great Explorations in Math and Science" (GEMS) developed in Berkeley. While no empirical studies exist specifically examining the impact of the interdisciplinary nature of GEMS on student achievement, Granger et al. (2012) used GEMS as a "student-centred" experimental intervention contrasted with a "teacher-centred" control group in a large-scale, randomized-cluster experimental design. They found significant gains in science learning for the experimental (GEMS) group.

In conclusion, while there is evidence to support the positive impact on student achievement of an interdisciplinary approach to mathematics-science education, further work is required to support the evidence of significant gains in science learning through integration (Hurley, 2001) and, most importantly, to ground this work in the day to day of classroom practice, reflecting the constraints of timetabling, over-laden syllabi, assessment systems and so on.

INTERDISCIPLINARY IMPLICATIONS FOR TEACHING AND LEARNING OF SCIENCE

The sections preceding this indicate that interdisciplinary approaches to science education are implied from real world problems and the modern practice of science. This should include an integration not only of the traditional sciences and mathematics but also other disciplines of knowledge that allow for the multiple perspectives required to reflect decision-making and knowledge development related to science in a complex, modern society.

This perspective is reflected in policy documents such as "Science Education in Europe" published by the Eurydice (2011) which notes that the steering documents of most European countries recommend cross-curricular, contextual approaches in science learning, neatly summarised on p. 66. In part, this is advocated to awaken curiosity about the natural world in a manner that is relevant to student lives. Real-life contexts provide starting points for the development of scientific ideas (Bennett, Lubben, & Hogarth, 2007, cited in Eurydice, 2011, p. 64) and this "context-based science" teaching approach emphasises the philosophical, historical and societal aspects of science. In America, the National Research Council has recently released the "Next Generation of Science Standards" (NGSS, 2013) which Chowdhary et al. (2014) claim describes a new vision of science competence which they term "Interdisciplinary Science Inquiry" (ISI). ISI is rooted in project-based learning and is conceived as including the following attributes (p. 867)

- A contextualized nature of problems which establishes relevance to students' lives
- Incorporation of inquiry and engineering process skills or practices to learn science
- Creating connections within and across disciplines such as Mathematics, English Language Arts, Engineering, and Science, and
- Anchored within specific science disciplines.

Chowdhary et al. (2014) recognise that it is not just skills that are required to achieve ISI but a change in "mindset" (p. 866). In providing professional development (PD) to support teacher ISI practice they encountered many difficulties including teachers missing PD sessions, failing to complete documentation, expressing beliefs in child-centred approaches but remaining teacher focused in practice (in part deriving from low expectations of students) and limited conceptions of the purposes of ISI.

Czerniak and Johnson (2014) provide a useful review of the history of interdisciplinary science teaching and approaches to integrated curriculum design in the light of an increasing focus on disciplinary integration in STEM education. It is recognised that integration allows for a focus on important contexts and is inherently student-centred but presents significant problems in the move away from the traditional, discipline-centric school curriculum.

Few studies have examined teacher capacity to design and deliver an ISI curriculum with most focusing on teacher attitudes and beliefs towards such a curriculum but both Chowdhary et al. (2014) and Czerniak and Johnson (2014) highlight the affordances of project- and problem-based learning in realising ISI. While explicit beliefs significantly influence practice many researchers have indicated the overarching effects of (often subconscious) customs and habits. Korthagen (2010) provides a useful framework to understand teacher decision making in the classroom which he describes as immediate behaviour, driven by a "gestalt" i.e. unreflective, based on "momentarily triggered images, feelings, notions, values, needs or behavioural inclinations" (p. 101). Gestalts are developed through concrete, repeated experience and Nuthall (2005) argues that these take the form of "Ritualised Routines" where both students and teachers know what is expected of them in terms of their roles and classroom activities. For the latter these routines are formed early through an "apprenticeship of observation" (Lortie, 1975), meaning we teach how we were taught – the system reproduces itself, has significant inertia and is resistant to change, no matter how well supported. Complex factors influence routines to include power relations between students and teachers (Donnelly et al., 2014) which can either stifle or facilitate IBSE through for example, teacher monitoring of task completion rather than understanding, lack of student engagement in ownership of ideas and forms of questioning.

Crucially though, routines are necessary in managing the learning of large classrooms and are internally consistent within systems that place an emphasis on covering large volumes of content efficiently, to be evaluated by high stakes summative assessment. PD should certainly focus on improving the pedagogical skill of practitioners in the areas of project- and problem-based teaching and learning but this is only a necessary, not a sufficient condition for change. There is a need to develop new routines that are more coherent and internally consistent with ISI. This should include greater flexibility (with respect to timetables and class length in particular), less content to "cover" and learning outcomes that necessitate a child-centred approach with coherent modes of assessment.

This would suggest that any teaching and learning developments that support ISI must take a more comprehensive perspective, at the level of curriculum. Indeed, routines must be developed that encourage an inquiry orientation across the entire curriculum, not just within the sciences so that this "habit" can become the norm. Furthermore it is important that teacher identity as a subject expert is not undermined and that students develop appropriate levels of expertise in a range of disciplines. Lederman and Niess (1997) recommend that an interdisciplinary approach should be employed but in a way that preserves subject identity as each discipline has a unique epistemology that may not be fully engaged with in an integrated approach.

In the light of these considerations, Ireland provides an interesting current case of curriculum reform. The existing Junior Certificate (junior cycle education which takes place over three years from the age of (typically) 12–15 years and culminates with a state exam at the end of the third year) is being replaced by the Junior Cycle which places an emphasis on the development of key skills (including communication, working with others, critical and creative thinking and information processing, significant changes in the assessment system and a much greater emphasis on child-centred learning. The new science syllabus presents an interdisciplinary view of the sciences with the four core strands considered as overlapping bodies of knowledge that provide scope for integration, united by the overarching, Nature of Science strand (NCCA, 2015). Indeed, the theme of "Nature of Science" has been rather critical in the infusion of interdisciplinary perspectives in science education, for instance perspectives from philosophy of science (Erduran & Dagher, 2014; Erduran & Mugaloglu, 2014).

Students will be invited to conduct “Issues Investigations” which provides formal space for STS work but perhaps the most significant opportunity for ISI rests with the Short Courses (SCs). SCs allow space for more flexible learning and while some will be provided as exemplars, schools have been encouraged to design their own SCs to address issues of local concern/interest. One of the authors (O’Reilly) is currently conducting work with schools to design SCs using a Negotiated Integrated Curriculum (NIC) approach. What is to be learned and how it is to be learned is negotiated between teacher and student from the outset and on an ongoing basis. Students are encouraged to investigate issues that are of significant concern to them and generate their own questions and proposals for answering them. The work is inherently problem-based and radically alters the power dynamics of the classroom. Subjects are integrated as necessary to answer the questions posed and teachers are amongst those consulted in generating answers, but now as experts in their disciplines. This can re-invigorate subject learning and the extent of integration (and which subjects are more significant) depends on what is necessary to rigorously address the identified concerns. This neatly addresses issues about the extent and nature of integration discussed above.

Ongoing work has revealed the extent to which students are highly motived by this approach, enabling far more child-centred pedagogies and minimising classroom management issues. Power relationships are radically altered as students have a significant voice in their work, far more ownership of their learning and an inquiry orientation dominates classroom activities. A systematic approach to planning allows teachers to build confidence in developing new routines that are persuasive in terms of student engagement and enthusiasm for learning.

STS issues have emerged such as a focus on examining the introduction of charges for drinking water which has included the scientific consideration of the water purification process, including the addition of fluoride as a potential example of mass medication without consent. Other subjects integrated have included CSPE (equivalent to Civics or Politics) to examine the decision-making processes behind the introduction of the charges; Geography to consider the infrastructure for water delivery to homes; Business to compare the relative cost of water in other European countries and Technology to understand how water meters work. The culmination of student work was an evening where their findings were presented to, and discussed with parents and teachers. In this manner schools can become knowledge generators for issues relevant to their own communities and the focus on concerns makes it likely that STS issues will be highlighted by the students.

This represents one example of a structure that can provide the flexibility and scaffolding required for an ISI approach where integration is achieved, but not at the cost of subject identity. This can help to develop the sophisticated epistemic position described above where personal knowledge is developed in collaboration with others, assumptions are challenged and positions can be defended while remaining open to change. This is not to suggest that the path of change is easy and even in the best of circumstances difficulties will be encountered as teachers and students alike adjust to new routines. In addition to overcoming the obstacles to ISI noted above, space must be provided in the curriculum for the flexibility necessary to enact it or else there is a danger of tokenism where lip service is paid to interdisciplinary learning. Ideally the focus of STS investigations should be developed from the ground-up in schools, which will serve to define the boundaries and depth of subject contributions, and NIC provides one possible approach. Properly enacted schools can become knowledge generators rather than reproducers for society through ISI and this represents a radical reimagining of the purpose of schooling.

FURTHER READING

Duschl, R. A., Hamilton, R., & Grandy, R. (Eds.). (1992). *Philosophy of science, cognitive psychology and educational practice*. New York, NY: SUNY Press.

Haynes, C. (Eds.). (2002). *Innovations in interdisciplinary teaching*. Westport, CT: American Council on Education Oryx Press Series on Higher Education.

Matthews, M. (Ed.). (2014). *International handbook of history, philosophy and science teaching*. Dordrecht: Springer.

REFERENCES

Bauerle, C., Hatfull, G., & Hanauer, D. (2014). Facilitating STEM education through interdisciplinarity. In M. J. Curry & D. I. Hanauer (Eds.), *Language, literacy, and learning in STEM education* (pp. 167–179). Amsterdam: John Benjamins Publishing Company.

Chowdhary, B., Liu, X., Yerrick, R., Smith, E., & Grant, B. (2014). Examining science teachers' development of interdisciplinary science inquiry pedagogical knowledge and practices. *Journal of Science Teacher Education, 25*(8), 865–884.

Conant, J. B. (1948). *Harvard case histories in experimental science, 2 vols*. Cambridge, MA: Harvard University Press.

Connelly, F. M. (1969). Philosophy of science in the science curriculum. *Journal of Research in Science Teaching, 6*, 108–113.

Czerniak, C. M., & Johnson, C. C. (2014). Interdisciplinary science teaching. In N. G. Lederman & S. K. Abell (Eds.), *Handbook of research on science education* (Vol. 2, pp. 395–411). New York, NY: Routledge.

Donnelly, D. F., McGarr, O., & O'Reilly, J. (2014). 'Just be Quiet and Listen to Exactly what He's Saying': Conceptualising power relations in inquiry-oriented classrooms. *International Journal of Science Education, 36*(12), 2029–2054.

Duschl, R. A. (1985). Science education and philosophy of science: Twenty-five years of mutually exclusive development. *School Science and Mathematics, 87*(7), 541–555.

Duschl, R. A. (1994). Research on the history and philosophy of science. In D. Gable & D. Bunce (Eds.), *Handbook of research on science teaching and learning*. New York, NY: Macmillan.

Dagher, Z., & Erduran, S. (2014). The role of disciplinary knowledge in science education: The case of laws and explanations in biology and chemistry. In M. Matthews (Ed.), *Handbook of research on history, philosophy and sociology of science*. Dordrecht: Springer.

Erduran, S. (2007). Breaking the law: Promoting domain-specificity in chemical education in the context of arguing about the Periodic Law. *Foundations of Chemistry, 9*(3), 247–263.

Erduran, S., & Dagher, Z. (2014). *Reconceptualizing the nature of science for science education: Scientific knowledge, practices and other family categories*. Dordrecht: Springer.
Erduran, S., & Mugaloglu, E. (2013). Interactions of economics of science and science education: Investigating the implications for science teaching and learning. *Science & Education, 22*(10), 2405–2425.
Erduran, S., & Mugaloglu, E. (2014). Philosophy of chemistry in chemical education. In M. Matthews (Ed.), *Handbook of research on history, philosophy and sociology of science*. Dordrecht: Springer.
Eurydice. (2011). *Science education in Europe: National policies, practices and research*. Brussels, Education, Audiovisual and Cultural Executive Agency.
Furner, J., & Kumar, D. (2007). The mathematics and science integration argument: A stand for teacher education. *Eurasia Journal of Mathematics, Science and Technology, 3*(3), 185–189.
Golding, D. (2009). *Integrating the disciplines: Successful interdisciplinary subjects*. Australia: Centre for the Study for Higher Education, University of Melbourne.
Granger, E. M., Bevis, T. H., Saka, Y., Southerland, S. A., Sampson, V., & Tate, R. L. (2012). The efficacy of student-centered instruction in supporting science learning. *Science, 338*(6103), 105–108.
Hurley, M. M. (2001). Reviewing integrated science and mathematics: The search for evidence and definitions from new perspectives. *School Science and Mathematics, 101*(5), 259–268.
Irzik, G. (Ed.). (2013). Commercialisation and commodification of science: Educational responses. *Science & Education, 22*(10), 2375–2384.
Kaya, S., Erduran, S., & Birdthistle, N. (2015). *The role of entrepreneurship in nature of science and science education*. Proceedings of the International History, Philosophy and Science Teaching Biennial Conference, Rio de Janeiro, Brazil.
Klein, J. T. (1990). *Interdisciplinarity: History, theory and practice*. Detroit, MI: Wayne State University Press.
Korthagen, F. A. J. (2010). Situated learning theory and the pedagogy of teacher education: Towards an integrative view of teacher behaviour and teacher leaning. *Teaching and Teacher Education, 26*, 98–106.
Lederman, N. G., & Niess, M. G. (1997). Integrated, interdisciplinary or thematic instruction? Is this a question or is it questionable semantics? *School Science and Mathematics, 97*(2), 57–58.
Lortie, D. C. (1975). *Schoolteacher: A sociological study*. Chicago, IL: University of Chicago Press.
Matthews, M. (Ed.). (2014). *International handbook of history, philosophy and science teaching*. Dordrecht: Springer.
Nagle, B. (2013). Preparing high school students for the interdisciplinary nature of modern biology. *Life Sciences Education, 12*, 144–147.
NCCA. (2015). *Draft specification for junior cycle science*. Retrieved April 8, 2015, from http://juniorcycle.ie/NCCA_JuniorCycle/media/NCCA/Curriculum/Science/JC-ScienceSpec_ForConsultation.pdf
NGSS Public Release II. (2013). Retrieved from http://www.nextgenscience.org/next-generation-science-standards
Nuthall, G. (2005). The cultural myths and realities of classroom teaching and learning: A personal journey. *Teachers College Record, 107*(5), 895–934.
OECD. (2006). *PISA 2006. Assessing scientific, reading and mathematical literacy: A framework for PISA 2006*. Paris: Author.
Pang, J., & Good, R. (2000). A review of the integration of science and mathematics: Implications for further research. *School Science and Mathematics, 100*(2), 73–82.
Perry, W. G. (1981). Cognitive and ethical growth: The making of meaning. In A. W. Chickering (Ed.), *The modern American college* (pp. 76–116). London: Jossey-Bass.
Radder, H. (2010). *The commodification of academic research: Analyses, assessment, alternatives*. Pittsburgh, PA: University of Pittsburgh Press.
Wibble, J. R. (1998). *The economics of science: Methodology and epistemology as if economics really mattered*. London: Routledge.

SECTION II

THINKING AND LEARNING IN SCIENCE

SUDHAKAR AGARKAR AND RICHARD BROCK

7. LEARNING THEORIES IN SCIENCE EDUCATION

INTRODUCTION

Learning is a complex, multi-faceted phenomenon and it is therefore unlikely that a single model could explain all aspects of the process. The twentieth century has seen the development of many different models of learning, referred to as learning theories, some of which are discussed below. The theories emphasise different aspects of learning and are based on a range of different assumptions about knowledge. Though one of these theories, constructivism, is seen as the dominant way of thinking about learning in science education, other theories also contain insights for teachers and researchers. In this chapter, various theories will be grouped together under three broad headings, based on a similarity of assumptions: behaviourism, cognitive theory, and constructivism. As the theories outlined below make varied claims about the manner in which humans acquire new information, they lead to different recommendations for teachers' practice. The implications of each of the models in the science classroom are considered at the end of each section. Whilst considering the different theories it is worth holding in mind that each is a model of the learning process and, though no model is a complete description of the world, some are more useful than others. The reader may wish to consider the validity of the assumptions of each model, and the fruitfulness of the teaching approaches suggested.

BEHAVIOURISM

Behaviourism has its foundation in animal studies. Pavlov realised that if a bell was rung when the dogs in his laboratory were fed, the dogs salivary response was stimulated by the sound of the bell, even if no food was set out (Pavlov, 1927). He defined the development of an association between a stimulus (the sound of the bell) and a response (salivation) as conditioning. This stimulus-response link is central to behaviourist models of learning. The theory is concerned only with observable behaviours of learning, and avoids making claims about psychological processes or entities such as mental states or consciousness. These assumptions led to the development of one of the earliest sets of educational principles, Throndike's (1927) 'laws of learning.' For example, Thorndike proposed the law of effect, which suggests that events that occur after a stimulus-response pairing can alter the strength of the connection. The use of praise to reinforce on-task behaviour might be considered a classroom application of this law. Behaviourist principles came to

K. S. Taber & B. Akpan (Eds.), Science Education, 93–103.

dominate teacher education in the 1950's and saw the rise of models of teachers as dispensers of punishment and an emphasis on drill and practice approaches in the classroom. A key behaviourist thinker, Skinner (1958), proposed the idea of teaching machines, devices which would provide the best environment for conditioning. A student would enter their answer to a multiple-choice question into the machine, be given immediate feedback, and be directed to an appropriate follow-up question. These machines might be seen as precursors to contemporary adaptive learning tools that provide learners with personalised educational experiences.

Behaviourism fell out of fashion in science education research due to criticism on a number of fronts. Firstly, its principles were developed from experiments on animals and therefore might not adequately reflect the complexity of human learning (Stewart, 2012). Secondly, some psychologists argue that memory is more than a passive store of conditions and responses (Baddeley, 2000) but behaviourist models do not represent learning as an active process. Finally, certain interpretations of behaviourism have been associated with an authoritarian and teacher-centred model of teaching (Stewart, 2012). Despite these criticisms, some consequences of behaviourist theory still exert an influence on contemporary classrooms.

Implications for Practice

The behaviourist model of learning suggests a number of classroom strategies: the use of repetition as technique for learning skills and memorising factual knowledge; the introduction of classroom routines; the communication of clear learning objectives; the decomposition of complex tasks into a series of increasingly challenging steps; an emphasis on prompt feedback to promote or suppress behaviours; individualised learning programmes to enable students to work at different rates (Stewart, 2012). In the science classroom, behaviourism has been associated with drill-like practice and the transmission of facts and principles. For example, a teacher may focus on testing the verbatim recall of Newton's laws rather than students' ability to apply those principles to novel situations. However, the effectiveness of approaches that focus on rote-learning are contentious. Constructivists have described the highly successful performance on some measures of achievement by Asian science students, who are taught partly through drill-like approaches, as a paradox (Cheng & Wan, 2015). Though the success of Asian science education is a complex phenomenon, and factors beyond the classroom play a role, it might suggest that behaviourist educational strategies should not be lightly dismissed.

COGNITIVE SCIENCE

The second half of the twentieth century saw a reaction against behaviourist models of learning and the development of a novel approach, cognitive science. Cognitive scientists liken the human mind to a computer and therefore assume that learning can be modelled as information processing. This assumption has led to descriptions of learning as the action of a sequence of modules on incoming information. For example, some models of learning propose that a sensory interface initially processes signals from the world, then the preconscious interpreter filters the data before it is relayed to a conscious executive module (Taber, 2013, p. 66). Such models have explanatory power, for example, the action of the preconscious interpreter may explain how tacit conceptions of the physical world develop. Additionally, learning may be described using a model of memory which consists of the central executive (selectively channels information to other modules); the phonological loop (a short-term store of auditory information); the visuospatial sketchpad (a short-term store of visual information); and the episodic buffer (links visual and auditory information) (Baddeley, 2000). This model suggests that students might readily engage in a verbal and visual task at the same time, but completing two visual tasks simultaneously would be challenging.

The assumption that learning can be represented as a series of processes led to the development of models that distinguished different types of learning. For example, Robert Gagné (1965) assumed that different teaching approaches are appropriate for supporting different varieties of learning. Rote learning may be suitable for acquiring verbal information, such as the sequence of types of wave in the electromagnetic spectrum, but more complex approaches are required to help a learner acquire the skill of problem solving. A separate research programme was developed from the work of scientists attempting to create computer programmes that replicated human behaviours. Researchers such as Allen Newell and Herbert Simon began to study the behaviour of experts on simple tasks such as puzzles and games in order to catalogue the strategies they used. This work influenced researchers in science education and a number of approaches have sought to describe the differences between expert and novice learners or to catalogue the strategies students use when encountering new ideas in the classroom. The models of cognitive theory, developed by these and other research programmes, are diverse and too numerous to examine in detail in this chapter. For the interested reader Reif's (2008) book, *Applying Cognitive Science to Education*, presents a comprehensive discussion of ideas. Just as researchers have attempted to describe the strategies learners use, individuals may develop an awareness of their own learning processes. The study of metacogntion, that is thinking and knowledge about one's own cognitive processes (Flavell, 1979), is an established area of research on cognition in science education. A metacognitve model of learning would include systems that consciously regulate learning and, in general, such reflexive awareness is seen as beneficial to learning in the science classroom.

Cognitive models of learning have attracted criticism for their failure to address aspects of learning such as emotions, tacit skill acquisition and holistic phenomena like understanding (Haugeland, 1978). Critics have also highlighted that the narrow focus of cognitive science on the individual mind presents a limited model of learning. Though this criticism may be true of early research, a number of theorists, for example Bandura (1971), expanded the scope of cognitive theory to develop a social cognitive theory that has been widely used in research in science education. In Bandura's model a reciprocal relationship between the individual and the social environment is assumed. Changes in a student's understanding trigger responses from their teachers and peers, and vice versa.

Implications for Practice

A variety of models developed by cognitive scientists has led to a number of suggestions for classroom practice. As discussed, the approach of cognitive science is to represent cognition as a series of information processing systems. Therefore, the recommendations generated tend to be specific and focused on a particular aspect of cognition, rather than general approaches. Some examples are shown below:

- Research suggests that working memory is relatively limited and therefore 'chunking,' that is dividing content into smaller units, may assist with the retention of ideas (Taber, 2014, p. 166).
- Encoding memories in both visual and verbal forms may make them more stable (Reif, 2008, p. 95), so teaching with both visual and verbal transmission of information may be more memorable.
- Supporting students' ability to regulate their own thinking may make them more effective science learners.

CONSTRUCTIVISM

Constructivism is associated with a group of theories about learning that are a subset of cognitive theories, as they present propositions about information processing. However, constructivist thinkers go beyond the assumptions of cognitive science and, though a range of different varieties of constructivism exist, tend to share broadly similar axioms about learning:

- Learning science is an active process.
- Learners come to the science classroom with pre-existing ideas about many natural phenomena, which have an effect on their subsequent learning.
- It is possible to meaningfully model learner's knowledge as conceptual structures which have some commonalities but also display idiosyncratic features.
  Adapted from Taber (2009, p. 123)

Jean Piaget is often described as the founding thinker of constructivism. He proposed a view of learning, labelled genetic epistemology, which described the processes through which learners developed their knowledge of the world. He used earlier thinkers' notion of schemata, frameworks linking together concepts, memories and perceptions, which enable humans to make generalisations about the world, as the basis of his model. Piaget (1952) described two processes by which new information is added to existing schemata: assimilation occurs when novel information is reinterpreted in order to fit with pre-existing knowledge; accommodation involves the modification of an existing schema in order to accept new data. Learning is seen as proceeding through a cycle of assimilation, cognitive disequilibrium, accommodation, cognitive equilibrium and a return to assimilation. For example, a child may develop a schema that links metals with the solid phase at room temperature. Encountering mercury may lead to disequilibrium that causes the child to change their schema, an act of accommodation, before the next assimilation event occurs.

Piaget's ideas influenced David Ausubel (1963) who proposed that meaningful learning involves the integration of novel information into existing conceptual structure, resulting in an idiosyncratic reconstruction of the knowledge by each learner. Meaningful learning is contrasted with rote learning in which newly acquired concepts remain isolated from pre-existing conceptual structure. The repeated action of meaningful learning processes, over an extended period of time, leads to a model of expert knowledge as a highly inter-connected and well-integrated structure. It is assumed that this kind of knowledge organisation allows experts to apply ideas to novel contexts and adopt a flexible approach to problem solving. This model of learning as the development of conceptual structure, led Ausubel to suggest the idea of an advance organiser, information that is communicated to a student to help them integrate subsequent teaching. For example, a teacher may explain the nature of scientific models before teaching atomic bonding. An understanding of models may help a student interpret the nature of claims made about bonding, and to integrate the novel information into their conceptual structure.

Though his research was not influential outside of the Soviet Union till after his death, Lev Vygotsky has become an important constructivist thinker. He developed the construct of the zone of proximal development (ZPD), which he described as the set of tasks which are not accessible to a student working independently, but become achievable with support from teachers or peers (Vygotsky, 1978). This idea prefigured the development of scaffolding (Wood, Bruner, & Ross, 1976), an approach in which a teacher anticipates the difficulties a student may encounter with a task and provides appropriate information to bridge the gap between the required learning and the student's current position. Vygotsky described how a child's spontaneous concepts, or self-generated ideas, might conflict with the academic concepts they are being taught (Vygotsky, 1962). He suggested that spontaneous concepts are 'saturated with experience,' and therefore challenging to integrate with the abstract concepts of formal education (Vygotsky, 1962, p. 108). This idea was to become the foundation of the research programme into children's misconceptions that arose in the late 1970's (see below). A central assumption of Vygotsky's theory was that concepts appear first in the social realm and only subsequently in the psychological domain (Vygotsky, 1931/1981). A student is likely to encounter the scientific concept of force initially in social discussion in a classroom before internalising the notion. A group of constructivist theories, labelled social constructivisms, built on Vygotsky's claim to develop models of learning that emphasise the role of interactions between individuals. Therefore, rather than constructing learning as the result of processing in an individual's cognitive systems, social constructivists describe learning as the adoption of the behaviours and conventions of a particular culture or group. In this conceptualisation, the aim of science education is for students to adopt similar approaches to making sense of the world as experienced scientists. One of the most influential social constructivist learning theories arising out of the work of Vygotsky and others, is activity theory (Engeström, 1987). Activity theory suggests an individual needs to be understood within their relationship to a wider community and encourages sensitivity to the multiple understandings that may be possessed by different members of the community.

The constructivist research programme in science education began in the late 1970s and early 1980s. Keith Taber (2009, p. 113) has labelled five early papers in the field as a 'seminal corpus' for research into children's ideas in science. In the earliest of these papers, Rosalind Driver and Jack Easley (1978) suggested the term alternative frameworks to refer to students' constructions generated to understand the physical world, for example the idea that, during heating, the dimensions of particles themselves expand. These early papers founded a research programme initially directed at cataloguing students' alternative frameworks across a range of topics in science education. Research interest also focused on the manner in which students' initial understandings came to resemble those of experienced scientists, a process known as conceptual change. One of the earliest, and most influential, contributions to this research was the model of conceptual change proposed by Posner and colleagues (1982). The group suggested that a range of cognitive entities, of various types, for example analogies or beliefs about knowledge, is available to an individual at any given time. They labelled this set of resources a conceptual ecology. The assumption that learners have access to a variety of conceptual resources has become an axiom of a number of models of learning in science education (see, for example, diSessa's (1993) knowledge-in-pieces model).

Conceptual change has remained an area of much research activity in science education and many different approaches have been proposed, leading Tyson and colleagues (1997) to categorise the models into three groups: epistemological, ontological and social/affective approaches. Posner and colleagues (1982) argued conceptual change would occur when students became dissatisfied with an existing concept and a plausible and intelligible alternative was available. This kind of model has been classified as an epistemological (i.e. related to the nature of knowledge) approach to conceptual change. The second category of conceptual change contains the ontological model (Tyson et al., 1997) suggested by Chi and Slotta (1993). The central assumption of this approach is that learners categorise novel concepts into one of three major classes: matter, process or mental state. Students may initially misclassify scientific concepts into an inappropriate class, for example they may assume heat has substance-like properties, and conceptual change may occur through the reclassification of heat as a process. The final group of conceptual change theories are those that acknowledge the importance of social or emotional factors

(Tyson et al., 1997). For example, conceptual change may be driven by a desire to avoid the negative emotions associated with conflict between personal models and novel data and the development of coherence may be an emotionally agreeable and hence motivating experience.

Models of learning are necessarily dependent on how researchers conceptualise mental representations of knowledge. This is an area of intense discussion in science education and two major models of conceptual structures exist: knowledge-as-theories and knowledge-as-elements. The knowledge-as-theories position claims that, though they may differ from those of scientists, students' knowledge structures are relatively organised and coherent (Vosniadou, 2007). Conceptual change in this model is seen as a relatively slow process involving the gradual development of students' naïve theories. An alternative position, proposed by diSessa (1993), is that novice learners' thinking is characterised by a range of different elements that are relatively unstructured and inconsistently triggered across contexts. One of these elements, the phenomenological primitive, or p-prim, is seen as an intuitive knowledge structure that supports explanations of the physical world. For example, a proposed p-prim is the sense that, in general, more impetus or less resistance will lead to greater action. This intuition may underlie a number of alternative conceptions, for example, that objects with greater mass take less time to fall to the ground than lighter objects. P-prims are hypothesised to be contextually triggered and, rather than being discarded as the learner develops expertise, the conditions of their activation are adjusted. In this fragmented description of knowledge, learning is seen as the gradual organisation of knowledge elements into structured systems, known as coordination classes, which lead to a consistency in responses across a range of contexts.

The dominance of constructivism in science education may have obscured the arguments of other perspectives on learning (Solomon, 1994) and a number of critics have argued that the constructivist model is flawed (for a detailed summary of the various critiques of constructivism, see chapter 5 in Taber (2009)). Critics suggest that constructivist models of learning require teachers to accept all students' ideas as equally valid (Matthews, 1992) and the theory has been described as culturally imperialist (Bowers, 2007) due to the perception that the constructivist model has been imposed onto cultures which traditionally value knowledge transmission. Matthews (2002), argues that the strategies suggested by constructivism are well-known educational truisms and that the theory's jargon and underlying philosophical assumptions create confusion and inhibit communication. Proponents of constructivism have rebutted these attacks (Taber, 2009) and the assumptions of the programme are still widely accepted within science education.

Implications for Practice

The model of learning presented in constructivism has led to the proposal of a number of different approaches to teaching. The constructivist axiom that the individual student actively constructs knowledge underlies the discovery learning approach. The technique is often associated with Bruner (1961) but similar notions are found in the writing of Piaget and other thinkers. The central assumption of discovery learning is that 'discovering for oneself' leads to more effective learning (Bruner, 1961, p. 26). For example, students may be given a list of properties of a range of elements and asked to create a set of groupings based on physical similarities. It is argued that if students actively participate in the development of ideas, in contrast to simply being presented with facts, they will develop more meaningful understandings. Though the technique was popular in the 1960s and 1970s, in recent decades it has been criticised as an ineffective and potentially counterproductive strategy. The criticisms tend to be aimed at 'pure' discovery learning, in which a student receives little or no guidance, and Taber (2009, p. 205) argues that 'carefully guided discovery' may play a valid role in science education.

The assumption that learning is an active process performed by the individual, suggests a particular conceptualisation of the role of the teacher. Rather than acting solely as a source of knowledge, the teacher is seen as a facilitator of learning, that is they support students' personal acts of meaning making. Research on children's understanding of science has led to a general approach to science teaching that encourages teachers to assess students' pre-existing ideas and then develop, or alter students' alternative frameworks to facilitate the acceptance of novel material. One much discussed technique for modifying alternative frameworks is the triggering of conceptual (or cognitive) conflict, a recognition of incompatibility between the concepts a learner holds and novel material (Nussbaum & Novick, 1982). In the physics classroom, for example, a student who believes that a resultant force is required for motion at constant velocity might be shown a simulation of an object moving in space, in the absence of external forces. It is suggested that the dissonance between the simulation and the student's existing idea will cause conceptual change to occur. However, evidence from studies of the implementation of the approach suggest the technique produces mixed results and is challenging to implement in practice.

CONCLUDING THOUGHTS

The complexity and idiosyncrasy of human learning suggests it is unlikely that a single model will ever completely describe its nuances. Each of the theories above may be effective in describing some aspects of learning but other models, which there is insufficient space to examine in this chapter, may contribute to a fuller representation. The interested reader may wish to investigate, amongst others, semiotic, humanistic or neuroscientific theories of learning. It is important to recall, as with any model, the descriptions of learning presented here are incomplete and based on assumptions that are, to a greater or lesser extent, supported by evidence. Debates about the most appropriate manner to model learning may seem abstract and removed from practice. However, assumptions about the nature of knowledge and learning implicitly and explicitly influence the practice of teachers. The adoption of constructivist or behaviourist assumptions will lead to significantly different approaches to classroom practice, variation in teacher-student relationships and alternative interpretations of successful learning. It might seem that an evidence-based evaluation of the strategies suggested by the different learning theories would allow educators to decide which approach would be most effective in their context. However, the theories suggest different learning outcomes are desirable, for example the acquisition of factual knowledge is seen as central in behaviourist models, whereas constructivists emphasise understanding and meaningful learning. It is therefore difficult to develop a measure of effective learning that can be agreed on by the supporters of different models and hence it is challenging to compare the effectiveness of the approaches suggested.

Rather than seeing the multiple models of learning as existing in competition, the existence of a variety of perspectives may be more flexible and powerful than adherence to a single dominant approach. The richest account of learning is likely to involve models at a variety of different levels that account for the role of cultural pressures, social interactions, personal behaviours, cognitive systems, and the action of neurons. Classrooms are not uniform contexts; there is much variation between students in terms of prior knowledge, motivation, areas of interest and in preferred modes of social interaction. A teacher who has access to different approaches to conceptualising learning and a range of different strategies drawn from different traditions is likely to be well placed to cope with the diversity of classroom situations.

Whilst teachers and researchers may make an informed choice to conceptualise learning through a single framework, it is useful to remain sensitive to alternative interpretations. The dominance of the constructivist approach in science education might discourage new teachers and researchers from investigating models and strategies from other theories of learning. If nothing else, it is hoped that this chapter will encourage readers to engage with the different representations of learning that have been constructed and consider what they have to offer to practice or to research.

FURTHER READING

A comprehensive and accessible introduction to different learning theories and their application to teaching practice can be found in:

Schunk, D. H. (2014). *Learning theories: An educational perspective*. Harlow: Pearson Education Ltd.

A good overview of different learning theories in the context of science education can be found in the following chapter:

Duit, R., & Treagust, D. (1998). Learning in science – From behaviourism towards social constructivism and beyond. In B. Fraser & K. Tobin (Eds.), *International handbook of science education* (pp. 3–26). Dordrecht: Kluwer Academic Publishers.

Frederick Reif has written a broad introduction to the application of ideas from cognitive science to science teaching:

Reif, F. (2008). *Applying cognitive science to education*. Cambridge, MA: The MIT Press.

Keith Taber's book is a good source of discussion on constructivist models of learning in science education:

Taber, K. (2014). *Student thinking and learning in science*. New York, NY: Routledge.

REFERENCES

Ausubel, D. (1963). Cognitive structure and the facilitation of meaningful verbal learning. *Journal of Teacher Education, 14*(2), 217–222.

Baddeley, A. D. (2000). The episodic buffer: A new component of working memory? *Trends in Cognitive Sciences, 4*(11), 417–423.

Bandura, A. (1971). *Social learning theory*. Morristown, NJ: General Learning Press.

Bowers, C. A. (2007). *The false promise of constructivist theories of learning*. New York, NY: Peter Lang Publishing Inc.

Bruner, J. (1961). The act of discovery. *Harvard Educational Review, 31*, 21–32.

Cheng, M. H. M., & Wan, Z. H. (2015). Unpacking the paradox of Chinese science learners: Insights from research into Asian Chinese school students' attitudes towards learning science, science learning strategies, and scientific epistemological views. *Studies in Science Education, 52*(1), 29–62.

Chi, M. T. H., & Slotta, J. D. (1993). The ontological coherence of intuitive physics. *Cognition and Instruction, 10*(2–3), 249–260.

diSessa, A. (1993). Toward an epistemology of physics. *Cognition and Instruction, 10*(2–3), 105–225.

Driver, R., & Easley, J. (1978). Pupils and paradigms: A review of literature related to concept development in adolescent science students. *Studies in Science Education, 5*(1), 61–84.

Engeström, Y. (1987). *Learning by expanding: An activity theoretical approach to developmental research*. Helsinki: Orienta-Konsultit Oy.

Flavell, J. H. (1979). Metacognition and cognitive monitoring. *American Psychologist, 34*(10), 906–911.

Gagné, R. M. (1965). *The conditions of learning*. New York, NY: Holt, Rinehart & Winston.

Haugeland, J. (1978). The nature and plausibility of cognitivism. *Behavioral and Brain Sciences, 1*(2), 215–226.

Matthews, M. R. (1992). Constructivism and empiricism: an incomplete divorce. *Research in Science Education, 22*(1), 299–307.

Matthews, M. R. (2002). Constructivism and science education: A further appraisal. *Journal of Science Education and Technology, 11*(2), 121–134.

Nussbaum, J., & Novick, S. (1982). Alternative frameworks, conceptual conflict and accommodation: Toward a principled teaching strategy. *Instructional Science, 11*(3), 183–200.

Pavlov, I. P. (1927). *Conditioned reflexes* (G. V Anrep, Trans.) (Vol. 17). Oxford: Oxford University Press.

Piaget, J. (1952). *The origins of intelligence in children* (M. Cook, Trans.). New York, NY: International Universities Press.

Posner, G. J., Strike, K. A., Hewson, P. W., & Gertzog, W. A. (1982). Accommodation of a scientific conception: Toward a theory of conceptual change. *Science Education, 66*(2), 211–227.

Reif, F. (2008). *Applying cognitive science to education*. Cambridge, MA: The MIT Press.

Skinner, B. F. (1958). Teaching machines. *Science, 128*(3330), 969–978.

Solomon, J. (1994). The rise and fall of constructivism. *Studies in Science Education, 23*(1), 1–19.

Stewart, M. (2012). Understanding learning: Theories and critique. In L. Hunt & D. Chalmer (Eds.), *University teaching in focus* (pp. 3–20). Abingdon, Oxon: Routledge.

Strike, K. A., & Posner, G. J. (1982). Conceptual change and science teaching. *European Journal of Science Education, 4*(3), 231–240.

Taber, K. S. (2009). *Progressing science education: constructing the scientific research programme into the contingent nature of learning science.* Dordrecht: Springer.
Taber, K. S. (2013). *Modelling learners and learning in science education: Developing representations of concepts, conceptual structure and conceptual change to inform teaching and research.* Dordrecht: Springer.
Taber, K. S. (2014). *Student thinking and learning in science: Perspectives on the nature and development of learners' ideas.* New York, NY: Routledge.
Thorndike, E. (1927). The law of effect. *The American Journal of Psychology, 39*(1/4), 212–222.
Tyson, L. M., Venville, G. J., Harrison, A. G., & Treagust, D. F. (1997). A multidimensional framework for interpreting conceptual change events in the classroom. *Science Education, 81*(4), 387–404.
Vosniadou, S. (2007). Conceptual change and education. *Human Development, 50*(1), 47–54.
Vygotsky, L. (1931). The genesis of higher mental functions. In J. V. Wertsch (Ed.), *The concept of activity in Soviet psychology* (pp. 144–188). Armonk, NY: Sharpe.
Vygotsky, L. (1962). *Thought and language.* Mansfield, CT: Martino Publishing.
Vygotsky, L. (1978). *Mind in Society: The development of higher mental processes* (M. Cole, Trans.). Cambridge, MA: Harvard University Press.
Wood, D., Bruner, J. S., & Ross, G. (1976). The role of tutoring in problem solving. *Journal of Child Psychology and Psychiatry, and Allied Disciplines, 17*(2), 89–100.

CHRISTOPHER ANDERSEN AND MERCE GARCIA-MILA

8. SCIENTIFIC REASONING DURING INQUIRY

Teaching for Metacognition

As described in the US *Next Generation Science Standards*, scientific practices are the behaviors that scientists engage in as they investigate and build models and theories about the natural world. The use of the term "practices" has replaced the term "skills" to avoid the notion of rote mastery of a procedure. It also emphasizes the idea that engaging in scientific inquiry requires coordination of both knowledge and skills simultaneously. Part of the NGSS's intent is to better explain and extend what is meant by "inquiry" in science and the range of cognitive, social, and physical practices involved in it. The present chapter also strives in this direction. We provide an analysis of the reasoning challenges that the students face when they engage in scientific inquiry.

To answer the question "What is science?", we need to refer to: (1) a body of knowledge of concepts, laws, theories and ideas, (2) the 'process/method' as to what scientists do to develop/construct the body of knowledge and (3) the nature of science referring to the characteristics of scientific knowledge that are directly derived from the process/method used to develop the knowledge. Science is currently seen as tentative, empirical, theory-laden, creative, and social (see Chapter 2 for more information about the nature of science).

In spite of the science community recognition of the importance of theory in inquiry processes and the social nature of science, many science teachers continue to engage their students in theory-free data-gathering tasks. Many students believe in the distorted view of scientific inquiry as the application of 'The Scientific Method' as the single correct algorithm that students are expected to memorize, recite and follow as a recipe for success. In contrast, recent views of scientific inquiry provide no single fixed sequence of steps that all scientific investigations follow. The scientific questions guide the approach, and the approaches vary widely within and across scientific disciplines and fields. In spite of this, the Scientific Method is still posted on the walls of science classrooms as well as in science textbooks and laboratory manuals (despite it being discarded by philosophers, sociologists, and scientists), with teachers having their students memorize and structure their thinking along the rigid steps of the scientific method.

More than fifty years of psychological and educational research led to current definitions of scientific reasoning with three mains shifts from the classic recipe

K. S. Taber & B. Akpan (Eds.), Science Education, 105–117.

of scientific method to do inquiry. The first one calls our attention to the claim that scientific reasoning is not a content-free process. Processes of scientific reasoning develop hand in hand with the development of science content. The second shift refers to the fact that the processes that take place when applying the scientific method do not occur separately, but in integrated sets of inquiry where hypotheses are refined as observations are being made and variables operationalized. In addition, argumentation with peers plays an important role in the process of refining the strategies used in inquiry. What matters is the ability to coordinate the processes. The third shift refers to two dimensions that need to be paid attention to in order to understand how science is generated: epistemic and social. The epistemic dimension refers to being aware of the role of evidence in scientific practices and being aware of what counts as evidence. The social dimension refers to the science community providing recognition of the dialogic and dialectical practices that scientific reasoning involves.

Scientific reasoning is defined as the inquiry processes by which individuals revise and reconstruct their theories about the world; that is, the reasoning skills involved in experimentation, evidence evaluation, and inference making addressed to scientific understanding. It is essential to understand how students acquire scientific knowledge according to the procedures they use to discover, assess, revise, and communicate that knowledge. It is in this sense that cognitive developmental psychologists synergistically working with science educators should be very helpful.

Duschl and Grandy (2008) summarize the core of scientific inquiry as "acquiring data and transforming those data first into evidence and then into explanations" (p. 305). They propose a list of 30 scientific inquiry practices: posing questions, refining questions, evaluating questions, designing experiments, refining experiments, interpreting experiments, making observations, collecting data, representing data, analyzing data, relating data to hypotheses/models/theories, formulating hypotheses, learning theories, learning models, refining theories, refining models, comparing alternative theories/models with data, providing explanations, giving arguments for/against models and theories, comparing alternative models, making predictions, recording data, organizing data, discussing data, discussing theories/models, explaining theories/models, reading about data, reading about theories/models, writing about data, and writing about theories/models. The US *Next Generation Science Standards* propose a smaller set of science and engineering practices that are compatible with Duschl and Grandy: Asking questions (for science) and defining problems (for engineering); Developing and using models; Planning and carrying out investigations; Analyzing and interpreting data; Using mathematics and computational thinking; Constructing explanations (for science) and designing solutions (for engineering); Engaging in argument from evidence; and Obtaining, evaluating, and communicating information.

In the following section we make an analysis of the strategies involved in scientific reasoning and how they are used by students of different ages when they solve a scientific inquiry task.

HYPOTHESIS GENERATION

Asking Questions

For classroom scientific inquiry to be successful, teachers need to be able to help students transform their curiosity about natural phenomena into inquiry questions. Here is where the activity must be formulated in a way that transforms a given curiosity into an inquiry question. Task questions (what is the goal?) must be transformed into strategic questions (what can I do to achieve the goal?). Question-Asking is among one of the higher order thinking skills. It requires an explicit effort by science teachers to "challenge their students to ask relevant, in-context meaningful questions and, persistently, to exercise this capacity" (Zoller & Nahum, 2012, p. 211).

Student explorations of cause-and-effect relationships are an important part of classroom scientific inquiry. A familiar example for many elementary classroom teachers is to investigate the factors that effect the germination of seeds. A teacher who promotes student questioning by asking, "What do seeds need in order to start growing?" will receive responses about moisture, temperature, and other factors from virtually every child. Students often come into the classroom with a rich array of beliefs about the *causality* (e.g., "Water is needed for a seed to germinate") or *noncausality* ("The color of the pot doesn't make a difference") of factors that are investigated in classroom scientific inquiry, and these existing beliefs (correct or not) have an influence on the questions that students pose (and on other aspects of scientific inquiry, too).

The influence of cognitive biases. Scientific inquiry can be constrained by cognitive biases. For example, students' inquiry may be influenced by a *causal bias.* Initially, students' investigations of a phenomenon tend as focus on factors that they think will be causal. It may not be until after repeated encounters with the task (and a lack of causal results) that will they start to examine factors that they believe are noncausal. This bias toward causal variables is even more striking for young students, who may completely ignore testing their noncausal beliefs. Though the influence of causal bias may decrease with development, causal bias can affect inquiry at all ages.

Another important cognitive bias to consider in science learning is *confirmation bias,* in which students (and adults) tend to conduct their investigations in ways that support their existing beliefs. Both causal bias and confirmation bias can influence the entire process of scientific inquiry, from generating hypotheses to designing experiments to evaluating evidence.

At first glance, it may not be apparent why biases like these would be a concern for science learning. After all, for many problems in everyday life, there may be an enormous number of potential factors that could be at work, and it may be efficient first to investigate variables that are believed to be causal. But the goals of learning in the science classroom differ from the goals of problem solving in an everyday context in several ways. For example, in everyday problem solving, a goal is often just to solve the problem rather than understand the factors behind the problem. In contrast, understanding a scientific phenomenon often involves understanding all the factors involved, whether or not the student believes (perhaps incorrectly) that a given factor is causal or noncausal. Neglecting to understand factors that are believed to be noncausal can have serious repercussions; consider that for many decades, medical research focused on studies on males and failed to consider the possibility of gender differences in metabolism of pharmaceuticals, resulting in ineffective treatments and heightened toxicity.

What are the implications for teachers' practice in science class? Teachers need to be aware that when students are free to choose their own questions to pursue during inquiry, the questions they choose may reflect the cognitive biases described previously. A teacher could help counter students' cognitive biases by framing the inquiry in a manner that encourages students to continue to develop questions after they have they have confirmed their causal beliefs (e.g., "Find out as much as you can about…"). Teachers also need to help students become aware that these biases exist—unless students have explicit knowledge about their own thinking and its limitations, they will be hard-pressed to be able to change their thinking.

Making Hypotheses

Scientists use the term *hypothesis* for a tentative explanation for a scientific phenomenon. Hypotheses are tested to refine models and develop theories.[1] There are striking developmental differences in how children and adults use hypotheses during scientific inquiry.

Are students open to considering and testing alternative hypotheses? Compared to students age 11 and adults, younger students (age 8) tend to generate a single hypothesis and have difficulty considering alternatives. But even though older students and adults are willing to consider that alternative hypotheses exist, research shows that students of all ages tend to get stuck on testing only the hypothesis that they consider to be most likely. This confirmation bias causes students (and adults) to limit their investigation to the hypothesis that they believe to be correct (but may be actually incorrect) and interferes with their willingness to investigate other (possibly correct) hypotheses.

Are students aware of differences in the goal of experimentation? When students age 10–12 are asked to investigate a causal system, they tend to design experiments that are intended to result in desirable, interesting outcomes and avoid outcomes that are undesirable or less interesting (e.g., produce the fastest robot, change of color of a liquid, etc.) rather than design experiments to determine the causal relation among variables (e.g., which factors made a difference (and which factors did not make a difference) in the robot's speed). It is as if they have difficulty distinguishing between "producing" the phenomenon (an engineering approach) and "understanding" a phenomenon (a scientific approach). If students are focused on merely engineering a solution to a design problem, they may not come to an understanding of all of the variables in the phenomenon: they are likely to focus only on the variables that they believe to be causal (recall the discussion of confirmation bias and causal bias earlier in this chapter).

This contrast between the scientific inquiry approach ("Learn as much as you can about…") and engineering design approach ("Produce an outcome within these parameters") has important implications for teachers. The NGSS and STEM education more broadly have advocated for the increasing use of engineering approaches (e.g., design challenges) in science courses. There are benefits to framing a lesson as an engineering design problem: the content is applied in a real world context that may be more engaging than the traditional teaching of science. However, if the instruction is intended to foster students' scientific understanding of the variables that affect a phenomenon (and to foster students' scientific reasoning skills that are used to investigate those variables), then teachers need to be aware of how the goals of the learning task will promote students' use of a scientific versus an engineering approach to the activity. (See Schauble, Klopfer and Raghavan (1991) to learn more about scientific and engineering approaches).

EXPERIMENTAL DESIGN

In the previous section, we discussed some of the developmental changes in students' reasoning when they generate questions and hypotheses during scientific investigations. In this section, we will look at several aspects of how students design experiments to explore their scientific questions and hypotheses.

Designing Experiments and Investigating the Problem Space

Imagine that students are investigating the factors that affect the growth of plants within a simple causal system that has three factors to consider: (1) the type of seed *(Brassica* and *Rosette),* (2) the type of fertilizer (organic and chemical), and (3) the type of light (artificial and natural). To fully understand the causal and noncausal effects in scientific inquiry activities, students need to use two important scientific reasoning strategies when designing experiments: *control of variables* and *factorial combination of variables.*

Are students aware of the experimental value of controlled comparisons? When scientists design an experiment, they are careful to control for variables that are not the focus of the experiment. In our plant example, if a student intended to test whether the type of fertilizer made a difference in the growth of plants, a simple experiment could be designed in which one group of plants would receive organic fertilizer, another group of plants would receive chemical fertilizer, and the remaining variables (type of seed and type of light) would be identical across the two groups. If more than one variable differs between the two groups (for example, having the two groups differ both by the type of fertilizer and by the type of light), it will not be clear which variable (type of fertilizer or type of light) is responsible for any differences in growth.

Though the need to control variables may seem straightforward to a scientist (or a teacher), it is not always straightforward for students. When students investigate a simple causal system such as the plants task, they often do not begin designing controlled comparisons until around age 10.

What strategies do students use to control variables? Tschirgi (1980) summarized the different experimental strategies children use when they have to test the effect of a variable in a causal system and described three types of experimental strategies according to how many variables are changed and how many are held constant:

- *Vary-One-Thing-At-a-Time (VOTAT):* This strategy corresponds to the control of variables strategy, in which an unconfounded comparison is designed with one variable differing between the two comparison groups and the remaining variables kept constant. In our example that investigated the effect of the type of fertilizer (organic vs chemical) on the growth of plants, a student using VOTAT would change the type of fertilizer but the type of seed *(Brassica* vs *Rosette)* and type of light (artificial vs. natural) would be identical across the two groups.
- *Change-All (CA):* A student using CA to investigate the effect of the type of fertilizer would design a completely confounded experiment that did not control any of the variables. For example, one group might be organic fertilizer/Brassica seed/artificial light, and the corresponding comparison group would be chemical fertilizer/Rosette seed/natural light.
- *Hold-One-Thing-At-a-Time (HOTAT):* A student using HOTAT to investigate the effect of the type of fertilizer would keep the type of fertilizer the same across the two groups but change the remaining variables (type of seed and type of light).

Does the value of the outcome affect the choice of the experimental strategy? A teacher's lesson design can affect students' choice of inquiry strategy. Tschirgi found that age 7–11 students' choice of experimental design strategy varied depending on how they defined the task outcome (in her study, students varied ingredients to produce a cake). If the students were asked to investigate how the ingredients affected whether a "good" cake would result, they tended to use the Hold-One-Thing-At-a-Time strategy: in an effort to preserve the same "good" outcome in the comparison cake, students would keep constant the ingredient believed to be the cause of the good outcome. In contrast, another group of students was asked to investigate how the ingredients affected whether a "bad" cake would result. The students in the "bad" cake condition were more likely to use the Vary-One-Thing-At-a-Time strategy to vary the variable believed to be the cause of the bad outcome.

Do students work systematically when they design experiments? In addition to the control of variables strategy just discussed, another important scientific reasoning strategy used by students when designing experiments is factorial combination of variables. In the plants task described earlier, the causal system to be investigated is relatively small: 2 levels of seed (Brassica and Rosette), 2 levels of fertilizer (organic and chemical), and 2 levels of light (artificial and natural). In the plant causal system, the total possible number of unique combinations of plant variables that students could produce (sometimes called the *problem space*) is 8 (i.e., $2 \times 2 \times 2$). In order to investigate all the possible main effects and interactions in the plant causal system, the entire problem space (8 combinations) needs to be examined.

However, students do not always explore the problem space in a systematic way. In a study using the plants task with students age 11–12 (Garcia-Mila, Andersen, & Rojo, 2011), participants had the opportunity to design 10 experiments. With a problem space of 8, these students could have designed experiments that examined the entire problem space, with two experiments to spare. Instead, fewer than half of the participants examined the entire problem space. However, once students reach adolescence, they are more likely to organize their experiments so that they systematically investigate the factors, often with the help of notes that they produce to keep track of the factorial combinations.

What affects students' systematic examination of the problem space? Even with the factorial combination of variables strategy, adolescents and adults may have difficulty in taking advantage of examining the available problem space. One factor affecting the use of the strategy may be the increasing size of the problem space. In the plants task, the number of possible combinations was 8. In a factorial combination task with 16 possible combinations, Siegler and Liebert (1975) found that only 10% of age 14 students succeeded in covering the whole problem space. In a causal reasoning task with similarities to the plants task but with a much larger problem space of 48 ($2 \times 2 \times 2 \times 2 \times 3$), Kuhn, Garcia-Mila, Zohar, and Andersen (1995) found that adults only investigated about 2/3 of the available problem space. In addition, the preadolescents in this study had inaccurate perceptions of the size of the problem space. For example, when asked about the size of the problem space that they were asked to investigate, one of the preadolescents answered that the size of the problem space was "infinite", and when asked to be more precise, he said "10,000."

When children and adults did not systematically investigate the problem space, they were repeating experiments that they had already conducted. These repetitions appear to be influenced by the causal bias that was discussed earlier in the chapter. Individuals have a preference for investigating causal beliefs vs. noncausal ones, so they tend to design more experiments that confirm their causal theories even if they repeat the same experiment several times. These repetitions increase when students encounter results that are discrepant to their belief (specifically, when they incorrectly believe that a variable is causal and the evidence shows the variable to be noncausal): the result is interpreted to be some kind of experimental error, so they tend to repeat the experiment in order to find the expected causal result.

EVIDENCE EVALUATION

In the previous section, we discussed some of the developmental changes in students' reasoning when they are designing experiments. In this section, we will look at students' reasoning about the data that they generate from experiments.

Interpreting Data, Evaluating Evidence and Inference Making

Once evidence is produced, it can be used to draw inferences. In order for an inference to be valid, the inference needs to be based on comparisons in which all of the factors are controlled for except for the factor under investigation. (The controlled comparison strategy was discussed in more detail previously in the Experimental Design section). However, students (and even adults) do not consistently draw inferences that are valid.

What types of invalid inference do students draw? Recall the plants task described earlier, which has three factors to consider (type of seed, type of fertilizer, and type of light). Imagine that a student has designed several experiments with the plants and is now looking over the results in order to draw inferences about how the type of light (artificial and natural) may affect plant growth. An inference could be validly based on experiments that were designed using the control of variables strategy (i.e., Vary-One-Thing-At-a-Time) (for example, comparing the growth of Brassica/chemical fertilizer/natural light plants with Brassica/chemical fertilizer/artificial light plans). But there are several ways in which students use evidence incorrectly and draw invalid inferences:

- The inference could be based on an *uncontrolled comparison* that resulted from experimental design strategies (such as Change-All and Hold-One-Thing-At-a-Time) that don't properly isolate the factor in question (for example, comparing the growth of Brassica/chemical fertilizer/natural light plants with Rosette/organic fertilizer/natural light plants).
- The inference could be based on a *single instance,* without having any comparison group (for example, drawing an inference about how the type of light affects plant growth by examining only the Brassica/chemical fertilizer/natural light plants).
- After students draw inferences about how the type of light affects plant growth, some will not refer to the results of their experiments, particularly if the results do not support their pre-existing beliefs. For example, many students believe that the plants will grow better in natural light (compared to artificial light). However, when the experimental results indicate that the plants grow better with artificial light, some students will try to preserve their pre-existing (and incorrect) belief

by ignoring the evidence and instead using a *theory-based justification* of their inference (for example, invoking an explanation that "natural" is healthier than "artificial.")

Students draw inferences without considering the complete problem space. Earlier in the chapter, we described how students tend to have biases that influence their hypothesis generation and experimental design when engaging in scientific inquiry. Because of these bias, students may limit their problem space exploration to investigations that confirm their pre-existing beliefs about factors that they believe to be causal. As a result, they may have an incomplete database of results to use to draw inferences.

One implication of using an incomplete database of results is that it may be difficult to detect causal interactions. We used a computerized causal system that had structural similarities to the plants example described earlier but which allowed students to investigate five factors that affected the speed of cars. Among the factors to investigate were the size of the engine (large and small) and the presence or absence of a tail fin. The size of the engine had a simple causal effect (the large engine made the car go faster than the small engine). However, the fin had an interactive effect: the fin only had an effect when the large engine was present and had no effect when the small engine was present. So if students only conducted controlled experiments of the fin using the large engine, they might conclude that the fin has a causal effect. But if students only conducted with controlled experiments of the fin using the small engine, they might conclude that the fin has no effect. In the absence of a complete database from exploring the entire problem space, students may not have sufficient evidence to discover interactive effects.

Recording Data and Reviewing Data Records

Teachers frequently have their students record information while engaged in scientific inquiry, often using laboratory notebooks. This notetaking can be useful for preserving the design of individual experiments and their outcome, for organizing a systematic search of the problem space using factorial combination, and for describing hypotheses and inferences. Despite the potential utility of notetaking, students may not be effective at taking and using notes during scientific inquiry.

Why don't students take notes? Our understanding of our own memory develops during the early elementary school years (age 5–10). As a result, preadolescent students may overestimate their ability to remember the conditions of the experiments that they design and the results they obtain, so they may see notetaking as unnecessary.

Even in tasks that a teacher may perceive as complicated enough to require notetaking, students may not have the same perception. In the cars task mentioned earlier in the chapter, the system has a large problem space (five factors, which produces 48 different cars to complete ($2 \times 2 \times 3 \times 2 \times 2$)) over a memory-challenging length of time (five weeks). Participants were encouraged to take notes but were not required to take notes. All of the adults in the study took notes, while only half of the students age 10 took any notes at all.

Why do students take poor notes? If students take notes, they may do it poorly. In the cars task, students designing experiments may record the levels of only some of the factors (perhaps only those believed causal). After completing an experiment, they may not record the result. These incomplete notes may be the result of students' poor *prospective memory,* which involves anticipating one's state of knowledge in the future and then recording appropriate information in the present in order to meet that future need. For the participants in the cars task, the number of notes taken by adults remained stable over the study, while the number of notes taken by the students age 10 dropped to one half of the notes that they took at the initial session, perhaps because the students saw little usefulness in their poor notes.

Why don't students review notes? In the cars task, the adults regularly looked back at their notes when engaged in experimental design, inference making, and other aspects of scientific inquiry. In contrast, even if they took notes, students age 10 rarely reviewed them. The ability of students to draw valid inferences and to effectively explore the problem space is correlated with reviewing notes rather than with simply recording notes: a recorded note is not useful unless it is actually used.

A comparison of the cars task and the plants task points out a strategy for teachers to encourage taking and reviewing notes. In the cars task, students design an experiment, carry it out, observe the results, and draw inferences all within one session. Rather than examine all the data that they have generated across several sessions, students tend to focus on the trial that is immediately at hand and analyse the new result in isolation from the trials from previous sessions. In contrast, the plants task separated the design of the experiment from the evaluation of evidence: the seeds required time to grow, so the result of the experiment could not be observed until the session after the seeds were planted. By incorporating deferred evidence evaluation into the task design, the proportion of students who took notes and who reviewed notes in the plants task was remarkably higher than in the cars task. (See Garcia-Mila, Andersen, and Rojo (2009) to learn more about the role of notetaking in science reasoning.)

ARGUMENTATION

Current views of the nature of science see science as a social construction not restricted to processes that result from inquiry practices (planning and performing experiments), but also from communication processes that lead to discussions addressed to resolve controversies and reach consensus. Science advances through a progression of theory confirmation-disconfirmation processes (Popper, 1965), with findings being submitted to social scrutiny. Hence, argumentation becomes a core process in such scrutiny because it allows debating confirmatory and disconfirmatory claims in relation to evidence.

Essential components in constructing scientific arguments are assessing claims and alternatives, weighing evidence, and evaluating the potential validity of scientific claims, all fundamental in the progression of scientific knowledge. Hence, students in a science course must not only develop essential inquiry skills, such as controlling variables, and designing and conducting experiments, but also must be able to decide between two competing alternative theories to explain a phenomenon based on the evidence generated. (See Chapter 12 and Garcia-Mila & Andersen (2009) to learn more about argumentation in science education.)

METACOGNITION TO EXPLAIN SCIENTIFIC REASONING PERFORMANCE

In the previous sections of this chapter, we have described a variety of strategies used during scientific inquiry and some of the difficulties that students (and even adults) have in using them. Many of these strategies are not difficult for students (even young children) to learn, yet they often don't use them even after learning them. Simply because a new strategy has been taught (for example, justifying inferences using controlled comparisons) does not mean that the previous strategies (uncontrolled comparisons, single instance, and theory-based justification) fall by the wayside. Instead of replacing the old with the new, the development of scientific reasoning should be thought of as adding to a repertoire of strategies, with the new strategy co-existing alongside the old. The rate of use of each strategy in the repertoire shifts over time, with the new strategy being used more and more often. The change in use of each co-existing strategy is more than simply of matter of practice. Even adults may invoke theory-based justifications for inferences when faced with evidence that contradicts a strongly-held belief, for example.

For teachers, this raises the question of how students choose between the array of co-existing strategies in their repertoire. When a student decides which strategy to use, *metacognition* plays an important mediating role. Sometimes characterized as "thinking about thinking," metacognition encompasses a wide range of cognitive activities. Because this chapter is focused on scientific reasoning strategies, we will limit our discussion to the metacognitive knowledge that students have about strategies *(metastrategic knowledge),* as opposed to metacognitive knowledge that students might have about concepts, epistemology, or other areas of science knowledge. Strategic knowledge can be thought of as "How to do a strategy" (such as control of variables) while metastrategic knowledge can be thought of as "Why to do a strategy" (particularly when other strategies co-exist in the repertoire).

We have already encountered meta-level issues in this chapter. Why don't students take notes? In our studies, the students knew *how* to write down their experimental design (e.g., organic fertilizer/Brassica seed/artificial light) and the resulting plant growth. In terms of *why* to write this information down in their notes, there are several meta-level considerations for students. Do students understand the limits of their own memory? Do they understand the cognitive demands of the task? (e.g., analyzing the results of experiments in a large problem space over many weeks) Do they know what information will be needed in the future in order to analyze the results? Do they understand the utility of notetaking to accomplish this goal?

We have also encountered teaching strategies that consider meta-level concerns. In the cars task, students saw the results of their experiment in the same session that it was designed. However, in the plants task, the results were not available until the next session. By separating the experimental design and the evidence evaluation phases in time, the task design highlighted a task demand that showed the need for notetaking. Fostering students' metastrategic understanding need not be limited to interaction and feedback during inquiry but may also include direct instruction. For example, even students in early elementary grades can understand the notion of a "fair test," which is a cornerstone of the validity of the control of variables strategy.

The development of scientific reasoning involves not only the acquisition of strategies but also the acquisition of meta-level knowledge. A student can be taught how to control variables (a strategy), but without the metastrategic knowledge of when to apply the strategy or why it is effective, it is unlikely to be used (especially if there are more familiar strategies available in a student's repertoire). Scientific reasoning does not always develop naturally and it can be fostered by teachers through prompts, scaffolds, didactic interventions, or practices that include metastrategic understanding. Along with teaching "what to do," we need to teach "why this is the strategy to apply" and especially "why other strategies are less adequate or valid." (See Kuhn & Dean (2004) to learn more about teaching to support metacognition.)

NOTE

[1] Psychologists use *theory* in a broader sense that includes students' beliefs about a particular phenomenon, and it is in this sense that it will be used in the chapter.

FURTHER READING

Duschl, R. A., Schweingruber, H. A., & Shouse, A. W. (2007). *Taking science to school: Learning and teaching science in grades K-8.* Washington, DC: National Academies Press. Retrieved from http://dx.doi.org/10.17226/11625

Garcia-Mila, M., Andersen, C., & Rojo, N. (2009). The development of scientific inquiry strategies and representational practices in preadolescents. In C. Andersen, N. Scheuer, M. P. Pérez Echeverría, & E. Teubal (Eds.), *Representational systems and practices as learning tools* (pp. 167–185). Rotterdam: Sense Publishers.

Klahr, D., Zimmerman, C., & Jirout, J. (2011). Educational interventions to advance children's scientific thinking. *Science, 333*, 971–975. Retrieved from http://dx.doi.org/10.1126/science.1204528

Kuhn, D. (2005). *Education for thinking.* Cambridge, MA: Harvard University Press.

NGSS Lead States. (2014). Appendix F: Science and engineering practices. In *Next generation science standards: For states, by states.* Washington, DC: The National Academies Press. Retrieved from http://nextgenscience.org/resources/ngss-appendix-f-science-and-engineering-practices

Rapanta, C., Garcia-Mila, M., & Gilabert, S. (2013). What is meant by argumentative competence? An integrative review of methods of analysis and assessment in education. *Review of Educational Research, 83,* 483–520. Retrieved from http://dx.doi.org/10.3102/0034654313487606

Zimmerman, C. (2007). The development of scientific thinking skills in elementary and middle school. *Developmental Review, 27,* 172–223. Retrieved from http://dx.doi.org/10.1016/j.dr.2006.12.001

REFERENCES

Duschl, R. A., & Grandy, R. E. (Eds.). (2008). *Teaching scientific inquiry: Recommendations for research and implementation.* Rotterdam, Netherlands: Sense Publishers.

Garcia-Mila, M., & Andersen, C. (2007). Developmental change in note-taking during scientific inquiry. *International Journal of Science Education, 29,* 1035–1058. Retrieved from http://dx.doi.org/10.1080/09500690600931103

Garcia-Mila, M., & Andersen, C. (2009). Cognitive foundations of learning argumentation. In S. Erduran & M. P. Jimenez-Aleixandre (Eds.), *Argumentation in science education: Perspectives from classroom-based research* (pp. 29–47). Dordrecht, Netherland: Springer. Retrieved from http://dx.doi.org/10.1007/978-1-4020-6670-2_2

Garcia-Mila, M., Andersen, C., & Rojo, N. E. (2011). Elementary students' laboratory record keeping during scientific inquiry. *International Journal of Science Education, 33,* 915–942. Retrieved from http://dx.doi.org/10.1080/09500693.2010.480986

Kuhn, D., Garcia-Mila, M., Zohar, A., & Andersen, C. (1995). Strategies of knowledge acquisition. *Monographs of the Society for Research in Child Development, 60*(4, Serial No. 245). Retrieved from http://dx.doi.org/10.2307/1166059

Kuhn, D., & Dean, D. (2004). Metacognition: A bridge between cognitive psychology and educational practice. *Theory into Practice, 43,* 268–273. Retrieved from http://dx.doi.org/10.1207/s15430421tip4304_4

Popper, K. R. (1965). *Conjectures and refutations: The growth of scientific knowledge.* New York, NY: Harper Torch Books.

Schauble, L., Klopfer, L. E., & Raghavan, K. (1991). Students' transition from an engineering model to a science model of experimentation. *Journal of Research in Science Teaching, 28,* 859–882. Retrieved from http://dx.doi.org/10.1002/tea.3660280910

Siegler, R. S., & Liebert, D. E. (1975). Acquisition of formal scientific reasoning by 10- and 13-year-olds: Designing a factorial experiment. *Developmental Psychology, 11,* 401–402. Retrieved from http://dx.doi.org/10.1037/h0076579

Tschirgi, J. E. (1980). Sensible reasoning: A hypothesis about hypotheses. *Child Development, 51,* 1–10. Retrieved from http://dx.doi.org/10.2307/1129583

Zoller, U., & Nahum, T. L. (2012). From the teaching to KNOW to learning to THINK in science education. In B. J. Fraser, K. G. Tobin, & C. J. McRobbie (Eds.), *Second international handbook of science education* (pp. 209–229). Dordrecht, Netherland: Springer. Retrieved from http://dx.doi.org/10.1007/978-1-4020-9041-7_16

KEITH S. TABER

9. THE NATURE OF STUDENT CONCEPTIONS IN SCIENCE

INTRODUCTION

Prior knowledge – what learners already know and understand – is a major determinant of what students will learn from their science classes (Taber, 2015). A great deal of research suggests that very commonly students may hold ideas about science topics that are different to, and indeed often inconsistent with, canonical scientific principles and theories (Duit, 2009). Studies have described learners' own ideas about science topics in various ways such as misconceptions, intuitive theories and alternative conceptual frameworks, although there are not widely agreed meanings for these different terms. Research also suggests that the ideas elicited from students vary on a number of dimensions that influence how significant student thinking is for learning canonical scientific ideas (Taber, 2009). The chapter explains how student conceptions in science can vary in terms of degrees of acceptance, connectedness, multiplicity and explicitness. The nature of each of these dimensions is described in the chapter drawing upon examples from research into student thinking about science topics, and the significance of each dimension for student learning is explored.

THE SIGNIFICANCE OF PRIOR LEARNING

The perspective informing this chapter is sometimes called personal constructivism (or psychological or pedagogic constructivism). This is a perspective on the nature of learning and human knowledge that suggests that knowledge is not the kind of thing that can be simply copied between minds (for example, from the teacher to the student) but rather has to be constructed anew by each knower. This perspective is informed by research on human cognition which suggests that the human mind acts to make sense of the world by recognising patterns in experience as a basis of constructing models that allow a person to anticipate the future. The young child knows (i.e. has developed an expectation based on patterns of past experience) that letting go of the toy will lead to it dropping to the floor, that kicking the ball harder will lead to it travelling further across the garden, and that crying usually leads to the appearance of mother.

This process of developing models in the form of expectations is quite conservative in the sense that once patterns are recognised and expectations established, they tend

K. S. Taber & B. Akpan (Eds.), Science Education, 119–131.

to be automatically used to filter and interpret new experiences. Letting go of a helium filled balloon that does not sink to the floor may be surprising, but does not lead to immediately expecting that other objects released into space will also float away. The human brain has evolved to be an apparatus that makes sense of the world in terms of the existing set of models. It can also augment and adapt those models in the light of new experience: but, once established, existing patterns of thought tend to dominate.

Much school level learning is language-based of course, but there are strong reasons to consider that the expectations of the world built up in early childhood are important foundations for all later learning (Vygotsky, 1934/1986), including the learning of abstract academic concepts presented in formal language (Lakoff & Johnson, 1980).

Learning is Interpretive, Incremental and Iterative

In effect, human learning tends to be interpretive, incremental and iterative (Taber, 2013b). Learning is interpretive as sensory data (including that deriving from teaching) is processed through the brain's sense-making apparatus to produce perceptions of the world. Our current models of how cognition works suggest that learning is incremental because a key part of our cognitive apparatus, working memory, only attends to a very small amount of new input at any time. This is indeed an *inbuilt bias* as previously learnt material can be 'chunked' into extensive complexes that can be processed in working memory, whilst only a few discrete items of novel 'input' can be considered at a time. Thus when teaching, we have to consider the material to be taught from the learner's perspective and organise it into what will be manageable 'learning quanta' at the learner's resolution. The nature of such learning quanta will depend on the extent and level of integration of the existing knowledge and understanding that a particular learner is able to draw upon to make sense of teaching.

Learning is iterative because it is interpretive – once a student has developed a particular understanding then they will interpret new information according to this way of thinking, and tend to learn it in a way that reinforces the existing interpretation. We certainly can, and do, develop new ways of thinking, but the brain has evolved to primarily seek to fit new information within existing ways of understanding, and that is what usually happens.

Given all this, a major determinant of learning is what is already believed or understood. Making sense of teaching requires the learner to process what they are hearing and seeing through their existing interpretive resources – something that is usually largely automatic and so the learner is not even aware it is occurring – and the particular meanings that they take from teaching depend upon the particular resources brought to bear.

WHAT CAN GO WRONG IN TEACHING?

If we understand that learning is an interpretive, incremental, and iterative, process then this means that science teaching is a process of guiding learners to construct understandings of the world that match scientific models as well as possible – given that they will always be relying on their existing knowledge and understanding to interpret the teacher's presentation. In designing lessons, a teacher needs to plan the presentation of material so that it makes good sense to students and will be interpreted as intended. To do that well requires knowledge of the learners' current state of knowledge and understanding (Driver & Oldham, 1986). If there is a mismatch between (a) how the teaching is intended to be understood, and (b) how it is actually made sense of, then the desired learning will not occur.

There are different kinds of basic mismatches that may occur, leading to different kinds of impediments to intended learning (Taber, 2005). Sometimes the teacher assumes prior knowledge that the students do not have. It is also possible that when students do have the expected and needed prerequisite knowledge they do not bring it to mind during teaching. So the teacher may assume that everyone in the class appreciates the relevant prior learning, but some students may not see its relevance and so are not interpreting teaching in the way intended. Teachers should always be explicit about how new teaching is intended to build upon prior learning. Sometimes it may be possible to ensure this by using a preliminary activity which both highlights essential pre-requisite knowledge and helps students re-organise it into the most useful form to support building the intended new learning from it – what has been called a scaffolding PlaNK – a 'platform for new knowledge' (Taber, 2003).

It is also possible that learners do relate teaching to existing knowledge and understanding, and so do interpret teaching in a way that makes sense to them, but *not* in the way intended. If they do not have, or do not recognise the relevance of, the prior learning they are expected to bring to mind, then they may well instead make links with other ideas they do have available. The links may be irrelevant or even inappropriate from the teacher's perspective, but the student will not realise that. Despite this being unfortunate in the context of that particular lesson, this shows that students are being active learners, and creatively seeking links between different features of their learning. This is something to be encouraged, as science itself relies on suggesting such creative possibilities (Taber, 2011). The concepts that students learn in science – magnetic fields, photosynthesis, oxidation – were once brave new ideas deriving from someone's creative imagination. Most such ideas do not survive extensive testing, so science relies on lots of imaginative suggestions to be scrutinised and selected from. Having wrong-but-creative ideas is therefore a positive trait in a science student, so the teacher should be encouraging this tendency even though many of the outcomes are unhelpful in the context of the particular lesson (Kind & Kind, 2007). Students will have such ideas whether they share them or not, and when the teacher dismisses them in a critical way this will only encourage learners to keep such ideas to themselves. That makes the teacher's job more difficult.

The other possibility is that the learner does relate teaching to their previous knowledge and understanding about the appropriate topics or concepts – but that existing understanding is not canonical. That is, the student may already hold alternative conceptions of the topic area at odds with the accepted scientific accounts. Research has revealed that learners at all levels (including some teachers) commonly hold notions that are inconsistent with the science that is taught in the curriculum.

LEARNERS' IDEAS IN SCIENCE

Research shows that even when a class meets a science topic for the first time in the curriculum, teachers cannot assume that students do not know anything about the topic, as quite often learners have developed their own ways of thinking about science topics before being formally taught about them. These ways of thinking are sometimes consistent with the scientific accounts met in school and college science – but certainly not always. Moreover, research also shows that teaching does not necessarily 'correct' students' ideas that do not fit with science. Students' alternative ideas will sometimes be changed by teaching, but not always. Moreover, teaching sometimes leads to the modification of existing ideas in unintended ways, and even the development of *new* 'wrong' ways of thinking. (There are some examples of unhelpful ideas acquired through science teaching in Chapter 4: '*Beliefs and Science Education*'.) So again we see that science teaching has to be seen as responding to, and channelling, students' existing thinking, not just passing on scientific knowledge.

There is a vast research base exploring students' ideas in science, and a range of terms have been used to label and categorise these ideas (Duit, 2009). Anyone reading research in this area (and much of it is fascinating, as well as being useful to inform teaching) will find references to a whole range of descriptors: alternative conceptions, misconceptions, intuitive theories, alternative frameworks (or alternative conceptual frameworks), minitheories, p-prims, knowledge facets, intuitive knowledge elements, preconceptions, etc. Although some of these different terms are intended to relate to genuine distinctions, sadly there is little consistency in how such terms are used across the literature. Here I will refer to learners' conceptions, many of which (those that are inconsistent with canonical science) are considered alternative conceptions.

LEARNERS' CONCEPTIONS

A conception is a way of making sense of something – a way of conceptualising. In order to write about conceptions we need to formulate them in verbal terms, but we should bear in mind that student thinking about science is not all verbal. Indeed scientists' ways of thinking about science is not limited to verbal language: Einstein is just one example of a major scientific thinker who visualised a lot of his scientific ideas, before reformulating those ideas in equations and words (Miller, 1986).

Examples of Students' Conceptions

As the research literature reports vast numbers of conceptions related to scientific topics, it is only possible here to discuss a few examples and invite readers to consider the possible implications of such ideas for teaching students related topics. Perhaps the reader will imagine they are teaching a class the relevant topics and ask themselves how they would adapt their teaching if students in the class held the following alternative conceptions:

- a ball that has been thrown into the air is subject to an upwards force during the period it is moving upwards;
- no force is required to make an object move in a circle as long as the speed of the object does not change;
- a woman cannot get pregnant the first time she has sexual intercourse;
- the product of a neutralisation reaction is always neutral;
- current diminishes around a series circuit;
- insects are not animals;
- mushrooms are plants;
- compounds such as SF_6, PCl_5 and IF_7 cannot exist as atoms can only have eight electrons in their outer shells;
- a hydrogen bond is a covalent bond to a hydrogen atom;
- intelligence is fixed at birth;
- cave men used to hunt dinosaurs
- the Earth is nearest the Sun in summer

It is very likely that readers who are experienced science teachers will have come across many other examples of ideas students have which are at odds with the science taught in schools and colleges.

Responding to Student's Conceptions

As suggested above, when students who hold alternative conceptions come to science lessons and are taught science that is inconsistent with their ideas, a number of things can happen (Gilbert, Osborne, & Fensham, 1982). These include:

- sometimes students shift their thinking to take on the scientific account in place of their previous ideas;
- sometimes students effectively ignore and soon forget the teaching and maintain their previous ways of thinking;

- sometimes students learn the new scientific ideas, but as something additional to their existing ways of thinking (even if these seem inconsistent with each other);
- sometimes students modify their existing thinking to some extent in the light of the teaching – to a new conception intermediate between what they thought before and what the teacher is intending them to learn;
- sometimes students modify their existing thinking to some extent in the light of the teaching – to a new conception that is a hybrid containing elements of previous and canonical ideas, but not quite matching either.

This list is simplistic in an important sense as it seems to imply that teaching and learning are nicely compartmentalised as occurring in a lesson and so lead to particular outcomes. In practice, the learning process continues long after the lesson and the students' thinking may only slowly shift (as there are ongoing brain processes which revisit memories of experiences and act over time to modify the way we think about things). Teaching is seldom organised into totally discrete lessons – usually teachers teach sequences of lessons on a topic, revisiting and developing points over several lessons. A student's understanding of, say plant nutrition, before the lesson where photosynthesis is introduced, immediately after that lesson, at the end of the full sequence of lessons in the topic, and a month beyond that, may all be different.

Given this complex situation, different recommendations have been made to teachers about how to best deal with learners' alternative conceptions. In particular, three common suggestions are to (a) ignore them; (b) challenge them; and (c) build on them. Such a diverse range of options seems unhelpful to the teacher – but actually each of these options may be sensible sometimes (as will be explained below).

It is sometimes suggested that teachers should just ignore students' ideas because often they are not significant for learning, and then paying attention to them in the classroom will simply reinforce them and confuse students about which ideas are being validated by the science lesson. However, there are also strong suggestions in the literature that students' alternative conceptions can be tenacious, and come to dominate their thinking about a topic, unless they are challenged. Challenging usually involves demonstrating or arguing why these ideas do not fit observations and other evidence, and are less useful than the scientific models.

An alternative argument, also often made in the literature, is that as students' existing conceptions are the (only) resources they have available for constructing new learning, they should be worked with rather than challenged. This argument suggests that what appear to be firm conceptions are often best understood as the result of the learner putting together fairly isolated knowledge facets or 'knowledge in pieces' which can be in effect dismantled and rebuilt to form scientific conceptions. Of course, the arguments made in research literature are being summarised and simplified here, and deserve more careful study.

Each of these arguments seems sensible, but they each rely on a different characterisation of students' conceptions. For the teacher to know how to best teach to take into account learners' ideas, they need to know which description of students' conceptions applies.

SIX DIMENSIONS OF LEARNERS' CONCEPTIONS IN SCIENCE

As the reader may suspect, the different views of the nature of learners' alternative conceptions are all supported by some research evidence. This suggests that learners' conceptions about scientific topics are not all of the same kind, but rather they vary considerably (Taber, 2009). The teacher therefore needs to be aware of the kind of variation that occurs so they know how to respond in particular cases.

A person's conceptions inherently vary in terms of degrees of acceptance, connectedness, multiplicity and explicitness. Each of these dimensions will be considered below. In addition to these dimensions, learners' ideas vary in terms of how consistent they are with scientific models, and how similar they are to those of other students.

Degree of Inconsistency with Scientific Models

One way in which learners' alternative conceptions vary, is in terms of just *how alternative* they are. The author once taught a student who referred to 'electron shields' in atoms for what are normally referred to as electron shells. This student seemed to have a reasonable grasp of the nature of electron shells in atoms, but just used an alternative label. This was a rather trivial form of alternative conception, and may have actually been helpful in some circumstances (for example when thinking about ionisation enthalpies). However, other alternative conceptions may be alternative at a much deeper, conceptual level. A chemistry teaching colleague of the author thought that any sample of a strong acid would have a pH of 1. This knowledge 'worked' in the context of the actual practical work carried out in his classes – but showed a lack of (or more likely, failure to bring to mind long-neglected) understanding of acid strength and concentration.

Degree of Explicitness of Student Knowledge

Sometimes we become aware of student conceptions because our students say or write something that is clearly not correct from a scientific perspective. Sometimes this is a statement that reflects a particular conception they hold as explicit propositional knowledge. That is, this specific idea is specifically represented ('stored') as part of their science knowledge.

However, many things that we elicit from students – as when we ask them questions in class – report ideas *generated* at that time in response to the specific question, rather than being the accessing and recollection of some specific notion that has previously been learnt and represented in memory. Consider a student who suggested copper was magnetic in response to a point posed in class. It may be that student had previously learnt (inappropriately) that copper was magnetic. However, it might be more likely the student had never considered this property of copper previously, but generated the suggestion from two ideas that had previously been learnt: copper is a metal (which it is) and all metals are magnetic (which they are not). It is even possible that the student had never *explicitly* considered the idea that all metals were magnetic before, but rather this was a tacit idea – something that the student had not even been aware they were primed to think until something in the classroom discussion provoked the thought.

This may seem fanciful, but there is strong evidence that much of our knowledge is tacit in this way – at the level of intuitions (diSessa, 1993). These intuitions, or intuitive knowledge elements, are represented in the brain at a level that is not directly open to conscious awareness, but which still influences how we perceive things, and understand them (see Chapter 10: '*Tacit Knowledge in Science Education*'). More research is needed into these kinds of aspects of knowledge, sometimes called phenomenological primitives, or p-prims. It is believed that we all have a large repertoire of these intuitions that we apply without even realising it when making sense of the world. Sometimes students can tell us what they think in a particular situation, without seeming to be able to explain their reasoning – as they are just activating their intuitive knowledge (Watts & Taber, 1996).

Such intuitions are not verbal, but to talk and write about them we have to put them into words. So, for example, people tend to know, or expect, that getting close to some source leads to a stronger effect. So we are not surprised that the fire feels hotter as we get closer, or the sound is less intense as we move away from the loudspeaker (as early life experiences lead us to expect such a pattern). It is common when asking students to explain the seasons that many will suggest that Summer is the time when the earth gets closest to the Sun in its elliptical orbit. Probably students have not been taught that, and perhaps had never even thought it before, but when asked a question about the seasons this intuition or intuitive knowledge element may be triggered. The broader 'more effect when closer' intuition is *generally* sound, but in this case its application leads to an alternative conception.

Degree of Multiplicity of Student Knowledge

At first sight it may seem foolish to think two inconsistent things. Indeed the human brain seems to have evolved to prefer to maintain coherence between our different ideas. However, it is often possible to elicit from a student several ideas which seem to be inconsistent or even directly contradict. Of course, things that seem contradictory from the teacher's perspective do not always seem so from the student's perspective. For example, sometimes students will see as quite different phenomena what the science teacher conceptualises as different examples of the same basic phenomenon. So students may seem to change their minds about some science concept according to the context in which we ask the question (Palmer, 1997) – but from the students' perspectives different principles apply in the different situations. To make good sense of student thinking, we need to explore their ideas in their own terms.

Students may also offer alternative explanations for the same phenomenon if they feel that several complementary explanations are allowed. This is not so strange, as in science we often deal with complex phenomena with multiple causes, and indeed may recognise different levels of explanation. For example, we might discuss an animal eating in terms of instinctive drives, overall energy requirements, and specific metabolic processes.

A particular feature that some research has uncovered is how students may be able to learn scientific ideas in class, and reproduce them in formal tests, whilst retaining quite different and alternative conceptions they use in everyday discourse (Solomon, 1983). So a student may talk in the playground about exercise giving them energy, even though in the classroom they 'know' that exercising requires an energy source.

Degree of Connectedness of Student Knowledge

As human beings we know about all kinds of things. Some of our knowledge is highly integrated, especially when we have expertise in a topic. As science teachers we are aware that certain ideas, maybe conservation of energy or natural selection, apply across a great many topics, and we tend to see the links between different things we teach. Students have spent less time studying and thinking about the science, and their knowledge tends to be less well organised and integrated than that of an experienced teacher. Indeed some of their conceptions may be more or less isolated fragments of knowledge – little 'islands of knowledge' – that they do not see as significantly linked to anything else.

Other ideas, however, become firmly linked into networks that can become mutually reinforcing. One common alternative conception that relates to chemistry learning is the idea that atoms want, or need, to have full outer shells (or octets of electrons, see Chapter 24: '*Teaching and learning chemistry*'). Students relate this idea to a wide range of notions about chemical reactions, chemical bonding, stability of different chemical species, patterns of ionisation enthalpies and so on (Taber, 2013a). Students usually acquire the conception quite early in their chemistry learning, and it may influence (and distort) later learning in a range of core topics. Students often find it difficult to reject this conception, even when they are taught that it is not a helpful notion, because they have already constructed an extensive conceptional framework around it.

Degree of Commitment to Student Knowledge

Students are strongly committed to some of their alternative conceptions. In effect they have come to 'believe' the conception as an accurate fact about the world (see Chapter 4: '*Beliefs and Science Education*'). However, many student conceptions are more tentative. Sometimes students are perfectly happy to be told they have got something wrong, and to modify this aspect of their thinking. Many ideas elicited from students are not so much stable and committed conceptions but better considered conjectures – notions they are exploring and testing out. That of course is something to be encouraged in future scientists. Students may readily abandon many of these conjectures when they see, perhaps with the teacher's help, that they have limitations.

However, some other ideas may be committed to so strongly that it is very unlikely that any amount of telling or presentation of argument will persuade the students that their conceptions are wrong. People are generally very good at finding evidence that seems to support their ways of thinking and reasons to dismiss or see as flawed any counter arguments (Nickerson, 1998).

When students do not have any strongly committed conceptions relating to a particular topic, they may well consider and explore a range of alternative possibilities – perhaps shifting from one to another. This is one basis for finding students sometimes have manifold conceptions of the same topic as suggested above.

Degree of Commonality of Student Knowledge

A final dimension to consider is how common students' conceptions are. At one level each student has a unique personal history of learning experiences. However, all students live in the same physical world, and have similar biological apparatus for exploring and finding out about it. Moreover, in any school or college many of the students will have been brought up in the same cultural environment and they will often share the same language.

Not surprisingly, then, research suggests that some alternative conceptions are very common. The majority of people have difficulty learning about the physics of force and motion, because most people think that an object that is moving must be subject to some kind of force in the direction of motion (Watts & Zylbersztajn, 1981). Even after learning about Newton's principle of inertia (the first law of motion), people often get questions about this topic wrong because intuitively they reject the physics. In most science topics explored, researchers have found examples of common alternative conceptions – that is where many students of a certain age seem to have much the same alternative conceptions.

However, it is also possible to find students reporting ideas that do not seem to be reported in the literature, and are not shared by any of their classmates. One student the author taught had her own understanding of what the charge symbols (such as the '+' in Na^+) were intended to mean which was not only quite different from the understanding of her teachers, but also – it seemed – all her classmates (Taber, 1995). As learning is interpretive, incremental, and iterative, there is always potential for individuals to form idiosyncratic ideas based on their own unique history of experiences – but the common experience and discourse of the community tends to act as something of a brake on this, and channels most of our thinking to be aligned with those around us that we regularly talk and listen to. Consequently, as teachers, we face both common alternative conceptions, many of which are discussed in published studies, and more rare examples – which often have not been reported in the literature.

CONCLUSION: WHAT'S A TEACHER TO DO?

This discussion of the diverse nature of learners' conceptions may not seem very helpful in deciding whether teachers should (a) ignore, (b) challenge, or (c) seek to develop, learners' alternative conceptions. All three of these options seem sensible *sometimes*.

The teacher therefore needs to read about the literature on the topic being taught, and find out more about the particular conceptions that have been discovered in research and what is recommended in various cases. However, even more importantly, the teacher needs to maintain a dialogue with her students, to find out just what ideas her students are using to make sense of teaching (see Figure 1).

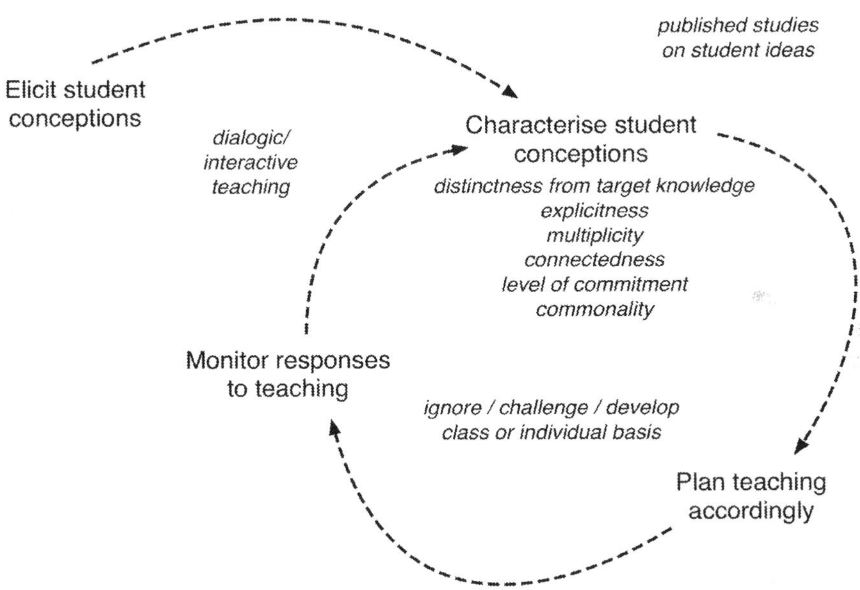

Figure 1. Effective teaching is informed by being familiar with pedagogic research and listening to learners' ideas

A major part of effective science teaching is making judgements about how to respond to learners' ideas – which ones can be best ignored, which are worth spending time challenging with demonstrations and arguments, and which should be seen as useful starting points for moulding towards more scientific accounts of the world. The best response to the 'same' alternative conception may even be different in different cases. Teaching science well is very challenging as it can only be planned so far in advance – much of the expertise relates to being able to make decisions within class about how to respond to particular ideas we elicit from students. The decisions we make in such work should not be seen as definitive either, but seen more as based on hypotheses to be tested in the classroom, where – like good scientists – we collect evidence to evaluate our conjectures about student thinking and revise them when indicated by new evidence.

A good science teacher is therefore not just an expert on the science to be taught, but is a clinician in the classroom, a science learning doctor: constantly diagnosing student thinking, responding to it, evaluating this process, and revising the treatment of the topics as needed. This is difficult and highly skilled work. It can also be extremely fascinating, highly motivating, and – when we start to see progress in students' scientific thinking and understanding – intensely satisfying.

FURTHER READING

Driver, R., Rushworth, P., Squires, A., & Wood-Robinson, V. (2013). *Making sense of secondary science: Research into children's ideas* (2nd ed.). London: Routledge.

Taber, K. S. (2014). *Student thinking and learning in science: Perspectives on the nature and development of learners' ideas.* New York, NY: Routledge.

REFERENCES

diSessa, A. A. (1993). Towards an epistemology of physics. *Cognition and Instruction, 10*(2&3), 105–225.

Driver, R., & Oldham, V. (1986). A constructivist approach to curriculum development in science. *Studies in Science Education, 13,* 105–122.

Duit, R. (2009). *Bibliography – Students' and teachers' conceptions and science education.* Kiel, Germany: IPN – Leibniz Institute for Science and Mathematics Education.

Gilbert, J. K., Osborne, R. J., & Fensham, P. J. (1982). Children's science and its consequences for teaching. *Science Education, 66*(4), 623–633.

Kind, P. M., & Kind, V. (2007). Creativity in science education: Perspectives and challenges for developing school science. *Studies in Science Education, 43*(1), 1–37. doi:10.1080/03057260708560225

Lakoff, G., & Johnson, M. (1980). The metaphorical structure of the human conceptual system. *Cognitive Science, 4*(2), 195–208.

Miller, A. I. (1986). *Imagery in scientific thought.* Cambridge, MA: MIT Press.

Nickerson, R. S. (1998). Confirmation bias: A ubiquitous phenomenon in many guises. *Review of General Psychology, 2*(2), 175–220.

Palmer, D. (1997). The effect of context on students' reasoning about forces. *International Journal of Science Education, 19*(16), 681–696. doi:10.1080/0950069970190605

Solomon, J. (1983). Learning about energy: How pupils think in two domains. *European Journal of Science Education, 5*(1), 49–59. doi:10.1080/0140528830050105

Taber, K. S. (1995). Development of student understanding: A case study of stability and lability in cognitive structure. *Research in Science & Technological Education, 13*(1), 87–97.

Taber, K. S. (2003). Responding to alternative conceptions in the classroom. *School Science Review, 84*(308), 99–108.

Taber, K. S. (2005). Learning quanta: Barriers to stimulating transitions in student understanding of orbital ideas. *Science Education, 89*(1), 94–116.
Taber, K. S. (2009). *Progressing science education: Constructing the scientific research programme into the contingent nature of learning science*. Dordrecht: Springer.
Taber, K. S. (2011). The natures of scientific thinking: Creativity as the handmaiden to logic in the development of public and personal knowledge. In M. S. Khine (Ed.), *Advances in the nature of science research: Concepts and methodologies* (pp. 51–74). Dordrecht: Springer.
Taber, K. S. (2013a). A common core to chemical conceptions: Learners' conceptions of chemical stability, change and bonding. In G. Tsaparlis & H. Sevian (Eds.), *Concepts of matter in science education* (pp. 391–418). Dordrecht: Springer.
Taber, K. S. (2013b). *Modelling learners and learning in science education: Developing representations of concepts, conceptual structure and conceptual change to inform teaching and research*. Dordrecht: Springer.
Taber, K. S. (2015). Prior knowledge. In R. Gunstone (Ed.), *Encyclopedia of science education* (pp. 785–786). Berlin-Heidelberg: Springer-Verlag.
Vygotsky, L. S. (1934/1986). *Thought and language*. London: MIT Press.
Watts, M., & Taber, K. S. (1996). An explanatory gestalt of essence: students' conceptions of the 'natural' in physical phenomena. *International Journal of Science Education, 18*(8), 939–954.
Watts, M., & Zylbersztajn, A. (1981). A survey of some children's ideas about force. *Physics Education, 16*(6), 360–365.

RICHARD BROCK

10. TACIT KNOWLEDGE IN SCIENCE EDUCATION

The Role of Intuition and Insight in Teaching and Learning Science

INTRODUCTION

Tacit knowledge, that is knowledge that cannot be expressed directly in words (Polanyi, 1966, p. 4), might appear to be an obscure concept for science teachers to be interested in. It may seem that inexpressible knowledge is unlikely to be of great importance to teaching and learning science. In this chapter, I will argue that the reverse is true: tacit knowledge has a manifest impact on learning in science and teachers and researchers should be alert to the presence of implicit understandings when describing the state of students' knowledge. There are two situations, which might be familiar to teachers, which hint at the presence of tacit knowledge. Firstly, intuitions can be thought of as the source of statements such as: 'I just know it's the right answer but I can't explain it in words. I just feel it's right.' These kinds of claims indicate the presence of inexpressible knowledge. Secondly, moments when students experience sudden clarity, the so-called 'A-ha' experience or insight, suggest that the processes that lead to understanding are not always available to consciousness. These two concepts, intuition and insight, are the foci of this chapter as they are conceptualised as processes in which the tacit interacts with the conscious. The structure of this chapter is as follows: after discussing the nature of tacit knowledge, the concept of intuition and its role in science education is examined leading to the proposal of strategies that may support intuition in the classroom. The second half of the chapter presents a similar discussion of insight in the context of science education. It is hoped that the chapter will encourage readers to be sensitive to the existence of tacit knowledge in the science classroom.

THE NATURE OF TACIT KNOWLEDGE

Tacit knowledge has been defined as knowledge that 'cannot be explicitly stated' (Polanyi, 1966, p. 4, italics removed). This assertion raises the question of how the presence of something that is inexpressible can be inferred. Consider the following examples taken from a series of interviews conducted with a 17-year-old student, about concepts in physics:

> [On explaining weightlessness] I understand what this is trying to do but I don't know how it say it out in words…

K. S. Taber & B. Akpan (Eds.), Science Education, 133–142.

> Yeah, in potential difference, yeah,... when I got it then I just knew how to do the rest of it and everything...

In the first quotation, the student appears to be aware of an understanding that is inexpressible in words; in the second excerpt, the student describes a moment of sudden clarity in their understanding which appears to arrive without conscious control.

There are a number of ways in which the existence of this kind of tacit knowledge can be explained. Firstly, the impression we have of being able to access all of our thoughts and knowledge directly may be illusory; tacit knowledge and processing may take the form of knowledge that exists, or processing that occurs, beyond the limits of our conscious awareness of cognitive functioning (Nisbett & Wilson, 1977; Taber, 2014). Secondly, certain kinds of knowledge, for example about the motion of objects, may be encoded in a non-verbal form, for example kinaesthetically or visually, and so be difficult to express through language (diSessa, 2000, p. 96; Clement, 2008, p. 209). Finally, tacit knowledge may exist as abstracted rules, or heuristics, that guide a learner's responses whilst they remain unaware of the principles organising their thought (Kahneman, 2011, p. 98). If tacit knowledge were isolated and inert in conceptual structure, it would be impossible to detect, however, as illustrated in the excerpts above, though it is not directly expressible, tacit knowledge exerts an influence on conscious thought. The processes of intuition and insight can be seen as mechanisms in which tacit and explicit knowledge interact and will, therefore, be used as means to examine tacit knowledge in science education.

THE LINK BETWEEN TACIT KNOWLEDGE AND UNDERSTANDING

A number of commentators have observed that the acquisition of explicit knowledge does not necessarily lead to expert-like understanding (Bransford, Brown, & Cocking, 2000, p. 9; Kosso, 2002). Richard Feynman expressed a frustration that is experienced by many teachers: '...the students had memorised everything, but they didn't know what anything meant' (Feynman, 1985, p. 212). In order to illustrate the difference between knowing and understanding, consider the Chinese room thought experiment proposed by John Searle (1989). Imagine a person, who doesn't speak Chinese, is locked in a room. Inside the room is a set of instructions, written in English, which link sets of characters in Chinese to other groups of Chinese letters. If questions in Chinese are fed into the locked room, the imprisoned person can follow the instructions and pass seemingly fluent answers in Chinese out from the room. To an observer outside, the room's occupier appears to be fluent in Chinese, but the person is merely following a series of procedures, without comprehending the meaning of the output they produce. Though Searle's thought experiment was originally posed to make an argument concerning artificial intelligence, it is resonant in thinking about the processes students use in the classroom.

It may be that some students experience the classroom in a similar manner to the person in the Chinese room: they follow rules to complete tasks, but with little understanding of the nature of the processes, a phenomenon Airey and Linder (2008) have described as discourse imitation. That is, students have acquired the ability to perform a set of procedures, as one might mimic the sounds of a foreign language, without being aware of the meaning of their actions. A number of different factors could account for how it is possible to acquire the ability to preform tasks without understanding, but one plausible explanation is that the students lack certain kinds of tacit knowledge. Some forms of understanding may be tacit (Lipton, 2009), for example, observing the motion of an orrery may lead to the development of an understanding of the retrograde motion of Mars that is hard to express in words. Indeed, expertise, in a range of professions, has been linked to the acquisition of tacit knowledge (Sternberg & Horvath, 1999).

The link between tacit knowledge and expert understanding, may explain the reported frustration of teachers who discover that encouraging the acquisition of additional facts does not necessarily remedy deficits in understanding:

> I don't know how to tell you something that will transform you from a person who *can't* analyze new situations or solve problems, to a person who *can* ... But in the case of physics, I can't transform you from somebody who *can't* to somebody who *can*, so I don't know what to do.
>
> Because I *intuitively* understand what's going on physically, I find it difficult to communicate. (Feynman, 2013, p. 69, italics in original)

At least some of the knowledge of expert scientists is tacit and therefore difficult to transfer directly to students. Science curricula are, by necessity, a set of articulable statements of content. A student who rote learned the factual knowledge listed in a curriculum might resemble the person in the Chinese room. They may be able to recall knowledge and follow procedural steps but are likely to struggle to transfer their knowledge to novel situations and lack a sense of the relationship between ideas. By definition, the difference in tacit knowledge of experts and novices is difficult to describe, but might involve such elements as knowledge related to the kind of contexts in which a particular approach will be successful, a sense of the related underlying structures of situations or kinaesthetic models of how particular systems will behave. The process of good science teaching should foster students' awareness of the manner in which tacit knowledge affects their thinking and assist students in attaining expert tacit knowledge.

INTUITION IN SCIENCE EDUCATION

It might be assumed that learning about science is an entirely rational and explicit process. In this section, I will challenge this claim by examining the role of tacit knowledge in learning through the concept of intuition. Though intuition has been described as difficult to define, for the purposes of this section I will take it to mean: 'a tacit hunch or feeling that influences thought with little conscious effort'

(Brock, 2015, p. 2). A review of the diverse definitions and models of intuition can be found in an article in *Studies in Science Education* (Brock, 2015). Two plausible explanations for why students can develop knowledge about the physical world, which they are not able to express in words, are: (a) the knowledge is stored in a kinaesthetic or imagistic form; (b) the knowledge is stored as abstracted rules.

The embodied cognition hypothesis suggests that thought is influenced by features of the physical body beyond the brain (Shapiro, 2011). Young children begin to develop an understanding of the physical world before they develop language, and hence the basis of some types of knowledge, for example, an awareness of how objects move or understandings of which kinds of objects are agents, may be nonverbal. These early intuitions can initially be powerful routes to understanding the world but may interfere with the subsequent acquisition of the formal and explicit principles of scientific knowledge. This kind of non-verbal knowledge has been modelled as phenomenological primitives (or p-prims). P-prims encode a 'sense of mechanism' (diSessa, 1993, p. 106) and are inarticulate due to their visual or kinaesthetic form (diSessa, 2000, p. 96). In this construction of cognition, diSessa (1993, p. 114) argues experts do not discard their initial p-prims, rather they become integrated into their more developed understanding.

The importance of embodied cognitions in developing understanding is also supported by Clement (2008), who argues that experts may run mental simulations, based on perceptual data and motor sensations, to make sense of a situation. Indeed, it has been proposed that the kind of abstract concepts used in science, such as force or heat, can only be developed out of resources gained from experiencing the world (Lakoff & Johnson, 1980; Barsalou, 1999). In this manner, there exists an inextricable entanglement between explicit and tacit knowledge of scientific concepts. Though an experienced physicist may be able to clearly articulate Newton's laws, and other explicit knowledge concerning the nature of force, it is likely some of their understanding of the concept will take the form of inarticulable knowledge, or intuitions. The compounding of tacit and explicit knowledge may partly explain Feynman's observation that students can possess a significant body of factual knowledge concerning a topic, and yet fail to develop an expert understanding.

A different model of tacit knowledge suggests we learn about the world through developing abstracted rules, which are applicable across a range of situations, without conscious awareness of the rules' existence. Daniel Kahneman (2011, pp. 88–89) describes such rules as heuristics, that is, rapid and non-conscious routines for finding solutions to problems. For example, we tend to assume that emotionally charged events, such as air crashes, are more likely to occur than they do in reality: a strategy that might be useful for avoiding extreme events, but may lead to an overestimation of risk in some contexts. Similar implicit rules may underlie some kinds of thinking in science, one conceptualisation of such rules is the organising gestalt. For example, Andersson (1986) proposed the experiential gestalt of causation, a tendency to assume situations involve: a causal agent, an object and an instrument, through which the agent acts. The gestalt of causation may underlie students' understandings across a range of contexts, for example, they may assume bulbs that are closer to a battery will be brighter than those further away, the eye emits rays of light, or a wire must be in physical contact with the iron core of an electromagnet to be effective. A similar kind of abstraction is the explanatory gestalt of essence (Watts & Taber, 1996), an assumption some phenomena occur because they are 'natural'. The next section examines ways in which students' use of intuition in the classroom can be supported.

STRATEGIES TO SUPPORT INTUITION

Although tacit knowledge is often seen as a hindrance to novices' thinking, as their intuitions can differ from the accepted scientific models, expert intuitions can be powerful tools for rapid problem solving (Baylor, 2001). This section discusses how the transition from novice to expert intuition may be facilitated in the science classroom. As has been discussed, intuitions may develop through interactions with the physical world. Therefore two plausible suggestions to scaffold students' intuitive knowledge are engagement with appropriate practical work and the use of computer simulations. The implicit nature of intuitions may arise from the existence of knowledge that is stored as kinaesthetic (knowledge that allows the learning of complex muscle movements, for example, how to walk without conscious awareness) imagistic (memories stored as pictures) or other non-verbal forms.

It is possible that practical work can develop this kind of knowledge that is not expressible in words (Wellington & Osborne, 2001, p. 7). However, care must be taken in the design of practical work as there is a tendency for students' intuitive ideas to change their perception of what they have observed to match their preconceptions, reinforcing existing intuitions rather than causing change (Champagne, Gunstone, & Klopfer, 1985). An approach that enables greater control over the nature of events that students observe is the use of computer-based simulations. For example, diSessa (1986, p. 210) describes the development of a simulation of a drag-force free environment, dynaturtle. He argues that experiences with simulated worlds are, in principle, no different from engagement with the physical world and can therefore be used to develop students' intuitions about motion.

An obstacle to the development of students' tacit knowledge in science may be the notions they have regarding the nature of the subject. Much has been written about students' understanding of the nature of science (see Section I: Nature of Science and Science Education) and assumptions about the nature of knowledge may impact the manner in which students approach learning (Songer & Linn, 1991). For example, diSessa (1985) described the cases of two students, A and B, with differing understandings of the nature of science. Student A linked knowledge with equations and numbers, and produced classwork described as not developed beyond what was taught. Contrastingly, student B believed that qualitative models were significant in developing understanding and was sensitive to the presence of intuitions; their work is reported as containing novel ideas. It is unlikely that the development of an epistemology of science that values intuition will necessarily lead to higher achievement in science, rather it is likely that students' understanding of the content and nature of science co-develop. However, if a large number of students view school science as an exercise in the acquisition of facts (Osborne & Collins, 2000, p. 26), they may not be sensitive to the role of tacit knowledge and processes in their learning.

It might therefore be beneficial for teachers to highlight to students the role of tacit knowledge in learning. Several approaches (for example, Hammer and Elby (2003)) suggest making students aware of their own intuitions in order to support the transition to expert understanding. It may be useful to introduce students to the two systems model of thinking (Kahneman, 2011), which proposes people can use both a rapid but tacit and a deliberate, explicit system of thinking, and encourage reflection on when each process is used. Such a discussion should not define one system as superior to the other, but encourage reflection on the nature and limitations of the two processes. Historical reports of the use of intuition by scientists may act as useful exemplars for students. For example, Rutherford (Oliphant, 1972, p. 19) and Watson (1980, pp. 99–101) recount how their intuitions lead them to make discoveries, whilst Newton and Schrödinger are described as having to overcome their initial intuitions to develop novel theories (Rohrlich, 1996). Fostering in students an awareness of the processes they are using to reason, will aid their thinking in, and beyond, the science classroom.

INSIGHT IN SCIENCE EDUCATION

Insight can be thought of as '…an explicit awareness of novel relations between concepts that arrives with apparent suddenness and little conscious awareness of processing' (Brock, 2015, p. 2). For insights, the tacit element is that the processes leading to understanding are obscure, though the resulting knowledge is consciously available. In deliberate reasoning, a learner might consciously link concepts, overcome contradictions and clarify relationships before reaching a coherent understanding. In the case of insight, processing happens without conscious awareness and the final understanding, which is explicit, appears to arrive without deliberate effort. The moments when teachers cause sudden clarifications in understanding have been described as some of the most powerfully rewarding experiences in science teaching (Fuller, 1993, p. 300; Halpern, 2005, p. 141).

Insight in science education might be conceptualised as a kind of rapid conceptual change. Conceptual change is generally seen as a gradual process and sudden changes in understanding are assumed to be rare events (Vosniadou, 2008, p. xvi). Nevertheless, there are a number of models of conceptual change that allow for the kind of discontinuous change seen in a moment of insight. For example, in Chi's (1997) model of conceptual change as ontological reclassification, certain categorisations are seen as ontologically distinct, for example, the attributes associated with entities classified as matter are necessarily separate from the properties of those labelled as processes. The 'aha' phenomenon then, is linked to the transfer of a concept between ontologies (Chi, 1997, p. 230), a transition that might explain the apparent suddenness of insight as there can be no intermediate categorisation. For example, a student may experience a moment of insight when they reclassify electrical potential difference from its original association with physical objects to an understanding of it as an abstract process.

Humans have a tendency to resist changes to understandings they have put effort into developing (Heine, Proulx, & Vohs, 2006). For example, Duncker (1945) developed the notion of functional fixedness: problem solvers tend to see the role of an object as limited to its usual function. Such fixedness has been described in the context of science education. Furió, Calatayud, Barcenas and Padilla (2000) report how students' commitment to a rote-learned theory, Le Chatelier's principle, led to a difficulty in problem solving. The ability to move way from an initial conceptualisation to a novel understanding of a problem might underlie some moments of insight (Clement, 1989, p. 350). One approach, therefore, that might be used to encourage students to experience moments of insight, is to support the development of multiple problem conceptualisations. For example, students might be encouraged to solve problems in dynamics by considering approaches that focus on forces and conservation of energy. Other approaches to encourage students to experience moments of insight are discussed in the section below.

STRATEGIES TO FOSTER INSIGHT

Before suggesting approaches to fostering insight in students, it is worth considering whether moments of insight are a desirable aim in the science classroom. Given the definition of insight as awareness reached without conscious control, the process might seem undesirable in classrooms in which time is limited and teaching aims to explicate the processes of problem solving. However, moments of insight are described as productive in the work of scientists (Ramsland, 2012) and may be powerful emotional events that students find motivating (Metcalfe & Wiebe, 1987, p. 238). Therefore, whilst moments of insight should not be expected to occur frequently in science lessons, they should be noted and reflected upon when they do.

After a period of unfruitful, conscious engagement with a task, it is reported that insight may occur during engagement with an unrelated task. The interval of working on a different activity is called an incubation period (Smith, 1995, p. 241). Students in the United Kingdom reported being rushed through the content in science lessons (Osborne & Collins, 2000, p. 25); haste which may come at the expense of incubation necessary for the development of insights. Allowing students sufficient time to engage in problem solving, or to think of answers to questioning in class, may allow their tacit processing to reach a conscious outcome. Certain kinds of problems, ones that require a change in problem representation to be solved, have been classified as insight problems (Weisberg, 1995). The kind of problems set in the science classroom are commonly ones which involve the recall of factual knowledge or the application of a set of routines to solving a problem. Such exercises are important but it may benefit students if they are, occasionally, given opportunities to solve insight problems (for sources of such problems in physics see, for example, Epstein, 2009). Insight problems may help students to understand that problem solving in science is not only a matter of following a set of predetermined steps and the positive emotions associated with successful solution may be motivating.

Though it may be a difficult concept to describe, tacit knowledge plays a role in how students come to understand scientific concepts. It is hoped conceptualising the concepts of intuition and insight as links between the tacit and the explicit will enable teachers and researchers to reflect on the role of tacit knowledge in the learning experiences of students.

SUMMARY

- There exists a kind of knowledge, tacit knowledge, that is not directly expressible in words yet has an impact on learning in science education.
- Although tacit knowledge is not directly accessible, its influence on conscious thought can be detected through the action of the processes of intuition and insight.
- Intuitions are pieces of tacit knowledge that influence conscious thought. They are a feature of both novice and expert thinking.
- Intuitions may be knowledge that is encoded kinaesthetically, as images, or as abstracted rules.
- Insights occur when a sudden awareness of connections between concepts arises without a conscious understanding of the processes leading to that awareness.
- Students should be encouraged to be aware of the existence of tacit knowledge and to develop sensitivity to the manner in which it impacts on their learning in science.

RECOMMENDED RESOURCES

For an extended version of the ideas discussed in this chapter see:

Brock, R. (2015). Intuition and insight: Two concepts that illuminate the tacit in science education. *Studies in Science Education. 51*(2) 127–167.

Another good source of information on the role of tacit knowledge in the context of chemistry education is Keith Taber's article:

Taber, K. S. (2014). The significance of implicit knowledge for learning and teaching chemistry. *Chemistry Education Research and Practice, 15*(4), 447–461.

For general reading on the nature of intuition and insight, I can recommend the following books:

Kahneman, D. (2011). *Thinking, fast and slow*. London: Penguin.
Kounios, J., & Beeman, M. (2015). *The Eureka factor: Aha moments, creative insight, and the brain*. New York, NY: Random House.
Myers, D. G. (2004). *Intuition: Its powers and perils*. London: Yale University Press.

REFERENCES

Airey, J., & Linder, C. (2008). A disciplinary discourse perspective on university science learning: Achieving fluency in a critical constellation of modes. *Journal of Research in Science Teaching, 46*(1), 27–49.

Andersson, B. (1986). The experiential gestalt of causation: A common core to pupils' pre- conceptions in science. *European Journal of Science Education, 8*(2), 155–171.

Barsalou, L. W. (1999). Perceptual symbol systems. *Behavioral and Brain Sciences, 22*(4), 577–660.

Baylor, A. L. (2001). A U-shaped model for the development of intuition by level of expertise. *New Ideas in Psychology, 19*(3), 237–244.

Bransford, J., Brown, A. L., & Cocking, R. (Eds.). (2000). *How people learn: Brain, mind, experience, and school.* Washington, DC, US: National Academies Press.

Brock, R. (2015). Intuition and insight: two concepts that illuminate the tacit in science education. *Studies in Science Education, 51*(2), 127–167.

Champagne, A. B., Gunstone, R. F., & Klopfer, L. E. (1985). Instructional consequences of students' knowledge about physical phenomena. In L. H. T. West & A. L. Pines (Eds.), *Cognitive structure and conceptual change* (pp. 61–90). Orlando, FL: Academic Press.

Chi, M. T. H. (1997). Creativity: Shifting across ontological categories flexibly. In T. Ward & S. M. Smith (Eds.), *Creative thought: An investigation of conceptual structures and processes* (pp. 209–234). Washington, DC: American Psychological Association.

Clement, J. (1989). Learning via model construction and criticism. In J. A. Glover, R. Ronning, & C. Reynolds (Eds.), *Handbook of creativity: Assessment, theory and research* (pp. 341–381). New York, NY: Plenum.

Clement, J. J. (2008). *Creative model construction in scientists and students.* Dordrecht: Springer.

diSessa, A. A. (1985). Learning about knowing. In E. L. Klein (Ed.), *New directions for child development: Children and computers* (Vol. 28, pp. 97–124). San Francisco, CA: Jossey- Bass.

diSessa, A. A. (1986). Artificial worlds and real experience. *Instructional Science, 14*(3), 207–227.

diSessa, A. A. (1993). Toward an epistemology of physics. *Cognition and Instruction, 10*(2/3), 105–225.

diSessa, A. A. (2000). *Changing minds: Computers, learning, and literacy.* Cambridge, MA: MIT Press.

Duncker, K. (1945). On problem solving. *Psychological Monographs, 58*(5), i–113.

Epstein, L. C. (2009). *Thinking physics: Understandable practical reality.* San Francisco, CA: Insight Press.

Feynman, R. (1985). *Surely you're joking Mr Feynman.* New York, NY: Bantam Books.

Feynman, R. P. (2013). Laws and intuition. In R. P. Feynman, M. A. Gottlieb, & R. Leighton (Eds.), *Feynman's tips on physics: Reflections, advice, insights, practice* (pp. 61–89). New York, NY: Basic Books.

Fuller, R. G. (1993). Millikan lecture 1992: Hypermedia and the knowing of physics: Standing upon the shoulders of giants. *American Journal of Physics, 61*(4), 300–304.

Furió, C., Calatayud, M. L., Barcenas, S. L., & Padilla, O. M. (2000). Functional fixedness and functional reduction as common sense reasonings in chemical equilibrium and in geometry and polarity of molecules. *Science Education, 84*(5), 545–565.

Halpern, D. F. (2005). That aha moment when understanding happens-that is why I teach. In T. A. Benson, C. Burke, A. Amstadter, R. Siney, V. Hevern, B. Beins, & W. Buskist (Eds.), *Teaching psychology in autobiography: Perspectives from exemplary psychology teachers* (pp. 135–140). Society for the Teaching of Psychology. E-Book available from the Society for the Teaching of Psychology. Retrieved August 24, 2016, from http://teachpsych.org/ebooks/tia2005/index.php

Hammer, D., & Elby, A. (2003). Tapping epistemological resources for learning physics. *Journal of the Learning Sciences, 12*(1), 53–90.

Heine, S. J., Proulx, T., & Vohs, K. D. (2006). The meaning maintenance model: On the coherence of social motivations. *Personality and Social Psychology Review, 10*(2), 88–110.

Kahneman, D. (2011). *Thinking, fast and slow.* London: Penguin.

Kosso, P. (2002). The omniscienter: Beauty and scientific understanding. *International Studies in the Philosophy of Science, 16*(1), 39–48.

Lakoff, G., & Johnson, M. (1980). *Metaphors we live by*. London: University of Chicago Press.
Lipton, P. (2009). Understanding without explanation. In H. de Regt, S. Leonelli, & K. Enger (Eds.). *Scientific understanding: Philosophical perspectives* (pp. 43–63). Pittsburgh, PA: Pittsburgh University Press.
Metcalfe, J., & Wiebe, D. (1987). Intuition in insight and noninsight problem solving. *Memory & Cognition, 15*(3), 238–246.
Nisbett, R. E., & Wilson, T. D. (1977). Telling more than we can know: Verbal reports on mental processes. *Psychological Review, 84*(3), 231–259.
Oliphant, M. (1972). Some personal recollections of Rutherford, the man. *Notes and Records of the Royal Society, 27*(1), 7–23.
Osborne, J., & Collins, S. (2000). Pupils' and parents' views of the school science curriculum. *School Science Review, 82*(298), 23–31.
Polanyi, M. (1966). The logic of tacit inference. *Philosophy, 41*(155), 1–18.
Ramsland, K. M. (2012). *Snap: Seizing your aha! moments*. Amherst, NY: Prometheus Books.
Rohrlich, F. (1996). The unreasonable effectiveness of physical intuition: Success while ignoring objections. *Foundations of Physics, 26*(12), 1617–1626.
Searle, J. (1980). Minds, brains and programs. *Behavioral and Brain Sciences, 3*(3) 417–457.
Shapiro, L. (2011). *Embodied cognition*. Abingdon, Oxon: Routledge.
Smith, S. M. (1995). Getting into and out of mental ruts: A theory of fixation, incubation, and insight. In R. J. Sternberg & J. E. Davidson (Eds.), *The nature of insight* (pp. 229–251). Cambridge, MA: MIT Press.
Songer, N. B., & Linn, M. C. (1991). How do students' views of science influence knowledge integration? *Journal of Research in Science Teaching, 28*(9), 761–784.
Sternberg, R., & Horvath, J. (1999). *Tacit knowledge in professional practice*. London: Laurence Erlbaum.
Taber, K. S. (2014). The significance of implicit knowledge for learning and teaching chemistry. *Chemistry Education Research and Practice, 15*(4), 447–461.
Vosniadou, S. (2008). Conceptual change research: An introduction. In S. Vosniadou (Ed.), *International handbook of research on conceptual change* (1st ed., pp. xiii–xxviii). New York, NY: Routledge.
Watson, J. D. (1980). *The double helix: A personal account of the discovery of the structure of DNA*. New York, NY: Norton.
Watts, M., & Taber, K. S. (1996). An explanatory gestalt of essence: students' conceptions of the 'natural' in physical phenomena. *International Journal of Science Education, 18*(8), 939–954.
Weisberg, R. W. (1995). Prolegomena to theories of insight in problem solving: A taxonomy of problems. In R. J. Sternberg & J. E. Davidson (Eds.), *The nature of insight* (pp. 157–196). Cambridge, MA: MIT Press.
Wellington, J., & Osborne, J. F. (2001). *Language and literacy in science education*. Buckingham: Open University Press.

ALAN J. MCCORMACK

11. DEVELOPING VISUAL/SPATIAL THINKING IN SCIENCE EDUCATION

Visual/spatial thinking (VST) involves purposeful use of your mind's eye to develop mental pictures or images. At higher levels, it is characterized by both logical and creative processing of mental images to solve problems, create new ideas, improve physical skills, and/or even quiet tumultuous emotional states.

Visual/spatial thinking pervades all human experience, from the frontiers of professional science and art to everyday tasks such as packing a suitcase. Chemists visualize molecular configurations and changes in shape; a science fiction writer imagines Klingons passing through a time warp. All are thinking visually. Everyday each of us uses visual/spatial thinking skills to accomplish pedestrian tasks: using road maps, rearranging furniture, reading books, etc.

Visual/spatial solutions to problems have frequently been recounted by brilliant inventors and scientists. In one historically important episode, the structure of DNA was ferreted out by Watson, Crick and Franklin using a variety of VST techniques: thought experiments, drawing, and three-dimensional modeling.

Most of what is studied by present-day scientists cannot be directly observed or photographed. They are ideas based on indirect evidence and expressed as visual metaphors. Drawings of molecular structures, genes, distant galaxies, and electrons flowing through computer chips are probably *not* absolutely accurate representations of reality, but they *are* indispensable aids to our understanding. Without these imagined visual metaphors, science would stagnate and communication of scientific discoveries would nearly be impossible. According to E. S. Ferguson (1977), many scientific and engineering problems simply cannot be described verbally, and VST is essential to scientists as an aid to thinking, a means of problem formulation, and as the essence of problem-solving. Perhaps MacFarlane-Smith (1964) is correct in his suggestion that after humans have attained a certain minimal verbal activity, it is visualization skills that determine how far a person will progress in the sciences.

Recent research shows that higher spatial ability is an excellent predictor of interest and success in STEM disciplines (Science, Technology, Engineering, and Mathematics). Jonathan Wai and his colleagues conducted a huge, long-term study (400,000 students tracked for 11+ years) and found spatial ability to be a significant factor in predicting success in advanced educational STEM-oriented degrees and STEM careers (Wai et al., 2009). Similarly, Elizabeth Gunderson and coworkers found preschoolers who performed well on spatial tests are better than others at

K. S. Taber & B. Akpan (Eds.), Science Education, 143–156.

mathematics as 8 year olds (Gunderson et al., 2012). Other investigators report that professionals in STEM disciplines such as geosciences are better able to navigate their environments than workers in non-STEM fields (Hegarty et al., 2010).

Strong arguments have been put forth for broad use of one aspect of VST by John Gilbert and colleagues in *Visualization: Theory and Practice in Science Education* (2008). Gilbert champions the importance of having students develop *visualizations* of their science experiences and be able to interpret and use visualizations produced by others, such as photographs, diagrams, graphs, and structural representations. Gilbert categorizes these visualizations into several levels as 3D, 2D, and 1D representations in terms of levels of abstraction. His coauthors Uttal and O'Doherty of the same volume build arguments for the usefulness in science education of visualizations, pointing out that these mental constructions "make complex information accessible and cognitively tractable." They go on further: "*Visualizations* highlight the portions of the information that the designer intends for the learner to see and hence support both learning among novices and new discoveries among experts. They allow us to perceive, and to think about, relations among items that would be difficult to comprehend otherwise" (Uttal & O'Doherty, 2008, p. 53).

A somewhat similar call for more attention to visual representations is made by Finson and Pedersen in *Visual Data and Their Use in Science Education* (2013). Their book focuses mainly on supporting the use of textbook and computer simulation visuals as boosts to concept development, but also favors the use of mind-mapping, concept maps, and web diagrams as visual learning aids for science students (Finson & Pedersen, 2013). The limitation of both the Gilbert et al. and Finson and Pedersen compendiums is that their focus is mainly on student interpretation of various visuals, rather than on the many other dimensions of VST, including variations of logical processing and creative production of images. Current publishers of textbooks and supporting learning materials in science include heavy doses of visual representation. In an important study of this issue, a visual-arts educator observed that current college students "could view pictures but could not craft images. They could read a map but could not map data" (Metros, 2008. p. 103).

VST researcher Newcombe notes that teachers assume that students understand their textbooks' visuals, but they frequently do not. She observed that "many students have little idea what the arrows in diagrams may mean, or how the zoom-outs or cutaways relate to the main diagram, and they often fail to read the captions and legends. Some students may rarely consult the diagrams at all, despite the fact that diagrams frequently present information that is not also presented in the verbal text" (Newcombe, 2013, p. 30). Newcombe suggests that teachers explicitly work with students on how to understand diagrams, and to involve students in sketching activities, pointing out that practicing scientists often make sketches to communicate their ideas to others. Ainsworth and colleagues agree, reporting in the respected journal *Science* that sketching by students is beneficial for many reasons: it promotes engagement, helps develop deeper understanding of ideas, activates reasoning, forces ideas to crystallize, and facilitates communication within working groups (Ainsworth et al., 2011).

The highly science-significant and politically-influential National Research Council realized the importance of VST in 2006 and constituted a blue ribbon committee – the Committee on Support for Thinking Spatially – resulting in publication of their excellent report: *Learning to Think Spatially* (NRC, 2006). The committee developed a position statement citing "the educational necessity for teaching and learning about spatial thinking" (NRC, 2006. p. 5). Here are just some of their position statements:

- Spatial thinking is integral to everyday life. People, natural objects, human-made objects, and human-made structures exist somewhere in space, and the interactions of people and things must be understood in terms of locations, distances, directions, shapes, and patterns.
- Spatial thinking is powerful. It solves problems by managing, transforming and analyzing data, especially complex and large data sets, and by communicating the results of those processes to one's self and to others.
- Spatial thinking is integral to the everyday work of scientists and engineers, and it has underpinned many scientific and technical breakthroughs.
- Spatial thinking is a skill that can – and should – be learned by everyone.
- Expertise in spatial thinking develops in the context of specific disciplines and becomes transformed through training and extensive practice.
- Spatial thinking is currently not systematically instructed in the K-12 (ages 5–18 years) curriculum despite its fundamental importance and despite its significant role in the sets of national standards for science, Mathematics, geography. This is a major blind spot in the American educational system (NRC, 2006, pp. 5–6).

VISUAL/SPATIAL THINKING IS MULTIDIMENSIONAL

Visual/spatial thinking is not a single, unidimensional mental skill. Instead, it is a complex set of interrelated abilities involving perception, memory, logic, and creativity. Certainly, it involves far more than only processing visual diagrams and images found in textbooks and computer simulation programs. There exist a plethora of VST skills that need to be mastered as a basis for understanding and learning from textbook and digital images. VST involves more than consuming visualizations, but in learning many strategies to perceive, remember, manipulate and produce images.

In the early years of schooling, children should improve visualization skills through hands-on experiences allowing them to shrink, expand, turn, and deform two- and three-dimensional shapes. They should begin to master ideas of relative position of objects, learning the meaning of above, below, near, left, and right. They should build structures with blocks, plastic straws, clay and Legos. Later, they should become comfortable drawing views from perspectives of different observers, inferring shapes of objects from the shadow they cast, and inferring qualities of objects that cannot be directly observed but can be imagined. They can learn to draw maps of their bedrooms and graduate to mapping routes to navigate from home to school. Through the elementary and middle school years they can learn some of the design tricks of engineers and build their own inventions to solve some of everyday life's annoying challenges. There are a myriad of VST skills and activities circulating through the veins of all school subjects, so we need some sort of a map to begin to sort these out.

In order to study and develop children's visual/spatial thinking skills, the need for a table of specifications describing the visual mental complex became apparent. *The Taxonomy of Visual/Spatial Thinking Skills* (McCormack, 1988) was developed for this purpose. The *Taxonomy* was derived from and extensive search of the research literature and studies of all existing VST measuring instruments. According to this classification system, VST can be organized into four hierarchical domains:

- *Visual/Spatial Perception*: includes rudimentary physiological ability to observe objects and form representative mental images.
- *Visual/Spatial Memory*: comprised of abilities to mentally store images and retrieve them at a later time and to communicate descriptions of images through drawings and language.
- *Logical Visual/Spatial Thinking*: consists of operations involving mental images where the operations are based on a set of rules and analytical, convergent thinking. Most of these operations involve processes of logical inference.
- *Creative Visual/Spatial Thinking*: involves production of rare, unique, or original mental images. Images may be produced in realms of fantasy, invention, design, aesthetics, humor, and metaphor.

A pyramid is a good model showing a hierarchical arrangement of the major domains of visual/spatial thinking, illustrating that visual memory depends on functioning of visual perceptions, logical visual thinking depends on use of both perception and visual memory, while creative visual thinking functions effectively only by being based firmly on three underlying domains.

Each domain, of course, is comprised of numerous skills and cognitive processes and organizing them is a good first step toward building VST into the STEM curriculum. Following are key behaviors or cognitive skills characteristic of each of the four domains, and some brief sample learning activities that might effectively be used in developing skills of the domain:

DOMAIN ONE: VISUAL/SPATIAL PERCEPTION

Key Skills or Behaviors

- Using our visual sense to form images and record data using drawings or verbal descriptions
- Ability to observe fine details of objects

- Forming mental images based on limited sensory input
- Differentiating between figure and ground in complex drawings or photographs
- Forming visual images from perceptual data *other than* sight
- Making understood connections between real (actual) objects and drawings, photographs, or media images that represent the objects

Sample Learning Activities

VST Perception I: Alphabets in nature (observing fine details of natural objects). A wonderful book entitled *The Butterfly Alphabet* by Kjell B. Sandved (1996) captures a decades-long search by the author for images of letters of the alphabet on the wings of butterflies and moths. After introducing Sandved's great adventure, challenge your students to find as many naturally-occurring letters of the alphabet as they can. Discovered letters may be captured in many creative ways: digital or Polaroid photos, drawings, clay impressions, crayon rubbings, or actual samples.

VST Perception II: Blurry photographs or drawings (forming mental images based on limited visual input). Choose some photographs or line drawings of items relevant to a current unit of study in science. (Images of animals, biological cell-types, machines, etc.) Using a projector, project very blurry images onto a screen so that most detail is lost, leaving only highlights or generalized shapes. Have students search for visual clues, and infer what image they may be observing.

DOMAIN TWO: VISUAL/SPATIAL MEMORY

Key Skills or Behaviors

- Storage and retrieval of mental images representing observed objects or situations.
- Bringing back mentally-recorded images to a conscious level
- Visualizing objects or situations based on verbal descriptions
- Retrieval of images of visually-observed patterns

Sample Learning Activities

VST Memory I: Fly's eye view – what my bedroom looks like to a fly on the ceiling (communication of previously perceived objects through drawing). Most kids are ready to use simple mapping skills as early as age 3 or 4. For example, preschoolers have been shown capable of interpreting a map of their living room floor plan and the to use a map to show where in the room they have hidden a toy (Shusterman et al., 2008).

For this activity, ask students: "Are you aware of your surroundings? Do you really *observe* your environment? Let's find out – pretend there is a fly in the middle of your bedroom ceiling. The fly can see everything in the room. What would the fly see?"

Challenge kids to make a line drawing (strictly from memory) of a view from above of their bedroom. When drawings are complete, invite them to share their masterpieces with the class. Then, have them take the drawings home and judge how accurate their memories were – have them list all items that were *not included in their drawings*, and bring these lists back for discussion.

Visual Memory II: Quick Draw (Retrieval of images of visually-observed patterns) Line drawings of patterns are "flashed" on a screen for a few seconds of observation (using a projector of your choice.) – Ask: "Can you remember images you have observed? Draw the target drawings as best you can." This improves with practice, and could be done as a routine classroom activity. When first attempting visual/spatial activities such as these, be prepared for slow progress in the beginning stages. Researcher Nora Newcombe (2010) notes that students with poor spatial skills can be slow to improve – she found it takes six sessions or more before skills are found to get better.

DOMAIN THREE: LOGICAL VISUAL/SPATIAL THINKING

Key Skills or Behaviors

- Applying rules of logical analysis to mental images
- Visualizing objects as observed from different points of view
- Visualizing sections of objects (cross-sections and longitudinal-sections)
- Figure completions (completing incomplete drawings of patterns or objects)
- Identifying hidden shapes in complex figures (embedded figures)
- Recognizing patterns in both the natural and human-made world
- Visualization of repeating elements in a pattern
- Interpreting 2-dimensional representations of 3-dimensional objects
- Visualizing motion of objects and patterns created by motion of objects
- Visualizing rotations of 2-dimensional objects
- Visualizing 3-dimensional shapes that may be formed by rotation of 2-dimensional objects
- Visualizing rotations of 3-dimensional objects
- Identifying and constructing surface patterns of 3-dimensional objects (surface "peels")
- Making inferences regarding shapes of objects based on the shadows they cast.
- Visualizing shapes of hidden objects based on non-visual data (data obtained through senses other than vision)
- Developing inferential models to represent objects that have not been directly observed
- Visualizing a predicted transformation of an object
- Visualizing a predicted reorganization of the relative locations of a set of objects
- Wayfinding – following a designated route to a destination; accurately reading maps

Sample Learning Activities

Logical VST I: Visualizing sections of objects. In science, "sections" are straight cuts through objects, revealing a planar view of what is inside the object and the shape of the outer circumference of the object. Textbooks of biology are ordinarily filled with various sections of organisms. Earth science books use sections of the entire Earth and parts of the Earth's crust to clarify concepts of geology. Engineers frequently employ sections of buildings and machines to help clarify various structures. Clearly, the idea of "sectioning" living and nonliving things is an important tool in the sciences. A survey of K-12 science programs, revealed no instances where the section idea is actually taught. Somehow, curriculum developers assume that students will naturally comprehend images of sections! Students often don't understand. Spatial thinking researcher Nora Newcombe (2013) found that students frequently have little understanding of drawings such as cross sections, and they often ignore them while reading text books. She suggests that teachers explicitly teach the meaning of sectional drawings, and to point out that these drawings often contain information that is not provided otherwise in the text. Here is an activity that can help build understanding of cross- and longitudinal sections.

Sectioning play dough shapes – prepare small solid play dough shapes: a cube and a sphere about the size of game board dice and toy marbles. Use the same color of play dough for both of these. Take larger amounts of a contrasting color of play dough, and enclose each small cube and sphere within a round ball (about 5cm in diameter). Be careful to maintain the shapes of the encased small objects as you create the larger spheres. Each working group of students should be given two of the clay balls (one containing a cube shape, the other the spherical shape), and a safe, plastic picnic knife. Invite the groups to make a cut through each clay ball, right through its center. Have students observe the exposed slices through the imbedded objects, and make inferences about their overall (3-dimensional) shapes.

Logical VST II: Observing from different points of view with Mr. Observer. All of us are called upon occasionally to describe where a place or object is located: "Where is the nearest gas station?" and, "Where is your house?" are common questions we all encounter. We answer with directions such as, "The gas stations is six blocks ahead and is located next to the super market." In such situations, it is not possible to describe the location of an object without using other *reference objects.* We can only locate the position of an object relative to the position of other objects or to an observer. In science, the identification of an observer or a set of objects from which the position of another object is described as a *frame of reference.* Elementary level kids tend to be fixated on themselves as primary frames of reference. They need to learn how comprehend observations made from alternative points of view.

To facilitate this process, a class can be introduced to "Mr. Observer." He can be represented by a cut-out cardboard stand-up stick-figure puppet. Students can be challenged to report the location of various objects in the classroom from

Mr. Observer's point of view, and use him as a tool for observing orientations of drawings, block constructions, and pathways of moving objects.

DOMAIN FOUR: CREATIVE VISUAL/SPATIAL THINKING

Key Skills or Behaviors

- Fluency, flexibility, and originality in production of divergent mental images related to a challenge or other stimulus
- Producing and inventing visual images beyond what has actually been experienced
- Visual invention (via drawings or actual construction) of devices, imaginary creatures, or other novel solutions to problems
- Imagining alternate explanations for the operation of a physical system, based on partial observation of the system
- Creating fantasy images of persons, places, objects, etc.
- Ability to create novel scientific models to explain observed data
- Willingness to risk experiences with fantasy
- Ability to create visual metaphors
- Ability to create visual humor through cartoons, pantomime, drawings, or constructed objects
- Ability to tap subconscious states through relaxation, meditation, and dreams
- Imagining alternative future events
- Janusian Imagery – holding two different mental images in the mind at the same time

Sample Learning Activities

Creative VST I: Anti-Coloring Book-style drawings. In the early 1980's, elementary art teacher Susan Striker had an epiphany – she realized that the traditional coloring books supplied to children were no use in stimulating children's creativity – in fact, these books tend to stifle imagination. Striker invented "The Anti-Coloring Books" as a result of her insight (Striker, 1984). These break the mold of the ordinary coloring book: instead of each coloring book page being completely filled with adult-drawn figures to be merely decorated with waxen colors, each of the Anti-Coloring Book pages is mostly blank space, inviting kids to add their own drawings. But the real key the Anti-Coloring book pages is that each page presents an intriguing challenge by way of a small beginning drawing and a challenge. One page depicts the dashboard and windshield of a spacecraft as viewed by the pilot of the craft. The student peers (in his/her mind's eye) through the windshield and is presented with the challenge: "Your spaceship is landing on Mars, and the Welcoming Committee is coming to greet you." Kids invent, draw, and color an amazing array of Martians. Dozens of other science-oriented challenges can be found in Striker's books.

Creative VST II: Create-a-Creature. Inventions are not necessarily inherent only to the domains of physics and engineering. Create-a-Creature will give student an opportunity to apply what they have learned about animal adaptations, as well as being enjoyably involved in collaborative development of a creature never before seen on this planet.

For "Create-a-Creature," make an overhead projector transparency for each group of 3 students. Each transparency has been divided into 3 sections, with "HEAD," "BODY," and, "TAIL," as labels on the three segments. Use scissors to cut the transparency into separate "head," "body," and, "tail" sections, and place the pieces into an envelope. Each working group of three students gets one prepared envelope and marking pens. They are directed to randomly reach into the envelope, select a segment, and each draw one section of the transparency. They are then challenged to draw (out of sight of their other group members), a portion of an actual or fantasy animal as labeled on their transparency part. Kids should be encouraged to think "far out" and be reminded they are trying to invent a creature that is brand new to the universe, not duplicate an existing animal.

After completing their independent animal segments, groups reconvene and match-up their body sections to form some amazing creatures. Some amazing creations will result. Now, you are staged to bring the lesson back to its biological roots: "You have invented some wonderful creatures. Now, let's think about the creature scientifically." Give each group a set or guiding questions from which they are challenged to develop a "Scientist's Guide" for their creature:

- Invent a good name for your new creature.
- Examine the body of your creature carefully – How large is it? (Insect size? Elephant size?)
- Based on the body characteristics of your creature, what is its likely habitat (Swamp? Trees? Desert? Or?)
- What does your creature eat?
- How does your creature move?
- How does your creature defend itself from enemies?
- What is the life cycle of your creature? Does it change shape during its life?
- How is your creature important in the environment where it lives? Is it harmful or helpful?

Time now for a scientific conference: Have each group, in turn, assemble their creature on the stage of your overhead projector and present it to the world. This pageant is likely to be both humorous and scientifically enlightening, as students will have amazingly clever names for their animals and will likely have some brilliant insights linking structure and function of their creature's anatomy.

GENDER DIFFERENCES IN VISUAL/SPATIAL THINKING

Much research evidence exists that males outperform females on various measures of visual/spatial thinking. That this difference between the sexes becomes most pronounced in early adolescence was established by a synthesis of research by Maccoby and Jacklin (1974). But, a number of studies have shown that sex differences are apparent even in preschool children. Siegel and Schadler (1977), for example, had 4 1/2 year olds construct scale models of their classroom by placing representations of furniture in their correct places on the model. Boys accomplished this with far greater accuracy than girls. In a second study, Herman and Siegel (1978) had kindergarten, second, and fifth grade children walk repeatedly through a large-scale model town. After considerable experience with the model, children were asked to reconstruct the arrangement of roads, buildings, and a railroad track from visual/spatial memory. The researchers found that boys significantly outperformed the girls on this task, and that accuracy improved with age.

According to McGee (1982), sex differences in visual/spatial thinking skills are among the most persistent of all individual differences cited in psychological research literature. For example, two of the most widely-used standardized tests for visual/spatial skills are Thurstone's Primary Mental Abilities (PMA) Test and the Differential Aptitude Test (DAT). Herzberg and Lepkin (1954) administered the PMA to more than 1,000 high school students and found boys significantly superior to girls on spatial tasks at every age studies. Girls tended to excel in language tasks such as word fluency. Studies by Wesman (1949) using the DAT found that males score significantly higher on spatial relations and mechanical reasoning tasks, while girls excel at tasks involving clerical speed, spelling, and language usage.

VISUAL/SPATIAL THINKING CAN BE IMPROVED

A number of significant successes in development of visual/spatial thinking abilities have been reported. Barbara Moses (1980) developed a carefully designed sequence of visual/spatial thinking activities for ninth grade classes. She found that the ability to think visually is not innate and can be improved. Her activities also revealed a strong positive relationship between problem-solving and VST skills. Moses found improvement most dramatic among female students who had few previous direct experiences with VST.

A key study by Smith and Schroeder (1979) with fourth graders found that those receiving training in visual thinking outperformed those who did not, and *there was no difference between the sexes in how training affected their abilities to perform visual/spatial tasks*. Girls performed as well as boys when they had similar previous experiences.

Herbert Cohen (1983) used hands-on science materials to examine effects of science manipulatives on VST abilities and whether effects were different for males and females. Using fifth grade students in a controlled study, he found that both males and females significantly improved in logical visual thinking abilities as a result of science experiences involving physical manipulatives. The research studies reviewed for this project tend to support five important observations:

1. Boys tend to outperform girls on visual/spatial tasks.
2. Visual/spatial abilities can be improved through appropriate training.
3. Both sexes benefit from planned visual/spatial exercises, but girls profit most in that these experiences narrow the gender gap in visual/spatial thinking.
4. Many children from diverse/underrepresented ethnic backgrounds would benefit tremendously from increased attention to more visual approaches to learning science.
5. Difficulties with visual/spatial skills tend to be a consequence of limited social and scholastic opportunities to develop these skills.

Humans of all ages need experiences with games requiring spatial orientation and visualization, constructing models and other devices from simple parts, and with drawings in two and three dimensions that traditionally have been reserved only for students who elect to take mechanical drawing courses. In order to establish true freedom of career choice, people of both sexes and from all ethnic groups should be given opportunities to hone their visual/spatial skills. I can think of no more obvious place to begin than with modification of elementary/middle level science programs.

SCIENCE ACTIVITIES BUILD VISUAL/THINKING SKILLS

Though development of VST seems an obvious and worthy objective for any science program, curriculum developers have failed to give direct attention to this set of skills. A study by the author of all presently available U.S. K-8 science programs (for ages 5–12 years) failed to find a single program which specifically mentions development of VST as one of its objectives. Some of these skills are sometimes casually included occasionally as part of an activity, but never given any clear focus or emphasis. If they are there, it is almost as if accidentally or peripherally.

As has always been true historically, VST continues to be treated as something people learn automatically or naturally. Not true. VST consists of an ultimately essential set of skills that is too important to be left to development by chance (Newcombe, 2013. p. 29).

To test the feasibility of improving VST on purpose through carefully constructed science experiences, Project VISTA at San Diego State University was initiated and funded by the National Science Foundation. VISTA enlisted the aid of 30 K-8 teachers and a group of university science educators to develop VST-oriented science activities, implement them with K-8 students, and assess any impact on VST skills.

Science activities involving VST parameters were used as enrichment to school district science programs with 537 K-8 students over a full academic year. Another 457 students in the same schools served as control subjects. A battery of age appropriate VST tests were borrowed from psychological test banks and used pre- and post. Tests of spatial memory, positions in space, spatial relations, spatial rotation, and spatial mechanical reasoning were employed. Data analysis showed a clear statistically significant advantage on mean test scores by treatment groups.

It is almost a cliché in the VST educational/psychological literature that males outperform females at all age levels in visual/spatial thinking. Our VISTA research found that, indeed, female subjects scored lower on pretest measures. But, at the end of the project, females mean scores equaled those of males, and many female individuals outscored males on many subtests! This seems highly significant as it indicates the traditional "gender gap" in VST skills can be closed if females receive direct experiences with VST activities.

SIGNIFICANCE FOR SCIENCE EDUCATION

VST is essential for professional scientists in *doing* science and for students of science in *learning* science. Current research supports the following premises:

1. VST has traditionally been overlooked and underused in school science curricula.
2. Teachers can learn to value VST. They can develop exciting and important VST activities and enthusiastically incorporate them into existing science programs.
3. Multiple VST skills can be assessed using paper and pencil instruments. The instruments are considered to be enjoyable "puzzles" by kids.
4. Summer professional development programs for teachers appear to be successful in producing teachers who are positively-oriented toward VST and who can produce sound VST science activities and implement them in their classrooms.
5. New and exciting VST science activities can be developed in all conceptual areas of science (life, earth, space, physical, technological) using a wide variety of activity formats: hands-on activities, paper and pencil challenges, games, simulations, storylines, and Imagineering.
6. Appropriate VST activities conducted on a regular basis can result in improved VST abilities of students.
7. Female students seem to especially benefit from increased exposure to VST science activities.
8. VST-oriented science activities appear to have special attitudinal benefits for students – they enjoy them!

FURTHER READING

National Research Council. (2006). *Learning to think spatially.* Washington, DC: The National Academies Press.

Shah, P., & Miyake, A. (2005). *The Cambridge handbook of visuospatial thinking.* New York, NY: Cambridge University Press.

Yenawine, P. (2014). *Visual thinking strategies.* Cambridge: Harvard Education Press.

REFERENCES

Ainsworth, S., Prain, V., & Tyler, R. (2011). Drawing to learn in science. *Science, 333*(6046), 1096–1097.

Cohen, H. (1983). A comparison of the effects of two types of student behaviors with manipulatives on the development of projective structures. *Journal of Research in Science Teaching, 20*(9), 875–883.

Ferguson, E. S. (1977). The mind's eye: Nonverbal thought in technology. *Science, 197*(4306), 826–836.

Gilbert, J. K., Reiner, M., & Nakhleh, M. (Eds.). (2008). *Visualization: Theory and practice in science education.* Dordrecht: Springer.

Gunderson, E. A., Ramirez, G., Beilock, S., & Levine, S. C. (2012). The relation between spatial skill and early number knowledge: The role of the linear number line. *Developmental Psychology, 48*(5), 1229–1241.

Hegarty, M., Crookes, R. D., Dara-Abrams, D., & Shipley, D. F. (2010). Do all science disciplines rely on spatial abilities? Preliminary evidence from self-report questionnaires. *Spatial Cognition, VII*, 85–94.

Herman, J. F., & Siegel, A. W. (1978). The development of cognitive mapping of the large-scale environment. *Journal of Experimental Child Psychology, 26*, 389–406.

Herzberg, F., & Lepkin, M. A. (1954). A study of sex differences in the PMA test. *Educational and Psychological Measurement, 14*, 687–689.

Kosslyn, S. M. (1985). Stalking the mental image. *Psychology Today, 19*(5), 22–28.

Maccoby, E., & Jacklin, C. (1974). *The psychology of sex differences.* Stanford, CA: Stanford University Press.

MacFarlane-Smith, I. (1964). *Spatial ability: Its educational and social significance.* San Diego, CA: Knapp Publishing Co.

Mason, C. L., & Kahle, J. B. (1989). Student attitudes toward science and science-related careers: A program designed to promote a stimulating gender-free learning environment. *Journal of Research in Science Teaching, 26*, 25–39.

McCormack, A. J. (1981). *Inventors workshop.* Belmont, CA. Fearon Publishing Co.

McCormack, A. J. (1988). *Visual/Spatial thinking: An essential element of elementary school science.* Washington, DC. Council for Elementary Science, International.

McGee, M. G. (1982). Spatial abilities: the influence of genetic factors. In M. Potegal (Ed.), *Spatial abilities: Development and psychological foundations* (pp. 199–216). New York, NY: Academic Press.

Metros, S. E. (2008). The educator's role in preparing visually literate learners. *Theory into Practice, 47*, 102–109.

Moses, B. (1980, April 7). *The relationship between visual thinking tasks and problem-solving performance.* A paper presented at the Annual Meeting of the American Educational Research Association, Boston, MA.

Newcombe, N. S. (2013). Seeing relationships: Using spatial thinking to teach science, mathematics, and social studies. *American Educator, 37*(1), 26–31.

NRC. (2006). *Learning to think spatially.* Washington, DC: The National Academies Press.

Piaget, J., & Inhelder, B. (1956). *The child's conception of space.* New York, NY: Humanities Press.

Rheingold, H.L., & Cook, K. V. (1975). The content of boys' and girls' rooms as an index of parent's behavior. *Child Development, 46*, 459–463.

Siegel, A. W., & Schadler, M. (1977). The development of young children's spatial representations of their classrooms. *Child Development, 48*, 388–394.

Smith, W. S., & Schroeder, C. (1979). Instruction of fourth grade boys and girls on spatial visualization. *Science Education, 63*(1), 61–66.

Sommer, R. (1978). *The mind's eye: Imagery in everyday life.* Palo Alto, CA: Dale Seymour Publications.

Uttal, D. W., & O'Doherty, K. (2008). Comprehending and learning from 'visualizations': A developmental perspective. In J. K. Gilbert, M. Reiner, & Nakhleh, M. (Eds.), *Visualization: Theory and practice in science education* (p. 53). Dordrecht: Springer.

Wai, J., Lubinski, D., & Benbow, C. P. (2009). Spatial ability for STEM domains aligning over 50 years of cumulative psychological knowledge solidifies its importance. *Journal of Educational Psychology, 101*(4), 817–835.

Wesman, A. G. (1949). Separation of sex groups in test reporting. *Journal of Educational Psychology, 40*, 223–229.

AGUSTÍN ADÚRIZ-BRAVO AND ANDREA REVEL CHION

12. LANGUAGE, DISCOURSE, ARGUMENTATION, AND SCIENCE EDUCATION

INTRODUCTION

In didactics of science (i.e., science education understood as a scholarly discipline) as well as in other academic fields – such as the philosophy of science, cognitive science, or classroom ethnography – there is nowadays extended consensus around the recognition that scientific knowledge is "dependent inextricably on language and language is also central to our ability to think [scientifically]" (Evagorou & Osborne, 2010, p. 136).[1] Language constitutes a key element in science education: it can be seen as a tool that allows us to understand and apprehend the natural world, to shape and express our ideas and reasoning on it, and to develop, share, transmit and perpetuate scientific knowledge.

The paramount role of language in scaffolding and configuring science learning processes started to be widely acknowledged in the 1960s; such acknowledgment can be at least partially attributed to the seminal works of Jerome Bruner and the dissemination of Lev Vygotsky's ideas in the English-speaking academic community.[2] However, it was not until the late 1980s, and following conceptual developments in the philosophy of science, linguistics, and educational studies, that science education research began to pay consistent attention to the linguistic aspects inherent to science teaching and learning. As a consequence of this new focus, a very active and rapidly expanding research line has emerged and consolidated in the last two decades. Such line comprises several theoretical perspectives focussing on different aspects of the nature and use of the scientific language in the classroom (e.g., Lemke, 1990; Driver et al., 1994; Candela, 1995; Sutton, 1996; Sanmartí et al., 1998; Osborne, 2010), with studies looking into all educational levels, from Kindergarten to University.

Clive Sutton (1996), in his now classic paper "Beliefs about science and beliefs about language", portrays two distinct *epistemic* (i.e., knowledge-construction) functions that language can perform in science: language can serve as a *labelling* system, to tag and transmit established pieces of knowledge, or as an *interpretive* system, actively used to generate and consolidate new understandings. In his text, Sutton is advocating for a shift from the positivistic emphasis on language as a means of merely conveying conceptual information towards the constructivist idea of understanding language as a way of making and negotiating meanings. Adhering to this twofold characterisation of scientific language for the science classes would

K. S. Taber & B. Akpan (Eds.), Science Education, 157–166.

import that students need to be introduced, in a coordinated way, into the modes of reasoning and in the patterns of language that are deployed in the context of doing science. Along this line, Evagorou and Osborne (2010, p. 138) claim:

> [B]ecause reading and writing are activities that are constitutive of science, and because the language of science is complex and foreign to many students, we see teaching science as fundamentally a process of teaching a language – one in which the teacher has both to help students to interpret and construct meaning from scientific text and one in which they must provide opportunities to develop their fluency and capabilities with that language. In the classroom, three main forms of language are used as tools for understanding, communicating, and developing knowledge: talk, writing and reading.

In the same spirit of the previous paragraph, Jay Lemke (2001) argues that we could understand science education as a "second socialisation": an enculturation into a sub-community – science – that has its own representations, methods, ethos, and jargon. This theoretical approach should motivate us to examine how people learn to talk and write the language of science *while engaging in specific scientific activities*, such as observing, experimenting, hypothesising, debating, or publishing. In his renowned book *Talking science*, Lemke (1990) equates science learning – at least in some aspects – to learning to "talk science". This implies moving away from science lessons dominated by a "triadic dialogue" centred on teachers' authoritative talk – as in the classical IRF (initiation-response-feedback) sequences – towards designing and implementing classes in which students actively (re)construct language and flexibly use it to make sense of natural phenomena.

Lemke introduces a very suggestive idea: talking science could be considered a very elaborate and controlled social process; his description of this process is modelled on the metaphor that science is a foreign language that students have to learn. In his own words:

> Learning science means learning to talk science. It also means learning to use this specialized conceptual language in reading and writing, in reasoning and problem solving, in guiding practical action in the laboratory and in daily life. It means learning to communicate in the language of science and act as a member of the community of people who do so. "Talking science" means observing, describing, comparing, classifying, analyzing, discussing, hypothesizing, theorizing, questioning, challenging, arguing, designing experiments, following procedures, judging, evaluating, deciding, concluding, generalizing, reporting, writing, lecturing, teaching in and through the language of science. (Lemke, 1990, p. 1)

Lemke asserts that we learn to speak the language of science in much the same way in which we learn any other language: practicing it with people who master it and using it in a variety of *pragmatic* communicational situations, where it should be employed in its most frequent typologies, genres, and text formats. In accordance

14

SCHOOL SCIENTIFIC ARGUMENTATION

Many scholars in didactics of science have converged in the recognition that the production of argumentative texts is particularly powerful for science education;[4] in addition, the function that mastery of argumentation has in different aspects of learning has been extensively investigated *outside* science education, and the results of such research have reinforced conviction within our community to adopt argumentation as a privileged strategy to teach science. For instance, Greg Kelly and Charles Bazerman (2003) – within the framing of ideas known as "writing across the curriculum" (WAC) – propose to engage students in instances in which they produce arguments in the scientific disciplines and beyond them. From these arguments, students learn to talk and write the language of the different academic fields.

In their proposal, these authors indicate that argumentative discourse would be one of the communicational functions *par excellence* in the development of scientific knowledge, and hence its importance in a science learning that seeks "authenticity". Basing on scientific models, students can construct a special kind of evidence-based explanation to make sense of the world around them; in this construction, we can help them identify that the sound connection between data (evidence) and conclusions (explanation) counts as argumentation:

> [S]tudents not only need to write in order to master the concepts and work of a field, but more particularly to develop competencies in the specific argumentative practices of their fields [...]. In addition to the genre-specific writing competencies, with associated argumentative patterns, students must begin to gain a feel for the argumentative forums and dynamics of their fields. They must learn the kinds of claims people make [and] what kind of evidence is needed to warrant arguments [...]. (Kelly & Bazerman, 2003, pp. 29–30)

Scientific argumentation demands a style of monitoring that enables student-writers, among other things, to: recognise the presence of adequately derived conclusions; revise the strong connection between these and the elements providing and justifying the *transition* from data – what Stephen Toulmin calls "warrants" and "backings" (cf., Erduran et al., 2004); and seek for strong structural coherence between propositions. The hybrid nature of argumentation – as a tight welding of the epistemic operations of *explaining* and *convincing* (cf., Kelly & Takao, 2002; Sampson & Clark, 2008) – provides both a "window" for the teachers to assess students' understandings, and an arena for the students to test their own appropriation of what they are learning.

As we have suggested, understanding science accomplishments as products *and* processes – i.e., what we know and how we know it – requires systematic discursive exploration of the intricate *interaction* between theoretical ideas and the evidence (empirical and other) that they rely on. In other words, students should become aware that scientific constructs divert in many aspects from common sense, and are often far from transparent; they should rather see those constructs as the laborious *inferential* products of one of the most intricate cognitive and social activities of humankind. All of this means sustained argumentative work in the science classroom.

Nowadays, a broad range of theoretical conceptions on the nature of scientific argumentation is available in the literature of didactics of science;[5] these conceptions are mostly imported from classical and renewed positions in linguistics (including here argumentation theory, dialectics, and rhetoric) and – to a lesser extent – in the philosophy of science. Conceiving argumentation as a "cognitive-linguistic ability"[6] that links phenomena, models, evidences and explanations through discourse would locate it – together with explanation – at the very vertex of the "scientific pyramid" (cf., Duschl, 1990): argumentation would be one of the most inclusive and elaborate scientific processes, in which models are put into action in order to give meaning to the world.

Didacticians of science Sibel Erduran and Marilar Jiménez-Aleixandre are amongst the most cogent advocates for the indispensability of school scientific argumentation:

> This competence is instrumental in the generation of knowledge about the natural world [and] plays a central role in the building of explanations, models and theories [...] as scientists use arguments to relate the evidence they select to the claims they reach through use of warrants and backings. [...] [A]rgumentation is a critically important discourse process in science, and [...] it should be promoted in the science classroom. (Jiménez-Aleixandre & Erduran, 2007, p. 4)

These authors propose that there are at least five intertwined "potential contributions" that arise from the introduction of argumentation in the science classrooms (cf., Jiménez-Aleixandre & Erduran, 2007, p. 5):

1. Through arguing, students could get access to the cognitive and meta-cognitive processes characterising expert performance and enabling modelling.
2. They could develop the kind of communicative competencies that foster critical thinking.
3. Argumentation could support students' achievement of scientific literacy and their empowerment to talk and write the language of science.
4. Students could be introduced into the epistemic practices pertaining to the scientific culture and they could consequently construct epistemic criteria for knowledge evaluation.
5. Argumentation could accompany the development of scientific reasoning among students, particularly the choice between theories or positions based on rational criteria.

The research group LIEC (Llenguatge i Ensenyament de les Ciències/Language and Science Teaching, at the Universitat Autònoma de Barcelona in Spain) defines argumentation – following Dutch authors inscribed in the so-called "pragma-dialectics" – as a social, intellectual, and verbal activity directed to support or rebut a claim. When arguing, in addition to the sheer content of the claim, its *purpose* and *recipients* are important: the "arguer" has to choose between explanations and to provide reasoned criteria to give validity to the most appropriate choices (Sanmartí, 2003). In order to be able to construct operative models about the natural world, students need, besides meaningfully learning the involved concepts, to be able to distinguish between different kinds of explanations and to apprehend criteria that enable critical evaluation. In the scientific community, such choice (usually referred to as "scientific judgment") occurs in a context of debate or controversy; in the classroom, argumentative dialogue is generally enacted through the presentation of opposing positions and the discussion of reasons and evidence supporting them. School scientific argumentation thus establishes a very specific and sophisticated kind of oral communication and of written text production.

In our own work, we identify to some extent the skills of explaining, justifying and arguing with one another, though some authors from the field of linguistics make sharper distinctions between them based on formal or pragmatic considerations (for instance, it is usually pointed out that arguing as a "rhetorical move" implies a strong will to make the audience *see* something in a particular way). In particular, we define school scientific argumentation as the production of a text in which a natural phenomenon is *subsumed* under a theoretical model by means of an analogical procedure (Adúriz-Bravo, 2011, 2014; Revel Chion & Adúriz-Bravo, 2014). Argumentation can therefore be considered as the "textual concretion" of a fully-deployed scientific explanation.

In a "complete" school scientific argumentation, we recognise the following overlapping dimensions, which we call *components*:

1. The *theoretical* component, meaning that there must be a scientific model (Giere, 1988) as a reference, allowing explanation of a phenomenon by its "similarity" to that model.
2. The *logical* component, meaning that arguments have a rich syntactic structure and can be formalised as reasoning patterns (for instance: deductive, abductive, analogical, relational, causal, functional…).
3. The *rhetorical* component, meaning that arguments have persuading and convincing as constitutive aims (Osborne, 2001).
4. The *pragmatic* component, meaning that arguments are situated in a particular communicational context that configures them and from which they take meaning.

We use this four-component analysis of school scientific argumentation with three purposes that support one another. In the first place, to guide the design and implementation of instructional activities aimed at teaching students how to argue (see Adúriz-Bravo, 2011). Then, to communicate to those students which are the key characteristics of good-quality argumentation. And finally, to perform evaluative analyses and give feedback on the texts they produce.

SUMMARY

In this chapter, we have examined the following key ideas:

1. Research on language in science teaching has consolidated in the last three decades, and it is undertaken from didactics of science and from other disciplinary approaches.
2. Language plays a fundamental role in science education due to its epistemic function of meaning-making. Such function is accomplished through the construction and re-signification of terms, the implementation of metaphors and analogies, and the utilisation of complex linguistic procedures such as explanation and argumentation.
3. Discourse (oral and written, in different genres and formats) permits teachers to look into students' science learning, and allows students to test their own understanding of scientific ideas.
4. Argumentation is a key act of speech in the development of science; thus, science teaching that seeks to convey clear messages on how science works should give prominence to this competency.

NOTES

[1] See Wellington and Osborne (2001) for a collection of references that attest the importance conceded to language in the science classrooms.

[2] See, for instance, Bruner (1966, 1990) and Vygotsky (1934/1962).

[3] This condition of being theoretical is independent of being "observable" or not in instrumental terms; it has to do with the conceptual "ladenness", and the hypothetical and inferential nature of these entities.

[4] See the literature compiled in Erduran and Jiménez-Aleixandre (2008).

[5] Jonathan Osborne (2001) has reviewed a number of educational definitions of scientific argumentation and examined their epistemological foundations.

[6] These abilities reflect high-order cognitive capacities but at the same time imply the production of very elaborate oral and written texts (Sanmartí, 2003).

FURTHER READINGS

Driver, R., Newton, P., & Osborne, J. F. (2000). Establishing the norms of scientific argumentation in classrooms. *Science Education, 84*(3), 287–312.

Duschl, R. A., & Osborne, J. (2002). Supporting and promoting argumentation discourse in science education. *Studies in Science Education, 38*, 39–72.

Erduran, S., & Jiménez-Aleixandre, M. P. (Eds.). (2008). *Argumentation in science education: Perspectives from classroom-based research.* Dordrecht: Springer.

Jiménez Aleixandre, M. P. (2010). *Diez ideas clave: Competencias en argumentación y uso de pruebas.* Barcelona: Graó.

Newton, P., Driver, R., & Osborne, J. (1999). The place of argumentation in the pedagogy of school science. *International Journal of Science Education, 21*(5), 553–576.

Ogborn, J., Kress, G., Martins, I., & McGillicuddy, K. (1996). *Explaining science in the classroom.* Buckingham: Open University Press.

Wellington, J., & Osborne, J. (2001). *Language and literacy in science education.* Buckingham: Open University Press.

REFERENCES

Adúriz-Bravo, A. (2011). Fostering model-based school scientific argumentation among prospective science teachers. *US-China Education Review, 8*(2), 718–723.

Adúriz-Bravo, A. (2014). Revisiting school scientific argumentation from the perspective of the history and philosophy of science. In M. R. Matthews (Ed.), *International handbook of research in history, philosophy and science teaching* (pp. 1443–1472). Dordrecht: Springer.

Bailer-Jones, D. M. (2000). Scientific models as metaphors. In F. Hallyn (Ed.), *Metaphor and analogy in the sciences* (pp. 181–198). Dordrecht: Kluwer Academic Publishers.

Bruner, J. (1966). *Toward a theory of instruction.* Cambridge, MA: Harvard University Press.

Bruner, J. (1990). *Acts of meaning.* Cambridge, MA: Harvard University Press.

Candela, A. (1995). Consensus construction as a collective task in Mexican science classes. *Anthropology and Educational Quarterly, 26*(special issue "Vygotsky's theory of human development: An international perspective"), 458–475.

Charolles, M. (1986). L'analyse des processus rédactionnels: Aspects linguistiques, psychologiques et didactiques. *Pratiques, 49*("Les activités rédactionnelles"), 3–21.

Driver, R., Asoko, H., Leach, J., & Scott, P. (1994). Constructing scientific knowledge in the classroom. *Educational Researcher, 23*, 5–12.

Duschl, R. A. (1990). *Restructuring science education: The importance of theories and their development.* New York, NY: Teachers College Press.

Erduran, S., & Jiménez-Aleixandre, M. P. (Eds.). (2008). *Argumentation in science education: Perspectives from classroom-based research.* Dordrecht: Springer.

Erduran, S., Simon, S., & Osborne, J. F. (2004). TAPping into argumentation: Developments in the application of Toulmin's argument pattern for studying science discourse. *Science Education, 88*(6), 915–933.

Evagorou, M., & Osborne, J. (2010). The role of language in the learning and teaching of science. In J. Osborne & J. Dillon (Eds.), *Good practice in science teaching: What research has to say* (2nd ed., pp. 135–157). New York, NY: Open University Press/McGraw-Hill.

Giere, R. N. (1988). *Explaining science: A cognitive approach.* Chicago, IL: The University of Chicago Press.

Harré, R. (2004). *Modeling: Gateway to the unknown.* Amsterdam: Elsevier.

Jiménez-Aleixandre, M., & Erduran, S. (2008). Argumentation in science education: An overview. In S. Erduran & M. Jiménez-Aleixandre (Eds.), *Argumentation in science education: Perspectives from classroom-based research* (pp. 3–27). Dordrecht: Springer.

Kelly, G., & Bazerman, C. (2003). How students argue scientific claims: A rhetorical-semantic analysis. *Applied Linguistics, 24*(1), 28–55.

Kelly, G., & Takao, A. (2002). Epistemic levels in argument: An analysis of university oceanography students' use of evidence in writing. *Science Education, 86*(3), 314–342.

Kuhn, D. (2010). Teaching and learning science as argument. *Science Education, 94*, 810–824.

Langer, J., & Applebee, A. (1987). *How writing shapes thinking: A study of teaching and learning.* Urbana, IL: National Council of Teachers of English.

Lemke, J. L. (1990). *Talking science: Language, learning, and values.* Norwood, MA: Ablex Publishing Corporation.

Lemke, J. L. (2001). Articulating communities: Sociocultural perspectives on science education. *Journal of Research in Science Teaching, 38*(3), 296–316.

Navarro, F., & Revel Chion, A. (2013). *Escribir para aprender: Disciplinas y escritura en la escuela secundaria.* Buenos Aires: Paidós.

Osborne, J. (2010). Arguing to learn in science: The role of collaborative, critical discourse. *Science, 328*, 463–466.

Osborne, J. F. (2001). Promoting argument in the science classroom: A rhetorical perspective. *Canadian Journal of Science, Mathematics, and Technology Education, 1*(3), 271–290.

Revel Chion, A., & Adúriz-Bravo, A. (2014). La argumentación científica escolar: Contribuciones a una alfabetización de calidad. *Pensamiento Americano, 7*(13), 113–122.

Sampson, V., & Clark, D. (2008). Assessment of the ways students generate arguments in science education: Current perspectives and recommendations for future directions. *Science Education, 92*(3), 447–472.

Sanmartí, N. (Ed.). (2003). *Aprende ciències tot aprenent a escriture ciència.* Barcelona: Edicions 62.

Sanmartí, N., Izquierdo, M., & García, P. (1999). Hablar y escribir: Una condición necesaria para aprender ciencias. *Infancia y Aprendizaje, 281*, 54–58.

Strange, R. L. (1986). *An investigation of the ability of sixth graders to write with sense of audience* (Dissertation). Indiana University, Bloomington, IN.

Sutton, C. (1996). Beliefs about science and beliefs about language. *International Journal of Science Education, 18*(1), 1–18.

Tynjälä, P., Mason, L., & Lonka, K. (2001). Writing as a learning tool: An introduction. In P. Tynjälä, L. Mason, & K. Lonka (Eds.), *Writing as a learning tool: Integrating theory and practice, "Studies in writing"* (Vol. 7, pp. 7–22). Dordrecht: Springer.

Vygotsky, L. S. (1962). *Thought and language.* Cambridge, MA: The MIT Press. (Russian original of 1934.)

Wellington, J., & Osborne, J. (2001). *Language and literacy in science education.* Buckingham: Open University Press.

SECTION III

THE SCIENCE CURRICULUM

INGO EILKS AND AVI HOFSTEIN

13. CURRICULUM DEVELOPMENT IN SCIENCE EDUCATION

INTRODUCTION

Science curriculum development can involve changes in what is taught (the content and its related applications), to whom (target audiences, namely the learners), and how (ways of teaching and learning, different instructional interventions). The chapter is concerned with the following key questions: *Why* develop the science curriculum? *How* and *by whom* is the developmental process initiated and sustained? *In which way* should it be developed? In addition, we deal in this chapter with issues related to justifying change in the current curriculum and about models that can be used to guide the process of curriculum development in science education.

DRIVING FACTORS FOR DEVELOPING THE SCIENCE CURRICULUM

Throughout the last 60 years, the goals and related objectives for science teaching and learning have undergone changes several times; often leading to reforms in the way the curriculum was taught and science was learned. The five key factors that have influenced a change in curriculum goals for teaching science are: the learners, the teachers, the content, the pedagogy of teaching and learning both in and out of school, and the assessment of students' achievement. In this chapter we will address the first three of them.

The Learner

A long tradition of research on learning and teaching science suggests that (NRC, 1996):

> Learners are goal-directed agents who actively seek information. They come to formal education with a range of prior knowledge, skills, and beliefs. In addition, they are directed by their concepts, interest, motivation, and attitudes that significantly influence what they notice about the environment and how they organize and interpret it. This, in turn, affects their abilities to remember, reason, solve problems, and acquire new knowledge. (p. 10)

K. S. Taber & B. Akpan (Eds.), Science Education, 169–181.

Studies indicate that affective (interest, motivation and attitudes), meta-cognitive, and socio-cultural aspects play an important role in the teaching-learning process (Linn, Songer, & Eylon, 1996). There is agreement among many science educators that the range (or repertoire) of learners' ideas and ways of making sense of the world should be a key factor in setting curricular goals and in developing teaching strategies and learning materials.

In the process of science learning, learners, either as individuals or as a group studying together, may grapple with a repertoire of ideas that are not necessarily consistent with each other. Science educators hold different opinions regarding the repertoire of learners' ideas. Some regard them as barriers to the process of learning and design strategies to eliminate them, while others regard the repertoire as an essential and useful resource enabling learners to build on their experience and intuitions in the means of conceptual development. Therefore the curricular goals, the teaching strategies, and the assessments differ in these approaches.

When reviewing the science curricula from the 1960s and early 1970s, one can see that at that time the main goal of science education was to give a limited portion of students a solid foundation of knowledge in the sciences. This foundation was thought to recruit and prepare these few students for future careers in science, engineering, medicine, and its related disciplines. Examples for such kind of curricula were, for example, PSSC (Physical Science Study Committee) in physics, BSCS (Biological Sciences Curriculum Study) in biology, or CHEMStudy in chemistry in the USA and the various Nuffield curricular projects in the UK (see e.g. Eilks, Rauch, Ralle, & Hofstein, 2013).

The results were that science curricula mainly focused on the learning of scientific facts and were structured analogous to science textbooks from the university. The focus was mainly on pure, not on applied science. Regarding teaching and learning school science, The American Association for the Advancement of Science in 1962 (based on Harms & Yager, 1981) summarized the goals of these curricular initiatives in the case of the USA as follows:

- Science education should present learners with a real picture of science, including theories and models;
- Science education should present an authentic picture of scientists and their method of research;
- Science education should present the nature of science (NOS);
- Science education should be structured and developed using the discipline approach (key concepts in each of the subjects).

About 20 years later, in the 1980s, there was a shift in many countries toward addressing the needs and abilities of all citizens. For example, an NSF sponsored project, *Project Synthesis* (Harms & Yager, 1981), which analyzed science curricula in previous years, led to a call to change the scope and goals for science teaching and learning, advocating that science education should:

- Include major concerns regarding science as a means of resolving current societal problems;
- Provide a means to attend to the personal needs of students; and
- Provide greater awareness of potential careers in science, technology, and its related fields.

These goals led, for example, to curriculum projects focusing on the interplay of Science, Technology, and Society (STS) around the world (Solomon & Aikenhead, 1994). These curricula were aligned with the developments in the late 1980s and early 1990s focusing a change in standards for new science curricula, i.e. the concept of *Scientific Literacy for all* students (Fensham, 1986; NRC, 1996; AAAS, 1993). New curricula had less focus on the preparation of individual students for their career in science and engineering. National science education standards worldwide started acknowledging that every future citizen needs a basic understanding of science and its related applications and ramifications. This re-orientation of the goals and objectives of science education led to intense debate about a potentially promising re-orientation and changed structures of the science curricula to fulfill the new set goals (Hofstein, Eilks, & Bybee, 2011).

The re-orientation of the curriculum towards science education *for all* became guiding educational policy in many countries. New standards reflected an expectation for science education to more thoroughly contribute to general educational objectives. The innovative work *Science for All Americans* (Rutherford & Ahlgren, 1989), and subsequent publications by the Project 2061, e.g., *Benchmarks for Science Literacy* (AAAS, 1993) and the *National Science Education Standards* (NRC, 1996) in the USA, influenced and directed similar national standards and policies in other countries such as the UK, Germany, or Israel (Hofstein et al., 2011). In parallel, the OECD (2006) in their framework for the *Program for International Student Assessment* (PISA) described the overriding target for any science education to allow all students to achieve scientific literacy in the means of:

> The capacity to use scientific knowledge, to identify questions and to draw evidence-based conclusions in order to understand and help make decisions about the natural world and the change made to it through human activity. (p. 3)

The Teacher

The role of the teacher in science education is described and detailed in section IV of this book. However, in this chapter we want to highlight the idea that very often the availability of science teachers, their ability to implement new science curricula in their schools, and their ability to cope with new scientific concepts influences the nature and content of the operated science curriculum.

One of the key factors regarding curriculum change are the teachers. In top-down approaches, teachers are generally reluctant to accept radical changes and often do not implement them in accordance with the rationale for the change suggested by the curriculum developers. Such changes may not be aligned with teachers' existing views and practices, and may require new knowledge, perhaps content knowledge (CK), its related pedagogical content knowledge (PCK), and as being part of its revised curricular knowledge

Important factors influencing teachers' response to change include among others: personal experiences, cultural norms (e.g., questioning behavior), the professional status of the teacher, the teacher's understanding of the proposed change and its rationale, and teachers' views on students' potential future career opportunities. What teachers think, believe, and know has an effect on their teaching and classroom behaviour. These factors are therefore important when it comes to effectively and successfully reforming the curriculum. Any educational reform and implementation of changed curricula can only be successful if teachers` beliefs, their prior- knowledge and attitudes are acknowledged and are taken into account seriously in the reform (Haney, Czerniak, & Lumpe, 1996).

The Content and its Organization

Scientific content, skills, or scientific practices to be learned constitute the major fabric of the curriculum. Criteria for choosing scientific key ideas may relate to: the importance of concepts within and across disciplines, the provision of key tools for understanding, investigating and problem-solving, enhancing interest, the meaning for life experiences and the connection to personal and societal concerns. In addition it should be teachable and learnable over multiple grades at increasing levels of depth and sophistication (e.g. "learning progressions"). Changes in conceptions about how topics should be organized have influenced curricular change. For example, 'context-based science' (e.g. in the PLON curriculum in the Netherlands and the Salters' projects in the UK) and 'knowledge for use' approaches depart significantly from the traditional 'structure of the discipline' approach that previously often underpinned science curriculum developments (Pilot & Bulte, 2006).

Aligning school science with contemporary scientific knowledge should also be considered in areas that change at a very rapid pace such as molecular biology or nano-sciences, as well as topics that are interdisciplinary in nature such as brain science and climate change. Changes of this kind in the fields of science and technology are the driving force behind many innovations and reforms in school science curricula.

Another central issue is the methodologies used for enhancing the acquisition of skills in science curricula. There is a consensus that skills should be developed in the context of content, and that in order to develop a generalizable skill (transfer), it must be studied explicitly and practiced in different topics. This is not only a question of the pedagogy, but also about approaching and organizing (or structuring) the content in the curriculum.

THE QUESTION OF THE RELEVANCE OF SCIENCE LEARNING

In general, modern science education curricula emphasize both the learning of scientific theories and knowledge, and in the same time gaining science-related general educational skills, like problem solving (Hofstein et al., 2011). Skills development for relevant science learning in the context of recognizing and understanding questions about everyday life is needed, for decisions which pupils currently have to make on personal and societal issues, and for future actions and career choices (Stuckey, Hofstein, Mamlok-Naaman, & Eilks, 2013).

In order to theoretically operate effectively in these different fields that justify relevant science education, we need to examine what is meant by "relevant" (Newton, 1988). The word 'relevance' is currently present in many debates about why so many students do not like or do not opt to learn science. They often perceive their science learning as being irrelevant to them.

It has been shown that students attend more readily to their science studies if the subject matter presented to them is perceived as useful and relevant, than if it appears remote (Johnstone, 1981). However, the term 'relevance' is not a clear cut theoretical construct. For example the *ROSE – Relevance of Science Education Study* (Schreiner & Sjoberg, 2004) uses the word relevance as a synonym for students' interest but does not really differentiate between the two terms.

Relevance in science education can have a broader meaning than just meeting students' interest (Stuckey et al., 2013). In an early approach towards understanding relevance with respect to education, Keller (1983) defined relevance as the students' perception of whether the content they are taught satisfies their personal needs, personal goals, or career aims. In this set of needs, one has to keep in mind that students' future needs, goals, and career aims might not be conscious to them at the time they are having science lessons in the compulsory phase of their school education. Therefore, the question of relevance is not an easy one. It is suggested that the issue of relevance is always related to a further set of questions, namely; relevant to whom, for what something should be considered to be relevant, or who is deciding about that.

Since the 1980s there were different suggestions for organizers regarding the question of relevance in science education (e.g. Newton, 1988; Harms & Yager, 1981). Among these ideas there are different aspects of potential relevance that can found in several papers. According to Stuckey et al. (2013), these aspects can be summed up in three dimensions of potential relevance of science education of which all three have an actual component (connected to the students' interest today) and a future component (of which the student might not be aware today), as well as an extrinsic to intrinsic dimension:

- *Relevance for the individual*: meeting students' curiosity and interest, giving them necessary and useful skills for coping in their everyday life today and in future, or contributing the students' intellectual skill development.
- *Relevance for a future profession*: offering orientation for future professions, preparation for further academic or vocational training, or opening formal career chances (e.g. by having sufficient courses and achievements for being allowed to study medicine).
- *Relevance for the society*: understanding the interdependence and interaction of science and society, developing skills for societal participation, or competencies in contributing society's development.

It is suggested that modern science curricula should provide a balanced consideration of these three dimensions to come up with a holistic innovation of the science curriculum to provide relevant science education (Stuckey et al., 2013).

MODELS FOR DEVELOPING SCIENCE CURRICULA

Ideally, a curriculum-development process should be a holistic, continuous, and long-term endeavor involving several components, and respecting the different dimensions of relevant science education often carried out in parallel. Key components include: initial setting of goals, analysis and selection of the topics aligned with official syllabi; diagnosis of students' ideas as well as analysis of the inherent characteristics of the science concepts; design of learning, teaching and assessment materials (e.g. crafting tasks, uses of representations and didactical aids); small scale implementation and teacher development cycles accompanied by research (teaching experiments). This process often leads to reconsideration of goals, the pedagogical resources and the teacher development activities. Advanced stages of the process can lead to large-scale implementation and evaluation studies.

Over the years, the need for changes in the curriculum has been raised by different stakeholders in science education, such as policy makers, scientists, science educators, curriculum developers, teachers, local initiators (e.g., a school, a school district, or schools networks), or parents. Pressure for change has also come from societal or socio-economic sources. In recent years, in many countries, curriculum change was often also initiated and influenced by national and international standards, large scale assessments, and frameworks that characterize desirable change and are prepared by national academies, ministries of education and other influential organizations. Examples of such initiatives include the *National Standards in Science Education* developed by the US National Research Council in 1996 and revised 2012 in the Next Generation Science Standards, and the *Benchmarks of Science for all Americans* arising from Project 2061. The suggested frameworks have been used for changing curricula and evaluating their effects. Teacher associations have been very influential in initiating curriculum change through the development of corresponding frameworks (e.g., the NSTA in the USA, the ASE in the UK, or the

ASTA in Australia). Another mechanism for initiating change has been through influential reports discussing goals, methods, and recommendations related to teaching and learning science, an example for such a report from the European Union is *Beyond 2000* by Millar and Osborne (1998).

Calls for change have led to two key models of curriculum development efforts that differ in their methods of design and implementation and in the constituents involved in the developmental process: a centre-periphery top-down model in which a central development group tries to influence those on the periphery; and a bottom-up model, responding to local needs through school-based (or teacher-based) curriculum development or where change is instigated and implemented by leading teachers and then adopted by others.

These two key models differ in the nature and magnitude of teacher involvement in the developmental process, in the activities of implementation, and in teachers' professional development. The change processes associated with each of these models differ also in the scope of curriculum adoption, in the relationship between the intended and implemented curriculum, in teacher ownership and ways of adaptation, and in the degree of sustainability. In both models, a major concern is how to prepare 'educative materials', namely materials that promote teacher professional growth in addition to student learning, and how to ensure effective implementation and sustainability.

CURRENT TRENDS IN CURRICULUM REFORM

The theory of situated cognition points out that developing the ability to apply the learned science theory only takes place if the learning process is embedded into the learner's life. Therefore it is better to start the science curriculum units from contexts that make sense to the learner (Figure 1). Science learning should start

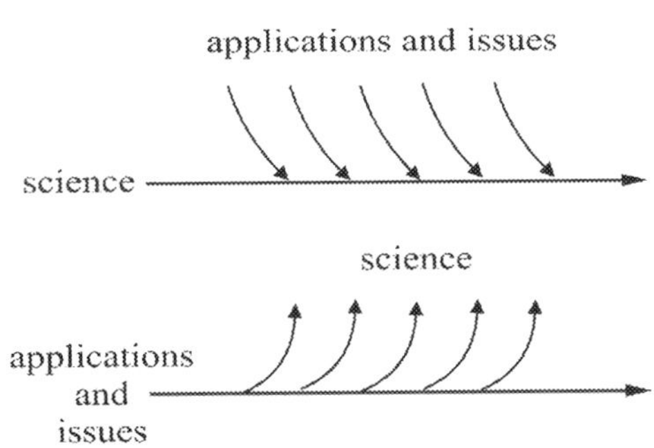

Figure 1. Traditional curricula driven by the structure of the discipline vs. curricula driven by applications and issues (based on Holman, 1987). Traditionally, science is taught and applications and issues come in for illustration. Context-led curricula are structured by different applications and issues that lead into science learning

from contexts and issues that are connected to the life of the students, their prior experiences, their interests, and therefore it should have a meaning to them. But, contexts also have to be chosen in such a way that they relate to the application of the learned knowledge. For the majority of the students who will not embark on a career as a scientist such a context will not originate from academic science. As such the everyday lives of students and the society which they live in have the potential to offer more meaningful contexts to the students.

Context-based Science Education

Since the 1980s' projects were launched in many countries with the goal of teaching science through a context-based approach. A common characteristic of these approaches was described by Bennett and Lubben (2006) as:

- The use of everyday contexts and applications of science as the starting point for developing scientific (in our case science) understanding,
- The adoption of student-centred approaches,
- Introducing and developing scientific ideas via a "spiral curriculum" (a curriculum where a scientific concept is dealt with repeatedly on different age levels leading to a more and more elaborated understanding), and
- Using a "need to know" approach.

When we use the word context today, it has many different educational meanings and connotations (Gilbert, 2006; Eilks et al., 2013):

- *Context as a direct application of concepts*: An application is operated to illustrate a science concept's use and significance. Topics are chosen from the presumed personal/social everyday life of the students to which the concepts of science are taught as abstractions. The concepts are then applied so that the students understand the applicability of the concept. This approach is strictly about how the concepts are used in the applications, almost as an afterthought, to the end of the theoretical treatment of concepts and often without a consideration of their cultural significance. As a post-hoc illustration, it is only an attempt to give meaning to a concept after it has been learnt and is therefore hardly meets the idea of situated learning.
- *Context as reciprocity between concepts and applications*: In this approach, applying contexts affects the meaning attributed to the concepts. Viewing concepts from different perspectives (the scientist, the engineer, the politician) implies different meanings for one concept. This model provides a better basis for context-based science education than the first one, although there is no obvious need for students to value the setting as the social, spatial, or temporal framework for a community of practice. But the behavioural environment may be of higher quality, dependent on the teacher's understanding of the setting being used. The risk is that students do not see the relationship between a certain problem and why

they should use some science to deal with it, because the context of an expert does not automatically become a context of the learner.

- *Context provided as personal mental activity*: A specific person fixed in time and space who was seeking to explain a specific topic using science is employed as context for learning science. The model seems to be of greatest value when applied to cases of recent major events in science. But, the use of this kind of events in science will only be successful if students see the value of it. This is not always the case if the major events are historic, and as such took place long ago and have less meaning to the student. Also the chance for students to become actively involved is limited and the social dimension, through interaction within a community of practice, is missing.
- *Context as a social circumstance*: The social dimension of a context is put in focus as a cultural entity in society. This kind of context considers the importance of the context to the life of communities within society. Here, meaning-making can take place from two different perspectives, from a context as social surrounding or by a context as social activity. In science education, within this interpretation the context becomes intrinsic to student learning and fits most the ideas from situated learning and activity theory.

But, when trying to connect the science curriculum along meaningful contexts, one has to be aware that there is not only a discrepancy between the learner and the scientist, it is sometimes also a gap between the learner and the teacher. Not every context considered by a teacher as being meaningful will necessarily work. A meaningful context for the teacher does not always signify that it is also meaningful or relevant to the student.

Societal Driven Science Education

A more thorough approach in context-based science education is subsumed under the term of Socio-Scientific Issues (SSI)-based science education. SSI approaches focus a specific orientation of potential contexts for science education, namely societal issues and controversial scientific concerns. The idea for promoting more learning about the interrelatedness of science, technology and society in the means of SSI-based education has also its roots in the 1980s. Different acronyms were used and operated into whole curricula from that time. Examples are Science-Technology-Society (STS) from Canada and the US, Science And Technology In Society (SATIS) from the UK, or Scientific and Technological Literacy for All (STL) in the framework of the UNESCO project 2000+ (Eilks et al., 2013).

This more extended focus can be justified by applying *Activity Theory* (Holbrook & Rannikmae, 2007) or Allgemeinbildung (Hofstein et al., 2011) to science education. Activity Theory deals with the relationship of knowledge and learning with their use for societal practices. This link can be described as *"interlinking of knowledge and social practice through establishing a need (relevant in the eyes of students), identifying the motives (wanting to solve scientific problems and make socio-scientific decisions) leading to activity constituted by actions (learning in school towards becoming a scientifically literate, responsible citizen)"* (Holbrook & Rannikmae, 2007, p. 1353). Allgemeinbildung is a central European tradition of defining any goals of education with a view to the individual and its living within society (Hofstein et al., 2011). Today, Allgemeinbildung is seen as the ability to recognize and follow one's own interests and to be able to participate within a democratic society as a responsible citizen – and this is also more and more commonly seen as an important aim of science education.

Based on this understanding, SSI-oriented science education is more than solely a specific form of context-based science learning. Coming from the interplay of science, technology and society in recent years, e.g. Sadler and Zeidler (2009) in the US, or Marks and Eilks (2009) in Germany plead for more thorough developing STS curricula beyond using STS contexts to contextualize the learning of science. A step further is the thorough orientation of science lessons along with socio-scientific issues in order to promote general educational skills in communication and decision making as a contribution to participatory learning. Participatory learning means preparing students for a self-determined, responsible life and participation in a democratic society.

For Sadler (2004), the most fruitful settings for this kind of science teaching are those,

> which encourage personal connections between students and the issues discussed, explicitly address the value of justifying claims and expose the importance of attending to contradictory opinions. (p. 253)

For selecting corresponding issues with potential for participatory learning Marks and Eilks (2009), operated later by Stolz, Witteck, Marks and Eilks (2013), suggested five criteria: authenticity, relevance, being undetermined in a societal respect, potential for open discussion, and connection to a question of science and technology (Table 1).

SUMMARY

Curriculum development is an ongoing, dynamic process. It undergoes trends and is influenced by many stakeholders in the education field. Currently, a more thorough orientation on meaningful contexts and socio-scientific issues are suggested to reveal the full potential of science education to contribute to the general goals of education and school learning (Eilks et al., 2013). However, any curriculum development needs to be seen and reflected in the foreground of the cultural and socio-economic environments of the corresponding local or national educational system where it is to

Table 1. Criteria for selecting the most powerful socio-scientific issues for science learning and potential proofs by Eilks et al. (2013)

Criterion	*Explanation*	*Methodology for evaluating against the criterion*
Authenticity	The issue is authentic because it is – in fact – discussed in society.	It is checked for whether the issue actually is discussed in everyday life media (newspapers, magazines, TV, advertisings, …)?
Relevance	The issue is relevant, because societal decisions on the issue will have direct impact on students' life, today or in future.	Scenarios are outlined and reflected upon regarding the impact specific societal decisions will have on how the individual could potentially act, e.g. as a consumer.
Evaluation undetermined in a socio-scientific respect	The societal evaluation is undetermined, it allows for different points of view.	The public debate is analysed to whether there are – in fact – different, controversial points of view outlined (by lobbyists, media, politicians, …).
Allows for open discussions	The issues can be openly discussed.	Thought experiments are conducted in order to consider whether expressing different points of view will harm the feelings of persons and groups because of their socio-economic background or cultural and ethical concerns.
Deals with questions from science and technology	The issues center around scientific and technological questions, for which the understanding of science and technology is fundamental.	The discourse in the media is analyzed to examine whether basic concepts of science and technology are touched or used for argumentation – explicit or implicit.

be imbedded. An issue and development suggested in one environment might inspire others, but does not necessarily fit it.

There are still many open questions that require further research concerning the ways to enhance the development of useful practical and research-based knowledge relevant to curriculum development in specific topics, such as: How can one communicate detailed knowledge about teaching and learning sequences? How can one encapsulate and conceptualize practical knowledge of teachers? How can one develop cumulative research-based knowledge on the development of learning and teaching of specific topics and concepts? How is it possible to recognize the cultural and socio-economic issues of different educational environments for effective development and implementation of science curricula?

FURTHER READING

Eilks, I. (2015). On the transformation of research on teaching and learning about the sub-micro world in chemistry education into feasible classroom practice. *LUMAT: Research and Practice in Math, Science and Technology Education, 3*(3), 269–284. Retrieved from http://www.luma.fi/lumat-en/

Eilks, I., & Feierabend, T. (2013). Developing the curriculum by Participatory Action Research – An interdisciplinary project on climate change. In T. Plomp & N. Nieveen (Eds.), *Educational design research: Introduction and illustrative cases* (pp. 321–338). Enschede: SLO. Retrieved from http://international.slo.nl/edr/

Eilks, I., Rauch, F., Ralle, B., & Hofstein, A. (2013). How to allocate the chemistry curriculum between science and society. In I. Eilks & A. Hofstein (eds.), *Teaching chemistry – A studybook* (pp. 1–36). Rotterdam: Sense Publishers. Retrieved from http://www.sensepublishers.com

Fraser, B. J., Tobin, K. G., & McRobbie, C. J. (Eds.). (2012). *Second international handbook of science education* (Section VI (Curriculum and Reform), Vol. 2). Dordrecht: Springer.

Gunstone, R. (Ed.). (2015). *Encyclopedia of science education* (Section on curriculum). Dordrecht: Springer.

REFERENCES

AAAS – American Association for the Advancement of Science. (1993). *Benchmarks for science literacy.* New York, NY: Oxford University Press.

Bennett, J., & Lubben, F. (2006). Context-based chemistry: The salters approach. *International Journal of Science Education, 28*, 999–1015.

Eilks, I., Ralle, B., Rauch, F., & Hofstein, A. (2013). How to balance the chemistry curriculum between science and society. In I. Eilks & A. Hofstein (Eds.), *Teaching chemistry – A studybook* (pp. 1–36). Rotterdam: Sense Publishers.

Fensham, P. J. (1988). Familiar but different: Some dilemmas and new direction in science education. In P. J. Fensham (Ed.), *Development and dilemmas in science education* (pp. 1–26). London: Falmer.

Gilbert, J. K. (2006). On the nature of "context" in chemical education. *International Journal of Science Education, 28*, 957–976.

Haney, J. J., Czerniak, C. M., & Lumpe, A. T. (1996). Teacher beliefs and intentions regarding the implementation of science education reform strands. *Journal of Research in Science Teaching, 33*, 971–993.

Harms, N., & Yager, R. E. (1981). *What Research says to the science teacher* (Vol. 3). Washington, DC: NSTA.

Hofstein, A., Eilks, I., & Bybee, R. (2011). Societal issues and their importance for contemporary science education: A pedagogical justification and the state of the art in Israel, Germany and the USA. *International Journal of Science and Mathematics Education, 9*, 1459–1483.

Holbrook, J., & Rannikmäe, M. (2007). The nature of science education for enhancing scientific literacy. *International Journal of Science Education, 29*, 1347–1362.

Holman, J. (1987). Resources or courses? Contrasting approaches to the introduction of industry and technology to the secondary curriculum. *School Science Review, 68*, 432–437.

Johnstone, A. H. (1981). Chemical education research-facts, findings and consequences. *Chemistry in Britain, 17*, 130–135.

Keller, J. M. (1987). Development and use of the ARCS model of instructional design. *Journal of Instructional Development, 10*(3), 2–10.

Linn, M. C., Songer, N. B., & Eylon, B. (1996). Shifts and convergences in science learning and instruction. In D. C. Berliner & R. C. Calfee (Eds.), *scope* (pp. 438–490). Mahwah, NJ: Lawrence Erlbaum.

Marks, R., & Eilks, I. (2009). Promoting scientific literacy using a socio-critical and problem-oriented approach to chemistry teaching: Concept, examples, experiences. *International Journal of Environmental and Science Education, 4*, 231–245.

Millar, R., & Osborne, J. (1998). *Beyond 2000: Science education for the future*. London: King's College.
Newton, D. P. (1988b). *Making science education relevant*. London: Kogan Page.
National Research Council (NRC). (1996). *National science education standards*. Washington, DC: National Academy Press.
OECD. (2006). *Assessing scientific, reading and mathematical literacy.* Paris: OECD.
Pilot, A., & Bulte, A. M. W. (2006). Special issue: Context based chemistry education. *International Journal of Science Education, 28*, 953–1112.
Rutherford, F. J., & Ahlgren, A. (1991). *Science for all Americans: The project 2061*. New York, NY: Oxford University Press.
Sadler, T. D. (2004). Informal reasoning regarding socioscientific issues: A critical review of research. *Journal of Research Science Teaching, 41*, 513–536.
Sadler, T. D., & Zeidler, D. (2009). Scientific literacy, PISA, and socioscientific discourse: Assessment for progressive aims of science education. *Journal of Research in Science Teaching, 46*, 909–921.
Schreiner, C., & Sjøberg, S. (2004). *Sowing the seeds of ROSE. Background, rationale, questionnaire development and data collection for ROSE (The Relevance of Science Education): A comparative study of students' views of science and science education* (Acta Didactica 4/2004). Oslo: University of Oslo.
Solomon, J., & Aikenhead, G. (Eds.). (1994). *STS education: International perspectives on reform.* New York, NY: Teachers College Press.
Stolz, M., Witteck, T., Marks, R., & Eilks, I. (2013). Reflecting socio-scientific issues for science education coming from the case of curriculum development on doping in chemistry education. *Eurasia Journal of Mathematics, Science and Technological Education, 9*, 273–282.
Stuckey, M., Mamlok-Naaman, R., Hofstein, A., & Eilks, I. (2013). The meaning of 'relevance' in science education and its implications for the science curriculum. *Studies in Science Education, 49*, 1–34.

BEN AKPAN

14. SCIENCE CURRICULUM DEVELOPMENT INITIATIVES

INTRODUCTION

This chapter explores the various curriculum development efforts from the post-sputnik1 era to present. Sputnik1, according to Wikipedia, was the first artificial Earth satellite (measuring 58 cm in diameter) launched by the then Soviet Union into an elliptical low Earth orbit on 4 October, 1957. The chapter is focused on developments in the United States of America and the United Kingdom as typical of efforts in many countries. The review begins with post-sputnik reforms in both US and UK, before turning to consider the situation in Nigeria to demonstrate what occurred in the developing world. This is followed by a comparison of traditional and new curricula, and a review of recent science curriculum projects. The chapter ends with consideration of those factors that determine curriculum change.

POST-SPUTNIK CURRICULUM DEVELOPMENT EFFORTS

Globally, the science curriculum has always undergone changes. However, a major milestone was reached in the 1960s and 1970s. While the causes of the intense science curriculum development efforts differed from country to country, in the United States of America it is contended that the launch of the satellite Sputnik1 in 1957 by the USSR provided the much needed catalyst. In this section, an attempt is made to review some of the curriculum projects as examples of what occurred during the two-decade post-sputnik era, spanning the 1960s and '70s.

Curriculum Projects in the United States

Following the launch of Sputnik on 4 October, 1957, there were agitations in the United States for the reorganization of science curricula. These resulted in the following curriculum projects:

Physical Science Study Committee (PSSC): This project commenced in 1956 and was based in Cambridge Massachusetts. Through the development of a textbook, laboratory manual, teachers' guide and other teaching aids, PSSC aimed to modernise physics at the high school level. The resulting PSSC course placed emphasis on

K. S. Taber & B. Akpan (Eds.), Science Education, 183–198.

laboratory work, treated fewer topics at greater depth, and placed a high premium on basic physics when compared to the traditional physics course.

Biological Sciences Curriculum Study (BSCS): The Biological Sciences Curriculum Study was based in Colorado Springs, Colorado. It commenced in 1960. BSCS developed new programmes in Biology with emphasis on concepts and investigations in contrast with the supply of facts through lecture method of teaching. Additional programmed materials and inquiry slides were developed as well as materials for the intellectually disabled. Now in its 6th decade, BSCS has continued to develop several inquiry-based curricula including textbooks.

Chemical Education Materials Study (CHEM Study): Established in 1959 by the American Chemical Society which is based in Washington DC, CHEM Study sought to groom high school students for college chemistry, among other objectives. It engaged in the production of chemistry textbook, manual for laboratory work, achievement tests, and programmed instruction guides.

Conceptually Oriented Programme in Elementary Science (COPES): COPES was established in 1965 at New York University. It developed K – 6 programme using the following themes which were taught in spiral form according to the grade of the children: the structural units of the universe, interaction and change, conservation of energy, degradation of energy, and statistical view of nature.

Science Curriculum Improvement Study (SCIS): SCIS commenced in 1965 at the University of California at Berkeley. It adopted an investigative approach to the study of several topics through the use of kits. Each kit was adequate for use in a class of 32.

Elementary School Science Project (ESSP): The ESSP revised and updated the astronomy programme in line with knowledge at the time. It began in 1960 at the University of Illinois.

Elementary Science Study (ESS): ESS was based at the Education Development Center, Newton, Massachusetts. It began in 1960 and sought to develop meaningful science materials for use by children in grades K-9. With more than 50 units, it was designed to enhance children's curiosity about nature.

Science – A Process Approach (SAPA): Established in 1962 by the American Association for the Advancement of Science, SAPA worked on the notion by Robert Gagne' that learning should proceed from simple to complex ideas. It placed considerable emphasis on the development of skills of observation, classification, measuring, communicating, inferring, predicting, and space-time relationships at the primary grades. At the intermediate grade level, emphasis was on hypothesizing, controlling variables, interpreting data, defining operationally, and experimentation.

Minnesota Mathematics and Science Technology Project (MINNEMAST): MINNEMAST started in 1961 at the University of Minnesota. It developed coordinated science and mathematics programme for K-6 grades. The programme emphasized the processes of science. In addition, materials were designed for in-service training as part of the programme.

Curriculum Projects in the UK

In the United Kingdom, many curriculum projects also sprang up. Among the projects were the Scottish Integrated Science, Nuffield Combined Science, Nuffield Secondary Science, Nuffield O Level Science, Nuffield A Level Science, and Scottish Science.

Scottish Integrated Science: This took off in 1966 with a syllabus for the first two years of secondary education. Textbooks were also published in line with the syllabus. Lucas and Chisman (1973) report that the integrated science programme placed emphasis on areas of science that were pivotal to general education since the first two years of secondary education made it mandatory for all students to study science. The programme enabled students to be familiar with scientific language, experimental procedures, scientific apparatus, and how to draw conclusions from experiments.

Nuffield Combined Science: The Nuffield Combined Science commenced in 1966. It was for children in their first two years of secondary education. The content of the project featured the world around us, looking for patterns, how living things began, air, electricity, water, small things, earth, insects, and energy. In terms of teaching methodology, Combined Science relied on students' first hand interaction through laboratory investigation. Experiments were also recommended for performance outside the laboratory,

Nuffield Secondary Science: While both the Scottish Integrated Science and Nuffield Combined Science were sequentially written in terms of content, the Nuffield Secondary Science was meant to be a pool from which one could choose suitable content to obtain a course of study. The target population for secondary science was for those not likely to take examinations in science at the General Certificate of Education Ordinary Level. These were students aged 13 to 16 in the lower 75 percent ability range. Nuffield Secondary science sought to facilitate the scientific processes of observation, deduction, hypothesizing, and design of experiments. Emphasis was placed on contents which were considered pivotal to the realization of these objectives. The theme for secondary science were interdependence of living things, continuity of life, biology of man, harnessing energy, extension of sense perception, movement, using materials, and the earth and its place in the universe. The programme was supported with visual aids, background readers, apparatus guide, and teachers' guide.

Nuffield O level Science: On the other hand, the Nuffield O Level science programme in physics, chemistry, and biology which targeted the top 25–30 percent of ability groups in grades 7–11 was actually the first curriculum project to be supported by the Nuffield Foundation. The duration of the course was five years and led to the General Certificate of Education Ordinary level. The teaching methodology emphasized guided inquiry. In Biology, the following content was adopted: Introducing living things (Year 1), Life and living processes (Year 2), the maintenance of life (Year 3), living things in action (Year 4), and the perpetuation of life (Year 5). The chemistry programme was designed for those who might or might not go on to major in chemistry at A Level or University. It had the following themes: exploration of materials, the ideas that chemists use, and a course of options. In physics, the emphasis was on pupil investigation. It treated the following topics: Year 1: materials and molecules, making a microbalance, rough measurement, looking for a law of levers, investigation of springs, air pressure and molecules, measurement of a molecule, energy; Year 2: forces, electric circuits, electric currents, more forces, energy, heat, heat transfer; Year 3: waves, optics, motion and force, molecules in motion, electromagnetism, cells and voltage, electrostatics, a fruitful theory, (teaching) the use of theory; Year 4: physical basis of Newtonian mechanics, kinetic theory of gases, universal conservation of energy, power, electricity, electrons; Year 5: motion in orbit, electrons in orbits, the grand theory, oscillations and waves, interference of light waves, radioactivity, waves and particles. The physics courses were supported with Teachers Guide, a Guide to Experiments, and a Questions Book.

Nuffield A Level Science: The Nuffield A'Level Science project took off in 1966. It was established for Grades 12 and 13 as a two-year post O'level course in Biological Science, Chemistry, Physics, and Physical Science.

Scottish Science: The Scottish Science project was established by the Scottish Education Department. The project which covered physics, chemistry and biology comprised of schemes of work for the four years leading to the Scottish Certificate of Education, Ordinary Grade. An additional section provided for courses leading to the Higher Grade Certificate. The physics and chemistry schemes of work were operational in 1962 while those for biology followed a few years later.

Situation in Nigeria

In Nigeria, the Science Teachers Association of Nigeria (STAN) prompted by the curriculum reform movements in the United States and the United Kingdom, and with financial support from the Ford Foundation through the Comparative Education Study and Adaptation Centre (CESAC) and the Centre for Curriculum Development Overseas (CREDO), produced an integrated science course for use in the first two years of secondary education for the first time. A flow chart of the outline content is shown in Figure 1 (STAN, 1970). In liaison with the West African Examinations Council (WAEC) STAN also produced new syllabuses for chemistry, biology, and physics at the secondary school level.

Traditional Versus New Curricula

Traditional (pre-Sputnik) and new curricula (post-Sputnik) differed in various ways. These differences are enumerated in Table 1.

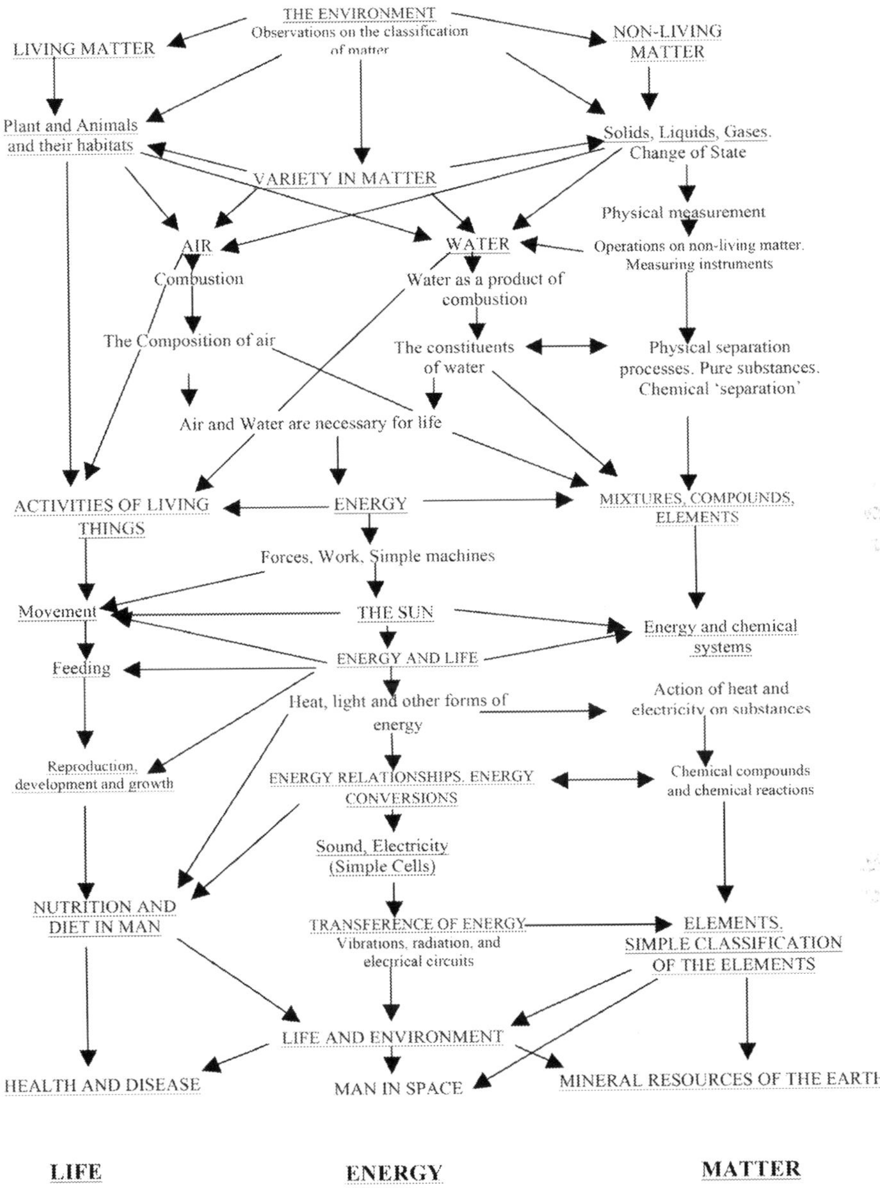

Figure 1. Integrated Science: A course for the first two years of Nigerian Secondary Schools – A flow chart of the outline content

Table 1. Differences between traditional (Pre-Sputnik) and new (Post-Sputnik) curricula

Traditional Curricula	*New Curricula*
Qualitative Techniques dominant	Emphasis on quantitative techniques such as data gathering, and drawing of graphs
Emphasis on applications and technology	Premium placed on abstractions, and theory
Teaching method is teacher-centred, lecture-demonstrations	Teaching is learner centred
Less emphasis on basic science	Emphasis on basic science
Less emphasis on laboratory work; emphasis on verification	Much emphasis on laboratory work; premium on investigation through experimentation

RECENT SCIENCE CURRICULUM PROJECTS

In this section, we review some projects and efforts in the USA and the UK aimed at modernizing the science curriculum. These efforts are typical of the global trend in science curriculum development. Just as the launch of the sputnik catalysed curriculum efforts in 1960 and 1970s, much of the recent science curriculum development efforts in the United States is traceable to the 1983 report of American President Ronald Reagan's National Commission on Excellence in Education entitled *'A Nation at Risk: The Imperative for Educational Reform'.* Thought to be one of the most important reform publications in the USA, *A Nation at Risk* warned that the educational foundations of the American society were being eroded and that this posed a serious threat to America's future as a nation and a people. The report led to several efforts to reform American science education. One of such efforts is Project 2061.

Project 2061

Project 2061 was initiated by the American Association for the Advancement of Science (AAAS). The project takes a long-term view of science education reform. It began in 1985 the year Halley's Comet was last seen from the Earth. The project is of the view that the young people then in 1985 will as adults greatly influence what life on Earth will be like in 2061, the year Halley's Comet is expected to return. The overarching goal of the project is to help all Americans become literate in science, mathematics, and technology. In furtherance of this goal, the following publications have served as reform tools:

Science for All Americans. Science for All Americans was produced by expert panels of scientists, technologists, and mathematicians. The book defines literacy and provides some principles for learning and teaching effectively. It outlines what all students should know and be able to do by the time they leave high school. It was published in 1990.

Benchmarks for Science Literacy. Benchmarks for Science Literacy was established in 1993. It provides what all students should know and be able to do in science, technology, and mathematics by the end of grades 2,5,8, and 12 thus helping science educators decide what to include in a core curriculum as well as when and why to teach it.

Atlas of Science Literacy. This is a joint publication between AAAS and the National Science Teachers Association (NSTA). The book comprises of conceptual strand maps that are organized into the same chapters as in *Science for all Americans* and *Benchmarks for Science Literacy*. The strand maps demonstrate how students' understanding of the ideas and skills that give rise to literacy in science, technology, and mathematics change as time progresses.

Designs for Science Literacy. Designs for Science Literacy offers different options for restructuring time (reallocating time for greater focus on facts, principles, etc), instructional strategies, and content thereby indicating how to approach the science curriculum design challenge. *Designs* has a companion CD-ROM – *Designs on Disk* – which provides examples of the kinds of functions a computer-based curriculum design system could perform.

Resources for Science Literacy. Resources for Science Literacy is predicated on the need for continual support for science literacy goals of Project 2061. One form of support being provided is the creation of a CD-ROM tool, *Resources for Science Literacy: Professional Development*, which contains six components. This is useful in pre-service as well as in-service training.

Blueprints for Reform. Blueprints for reform targets three themes: the foundation, school context, and support structure. Based on these themes, Blueprints provides papers anchored by experts on aspects of the education system that require changes if the goal of Project 2061 is to be attained.

Scope, Sequence, and Coordination of Secondary Science

The scope, sequence, and coordination (SS&C) project was instituted by the National Science Teachers Association (NSTA) in 1992 with its then Executive Director Bill Aldridge as founder. Aldridge (1992) contended that an analysis of the existing scope, sequence, and coordination of science subject matter in American secondary schools revealed serious deficiencies which if tackled could increase the numbers of children who study science. According to him, the changes were suggested partly by analyses of science courses in some countries especially the USSR and China. In comparing the USA with USSR and China, Aldridge stated:

> The secret of both countries' success in teaching science is multifaceted. First, they offer the subjects over four or five years instead of just one year, sometimes devoting only one or two hours a week on a subject for one or more of those years. Second, they sequence the content and approach starting with the concrete, phenomenological, and descriptive in the early years to the semi-quantitative and empirical in the middle years, then the theoretical and abstract in the later years. ...In the USSR and the People's Republic of China, each high school science course includes many practical applications, so that young people understand why they need to learn science. Finally, these countries closely coordinate the content from one subject to another...Here is an example of how the USSR coordinates its science. During a certain number of class hours in one year, students in biology class explore the human heart and the circulatory system. In chemistry class, the same students investigate chemical reactions that involve oxygen and, in particular, such metabolic reactions as oxygen transport by hemoglobin. At the same time in physics class they study the kinematics and dynamics of fluid flow. (pp. 4–5)

The SS&C project had four publications to support it: Volume I- *The Content Core*, Volume II- *Relevant Research*; *An Addendum to the Content Core Based on the 1994 Draft National Science Education Standards*, and *A High School Framework for Science Education.*

National Science Education Standards

The National Science Education Standards were designed by the United States National Research Council in 1996. They are a set of guidelines which together present a vision of science literacy and provide criteria for science education that will assure the attainment of that vision such as encouraging ideas that will facilitate coordination, consistency and coherence of science programmes. The standards are organized into six categories:

Science Teaching Standards. The planning of inquiry-based programmes, the actions taken to guide and facilitate student learning, the assessments made of teaching and student learning, the development of environments that enable students to learn science, the creation of communities of science learners, and the planning and development of the school science programme.

Professional Development Standards. The learning of science content through inquiry, the integration of knowledge about science with knowledge about learning, pedagogy, and students; the development of the understanding and ability for lifelong learning, and the coherence and integration of professional development programmes.

Assessment Standards. The consistency of assessments with decisions they are designed to inform, the assessment of both achievement and opportunity to learn science, the match between the technical quality of the data collected and the consequences of the actions taken on the basis of those data, the fairness of assessment practices, and the soundness of inferences made from assessments about student achievement and opportunity to learn.

Science Content Standards. Unifying concepts and processes in science, science as inquiry, physical science, life science, earth and space science, science and technology, science in personal and social perspective, and history and nature of science.

Science Education Programme Standards. The consistency of the science programme with the other standards and across grade levels; the inclusion of all content standards in a variety of curricula that are developmentally appropriate, interesting, relevant to students' lives; organized around inquiry, and connected with other school subjects; the coordination of the science programme with mathematics education, the provision of appropriate and sufficient resources to all students, the provision of equitable opportunities for all students to learn the standards; and the development of communities that encourage, support, and sustain teachers.

Science Education System Standards. The congruency of policies that influence science education with the teaching, professional development, assessment, content, and programme standards; the coordination of science education policies within and across agencies, institutions, and organisations; the continuity of science education policies over time, the provision of resources to support science education policies, the possible unanticipated effects of policies on science education; and the responsibility of individuals to achieve the new vision of science education portrayed in the standards.

The Next Generation Science Standards

By 2011, a decade and half following the publication of the National Science Education Standards, there were renewed agitations for America to step up efforts at improving on economic competitive edge, achievement of students in international competitions, and science literacy in the populace. The Next Generation Science Standards were designed and released in 2013 to assure attainment of these goals by the National Research Council (NRC) in consort with the National Science Teachers Association, the American Association for the Advancement of Science, and a non-profit organization – Achieve. The NGSS are arranged in a coherent fashion across grades following NRC's *Framework for K-12 Science Education.* Each standard comprises of three dimensions – practices, which refer to the methods used by scientists and engineers; crosscutting concepts having application across as well as links to different areas of science; and disciplinary core ideas consisting of specific content and subject areas.

In its position statement on the NGSS, the NSTA has recommended the adoption and implementation of the standards to assure that all students are provided with access to high quality science education which should enable them to be well-informed citizens. According to the NSTA, NGSS has seven conceptual shifts: interconnected nature of science, performance expectations clarify what students need to know so do not prescribe curriculum, provides coherent science programme across all grades, focus on deeper understanding and application of content, science and engineering are integrated across grades, prepares students for college, career, and citizenship; and aligns with standards in English language, arts, and mathematics. Many states in the US have adopted the NGSS.

In the United Kingdom, there have been efforts from several organizations and agencies aimed at reforming the science curriculum. In what follows, we review the efforts by Nuffield Foundation and the Welcome Trust to intervene in science curriculum development through the promotion of seminars that provided the much needed avenue for participants to contribute towards this topical issue. First we turn to a report that emanated from series of seminars conducted at the instance of the Nuffield Foundation.

Beyond 2000 – Science Education for the Future

Beyond 2000. Science Education for the Future is a report of seminars/meetings supported by the Nuffield Foundation in the United Kingdom in 1997 and 1998. Four focal points were addressed: successes and failures of science education to date, science education needed by young people of the day, content and structure of a suitable model for a science curriculum of all young people, and the problems and issues that would arise by the implementation of such a curriculum, and how these might be addressed (Millar & Osborne, 1998). The report noted the series of achievements in science education in England and Wales especially since the introduction of the National Curriculum in 1989. These included the fact that science had become a core subject in the curriculum of 5–16 year olds and the improved performances of pupils in the Third International Mathematics and Science Survey (TIMSS). The achievement notwithstanding, the report pointed to a multiplicity of problems on the science curriculum in use at the time – lack of relevance, lack of properly articulated set of aims, assessment being unrelated to situations where science knowledge may be required in later life, unhelpful separation of science and technology, and an insufficient emphasis on scientific issues that have bearing on modern life. The report made the following recommendations:

i) The science curriculum from 5 to 16 should be seen primarily as a course to enhance general scientific literacy by providing sufficient scientific knowledge and understanding that assures students are able, for example, to read and understand simple scientific articles and related programmes.

ii) At key stage 4 (ages 14–16), the structure of the science curriculum needs to differentiate more explicitly between those elements designed to enhance scientific literacy and those designed as the early stages of a specialist training in science, so that the requirement for the latter does not come to distort the former.

iii) The science curriculum needs to contain a clear statement of its aims – making clear why we consider it valuable for all young people to study science, and what we would wish them to gain from experience – these aims need to be clear, and easily understood by teachers, pupils, and parents; they also need to be realistic and achievable (see Box 1)

AIMS OF THE SCIENCE CURRICULUM

The purpose of science education, as a component of young people's whole educational experience, is to prepare them for a full and satisfying life in the world of the 21st century. More specifically, the science curriculum should:

- sustain and develop the curiosity of young people about the natural world around them, and build up their confidence in their ability to inquire into its behavior. It should seek to foster a sense of wonder, enthusiasm and interest in science so that young people feel confident and competent to engage with scientific and technical matters.
- help young people acquire a broad, general understanding of the important ideas and explanatory frameworks of science, and of the procedures of scientific inquiry, which have had a major impact on our material environment and on our culture in general, so that they can:
  - appreciate why these ideas are valued;
  - appreciate the underlying rationale for decisions (for example about diet, or medical treatment, or energy use) which they may wish, or be advised, to take in everyday contexts, both now and in later life;
  - be able to understand, and respond critically to, media reports of issues with a science component;
  - feel empowered to hold and express a personal point of view on issue with a science component which enter the arena of public debate, and perhaps to become actively involved in some of these;
  - acquire further knowledge when required, either for interest or for vocational purposes.

Box 1. Source: Millar & Osborne, 1988:12

iv) The curriculum needs to be presented clearly and simply, and its content needs to be seen to follow from the statement of aims (in Box 12.1) – scientific knowledge can best be presented in the curriculum as a number of key explanatory stories (Box 2), even as the curriculum should introduce young people to a number of important ideas about science.
v) Work should be undertaken to explore how aspects of technology and the applications of science currently omitted could be incorporated within a science curriculum designed to enhance science literacy – the cluster of sciences and technologies concerned with the transmission, storage, processing and replication of information provide the framework for understanding channels of communication in telephone lines, satellite systems, and fibre optic cables; they facilitate the understanding of the functioning of computers; yet they are almost wholly absent from the curriculum.
vi) The science curriculum should provide young people with an understanding of some key ideas-about-science, that is, ideas about the ways in which reliable knowledge of the natural world has been, and is being, obtained.
vii) The science curriculum should encourage the use of a wide variety of teaching methods and approaches-case studies of historical and current issues should be used to consolidate understanding of the explanatory stories and of key ideas-about-science, and to make it easier for teachers to match work to the needs and interests of learners.
viii) The assessment approaches used to report on pupils' performance should encourage teachers to focus on pupils' ability to understand and interpret scientific information, and to discuss controversial issues, as well as on their knowledge and understanding of scientific ideas.
ix) In the short term, the aims of the existing science National curriculum should be clearly stated with an indication of how the proposed content is seen as appropriate for achieving those aims.
x) In the medium to long term, a formal procedure should be established whereby innovative approaches in science education are trialled on a restricted scale in a representative range of schools for a fixed period – such innovations are then evaluated and the outcomes used to inform subsequent changes at national level.

Also in the UK, the Welcome Trust organized a high-level seminar in 2010 to examine the National Curriculum in science which was first introduced through the 1988 Education Reform Act. The objective of the seminar was to consider the lessons that could be learned from the history of the national curriculum in science so as to make informed decisions about its future. In terms of benefits, the group agreed that the national curriculum had led to a number of improvements – science education for all, raising of standards, increase in number of girls studying science, making science a core subject at primary level, and facilitating continuity and transferability of pupils. Even so, some drawbacks were observed – disempowerment of teachers, over- prescription of curriculum content, frequent changes to curriculum, and

From our point of view on the Earth, it seems that we are living on a flat stationary surface. However, imagine moving to a point in Space, well away from the Earth. Then we would see that it is roughly a sphere which is moving in two ways. First the Earth is spinning on an axis through its North and South Poles; this means that different parts of the Earth's surface point towards the Sun at different times, resulting in day and night. Second, it is also moving, roughly in a circle, round, the Sun, taking one year to make a complete orbit. The Earth is kept in its orbit by the gravitational force between the two masses of the sun and the Earth. Because the axis around which the Earth spins is tilted at an angle to the plane of its orbit, the relative lengths of day and night are different for the northern and southern hemispheres and, moreover, change as the Earth moves round its orbit. This is what causes the seasons.

In both our spinning and our orbital motion, we keep on going at a steady speed, unlike things here on Earth, because there is no friction to slow us down. We are not the only planet going round the Sun; there are others. Three of them (Mars, Venus and Mercury) are close to the sun like us. Then there are two really big ones (Jupiter and Saturn), very different from us and much further away. Finally there are the outer ones which are very much further away and really cold. Several of the planets, including the Earth, have moons which orbit around them.

Of the planets, the only one with life on it (so far as we know) is the Earth. It is possible that there is life on Mars and one of the moons of Jupiter, but we don't know. If we did find life there as well, it would make the possibility of other life elsewhere in the Universe much more likely.

Our planet is really quite unusual. Whilst most of the Universe consists of hydrogen and helium, we live on a tiny rocky planet made out of elements which together make up less than 2% of all the matter in the Universe. Moreover, we are just sufficiently far from the Sun for water to be a liquid on the majority of the surface. This has enabled life to begin. We are also big enough for there to be sufficient gravity to keep our atmosphere, unlike Mercury or the Moon.

Surprisingly, the Sun is a star – a fairly ordinary, middle-aged star half way through its lifetime and a wonderful example of a balanced nuclear fusion reaction. How do we know? Well firstly, this is the only mechanism that could possibly produce so much energy and, secondly, theoretical models based on this idea predict the behavior of the Sun quite accurately. The Sun looks bigger than all the other stars because it is much nearer. The Sun itself is just one star in a cluster of a hundred thousand million stars which we call a galaxy. You can see the cluster edge on in the night sky as a band of stars called the 'Milky Way'. There are hundreds of millions of galaxies and these are found in clusters as well. Distances to the stars are enormous – the nearest one would take four years to reach travelling at the speed of light, and the furthest known one is 12 billion years away. So our home, the Earth, is really just a tiny speck in an enormous Universe.

Box 2. An explanatory story: The earth and beyond
Source: Millar and Osborne, 1998:16

examination-focused teaching. The group had messages for any future review of the curriculum: base the aim of the curriculum on the contribution of science to the entire education of the child; core knowledge should be a balance of information, skills and concepts; ensure appropriate means of assessment of progress and achievement; new curriculum should build on strengths of existing curriculum; and teachers as well as other stakeholders must be supported to facilitate the implementation of any new science curriculum.

The Constant Thing in Science Curriculum Trends

If there is one thing that governs global science curricula, it is 'change'. It is instructive to note that successive generations although using a superior curriculum compared to those beforehand, are always advocating the modernization of the curriculum. Four decades ago, Hurd (1975) had this to say:

> A curriculum is needed that is oriented toward a period not yet lived, influenced by discoveries not yet made and beset with social problems not yet predicted. The need is for an education designed to meet change, to appreciate the process of change, and to influence the direction of change. (p. 22)

Although many strides have been taken and several curricula produced, yet forty years on, the world is still in search of a curriculum with the above description. So the search for a curriculum that will withstand the test of time has remained elusive and is destined to remain so at least in the foreseeable future. The reason is not far-fetched. The science curriculum has remained essentially dynamic, responding to changes in government policies, growth in scientific knowledge, thinking by teachers and associated groups, teaching strategies, technologies of teaching, physical world, and funding. Figure 2 is an illustration of this. So change remains the constant factor in science curriculum trends.

SUMMARY

In this chapter, we have discussed the following:

- Post-Sputnik Curriculum developments efforts such as the Physical Science Study Committee, Biological Sciences Curriculum Study, Chemical Education Materials Study, Conceptually Oriented Programme in Elementary Science, Science Curriculum Improvement Study, Elementary School Science Project, Elementary School Science Study, Science – A Process Approach, Minnesota Mathematics and Science Teaching Project, the Scottish Integrated Science, Nuffield Combined Science, Nuffield Secondary Science, Nuffield O Level Science, Nuffield A Level Science, Scottish Science, and the Nigerian Integrated Science Project.
- Differences between traditional (pre-Sputnik) and new (post-Sputnik) curricula.

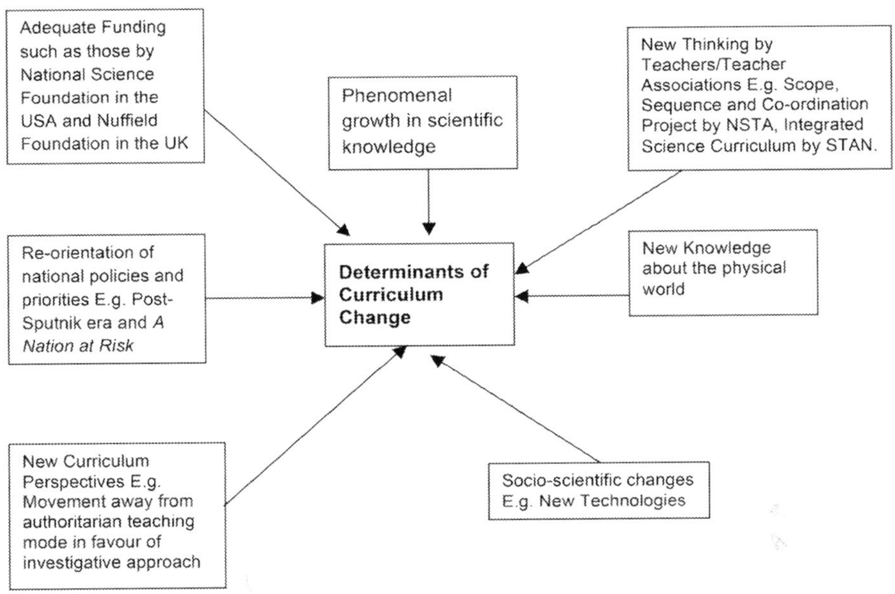

Figure 2. Determinants of curriculum change

- Recent Science Curriculum Projects – Project 2061 (Science for All Americans, Benchmarks for Science Literacy, Atlas of Science Literacy, Designs for Science Literacy, Resources for Science Literacy, Blueprints for Reform), Scope, Sequence, and Co-ordination of a Secondary Science, National Science Education Standards, The Next Generation Science Standards, and Beyond 2000: Science Education for the Future.
- Determinants of Curriculum Change

RECOMMENDED WEBSITES

National Research Council – www.nationalacademies.org/nrc/index.html
Project 2061 – www.aaas.org/program/project2061
Nuffield Foundation – www.nuffieldfoundation.org
Welcome Trust – www.welcome.ac.uk

REFERENCES

Akpan, B. B. (2015). The place of science education in Nigeria for global competitiveness. *Journal of the Science Teachers Association of Nigeria, 50*(1), 1–23.

Aldridge, B. G. (1992). Essential changes in secondary science: Scope, sequence, and co-ordination. In *Scope, sequence, and co-ordination of secondary school science* (Vol. 11). Washington, DC: National Science Teachers Association.

Hurd, P. D. (1975). Toward a theory of science education consistent with modern science. In E. Victor & M. S. Lerner (Eds.), *Readings in science education for the elementary school.* New York, NY: Macmillan Publishing Co.

Lucas, A. M., & Chisman, D. G. (1973). *A review of british science curriculum projects: Implications for curriculum developers.* Ohio: Center for Science and Mathematics Education, The Ohio State University, 244 Arps Hall, Columbus.

Millar, R., & Osborne, J. (Eds.). (1998). *Beyond 2000: Science education for the future*. London: King's College London.

Next Generation Science Standards. (n.d.). Retrieved May 12, 2015, from http://www.nextgenscience.org

STAN. (1970). *Curriculum Development Newsletter No 1: Integrated science*. Ibadan: Science Teachers Association of Nigeria.

RACHEL MAMLOK-NAAMAN

15. CURRICULUM IMPLEMENTATION IN SCIENCE EDUCATION

Rudduck and Kelly (1976) defined implementation as the process of moving from theory to practice – using ideas and materials in practice. Science educators and researchers suggest that in order to implement science curricula in upper secondary schools, several factors must be considered, e.g., the students' cognitive and affective aspects, the science content, cultural aspects, teachers' beliefs, school administration, preparation and professional development of science teachers and/ or guidance of teachers in schools, and the assessment of students' achievements. However, teachers are the key components in implementing any curriculum, and therefore, extensive, dynamic, and long-term professional development of science teachers should take place. This is thought to be one of the best ways of overcoming the teachers' anxiety regarding implementing a new curriculum. This chapter will deal with the problems and issues related to the implementation of science curricula in high schools by providing a theoretical framework that will involve the following: resource materials, models of teachers' professional development, students' needs, and assessment issues.

RESOURCE MATERIALS

Science educators hold different opinions regarding the variety of students' ideas elicited during teaching. Some educators regard them as barriers to the process of learning and therefore, design strategies to eliminate them, whereas others regard them as repertoires and as an essential and useful resource enabling students to build on their experience and intuitions. Therefore, the curricular goals, the teaching strategies, and the assessment tools differ in these approaches (Eylon & Hofstein, 2015).

During the 1960s and 70s, many reforms took place in pre-university science education worldwide. At the beginning of the reform in science education, more than sixty-five years ago, Jerome Bruner (1959) emphasized an idea that had begun to be used during that time by members of the science education community – the importance of effective implementation. He argued that teaching has to be carried out in such a way that it would represent "the structure of the discipline" integrated with an introduction to the discipline's concepts and its specific facts. "The structure of the discipline" was defined as a system of meta-principles around which the

K. S. Taber & B. Akpan (Eds.), Science Education, 199–210.

knowledge is organized. Bruner pointed out that some of these principles relate to disciplinary content knowledge and some relate to research methods. He explained that emphasizing the "structure of the discipline" in one's teaching is necessary for two reasons: (1) it enhances students' understanding of the conceptual essence (the nature) of the discipline and (2) it has pedagogical advantages, e.g., making information more understandable and more suitable for students' long-term memory.

In addition, the goals of the reform were strongly based on the view that science learning should serve students who plan to embark in the future on a career in the sciences, engineering, or medicine. In 1962 The American Association for the Advancement of Sciences summarized the goals of these curricular initiatives:

- Science education should present students with a real picture of science, including theories and models.
- Science education should present an authentic picture of scientists and their method of research.
- Science education should present the nature of science (NOS).
- Science education should be structured and developed using the discipline approach (key concepts of each of the subjects).

To attain these goals, a series of science curricula were developed, e.g., PSSC (Physical Science Study Committee), and HPP (Harvard Project Physics) in Physics; BSCS (*Biological* Sciences Curriculum Study) in Biology; CBA (*Chemical Bond Approach*) and afterwards CHEMStudy in Chemistry; and the Nuffield projects in the UK. The curricular materials were progressive in terms of learning materials, but they were generally not developed by educational personnel or institutions, but rather by researchers from the scientific academic community. One exception is the Nuffield project, which consisted of outstanding teachers who raised the interest of the academic world. However, the development of new curricula should be followed by intensive teacher workshops and courses, and formative evaluation (Ganiel, 1995). This involves interaction with the school system, the policy makers (e.g. the Ministry of Education, and the teacher educators, among others). Obviously, school systems are organized differently in different countries and school districts, and in particular they may be different in the extent to which they are centralized (i.e. with major decisions taken by a central authority) or decentralized.

Thirty years later, there was a new wave of science education reforms throughout the world. The concern about inadequate standards led to re-examining the goals in many countries, e.g., the National Curriculum in UK (1988), the California Framework (1990), and later, a project initiated by the National Research Council (NRC, 1996) in the U.S. The National Research Council established a National Committee on Science Education Standards and Assessment, to develop three groups of standards: Curriculum Standards, Teaching Standards, and Assessment Standards. Other countries have long had requirements that are dictated by a centralized examination system. This is common in many European countries, such as France, Italy, Belgium, Germany, France, and also in Israel.

Additional factors that influenced changes in curricular materials were as follows:

1. Teacher associations have been very influential in initiating curriculum changes through the development of networks such as the National Science Teachers' Association in the U.S., the Association for Science Education in the UK, the Science Teachers' Association in Ireland, and the Australian Science Teachers' Association.
2. Reports such as Beyond 2000 or ROSE (Sjøberg & Schreiner, 2010). For example, it has been suggested that East Asian students are more skeptical about the role of science and technology in society compared to other Western world students (Sjøberg, 2005). Furthermore, there are differences in the assumptions and norms in different educations systems, so for example, in some Far East contexts students are not encouraged to question their teachers and debate issues as much as in some other contexts.
3. Citizens' needs, which led to the development of different curricula, such as the "Science and Technology for All Curriculum", which was developed in Israel during the 1990s. In the U.S., for example, the NSF sponsored the Synthesis project, in which different curriculum materials were analyzed, and recommendations were made (Eylon & Hofstein, 2015): (1) To refer to current societal problems, (2) to attend to the personal needs of students, and (3) to provide a better awareness of potential careers in science and technology. Based on these recommendations, many curricular programs addressing science, technology, and society (STS) were developed all over the world. Eylon and Hofstein (2015) claim that *"Attempts have been made to make science more relevant to students and adjusted to their backgrounds"* (p. 260).
4. Changes and cutting-edge developments in science and technology, namely, contemporary scientific and technological knowledge, which constantly influence changes in the science curriculum in general, and in the high-school science curriculum in particular. The Science, Technology, Engineering and Mathematics (STEM) curriculum needs to be constantly realigned with the innovations in science, e.g., interdisciplinary topics or molecular biology.

A context-based curriculum will serve as an example of a curriculum that deals with scientific concepts that are related to real-world phenomena and everyday life problems. It is suggested that real-world problems emphasize the interdisciplinary nature of science and its relevance to students' lives. The following will serve as examples of context-based relevant curricula in chemistry:

1. *"I Have Chemistry with the Environment",* a teaching unit that focuses on teaching analytical chemistry together with environmental chemistry (Mandler et al., 2012). The unit consists of material to develop societal awareness of global warming, hazardous waste dumps, groundwater contamination, etc.,
2. *Chemistry in the Community* (*ChemCom*), a curricular unit designed by the American Chemical Society (American Chemical Society, 2006),

3. *PLON* curriculum in the Netherlands, and *"Salters"* Advanced Chemistry in the UK (Bennett, Grasel, & Parchmann, 2005).

In order to implement a context-based curriculum, changes should also be made in the school's learning environment, which allows students to interact physically and intellectually with instructional materials through: (1) Relevant hands-on experiences (Hofstein, Shore, & Kipnis, 2004), (2) design-based activities (Fortus et al., 2005), and (3) inquiry-oriented activities (Tobin, Capie, & Bettencourt, 1988), e.g., inquiry, argumentation, simulations, extracurricular activities, or updated technologies.

In addition, it is important to mention two additional aspects of curriculum development and implementation:

1. The meaning of relevance in curriculum development and implementation is controversial, due to: (1) The influences of different cultural backgrounds and different genders (Dkeidek, Mamlok-Naaman, & Hofstein, 2011), and (2) the different views of policy makers and educators compared with high-school science students regarding the concept of relevance. The personal context of relevance should be taken into account – topics that are relevant to science educators, to policy makers, or even to science teachers are not necessarily relevant to high-school students.
2. The "bottom-up" model versus the "top-down" model in implementing different curriculum materials, and especially the context-based relevant curricula (Eylon & Hofstein, 2015). The "bottom-up" (school-based) curriculum and its implementation model is in alignment with teachers and students' needs, and is usually developed by teams of leading teachers together with experts in curriculum development. It is suggested that traditional approaches to professional development are "top-down", with low interactivity, and have little less connection to teachers' needs and prior experiences. However, in many countries, and especially in Africa, it is not easy to introduce the "bottom-up" model of implementation. For example, studies conducted in Africa and Asia indicate that the lack of professional development and equipment, the socio-political priorities, the societal values, and the examination system, were inhibitors to introducing the "bottom-up" model (Sjøberg, 2005).

Last but not least, there is no one way to implement a certain curriculum. Should a curriculum be adopted exactly as it was recommended by the developers, or should it be adapted? There were times, especially during the 1970s and 80s, when policy makers in many countries adapted the curriculum materials and assessment methods of other countries. Needless to say, it did not always match the local needs, namely, teachers' cultural beliefs and their content knowledge or pedagogical content knowledge (PCK), students' cognitive and affective factors, the learning environment of schools in different cultures, the language capabilities, or the laboratory equipment and conditions. Adapting a new curriculum to suit the local needs takes into account all the above, including in-service professional development of the science teachers towards implementing the new curriculum and its assessment methods.

THE TEACHERS

Teachers in general and science teachers in particular, play a crucial role in curriculum implementation. A recent international policy document written by Osborne and Dillon (2008) reflects a consensus on the importance of good quality teachers and teaching:

> Good quality teachers with up-to-date knowledge and skills are the foundation of any system of formal science education. Systems to ensure the recruitment, retention, and continuous professional training of those individuals must be a policy priority in Europe. (p. 25)

There is constant growth in the body of knowledge in science as well as in science teaching. Therefore, in addition to textbooks and guide books related to the curriculum, teachers need to receive guidance and support throughout various teaching and implementation stages involving changes in the curriculum. They should be acquainted with new research and with changes in the science curricula, and attend life-long learning programs and workshops (Mamlok-Naaman et al., 2013).

Borko and Putnam (1995) have categorized the knowledge that teachers encounter as Subject Matter Knowledge (SMK), general Pedagogical Knowledge (PK), and Pedagogical Content Knowledge (PCK). Within this categorization, PCK is concerned with the teaching and learning of a particular domain: knowing how students learn within that domain, knowing their common misconceptions and the particular difficulties and challenges of that domain, and being able to apply this knowledge to teaching and learning within that particular domain (Shulman, 1987).

It is not easy for the teachers to undergo modifications that include changes in the content and in the way they teach. Teachers have different views and beliefs about teaching objectives, teaching strategies, their students' characteristics, or the classroom management. These beliefs are influenced by their personal experience, knowledge, social and cultural background, and many other different sources. However, independently of beliefs, it has been noted that teachers, in general, are excellent learners, and are generally interested in trying to teach a new curriculum, as well as in improving and enriching their teaching methods (Joyce & Showers, 1983). In addition, they should be encouraged to expand their repertoire of student assessment strategies to include techniques such as observation checklists, portfolios, and rubrics.

Teachers in general are however reluctant to accept radical changes and often do not implement them in accordance with the rationale for the change suggested by the curriculum developers. Such changes may not be aligned with teachers' existing views and practices and may require new knowledge, perhaps content knowledge (CK), its related pedagogical content knowledge (PCK), or curricular knowledge. Important factors influencing teachers' responses to changes include personal characteristics, curricular norms (e.g., the role of questioning), the professional status of the teacher, the teacher's understanding of the proposed change and its rationale, and systematic approaches to students' future career opportunities. An example would be the call for teaching case studies from the history of science (aimed at enhancing the nature of science education) that appeared in a work by Klopfer and Watson (1957). In the 1960s, a few curricula, such as the BSCS in biology, and Project Physics at Harvard University (Holton, 1967), incorporated history and philosophy (HPS) elements into their materials. However, apparently, teachers were not well prepared for this approach and they avoided those parts of the curriculum. Disappointment caused this approach to be abandoned for about twenty years.

One of the ways of overcoming the anxiety of teachers regarding reforms in science education is to encourage their active involvement in developing learning materials, instructional techniques, and related assessment tools (Loucks-Horsley et al., 1998). This was also what we have learned from the programs in the 1960s, that one of the keys in attaining the curricular goals should be involving teachers in the curriculum development process.

As mentioned above, the implementation of new content and pedagogical standards in science education necessitates intensive, life-long professional development of science teachers. In-service workshops conducted all over the world have usually been too short and sporadic to foster a lasting change in teachers' classroom practice and behavior. By attending continuous professional development (CPD) workshops, however, teachers will receive proper professional preparation in order to implement new curricular materials, receive the needed guidance and support while implementing them, and will become acquainted with new developments in science, as well as innovative curricular materials and innovative teaching strategies. Eylon and Hofstein (2015) claimed: *"The professional development of teachers, and providing them with opportunities and tools to customize instruction to their needs, is essential for effective implementation"* (p. 263).

There are many effective models of CPD of science teachers, such as "Involving teachers in the curriculum development", "Evidence-based professional development", or "Action Research". Loucks-Horsley et al. (1998) claimed that encouraging teachers' active involvement in developing learning materials, instructional techniques, and related assessment tools is an efficient way to help teachers overcome their anxiety regarding reforms in science education. "Evidence-based professional development" and "Action Research" focus on teachers' self-reflection, and are among the most promising strategies for teachers' CPD (Mamlok-Naaman & Eilks, 2012).

The 'evidence-based' approach consists of collecting artifacts in a particular science learning domain that presents teachers' work and students' learning, combined with written commentaries that explain the role of the artifacts within the learning context (Taitelbaum, Mamlok-Naaman, Carmeli, & Hofstein, 2008).

The artifacts should be turned into evidence during the CPD program, and they can be used to demonstrate evidence-based practice when teaching science. Based on findings from such a program that was implemented in Israel and in the UK, it was found that during the CPD, the teachers had gained more self-confidence in critiquing their own work, and in understanding their teaching strategies. As a result, their students' interest increased as the process of learning increased. Moreover, the students were more satisfied regarding the learning materials, the learning strategies, the assessment methods, and their ongoing dialog with their teachers.

Action Research programs and workshops deal with an inquiry into teachers' work and their students' learning in the classroom (Feldman & Minstrel, 2000). The teacher acts as a researcher of his or her own work in class, and reflects upon his or her own work in order to improve the teaching strategy. This may be followed by protocols assembled in a portfolio, which can be used to demonstrate evidence-based accomplished practice in science teaching, in an effort to achieve more effective teaching. The portfolio should document the activities, interactions, and behavior in the classroom. It can be viewed as a systematic and organized collection of evidence used to monitor the growth of a learner's knowledge, skills, and attitudes in a specific content area (Mamlok-Naaman & Eilks, 2012).

In summary, teachers who implement a new curriculum should receive sustained support in order to gain knowledge of different teaching strategies, instructional techniques, and assessment skills. This can be done by attending professional development workshops that deal with those topics, which will consequently stimulate their creativity and diversify their instructional strategies in the classroom. Such skills should improve their ability to teach as well as to understand and deal with their students' learning difficulties. Since the teachers will better understand the goals, strategies, and rationale of the curriculum, they will feel more qualified to modify the curriculum as needed. Such workshops help teachers become producers of curricular resources rather than just consumers. Such efforts and reforms in the way students are assessed (see "school-based and alternative assessment" in the "Students" section) necessitate support from other people not directly connected to the program, namely, school headmasters, science coordinators, and regional government consultants.

THE STUDENTS

The National Research Council (1996) in the United States suggests that students are:

> Good directed agents who actively seek information. They come to formal education with a range of prior knowledge, skills, beliefs. In addition, they are directed by their concepts, interest, motivation, and attitudes that significantly influence what they notice about the environment and how they organize and interpret it. This, in turn, affects their abilities to remember, reason, solve problems, and acquire knowledge. (p. 10)

Affective (interest, motivation, and attitudes), meta-cognitive, and sociocultural factors influence the learning-teaching process. Students' prior knowledge is an important aspect that should be taken into account when dealing with the cognitive and meta-cognitive factors that influence the learning-teaching process. A large body of knowledge exists regarding students' prior ideas, misconceptions, and alternative frameworks that develop in the process of learning. Students who learn the scientific disciplines from elementary grades to the upper high-school grades (grades 9–12), frequently encounter difficulties in understanding the abstract nature of scientific concepts and principles, as well as certain phenomena. Thus, they tend to build themselves alternative often erroneous conceptions and mental models (Taber, 2001a). Taber (2001b) claims that some students' alternative conceptions should be considered as pedagogic learning impediments, which reflect the way that they have been taught. Teachers as well as students use textbooks that tend to present science as a collection of true or complete facts and as generalizations and mathematical formulations (Nussbaum, 1989).

Some researchers have presented the idea that students' initial scientific knowledge is analogous to the knowledge of scientists in the ancient world, and consists of observations and conclusions that are often intuitive. Just as these scientists tended to personify objects, or describe processes and natural phenomena in emotional terms, so do children build a conceptual world that is adjusted to their own world of knowledge and emotions. They believe in what they sense and tend not to believe in what is out of the scope of their senses. However, the process of change is not simple, in particular, for those who encounter difficulties in grasping basic scientific concepts (Nussbaum, 1989).

Hofstein and Walberg (1995) suggested that instructional techniques in science should be matched with the students' characteristics, learning styles, and interests, in order to maximize the effectiveness of teaching and learning processes as well as to increase student motivation. A variety of instructional strategies may enable students to study science in ways that will match their learning styles and interests. In addition, the implementation of various instructional strategies should be in alignment with suitable assessment tools (National Research Council, 1996):

> Assessment policies and practices should be aligned with the goals, student expectations, and curriculum frameworks. Within the science program, the alignment of assessment with curriculum and teaching is one of the most critical pieces of science education reform. (p. 211)

The quality of available assessment tools tended to be quite limited in the past. Students' achievements were assessed mainly by paper-and-pencil tests. Other students' activities, such as group work, critical reading of scientific articles, or inquiry laboratory activities were seldom assessed. This may be due to the lack of appropriate assessment tools and to teachers' lack of experience in using them (Mamlok-Naaman, Hofstein, & Penick, 2007).

In addition to the increasing use of a diversity of learning and assessment strategies and tools, there is also a shift from the traditional way of teaching ("teacher-centered"), to "student-centered" approaches. "Student-centered" teaching and learning focus on students' activities, e.g., the students are active and not passive in class. They solve problems, answer questions, formulate their own questions, discuss, explain, and debate; they work in small groups and use a method of cooperative learning – they work in teams on problems and projects. The activities may consist of inquiry-based learning, problem-based learning, design-based or project-based learning. In this way, students can become involved in their own learning process, learn how to think better, and may become more motivated to study science. However, different cultures, traditions, norms, and social structures influence the teachers' ability to shift from "teacher-centered" to "student-centered" learning (Dkeidek et al., 2011).

EVALUATION

Evaluation of a new curriculum and its outcomes is a complex procedure which should raise questions such as: How do the teachers cope with the curriculum? To what extent did they address the goals and requirements of the curriculum? How do the learners perceive the curriculum materials? Were the learners motivated to study the new curriculum? What are the learners' achievements? What kind of assessment is used – summative or formative? What should be changed /added / revised regarding the curriculum materials?

As mentioned in the *Resource Materials* section, the United States National Committee on Science Education Standards and Assessment (National Research Council, 1996) has worked on establishing three groups of standards: Curriculum Standards, Teaching Standards, and Assessment Standards. In countries with a centralized educational system, policy decisions concerning the assessment of students may have a radical impact on what and how students learn (Eylon & Hofstein, 2015). As a result, there may be a need for making changes in the matriculation examinations (like in Israel), or decisions made by policy makers, such as implementing the "bottom-up" (school-based) continuous assessment conducted by teachers – a formative mode of assessment, in which students get involved in the assessment process, and may constantly improve their scores.

The implementation of a wide spectrum of instructional techniques in the science classroom necessitates selecting an appropriate assessment tool for each learning goal in order to effectively measure the students' achievements and progress. According to the National Research Council (1996):

> Assessment policies and practices should be aligned with the goals, student expectations, and curriculum frameworks. Within the science program, the alignment of assessment with curriculum and teaching is one of the most critical pieces of science education reform. (p. 211)

In the past, students often received final grades based mainly on their abilities in paper-and-pencil tests. Thus, only a small fraction of their activities in science were assessed. The reasons might have been linked to: (1) lack of appropriate assessment tools, (2) lack of valid and reliable criteria, (3) the fact that the teachers were less confident in using a variety of assessment methods, namely, school-based or alternative assessment methods, or (4) because teachers were not considered able to undertake objective marking of their own students. The range of available assessment tools can include tests, laboratory reports, quizzes, and assessment guides for carrying out mini-projects, writing essays, and for critical reading of scientific articles. In addition, each assessment tool should consist of detailed checklists (rubrics) and rating scales. The system of rubrics enables the students to correct their assignments and improve their scores. Moreover, it creates a continuous dialogue between the teachers and the students, and encourages the students to take responsibility for their learning processes. It is even recommended that students and their related teachers should be involved in designing the assessment methods and their respective weights. In order to effectively assess students with alternative assessment methods, teachers need to engage in a school-based alternative assessment process. Therefore, it is necessary that: (1) the teachers will be adequately prepared, and (2) the teachers will work in teams. Two studies conducted by Mamlok-Naaman et al. (2007), and Hofstein et al. (2004) may serve as examples.

Students' achievements and attitudes regarding a certain curriculum may be two of the aspects that should be considered in evaluating a new curriculum. A survey of teachers and students is needed in order to determine whether they feel the objectives of a curriculum are attained. The survey may use: (1) qualitative methods (teachers and students' self-reports and journals, interviews with teachers and students, observations in class, and others), (2) quantitative methods (questionnaires disseminated to teachers and students), (3) mixed methods – qualitative and quantitative (Tobin, 1995). In addition, in some countries, as part of the educational reforms, alternative assessment methods have been enacted, including observations by external raters.

SUMMARY

In summary, proper implementation of a new curriculum requires time, and many challenging aspects should be taken into account: The students' characteristics, the teachers' beliefs and pedagogical content knowledge, the scientific content, the cultural aspects, and the assessment of students' achievements. As Eylon and Hofstein (2015) claimed, the curricular process involves cycles of developing learning materials and pedagogical models, implementation, professional development, and research. Ganiel (1985) claimed:

> We used the words creation-implementation-evaluation-research in that order, and it may sound as if the activities occur in tandem, one after the other. That is

how it happened during our first trials, back in [the] 1960s and 1970s. However, as our approach matured, these components became more and more mixed, and they occur in interlocking cycles, feeding each other continuously. (p. 38)

Based on the above, it is suggested, that implementing a new curriculum requires both time and systematic efforts including continuous revisions of the curriculum based on national/local/cultural needs, teachers' professional development, and proper assessment of students' achievements.

FURTHER READING

Abell, S. K., & Ledemann, N. (Eds.). (2014). *Handbook of research on science education, Volume II* (Section V. Curriculum and assessment in science). New York, NY: Taylor & Francis eBooks,

Black, P., Harrison, C., Hodgen, J., Marshall, M., & Serret, N. (2011). Can teachers' summative assessments produce dependable results and also enhance classroom learning? *Assessment in Education, 18*, 451–469.

Krajcik, J. S., Mamlok, R., & Hug, B. (2001). Modern content and the enterprise of science: Science education in the 20th century. In L. Corno (Ed.), *Education across a century: The centennial volume* (pp. 205–238). Chicago, IL: National Society for the Study of Education.

Linn, M. C., Songer, N. B., & Eylon, B. (1996). Shifts of convergencesmin science learning and instruction. In D. C. Berliner & R. C. Calfee (Eds.), *Handbook of educational psychology* (pp. 438–490). London: Lawrence Erlbaum Associates.

Mamlok-Naaman, R., Franz R., Markic, S., & Fernandez, C. (2013). How to keep mysel fbeing a professional chemistry teacher? In I. Eilks, & A. Hofstein (Eds.), *Teaching chemistry – A studybook: A practical guide and textbook for student teachers, teacher trainees and teachers* (pp. 269–298). Rotterdam: The Netherlands: Sense Publishers.

Matthews, M. (2014). *Science teaching: The role of history and philosophy of science.* New York, NY: Routledge.

REFERENCES

American Chemical Society. (2006). *Chemistry in the community: ChemCom* (6th ed.). New York, NY: Freeman.

Bennett, J., Grasel, C., Parchmann, I., & Waddington, D. (2005). Context-based and conventional approaches to teaching chemistry: Comparing teachers' views. *International Journal of Science Education, 27*, 1521–1547.

Borko, H., & Putman, R. T. (1995). Expanding a teacher's knowledge base: A cognitive psychological perspective on professional development. In T. R. Guskey & M. Huberman (Eds.), *Professional development in education: New paradigm and practices* (pp. 35–66). New York, NY: Teachers College Press.

Dkeidek, I., Mamlok-Naaman, R., & Hofstein, A. (2011). Effect of culture on high school tudents' question-asking ability resulting from an inquiry-oriented chemistry laboratory. *International Journal of Science and Mathematics Education, 15*, 59–85.

Eylon, B., & Hofstein, A. (2015). Curriculum development. In R. Gunstone (Ed.), *Encyclopedia of science education.* Dordrecht: Springer.

Fortus, D., Dershimer, R. C., Krajcik, J., Marx, R. W., & Mamlok-Naaman, R. (2005). Design-Based Science (DBS) and real-world problem-solving. *International Journal of Science Education, 27*, 855–879.

Ganiel, U. (1995). Fostering change in science education: Creation, implementation, evaluation and research – the Israeli experience. In A. Hofstein, B. S. Eylon, & G. Giddings (Eds.), *Science education: From theory to practice* (pp. 31–39). Rehovot, Israel: The Weizmann Institute of Science.

Hofstein, A., & Walberg, H. J. (1995). Instructional strategies. In B. J. Fraser & H. J. Walberg (Eds.), *Improving science education* (pp. 70–89). Chicago, IL: National Society for the Study of Education.
Hofstein, A., Shore, R., & Kipnis, M. (2004). Providing high school chemistry students with opportunities to develop learning skills in an inquiry-type laboratory: A case study. *International Journal of Science Education, 26*, 47–62.
Holton, G. (1967). *Harvard project physics: A report on its aims and current status.* Cambridge, MA: Harvard University.
Joyce, B., & Showers, B. (1983). *Power and staff development through research on training.* Alexandria, VA: Association for Supervision and Curriculum Development.
Klopfer, L. E., & Watson, F. G. (1957). Historical materials and high school science teaching. *The Science Teacher,* 264–293.
Loucks-Horsley, S., Hewson, P. W., Love, N., & Stiles, K. (1998). *Designing professional development for teachers of science and mathematics.* Thousand Oaks, CA: Corwin Press.
Mamlok-Namman, R., & Eilks, I. (2012). Different types of action research to promote chemistry teachers' professional development: A joint theoretical reflection on two cases from Israel and Germany. *International Journal of Science and Mathematics Education, 10*, 581–610.
Mamlok-Naaman, R., Hofstein, A., & Penick, J. (2007). Involving teachers in the STS curricular process: A long-term intensive support framework for science teachers. *Journal of Science Teachers Education, 18*, 497–524.
Mandler, D., Mamlok-Naaman, R., Blonder, R., Yayon, M., & Hofstein, A. (2012). High-school chemistry teaching through environmentally oriented curricula. *Chemistry Education: Research and Practice in Europe, 13*, 80–92.
National Research Council. (1996). *National science education standards.* Washington, DC: National Academy Press.
Nussbaum, J. (1989). Classroom conceptual change: Philosophical perspectives. *International Journal of Science Education, 11*, 530–540.
Osborne, J., & Dillon, J. (2008). *Science education in Europe: Critical reflections.* London: Nuffield Foundation.
Rudduck, J., & Kelly, P. J. (1976). *The dissemination of curriculum development.* Slough, UK: N. F. E. R.
Project 2061: Science for all Americans. (1989). *American association for advancement of science.*
Shulman, L. S. (1987). Knowledge and teaching: Foundations of the new reform. *Harvard Educational Review, 57*, 1–22.
Sjøberg, S. (2005). How do learners in different cultures relate to science and technology? Results and perspectives from the project ROSE. *Asia-Pacific Forum on Science Learning and Teaching, 6*(12), 1–17.
Taber, K. S. (2001a). Building the structural concepts of chemistry: Some considerations from educational research. *Chemical Education: Research and Practice, 2*, 123–158. Retrieved from http://www.uoi.gr
Taber, K. S. (2001b). The mismatch between assumed prior knowledge and the learners' conceptions: A typology of learning impediments. *Educational Studies, 27*, 159–171.
Taitelbaum, D., Mamlok-Naaman, R., Carmeli, M., & Hofstein, A. (2008). Evidence-based Continuous Professional Development (CPD) in the Inquiry Chemistry Laboratory (ICL). *International Journal of Science Education, 30*, 593–617.
Tobin, K. (1995, April). *Issues of commensurability in the use of qualitative and quantitative measures.* Paper presented at the Annual Meeting of the National association for Research in Science Teaching, San Francisco, CA.
Tobin, K., Capie, W., & Bettencourt, A. (1988). Active teaching for higher cognitive learning in science. *International Journal of Science Education, 10*, 17–27.

AJEEVSING BHOLOA, GRÁINNE WALSHE
AND YASHWANTRAO RAMMA

16. CURRICULUM IMPLICATIONS OF THE INTEGRATION OF MATHEMATICS INTO SCIENCE

This section examines the curricular aspects of integrating mathematics into science. It looks at the importance of mathematics in science, the benefits of integrating mathematics and science, the variety of models of curriculum design that can underpin their integration and a case study of this integration in a teacher training institute.

THE IMPORTANCE OF MATHEMATICS FOR UNDERSTANDING SCIENCE

The close relationship between mathematics and science has long been recognized. Galileo once said that 'Nature is written in mathematical language'. Newton, who was as much a mathematician as a scientist, pioneered the use of calculus to support his understanding of motion. The study of genomics depends on quantifying massive amounts of biological data, while understanding changes in the environment, such as the relationship between carbon dioxide emissions and global warming, depends on the development of mathematical models by scientists for the prediction of climate change (Karsai & Kampis, 2010). This draws our attention to the fact that it is often not possible to carry out scientific inquiry without quantifying findings using the language of mathematics, its notations, equations and procedures. Almost all branches of science are becoming increasingly mathematical, ranging from bioinformatics to the development and testing of pharmaceuticals, and the investigation of the solar system in astronomy.

Science and mathematics provide the means for interdependent ways of knowing the world, as both are concerned with identifying patterns and relationships. Science provides contexts and applications for abstract mathematical principles, while mathematics provides the essential skills and processes necessary for science to make sense of the natural world. In recent times, the focus on STEM (science, technology, engineering and mathematics) education has led to growing interest in integrating all four of these subjects in various combinations in order to solve real life problems. This is said to promote students' capacities to adapt and transfer their scientific and mathematical knowledge and skills to other contexts and applications, for example, to other subjects, outside of school, and in their future workplaces and personal life. These capacities are essential for citizens in the rapidly changing globalized societies of the 21st century. At the secondary level, students are usually aware that

K. S. Taber & B. Akpan (Eds.), Science Education, 211–220.

mathematics is important to their study of physics, but they may not realize that it is also essential for understanding the statistical analysis of data in biology and for understanding proportional relationships between quantities of atoms and molecules in chemical reactions.

INTEGRATION OF MATHEMATICS INTO SCIENCE

There are numerous benefits to the integration of mathematics into science for the understanding of scientific concepts by secondary level students. While it is clear that the two subjects are undoubtedly related by nature, they are often compartmentalized into 'subject silos' within schools (Orton & Roper, 2000). This creates cognitive and affective barriers for students to apply their mathematical knowledge and understanding in the science class and *vice versa*. Also, since students have a tendency to memorise mathematics concepts and even procedures, it is an undeniable fact that the transfer of their mathematical knowledge to science is effected at a superficial level. Misconceptions start to develop the moment students perceive the same concept differently in two or more disciplines.

Many educational organizations have called for the integration of mathematics and science, and several countries have incorporated integration of science and mathematics into their second-level curricula in one form or another. Organisations that have highlighted connections between different school subjects include the American Association for the Advancement of Science, National Council of Educators of Mathematics and the National Research Council. Bossé, Lee, Swinson and Faulconer (2010) compared the learning standards from the National Council of Educators of Mathematics and the National Research Council (the American standards body) and concluded that science and mathematics should be integrated because the subjects should be learned in the same way. The 2012 *Framework for K-12 Science Education* in the US incorporates mathematical skills, as well as literacy and engineering into the study of science (Czerniak & Johnson, 2014). The *Next Generation Science Standards* identify the crosscutting concept of 'Scale, Proportion, and Quantity' for the study of scientific topics, in addition to developing student understanding of scientific practices, which include 'Using Mathematics and Computational Thinking' and 'Analyzing and Interpreting Data', all of which crucially depend on mathematical skills and processes (Next Generation Science Standards Lead States, 2013). Second-level curricula in Scotland, New Zealand and the Netherlands also advocate interdisciplinary and cross-curricular teaching and learning for these and other subjects (New Zealand Ministry of Education, 2007; Scottish Executive, 2004). However, even in discipline-based curricula, it is typical that science syllabuses refer to requirements for students to have an understanding of variables, to represent and analyse data, and to develop mathematical models taking account of errors and variation in measurement while drawing conclusions. None of these is possible in science teaching without incorporating knowledge, understanding and processes from mathematics.

This problem of students' inability to situate the links between mathematics and physics is generally accepted to be rooted in how educators provide instructions and facilitate the development of students' cognitive structures within an integrative framework. For example, the idea that a force produces motion is a common misconception, among even educators. While studying the case of a golf ball which, after being hit by a club, rolls on the grass, we found from our own observation that some pre-service secondary trainee educators, studying physics and mathematics (mechanics) consider a forward driving force acting on the ball to be responsible for sustaining motion. This is in keeping with research findings on force and motion (e.g. Besançon, 2013). As a result, students, learning from those educators – who teach according to their own alternative conceptions – can develop ideas which can be a barrier to conceptual development for future acquisition of concepts, be it in mathematics or in science. While some students can alter their ideas at a later stage in the face of instruction, some students unfortunately develop stable and tenacious ideas (Taber, 2002). It is the role of educators to develop their strategies so as to treat students' misconceptions as resources, rather than as obstacles, to influence and structure their learning (see Chapter 9).

MODELS FOR SCIENCE AND MATHEMATICS INTEGRATION

In principle, it is clear that mathematics is closely related to science and that integrating mathematical concepts and processes into students' learning of science has many potential benefits. However, defining what integration means and how to achieve it is not so simple. There are many different terms used in the literature to refer to integration, for example: interdisciplinary, multidisciplinary, trans-disciplinary, thematic, integrated, connected, nested, sequenced, shared, webbed, threaded, immersed, networked, blended, fused, coordinated and unified curricula (Czerniak & Johnson, 2014). Coupled with this, there are many different models for curriculum integration. For some, science and mathematics integration should occur as part of a wider project of curriculum integration which is based on social meaning and everyday significance for students. Some variants of STEM education adopt this approach whereby science, mathematics, technology and engineering are integrated to solve real life problems, such as the pollution of a local river. Others distinguish between integration as fully blending science and mathematics, and other forms of blending of the two subjects, such as interdisciplinary (preserving the disciplinary boundaries but making connections) and thematic education (Lederman & Niess, 1997). However, for many researchers, integration exists on a continuum starting from totally blended science and mathematics to separate subjects. Jacobs (1989) suggested ten options for curriculum design on a continuum ranging from discipline-based to multidisciplinary, interdisciplinary and completely integrated programs. Hurley (2001) found qualitative evidence revealing the existence of five forms of integration: sequenced (science and mathematics taught sequentially); parallel (science and mathematics planned and taught simultaneously through parallel concepts); partial (taught partially together and partially as separate disciplines in the same classroom); enhanced (either maths or science is the major discipline) and total (taught together in intended equality).

There are quite a number of models of curriculum design that can support differing levels of and differing purposes for mathematics and science integration. The national and local educational context will dictate to a great extent which model is most suitable. For example, if school structures encourage the development of lessons and activities based on interdisciplinary projects between different subjects, a thematic model would be appropriate. However, in a more subject-based educational system, educators of science and mathematics might collaborate locally to make the conceptual connections obvious to their students across both classes, in sequential, parallel or partial formats.

The problem related to the integration of concepts between disciplines is also found among educators. The discussion which will follow is based on a model of integration between science and mathematics (see Figure 1), which is based on the parallel mode of teaching (Hurley, 2001) adopted in a teacher training institution in Mauritius.

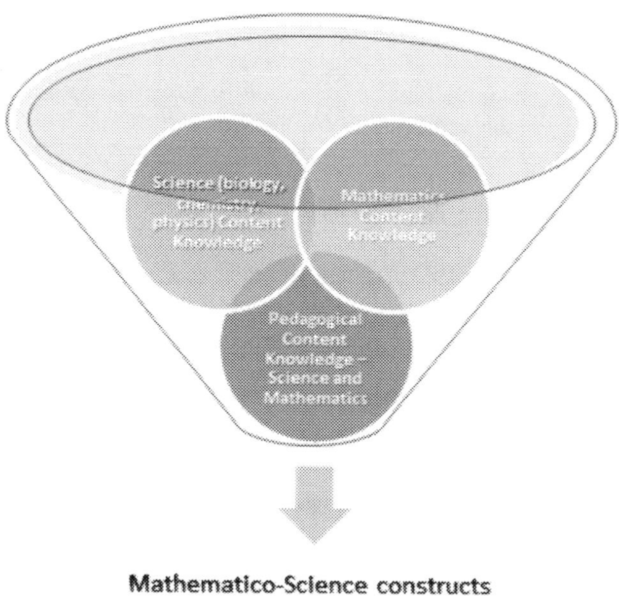

Figure 1. Integrating science and mathematics concepts

The intersection between science content knowledge, mathematical content knowledge and pedagogical content knowledge (science and mathematics) provides the basis for the integration of the concepts through a Thinking Process Model (TPM) (Ramma, Bholoa, Bessoondyal, & Thapermall-Ramasawmy, 2013). This model functions on the premise that there is adequate collaboration between science and mathematics educators leading to a common understanding of content and pedagogy during the preparation of the lessons.

The integration of mathematics into science (biology, chemistry and physics) through TPM requires that an identification of primary concepts and sub-concepts is carried out in a hierarchical arrangement in the first instance. A progressive development of the concepts (and sub-concepts) subsequently follows from concrete (known) to more general contexts and finally to abstraction (unknown). The link between mathematics and science concepts has to be made quite explicitly by educators so as to avoid misconceptions on the part of learners.

Implementation of a Model of Integration: A Case Study

At secondary level, most of the science concepts taught, starting from Grade 9 (age 14), rely heavily on mathematics concepts introduced in mathematics lessons as early as Grades 7 and 8 (ages 12 & 13). Educators from science and mathematics are required to work in collaboration with each other to implement an appropriate integrative strategy so that learners can make links with the mathematics-related concepts during science lessons, and likewise with the science-related concepts during the mathematics lessons. Inadequate acquisition of knowledge of integration by educators can lead students towards a poor understanding in mathematics and physics. For instance, in their study of secondary school students, Ramma and Bessoondyal (2001) showed the existence of a strong linear relationship between the performance in mathematics and physics. The computed correlation coefficient was 0.902.

Our investigation into the educators' perspectives in integration shows that both mathematics and science educators encounter difficulties, but it is more prominent on the side of the mathematics educators. For most of the mathematics educators, a knowledge of the content of physics syllabus is not deemed essential during preparation of their lessons, as they are of the opinion that their physics counterparts should be making the links – science with mathematics. Typical comments along these lines from mathematics educators include:

"I stick to my syllabus."

"The syllabus of math does not depend on physics."

We found that a lack of interest in, and an inadequate understanding of, integration among mathematics educators resulted in abstraction (and subsequently generalization) not taking place, thus leading to a dearth in mathematical thinking. Figure 2a shows the responses to an independent variable problem (Figure 2b) from the equations $y = mx + c$ (math) and $v = u + at$ (physics) by a sample of mathematics and physics educators. While the majority of physics educators could interconnect the independent variables in both situations, 7 out of 12 participating mathematics educators could not do so. The introduction of different symbolic representation of linear equations is meant to eventually consolidate students' conceptual understanding of that topic in mathematics. From this case study, it is found that most of the mathematics educators have learnt the formulae in the closed-form without attributing much physical interpretation to the variables and the constant terms. As such students' mathematical acquisition and reasoning will be partial and incomplete. The recognition of the distinction between dependent and independent variables is important as it enables students to understand the *cause and effect* relationship and also to visualise and analyse that relationship graphically.

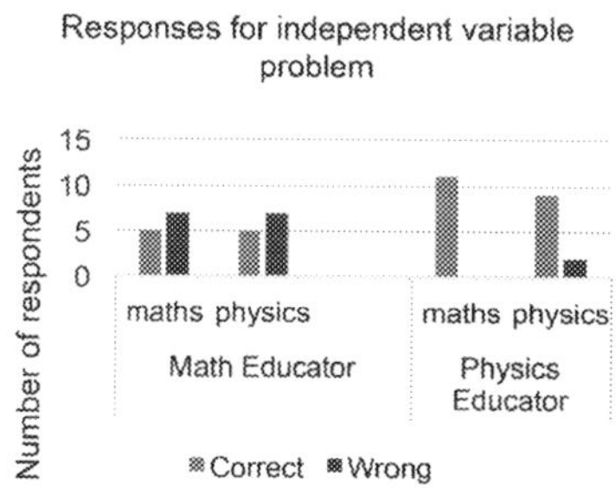

Figure 2a. Mathematics and physics educators' responses

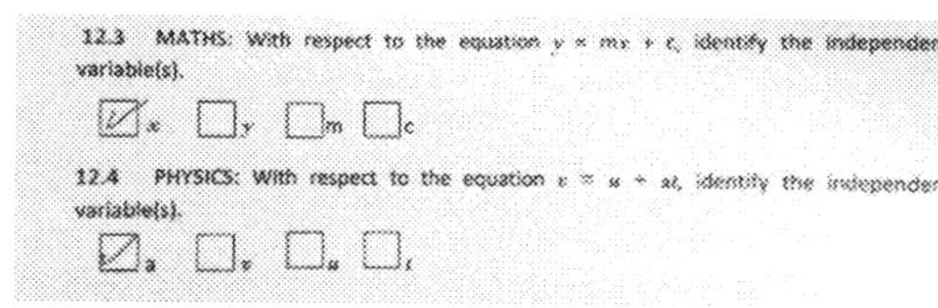

Figure 2b. Independent variable problem

In order to guide both science and mathematics educators to offer adequate pedagogical physico-mathematical support to students, the *Thinking Process Model* (Parmessur, Ramma, Ramdinny, & Bessoondyal, 2005) allows links to be formed between and among concepts of mathematics and physics using the concept map model. In this model, the *Thinking Stage* (TS) is an important element which ensures that thinking is done prior to retrieving a concept from memory within and across science, more specifically, physics(TS[P]) and mathematics (TS[M]). It is important that educators have a good notion of the prior knowledge of the students in the area that they wish to teach, as students may develop alternative ideas which are in contradiction with existing knowledge (Roschelle, 1995) in the course of the lesson.

The *Thinking Stage* (TS) incorporates a multimodal set of processes whereby learners are required to explore "a symbolic model of the task to determine a course of action that should be the best (or be at least satisfactory)" (Gilhooly, 1996, p. 1). The multimodal set of processes encapsulates, in addition to the process of thinking, a multitude of strategies, such as inquiry, questioning, concept formation, case studies, etc. The TS serves as a bridge, linking the content disciplines, pedagogy and teaching strategies into a coherent cluster to enable learners to create links towards developing appropriate cognitive structures since an 'integrated curriculum is apparently not a necessary condition for integrated understanding' (Badley, 2009, p. 118).

Figure 3 illustrates the *Thinking Process Model* (TPM) used during the integration of physics and mathematics concepts in a professional development course for physics educators. In this model, the core physics concept 'equation of motion' is directly mapped onto subsidiary physics concepts, assumed to be stored in the long-term memory in the form of *schema* (Clarke, Ayres, & Sweller, 2005), and onto the core (equation of straight line) and subsidiary mathematics concepts.

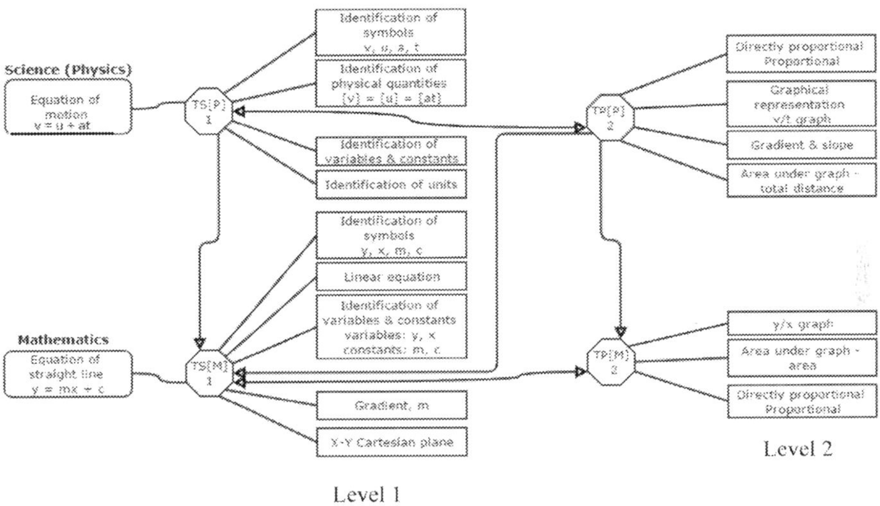

Figure 3. TPM – Integration of science and mathematics concepts

For learners to construct purposeful knowledge structures about 'Equation of Motion ($v = u + at$)', they need to critically retrieve, through *Thinking Stage 1* (TS[P]1), the required physics concepts which are directly related to the equation of motion and which include identification of symbols and their physical meaning, identification of variables, constants and units. Concurrently, learners should be able to relate their abstract mathematics knowledge of $y = mx + c$ and its concrete physical referents, $v = u + at$, through TS[M] 1. Such transitions are of paramount importance to maintain consistency and coherence in the relationship between mathematics and physics. It is also important that for any concept, be it in mathematics or physics, the learner undertakes the *Thinking Stage* before proceeding to a new stage (new level of thinking) to enable concrete building blocks to be developed in the mind. For instance, for constructive learning, if the learner is at Level 1, then to consider Level 2, the transition is done through TS[P]2 and TS[M]2. Such a multimodal approach to knowledge acquisition allows re-organization of existing knowledge structures rather than mere enrichment of knowledge. Such an approach concurs with Roschelle's (1995) suggestion for designers to, first, undertake to refine learners' prior knowledge, followed by anticipating a long term learning process and finally, to engage learners to interact with each other and with the teacher.

Both the mathematics and science teachers should engage in some collegial discussion to identify the connections in the area of integration between the subjects and then develop a common understanding of the concepts, followed by the design of a common plan of action. The lessons, although delivered independently of each other by the respective educators, should encourage learners to interconnect and understand mathematics and science concepts so as to internalize shared foundational meanings. With reference to the above-mentioned example, learners studying the mathematical equation $y = mx + c$ and the physics equation of motion $v = u + at$ would recognize linearity, nature of variables (dependent/independent), slope/gradient, coordinates, axes, rate of change, velocity, acceleration, constants, etc. as being complementary ideas for problem-solving.

DISCUSSION

For the understanding of a concept and to enable the construction of purposeful knowledge structures in relation to that concept, transition (thinking) from one level to another should be undertaken carefully. Reference to learners' existing knowledge or prior beliefs is an important stepping stone towards developing their critical thinking. In addition, Williams, Huang and Bargh (2009) point out that when learners are led to scaffold newer concepts with an already existing one, there is every opportunity for the concepts to become linked in the mind.

It is important to note that integration necessitates a collaborative culture to be established in such a way that there is flow of information between departments. The collaboration between the physics and mathematics departments is the key to adopting an integrated approach by educators, as stipulated by Carson (1999, p. 46):

> Mathematics courses can provide the tools needed to understand physics; physics courses can provide contexts in which mathematics can be applied. However, the connections between mathematics and science departments are often weak, and educators on both sides are unaware of where and why there are differences. Students too have difficulty transferring their knowledge between subjects. Where supplementary mathematics courses are offered to students they do not always relate well to the subjects they are designed to support. We need to work hard in schools and colleges to build bridges between departments and subjects, to look actively for opportunities to make connections, to have time to work and plan with colleagues.

Integration helps to overcome this division, not only between mathematics and physics, but also between mathematics and the other sciences, chemistry and biology.

It can help improve students' conceptual understanding in both subjects. In essence, the curriculum becomes coherent when educators make connections across different areas where there are conceptual overlaps (Geraedts, Boersma, & Eijkelhof, 2006). Moreover, integration has been shown to improve students' interest and motivation to engage in the learning of each subject. The issue here is about making the learning of both subjects more relevant to students by connecting the curriculum to their lives, for example to issues that may be of local significance. Even within the confines of school science, the integration of mathematics can support a more inquiry-based and problem-solving approach to teaching and learning, as students incorporate learning from different areas of the curriculum into their investigations. This encourages students to become critical thinkers rather than to simply absorb factual knowledge and disconnected skills from the syllabus, which they are then unable to apply to new and unfamiliar situations in their lives as citizens and/or in the workplace (Bybee, 2010; Czerniak & Johnson, 2014).

CONCLUSION

The Thinking Process Model (TPM) outlined in this chapter facilitates the synchronised approach to integrating mathematics into science. This approach relies on establishing common skills, knowledge or concepts that form part of teaching and learning in both mathematics and science. Although mathematics and science educators may teach their respective parts separately, they would have identified – prior to the delivery of the lessons in class– how the connections will be made so that concepts learned in one class can be transferred or revisited in the other, thus ensuring consistency in the acquisition of prior knowledge. Similar examples, tasks or assignments could be given to reinforce students' conceptual understanding. Our investigation in a teacher training institute however indicates that mathematics educators appear more indifferent to integration as they are content-driven in their teaching, with an emphasis on repetitive computational skills, instead of encouraging the connection between mathematical and scientific thinking. This calls for the attention of teacher training institutions to facilitate a common process of integration within their professional development courses for mathematics and science educators.

FURTHER READING

Pang, J., & Good, R. (2000). A review of the integration of science and mathematics: Implications for further research. *School Science and Mathematics, 100*(2), 73–82. doi:10.1111/j.1949-8594.2000.tb17239.x

Rennie, L., Venville, G., & Wallace, J. E. (2012). *Integrating science, technology, engineering, and mathematics: Issues, reflections, and ways forward.* New York, NY & London: Routledge.

Rennie, L., Wallace, J.,& Venville, G. (2012). Exploring curriculum integration: Why integrate? In L. Rennie, G. Venville, & J. Wallace (Eds.), *Integrating science, technology, engineering and mathematics: Issues, reflections and ways forward.* New York, NY: Routledge.

REFERENCES

Badley, K. (2009). Resisting curriculum integration. Do good fences make good neighbors? *Issues in Integrative Studies, 27*(8), 113–137.

Besançon, R. M. (2013). *Encyclopedia of physics* (3rd ed., p. 733). New York, NY: Springer Science and Business Media.

Bossé, M. J., Lee, T. D., Swinson, M., & Faulconer, J. (2010). The NCTM process standards and the five Es of science: Connecting math and science. *School Science and Mathematics, 110*(5), 262–276. doi:10.1111/j.1949-8594.2010.00033.x

Bybee, R. W. (2010). Advancing STEM education: A 2020 vision, *Technology and Engineering Teacher, 70*(1), 30–35.

Carson, S. (Ed.). (1999). *Shaping the future: Physics in mathematical mood.* Bristol: Institute of Physics.

Clarke, T., Ayres, P., & Sweller, J. (2005). The impact of sequencing and prior knowledge on learning mathematics through spreadsheet applications. *Educational Technology Research and Development, 53*(3), 15–24.

Czerniak, C. M., & Johnson, C. C. (2014). Interdisciplinary science teaching. In S. K. Abell & N. G. Lederman (Eds.), *Handbook of research on science education* (2nd ed., pp. 395–411). London & New York, NY: Routledge.

Geraedts, C., Boersma, K. T., & Eijkelhof, H. M. C. (2006). Towards coherent science and technology education. *Journal of Curriculum Studies, 38*(3), 307–325. doi:10.1080/00220270500391589

Gilhooly, K. J. (1996). *Thinking: Directed, undirected and creative* (3rd ed.). London: Academic Press.

Hurley, M. M. (2001). Reviewing integrated science and mathematics: The search for evidence and definitions from new perspectives. *School Science and Mathematics, 101*(5), 259–268. doi:10.1111/j.1949-8594.2001.tb18028.x

Jacobs, H. H. (1989). *Interdisciplinary curriculum: Design and implementation.* Alexandria, VA: Association for Supervision and Curriculum Development.

Karsai, I., & Kampis, G. (2010). The crossroads between biology and mathematics: The scientific method as the basics of scientific literacy. *BioScience, 60*(8), 632–638. doi:10.1525/bio.2010.60.8.9

Lederman, N. G., & Niess, M. L. (1997). Integrated, interdisciplinary or thematic instruction? Is this a question or is it questionable semantics? *School Science and Mathematics, 97*(2), 57–58. doi:10.1111/j.1949-8594.1997.tb17342.x

New Zealand Ministry of Education. (2007). *The New Zealand curriculum.* Wellington: Learning Media.

Next Generation Science Standards Lead States. (2013). *Next generation science standards: For states, by states.* Retrieved from http://www.nextgenscience.org/next-generation-science-standards

Orton, T., & Roper, T. (2000). Science and mathematics: A relationship in need of counseling? *Studies in Science Education, 35*, 123–153.

Parmessur, P., Ramma, Y., Ramdinny, A., & Bessoondyal, H. (2005). *Investigating the common core constructs in students' acquisition of logico-mathematical concepts in physics at HSC level.* Report of Unsolicited Research, Mauritius Research Council. Retrieved from www.mrc.org.mu

Ramma, Y., & Bessoondyal, H. (2001). The interrelationship between mathematics and physics. *Journal of Education, 1*(1), 21–28.

Ramma, Y., Bholoa, A., Bessoondyal, H., & Thapermall-Ramasawmy, S. (2013). *Integration of physics and mathematics concepts.* Mauritius: Mauritius Institute of Education. Retrieved from https://www.researchgate.net/publication/272567330_Integration_of_Physics_and_Mathematics_Concepts?ev=prf_pub

Roschelle, J. (1995). Learning in interactive environments: Prior knowledge and new experience. In J. H. Kalk & L. D. Dierking (Eds.), *Public institutions for personal learning: Establishing a research agenda* (pp. 37–51). Washington, DC: American Associations of Museums.

Scottish Executive. (2004). *A curriculum for excellence.* Edinburgh: The curriculum review group.

Taber, K. (2002). *Chemical misconceptions: Prevention, diagnosis and cure: Theoretical background* (Vol. 1). London: Royal Society of Chemistry.

Williams, L. E., Huang, J. Y., & Bargh, J. A. (2009). The scaffolded mind: Higher mental processes are grounded in early experience of the physical world. *European Journal of Social Psychology, 39*, 1257–1267.

MEHMET AYDENIZ AND GULTEKIN CAKMAKCI

17. INTEGRATING ENGINEERING CONCEPTS AND PRACTICES INTO SCIENCE EDUCATION

Challenges and Opportunities

With humanity's increasing reliance on the products of, and solutions through engineering, there is a high demand for professional workforce in engineering (Federal STEM Education-5-Year Strategic Plan, 2013; Gago et al., 2004). This in turn requires us to take a closer look at how we prepare our students for engineering careers through relevant science education programs. Engineering has recently become a topic of increasing interest in science education. In this chapter, therefore, we provide an in-depth critical review of current trends in engineering education and make recommendations for effective and meaningful integration of engineering concepts and practices in K-12 science curricula. The chapter consists of four sections: (i) importance of engineering education, (ii) core engineering concepts and practices, (iii) a review of programs that are designed to support teachers' pedagogical knowledge and skills in engineering education, and (iv) a list of curriculum programs related to engineering education and integration of engineering concepts and practices in K-12 science curricula.

AN INTERNATIONAL PERSPECTIVE ON SCIENCE, TECHNOLOGY, ENGINEERING AND MATHEMATICS (STEM) EDUCATION

Much of a country's economical and political leadership can be attributed to scientific and technological discoveries. Since there has been a significant decline of the number of scientists in some countries, heads of state and government across the globe have stressed the need to boost the number of people entering science, technology, engineering and mathematics (STEM) careers substantially (Gago et al., 2004). For example, in their report, *Europe Needs More Scientists*, Gago et al. (2004) claimed that there are serious shortcomings that stand in the way of increasing the number of science professionals in Europe. Concerns about STEM Education crisis vary across countries, yet a survey of related literature shows that the U.S reform discourse has invaded the STEM education space. The U.S based argument holds that not enough young people are inspired about pursuing an advanced degree in STEM (President's Council for Science and Technology (PCAST), 2010). Friedman (2005) for instance, argued "*The education in American junior high schools, in particular,*

K. S. Taber & B. Akpan (Eds.), Science Education, 221–232.

seems to be a black hole that is sapping the interest of young people, particularly young women, when it comes to the sciences" (p. 351). The common concern among US educators and public is not limited to students' declining interest in STEM fields. Stakeholders also complain about the quality and relevance of STEM programs in public schools. For instance, several government reports make the case that US schools are not well positioned to provide a quality education in STEM fields that will nurture students' problem solving, creativity and computational thinking skills. In response to this observation several stakeholders have made calls to address the issues. For instance, in its *Rising Above the Gathering Storm National Academies of Science* (2007), highlighted the need to (1) increase America's talent pool by improving K-12 mathematics and science education; (2) sustain and strengthen the nation's commitment to long-term basic research; (3) develop, recruit, and retain top students, scientists, and engineers from both the U.S. and abroad; and (4) ensure that the U.S. is the premier place in the world for innovation.

While a concern about lack of students in STEM is, in particular, a U.S specific concern, educators are concerned about the quality and relevancy of STEM programs across the globe. For instance, Turkish education system produces a surplus of STEM graduates, but the quality of STEM education provided in K-12 schools is continuously questioned (Aydeniz et al., 2015). The internationalization of STEM education and concerns over the quality of STEM programs have motivated several educators to develop innovative programs to address the quality of STEM education both in formal and informal settings. The underlying ideas are not only promoting STEM careers but also equip citizens to become critical consumers of STEM knowledge.

ENGINEERING AND SCIENCE EDUCATION

Our world has seen significant changes both in terms of the challenges posed and the opportunities present since the industrial revolution. While the science and engineering community has been able to come up with new discoveries and inventions that have revolutionized the ways in which we live, work, communicate and maintain our health in recent decades, the developments in STEM poses significant challenges to the environmental sustainability, national security, personal privacy, human health and world peace (National Academy of Engineering, 2004). Having diverse STEM skills may promote disruptive innovation. In particular, engineers play a significant role both in the advancement of communication, defense, educational and health-related technologies and addressing the challenges and risks posed by these developments. As we increasingly become dependent on science and technology, we face new and complex problems. To solve these new problems and to eliminate the potential risks posed by these developments we turn to engineers or think like an engineer to approach and solve the problems. Engineers are expected to develop new artifacts, processes and technologies to address these problems. In order for engineers to successfully deal with the emerging problems and demands, they will need to have new knowledge and a new set of skills. In addition, with our increasing reliance on the products and solutions of engineering, there is a high demand for professional workforce in engineering.

The changing nature and the increasing complexity of our problems, and the increasing need for engineers in the fields of national security, environmental sustainability, computing and communication technologies has necessitated new perspectives in engineering education. In this chapter we provide an in-depth critical review of current trends in engineering education and make recommendations for effective and meaningful integration of engineering in K-12 curriculum.

CORE ENGINEERING CONCEPTS AND PRACTICES

There are differences between pure science, applied science, technology and engineering. Technology and engineering are strongly associated with applied science and invention to the solution of technical problems. However, "technology [engineering] is much more than applied science and science is quite different from applied technology" (American Association for the Advancement of Science (AAAS), 1989, p. 26). One major way to distinguish among them is the intention and goals of the activity. Science is about a way of knowing things and aims to generate new knowledge and theoretical understanding. The purpose and goals of technology and engineering; however, are to control or to manipulate the physical world to produce an optimal solution to human needs and desires. To be more precise, while the end products of science are evidence-based explanations, the end products of engineering are solutions and artifacts/technology (NGSS, 2012).

Engineering is a unique field with its unique history, ways of knowing and practices. The National Academy of Engineering Committee on K-12 Engineering (2009) has identified several habits of minds (HoM) associated with engineering. According to the committee's report, habits of minds associated with engineering include: *(1) systems thinking, (2) creativity, (3) optimism, (4) collaboration; (5) communication; (6) attention to ethical considerations, and (7) finding solutions to problems that are based on scientific knowledge and models of material world.* HoM was being used to describe aspects of intelligence. For insistence Resnick (1999) argues that "intelligence is the habit of persistently trying to understand things and make them function better…Intelligence knows what one does (and doesn't) know, seeking information and organizing that information so that it makes sense and can be remembered" (p. 2).

System thinking is a core engineering practice because inherently engineering deals with designing, controlling, maintaining, analyzing and updating systems based on a set of goals. Such a way of thinking helps engineers to establish, evaluate and control complexity. Because engineering aims to solve complex problems or design systems to address a need, to propose a solution that best meets a need, it is inherently an innovative and creative endeavor. Developing innovative solutions requires constant engagement in critical thinking and problem solving. Engineers are charged not only to develop a solution to a problem but also asked to develop solutions that are most effective and efficient as well. Therefore, they need to be optimistic. National Academy of Sciences' (2009) report defines optimism as "a world view in which possibilities and opportunities can be found in every challenge and an understanding that every technology can be improved."(p. 5). Engineering is a practice that requires high level of collaboration. Collaboration improves the quality of engineering solutions and products because "*collaboration leverages the perspectives, knowledge, and capabilities of team members to address a design challenge*" (NAS, 2009, p. 6).

Engineers design solutions to address a need that has been brought up by customers. Designing the best solution requires an in-depth understanding of the problem or the need. Engineers need to identify the effectiveness, efficiency, and durability of design under different condition (NGSS, 2012). Similarly, communication is essential for the engineer to present prototypes, costs and other challenges and possibilities associated with a particular solution using multiple means of communication. Therefore, engineers must be effective communicators. Finally, engineers must pay special attention to ethical considerations when designing a solution. [The] Ethical aspect of engineering is concerned with the impacts of engineering on individuals, groups of people and the environment. Intentionally or unintentionally, engineering solutions can "disproportionately provide advantages or disadvantages for certain people or groups" (NAS, 2009, p. 9). Similar effects can be true for unintended consequences of a particular technology on human health, the environment or world peace. Preparing engineers who can think of these unintended consequences during the design process can help create a just society, peaceful world and a healthy environment.

ENGINEERING, TEACHER PREPARATION AND PROFESSIONAL DEVELOPMENT

While K-12 (kindergarten to grade 12 schools) educators and parents of school children have embraced the idea of engineering in schools, very few K–12 teachers are prepared to teach engineering practices (National Academy of Engineering, 2009). Preparation of quality teachers in engineering is a concern of multiple stakeholders. While some institutional bodies are concerned about the standards of teacher professional development (PD), others are worried about the structures that need to be put in place in order for school systems to provide quality-engineering education in the classrooms. While some stakeholders are concerned about the lack of standards for teaching engineering, others worry about the research-bases of proposed pedagogical recommendations. Both development of standards informed by research and knowledge produced through research programs provide guidelines for preparation of effective engineering teachers.

American Society for Engineering Education (ASEE)[1] has developed Standards for Preparation and Professional Development for Teachers of Engineering.

According to these standards, teachers of engineering should (1) focus on the fundamental nature, content, and practices of engineering" (p. 1), (2) engage teachers in authentic engineering practices and processes driven by a challenge through collaborative team work and (3) introduce participants to resources and tools that facilitate engineering design. Such tools may include simple tools such as rulers, or more "sophisticated technologies such as hardware and software" (p. 2), (4) PD programs designed for engineering teachers should focus on empowering teachers to acquire the fundamental knowledge of strategies that "enable success in engineering". Such strategies include "engaging in teams, asking questions, communication about design, and carefully documenting work" (p. 2), and (5) PD programs should give opportunities and encourage participants to reflect on their experiences with the engineering design process. Collectively these experiences, along with others that we did not report due to space limitations are believed to better prepare STEM teachers to teach core engineering concepts and practices in an effective manner.

While these standards are promising, stakeholders are concerned about the implementation of these standards. For instance, the National Association for Research in Science Teaching has declared a response in which it outlined its concerns for the implementation of engineering standards outlined in The National Research Council's publication of the *Framework for K-12 Science Education* (NGSS, 2012). The *NGSS* recommends integration of engineering into science curricula in two ways: (a) engineering as a pedagogical approach to teaching science content and (b) as a distinct and important content area in and of itself. The challenges to implementation of these two goals include: (1) lack of structures for preparation of pre-service teachers to teach engineering in K-12 classrooms, (2) many in-service science teachers do not feel prepared both attitudinally and professionally to teach engineering in K-12 classrooms (Yasar, Baker, Robinson-Kurpius, Krause, & Roberts, 2006) and (3) lack of access to quality curricular materials to teach engineering in K-12 classrooms.

These challenges can impact the quality of engineering in K-12 classrooms in many ways. First, teachers who have naive understanding of and limited experiences with engineering design process can misrepresent engineering in the classroom. These teachers can limit engineering to construction of physical objects with limited attention to problem solving, optimization and creativity in learning experiences that they provide to their students. Second, lack of engineering related quality PD to a great number of teachers can lead to inequitable implementation of engineering in K-12 classrooms. For instance, while schools with financial and intellectual resources may be able to provide significant PD and greater access to quality curriculum materials for their teachers, those who lack such resources may deprive their teachers of such PD. This in turn can create access to quality engineering education opportunities for different groups of students.

Despite these challenges, teacher pedagogical training is critical for engaging students in the engineering design process in an effective manner. The argument holds that a meaningful increase in teachers' content and pedagogical knowledge related to engineering design will improve their capacity to effectively develop and implement engineering lessons (Fransson, & Holmberg, 2012; Nadelson et al., 2015) Keeping this assumption in mind, STEM educators have designed and implemented several teacher PD programs (e.g., Nadelson, Seifert, & Moll, 2011: Guzey, Tank, Wang, Roehrig, & Moore, 2014; Lewis, 2006; Want et al., 2011) to prepare teachers to teach using design and engineering as contexts for teaching science and mathematics. The goal of these programs is to increase teacher content knowledge and comfort with teaching engineering in their classrooms. These programs aim to achieve this goal by engaging teachers in engineering design projects that adhere to student-centered instructional practices, focus on socially important and culturally relevant projects.

A research study conducted by Schnittka (2012) shows clearly how unprepared teachers can contribute to inequitable classroom implementation of engineering education. She reports the outcomes of a case study of a middle school science teacher who implemented an engineering-design based curriculum into two separate 8th grade science classes: one high achievers and one low achievers. The motivation for this case study came primarily from the claims by the NRC (2012). Specifically, the claim that engineering may benefit low-achieving students more than their peers, and thus level the playing field in education for students with different learning abilities. This engineering design based curriculum included one unit on heat transfer, which lasted 6–7 days. Even though the lessons for each class was based on the same goals and curriculum, Schnittka noticed there was a different ratio of teacher to student involvement in discussions. The class discussions in the high achiever class had much more student involvement and direction. In the exit interview the teacher admitted that she felt the need to give more direct instruction in the low achiever class due to concerns for classroom management. The pattern of questioning in the high-track class was also more open-ended compared to the questioning pattern in the low achiever class. The teacher also acknowledged a difference in ability for students to transfer practical knowledge to knowledge evident in traditional testing. She commented that more often in the low-track class students will demonstrate understanding of a topic but will not successfully answer questions about the same topic on a written exam the same week.

Nadelson et al. (2015) conducted a study with 142 K–5 elementary teachers who voluntarily participated in a STEM-focused PD. The goal of the PD program was to "enhance teachers' knowledge of engineering and the design process" (p. 1). Teachers participated in a three-day summer institute that consisted of presentations, hands-on learning activities, and curriculum planning and development. In this study, the authors "examined the elements of the design process that teachers emphasized in their instruction and the student-generated artifacts inspired by the lessons" (p. 23). The authors then evaluated and "classified the design assignments by the extent of responsibility taken by the teacher and student in terms of the structure of the elements in the design process."(p. 23). Through content analyses of observations of the lessons (169) of 142 K–5 elementary teachers who voluntarily participated in their PD, the authors reported the impact of PD on teachers' practice. The results indicate that teachers "experienced significant and sustained gains in their knowledge of the engineering design process." (p. 34). Authors reported that, "teachers implemented an array of design challenges, representing a diversity of creative expression and focusing on a range of topics" (p. 34). The authors concluded that even through a relatively short-term PD program, "teachers can develop lasting knowledge of engineering design." (p. 37). The authors attributed the success of the program to several factors. First, they argue that the program proved to be successful because "teachers were placed in situations where they actively interacted in design challenges in the context of the classroom" (p. 37). Second, the design activities conducted during the summer institute "provided the teachers with both knowledge of the design process and an instructional model for implementing an engineering design lesson" (p. 37). They argue that teachers' active participation in engineering design was instrumental both in terms of increasing their content knowledge and "enhancing their pedagogical knowledge of how to use design in teaching" (p. 37). The Authors pointed out, "modeling and engaging teachers in design activities appears to be a very effective way to increase both their procedural and content knowledge of engineering design and, therefore, their preparation to teach engineering design lessons" (p. 37). This is consistent with the recommendations in STEM education research for effective PD of STEM teachers (Loucks-Horsley et al., 2003; Penuel, Shear, Korbak, & Sparrow, 2005).

The two studies we report here highlight the importance of PD for effective implementation of engineering concepts and practices in an equitable manner. The challenge to the science education community is to embrace the standards proposed by the ASEEE and adopt best practices from the field to extend quality engineering education to diverse groups of students in different learning contexts such as museums and other informal learning centers as well as in the classrooms. We need to keep in mind that the demand for engineering from stakeholders is real and solutions to wider implementation of quality engineering education in K-12 classrooms awaits our contributions. Science educators can make such contributions through curriculum development, design of PD for teachers in all ladders of K-12 education and research.

Issue1: While the teacher PD programs in engineering design has primarily focused on preparation of secondary teachers (Burghardt & Hacker, 2007; Fontenot, Talkmitt, Morse, Marcy, Chandler, & Stennett, 2009; Tufenkjian & Lipton, 2007), in recent years there is an emerging interest in preparation of elementary science teachers to effectively teach engineering concepts and practices as well. This need is becoming more critical because of emphasis on engineering in the *Next Generation Science Standards* (NRC, 2013).

Issue2: While the number of PD programs aimed at preparing teachers to teach engineering are booming, we do not know if and how these programs are aligned with the standards proposed by NSEE. Second, these programs are often not offered in such ways that it will impact a significant number of practicing teachers.

Most implementation of engineering based PD programs occurs in isolated contexts and impact only a small number of teachers. We need to move beyond small-scale PD programs and expand opportunity to a greater number of teachers. However, we recognize that this is not possible without local school district leaders' and policymakers' support at the state education offices.

EXEMPLARY ENGINEERING EDUCATION PROGRAMS

We highlight three programs that we consider to be exemplary in terms of empowering teachers and students with knowledge and skills associated with engineering. The first program we highlight is CURENT at the University of Tennessee. Then, we highlight larger programs such as Project Lead the Way (Bottoms & Anthony, 2005) and Engineering is Elementary (EIE) (Cunningham & Hester, 2007). These two programs are particularly highlighted because they provide a well-defined structure and instructional resources for teachers to teach engineering content, concepts, and processes in K-12 schools in an effective manner.

The Center for Ultra-wide-area Resilient Electrical Energy Transmission Networks or CURENT is a center formed through collaboration of 16 higher education institutions in the US. The center seeks to meet the national and international energy sector needs for a skilled power systems workforce. CURENT performs research on electric power systems, with a focus on transmission, monitoring, and control of a modernized electric grid.

CURENT has several STEM education programs that focus on providing "multi-disciplinary, team-driven, and systems-oriented educational opportunities to pre-college and university students." The pre-college program focuses on inspiring young students through outreach programs. The overarching goals of pre-college programs are: (1) to inform K-12 students about current and anticipated energy-related challenges, (2) to encourage problem-solving, inquiry and promote creativity through hands-on learning activities and (3) to promote diversity from an early age with the goal of increasing enrollment of underrepresented populations in university engineering programs. The program conducts: (1) engineering family-nights at local K-12 schools in an effort to promote engineering literacy among elementary and middle school students, (2) provides teachers and high school students with research opportunities to advance their engineering knowledge and scientific research skills and (3) offers a girls only summer camp for middle school students to promote female middle school students' interest in engineering careers through hands-on learning activities.

Teachers attend a six-week summer institute where they work in research labs with university professors and graduate students and engage in authentic engineering related research experiences. The overall goal of the research experiences is to enhance science teachers' knowledge of electric circuits, power grids, renewable energy, and power systems and enhance their capacity to integrate science and engineering concepts in their curriculum. Teachers concurrently develop a unit plan consistent with the state science and engineering standards based on their research projects to implement in their classrooms. CURENT has impacted at least 7500 students, teachers, and parents since 2011 through its education outreach programs.

Project Lead the Way (PLTW) is a non-profit organization that is developing a STEM integrated program where the aim is to teach science, technology, engineering, and mathematics in one course. It started in 1987 in New York and has now reached 6,500 schools (PLTW, 2012). There are multiple studies that have investigated the impact of the program on students, teachers, and parents and principals (Kelly, 2008; Schenk, Rethwisch, Chapman, Laanan, Starobin, & Zhang, 2011; Bottoms & Uhn, 2007; Nathan et al., 2011; Daugherty, 2009; Rogers, 2007).

Several engineering education programs are offered in informal settings as well. For example, Boston's Museum of Science (BMOS) is one of the pioneers in this field. The museum offers workshops and develops resources in this field (http://www.mos.org/engineering-curriculum). The *Engineering is Elementary* (EiE) program has been widely used in primary schools throughout the U.S. Evaluations of EiE have found that incorporating engineering in science teaching, using inquiry-based pedagogic methods, results in highly desirable impacts on students and teachers, raising students' interest in science and engineering. BMOS is also expanded its influence to Europe by being a partner of European Union funded projects. ENGINEER project (www.engineer-project.eu) is among them. The project aims to develop some engineering design challenge units suited to European environments using *EiE's Engineering Design Plan* model. Each unit focuses on one engineering field and uses inexpensive materials for student-led design problem solving.

EiE has four main goals. These include: (1) Increase children's technological literacy; (2) Improve elementary educators' ability to teach engineering and technology; (3) Increase the number of schools in the United States that include engineering in their curricula; (4) Conduct research and assessments to further the first three goals and to develop a knowledge base on the teaching and learning of engineering at the elementary school level. To accomplish these goals, EiE has developed curricular materials, PD workshops and resources for teachers and teacher educators.

The EiE curriculum integrates engineering with core science and technology to prepare future generations of scientists and engineers. All EiE units have a common structure consisting of a preparatory lesson designed to prompt students to think about engineering, technology, and the engineering design process. The EiE unit guide provides teacher lesson plans, student duplication masters (worksheets), background resources for teachers, and assessment items.

The authors of the curriculum argue based on the feedback they have received from field-testing that "young children, are capable of much more complex engineering thinking than we originally anticipated." More specifically, "They can balance multiple constraints and criteria, compare the merits of designs, and represent their designs from different points of view" (NAE, 2009, p. 15). Not surprisingly, they also found that "contextualized design challenges appeal to children" (p. 15). Finally, they report that in comparison to their regular experiences in STEM classrooms, struggling students more frequently contribute, stay on task longer, and spend out-of-school time on the engineering challenges presented to them.

CONCLUSIONS

In this chapter, we attempted to achieve three goals. First, we wanted to establish the importance of engineering in K-12 classrooms. Second, we presented some of the habits of minds associated with engineering practices. Then, we discussed the importance of teacher preparation for effective teaching of engineering in the classroom. Finally, we presented several exemplary programs. Among all of these points highlighted in this chapter, we think that teacher preparation is the most critical element of engineering education agenda. While the literature on PD of STEM teachers is fairly consistent about what constitutes effective PD (Desimone, Porter, Garet, Yoon, & Birman, 2002; Loucks-Horsley et al., 2003; Penuel, Shear, Korbak, & Sparrow, 2005), there is an urgent need to investigate how STEM teachers develop and expand upon their existing pedagogical knowledge of specific content to teach engineering design. Incorporating engineering into K–12 schools on a large scale will challenge colleges of education and engineering to develop new models of teacher education. One such model has been developed by The University of Texas at El Paso. In partnership with colleges of education and science, the College of Engineering has developed a teacher preparation program for engineering. The objective of this program is to build the infrastructure for preparing future teachers of engineering. The project prepares engineering students for teaching careers by providing theoretical knowledge of inquiry-based instruction and first hand classroom experiences in partnership with high-needs school districts. More such programs are needed to address the demand for integration of engineering into STEM curriculum across grade levels.

Another point of discussion in engineering education literature is equity and diversity. The field appears to be doing well in terms of addressing diversity issues. The majority of the programs that we reviewed address diversity issues. Most programs have explicit statements of how their programs focus on recruitment of female students and other students coming from backgrounds that have historically been deprived of educational, social and economic opportunities. Finally, while engineering education programs are booming, assessment of learning in engineering practices is an area that teachers need support and guidance in. Similarly, more research is called for in the area of assessment of students' engineering knowledge and skills. We hope that the review provided here and the issues raised in this chapter will contribute to ongoing discussion and help elevate the status of engineering education in our community.

NOTE

1 https://www.asee.org/documents/papers-and-publications/papers/outreach/Standards_for_Preparation_and_Professional_Development.pdf

FURTHER READINGS

Massachusetts Department of Education. (2006). *Massachusetts science and technology/ engineering curriculum framework*. Retrieved from *http://www.doe.mass.edu/frameworks/scitech/1006.doc*

Boston's Museum of Science's Engineering Programs. Retrieved from http://www.mos.org/engineering-curriculum

Engineering design challenge units of ENGINEER Project. Retrieved from http://www.engineer-project.eu/about_2/index.html

Future Scientists and Engineers of America (FSEA). Retrieved from http://www.fsea.org

The National Action Council for Minorities in Engineering (NACME). Retrieved from http://guidemenacme.org/guideme

Pre K-12 Engineering. Retrieved from http://www.prek-12engineering.org

PTC Design & Technology in Schools Program. Retrieved from http://www.ptc.com/for/education/schools/index.htm

Teach Engineering Digital Library. Retrieved from http://www.teachengineering.com

Crismond, D. P., & Adams, R. S. (2012). The informed design teaching and learning matrix. *Journal of Engineering Education, 101*(4), 738–797.

Farmer, C., Klein-Gardner, S., & Nadelson, L. (2014). *Standards for professional development for K-12 teachers of engineering.* Washington, DC: The American Society for Engineering Education.

21Afterschool Alliance. (2014). *America After 3 PM: Afterschool programs in demand.* Washington, DC: Author. Retrieved February, 2015, from http://afterschoolalliance.org/documents/AA3PM-2014/AA3PM_National_Report.pdf

ENGINEERING EDUCATION JOURNALS

- Journal of Engineering Education
- Journal of Pre-College Engineering Education Research
- Journal of Research in STEM Education (J-STEM)

REFERENCES

Aydeniz, M., Cakmakcı, G., Cavas, B., Ozdemir, S., Akgunduz, D., Corlu, M. S., & Oner, T. (2015). *A report on STEM Education in Turkey: A provisional agenda or a necessity?* [In Turkish]. Istanbul, Turkey: Istanbul Aydin University. Retrieved from http://www.hstem.hacettepe.edu.tr/STEM_Rapor.pdf

Cunningham, C. M., & Hester, K. (2007). *Engineering is elementary: An engineering and technology curriculum for children.* Paper presented at the American Society for Engineering Education Annual Conference and Exposition, Honolulu, HI. Retrieved from http://eie.org/sites/default/files/research_article/research_file/ac2007full8.pdf

Daugherty, J. L. (2009). Engineering professional development design for secondary school teachers: A multiple case study. *Journal of Technology Education, 21*(1), 1–9.

Desimone, L., Porter, A. C., Garet, M., Yoon, K. S., & Birman, B. (2002). Effects of professional development on teachers' instruction: Results from a three-year longitudinal study. *Educational Evaluation and Policy Analysis, 24*, 81–112.

Friedman, T. (2005). *The world is flat.* New York, NY: Picador.

Federal STEM Education-5-Year Strategic Plan. (2013). *A report from the committee on stem education, National Science and Technology Council (NSTC)*. Retrieved from https://www.whitehouse.gov/sites/default/files/microsites/ostp/stem_stratplan_2013.pdf

Gago, J. M., Ziman, J., Caro, P., Constantinou, C., Davies, G., Parchmann, I., Rannikmäe, M., & Sjoberg, S. (2004). *Europe needs more scientists, Report by the High Level Group on Increasing Human Resources for Science and Technology in Europe 2004*, European Commission. Retrieved from http://ec.europa.eu/research/conferences/2004/sciprof/pdf/final_en.pdf

Loucks-Horsley, S., Love, N., Stiles, K. E., Mundry, S., & Hewson, P. W. (2003). *Designing professional development for teachers of science and mathematics* (2nd ed.). Thousand Oaks, CA: Corwin.

Nadelson, L. S., Pfiester, J., Callahan, J., Pyke, P., Hay, A., & Emmet, M. (2015). Constraints and criteria in the classroom: Engineering design as a context for teaching STEM. *Journal of Technology Education, 26*(2), 22–45.

Nathan, M. J., Atwood, A. K., Prevost, A., Phelps, L., A., & Tran, N. A. (2011). How professional development in project lead the way changes high school STEM teachers' beliefs about engineering education. *Journal of Pre-College Engineering Education Research, 1*(1), 15–29.

Schnittka, C. (2012). Engineering education in the science classroom: A case study of one teacher's disparate approach with ability-tracked classrooms. *Journal of Pre-College Engineering Education Research, 2*(1), 35.

Stohlmann, M. S., Moore, T. J., McClelland, J., & Roehrig, G. H. (2011). Year-long impressions of a middle school STEM integration program. *Middle School Journal, 43*(1), 32–40.

National Academies of Science. (2007). *Rising above the gathering storm*. Report from the Committee on Prospering in the Global economy of the 21st Century. Washington; DC: National Academies Press.

National Academy of Engineering (2004). *The Engineer of 2020: Visions of engineering in the new century.* Washington, DC: National Academies Press.

National Academy of Engineering. (2009). *Engineering in K–12 education: Understanding the status and improving the prospects*. Washington, DC: The National Academies Press.

National Research Council. (2011). *Successful K-12 STEM education: Identifying Effective Approaches in Science, Technology, Engineering, and Mathematics*. Committee on Highly Successful Science Programs for K-12 Science Education. Board on Science Education and Board on Testing and Assessment, Division of Behavioral and Social Sciences and Education. Washington, DC: The National Academies Press.

President's Council for Science and Technology (PCAST). (2010). *Prepare and inspire: K-12 science, technology, engineering, and math (STEM) education for America's future*. Washington, DC: PCAST.

Purzer, S., Moore, T., Baker, D., & Berland, L. (2014). *Supporting the implementation of the Next Generation Science Standards (NGSS) through research: Engineering*. Retrieved from https://narst.org/ngsspapers/engineering.cfm

Resnick, L. B. (1999). Making America smarter. *Education Week Century Series, 18*(40), 38–40.

Wang, H.-H., Moore, T. J., Roehrig, G. H., & Park, M. S. (2011). STEM integration: The impact of professional development on teacher perception and practice. *Journal of Pre-College Engineering Education Research, 1*(2), 1–13.

Yasar, S., Baker, D., Robinson-Kurpius, S., Krause, S., & Roberts, C. (2006). Development of a survey to assess K-12 teachers' perceptions of engineers and familiarity with teaching design, engineering, and technology. *Journal of Engineering Education, 95*(3), 205–216.

SECTION IV

SCIENCE TEACHING

JULIE A. LUFT AND SHANNON L. DUBOIS

18. ESSENTIAL INSTRUCTIONAL PRACTICES FOR SCIENCE TEACHING

ESSENTIAL INSTRUCTIONAL PRACTICES FOR SCIENCE TEACHING

On a daily basis, science teachers are called upon to provide rich learning experiences to their students. They are encouraged to use scientific inquiry or the practices of science to teach a concept (e.g., Abd-El-Khalick et al., 2004; NGSS Lead States, 2013). Fortunately, some early career teachers have experience using these methods of instruction (Davis, Petish, & Smithey, 2006; Luft et al., 2011), either through their teacher preparation program or during their induction program.

In order for early career science teachers to create a sound learning environment in science, we suggest that they focus on a cycle of instruction. As discussed by Zembal-Saul, Blumenfeld, and Krajcik (2000), a cycle of instruction consists of three phases: planning, instructing, and reflecting. The planning phase involves consideration of the required school standards or curriculum, and the students' current knowledge about the instructional topic. The instructing phase requires the use of materials and strategies in a way that supports the learning of science. The reflecting phase entails an assessment of the lesson, along with the goal of improving student learning. Figure 1 illustrates a cycle of instruction.

Within this cycle, specific instructional practices have been designed to promote student learning. When teachers emphasize the practices individually, they amount to little more than a set of procedures. Yet when the practices are applied collectively to a science topic, teachers can create a learning environment that allows students to construct their science knowledge. This collective orientation encourages the ongoing monitoring of student knowledge, so that teachers can continually adjust their instruction.

In order to guide early career science teachers as they are learning to teach, this chapter shares some essential instructional practices (EIPs) that should be included in a cycle of science instruction. Specific strategies are shared within each phase of a cycle of science instruction. As teachers learn about the strategies, they will come to realize that some of the strategies could exist in different phases. Use of the EIPs across different phases is entirely up to the early career teachers. With practice and over time, they will become more proficient, and will understand how the strategies can be used within and across phases.

K. S. Taber & B. Akpan (Eds.), Science Education, 235–245.

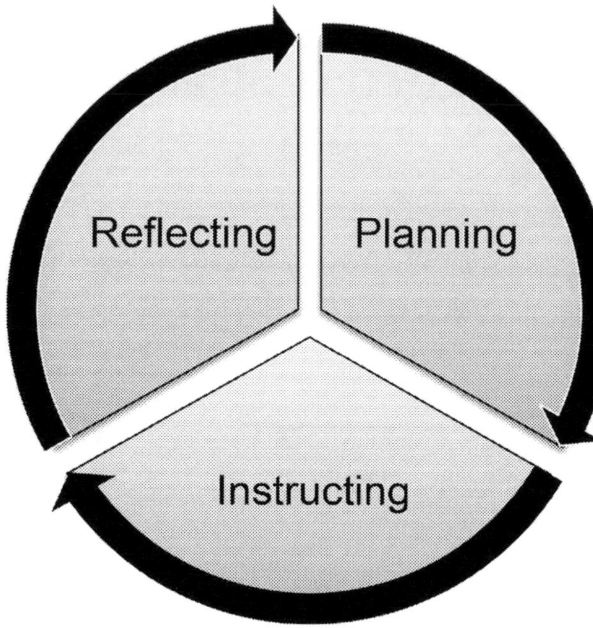

Figure 1. A cycle of instruction

TO BEGIN: PLANNING

The planning phase occurs before students arrive in the science classroom. During this phase, essential instructional strategies include: understanding the core/central ideas in science, identifying relevant instructional goals, determining students' prior knowledge, selecting appropriate instructional strategies, and considering the progression of lessons. The amount of time that is dedicated to each area will vary, depending on the lesson and the teacher's prior experience with the curriculum and students. The following sections discuss each of these EIPs.

Understand the Core Ideas in Science

Before a science teacher begins to plan for instruction, it is important that the teacher identify the core idea (s) of the unit or units. Core ideas in science are unique to each discipline. These ideas can vary by country, but they represent an overarching idea in a content area and are essential in answering fundamental questions about phenomena in nature. For instance, in the United States, in the physical sciences, core ideas in the *Next Generation of Science Standards* (*NGSS*) (NGSS Lead States, 2013) include: matter and its interactions, waves and their applications, energy, and motion and stability: forces and interactions.

Given the overarching nature of core ideas, early career teachers need to identify these ideas prior to planning a unit. For each lesson in the unit, the teacher should consider how the lesson supports the core idea and make this connection clear to students. Students will ideally develop depth of understanding in an area, as opposed to breadth. It should be noted that there could be a few core ideas in a unit, but only one core idea should be of interest in a lesson.

There are just a few core ideas in each disciplinary field. When trying to find a core idea, a new teacher should ask "What is the core idea that is essential for a student to know?" and "What is the core idea that I am addressing as a science teacher?"

Identify Instructional Goals

A science teacher should know the instructional goal(s) for the lessons or series of lessons. In some instances this goal may be predetermined, and in other instances it may be up to the science teacher to determine the goal. Countries with a national curriculum (e.g., Korea, Netherlands) may have specific goals with lessons that should be taught on predetermined days, or they may have concepts that are to be taught during a certain period of time. Countries without a national curriculum (e.g., Canada, United States) require that schools or regions identify content standards to be taught within broad or specific periods of time. In some instances, the content and pacing may even be left to the teacher.

Sound instructional goals are focused on scientific concepts, which allow for the utilization of specific scientific facts. They also support the use of scientific practices, process skills, or science as inquiry. For instance, the goal of a genetics lesson should not be for students to recite the phases of mitosis or meiosis. Instead, students should understand that genetic material is replicated during growth, and divided in half and recombined during reproduction. The phases of mitosis and meiosis would constitute supporting facts.

Ideally, to learn about the replication of genetic material, students could look at different cells that are dividing. They would notice that some cells contain the same genetic materials at the beginning and end of a division cycle. They would also notice that other cells contain half of the genetic material of a parent cell. These explanations would be the basis for understanding meiosis and mitosis.

In identifying an instructional goal, early career science teachers should ask "What is the central idea or concept that students should know and understand by the end of the lesson? How does this connect to the core idea?" This question encourages a teacher to focus on the concept, and not the facts that are embedded in the concept.

Determine Students' Prior Knowledge

Students bring their own ideas and experiences to science classrooms, and these experiences shape their science knowledge. Posner, Strike, Hewson and Gertzog

(1982) recognized that students held conceptions of science that could be intelligible to the student, but may be underdeveloped, or even at odds with the scientific community. In order to support students in learning scientifically correct concepts, they suggested that teachers try to determine the "prior knowledge" of their students.

For early career science teachers, it is essential to uncover a student's prior knowledge about the topic that will be taught in class. By understanding the prior knowledge of a student, science teachers can select appropriate instructional strategies to help students build upon the knowledge they bring to the classroom. For instance, students in a class could understand that temperature can cause a phase change, but they may have an incorrect notion about the molecular configuration within the different phases. In this instance, a teacher may want to plan instruction that supports a molecular understanding of phase changes.

In order to understand the knowledge that students have about a concept, early career teachers can look at research on misconceptions. This research reveals how students misunderstand certain topics, as well as their most common misconceptions. Early career teachers can also look at the work of their own students. Any sort of classroom artifact that is collected before formal instruction begins may provide insights into students' prior knowledge. Finally, early career teachers can talk to their experienced colleagues who have a good knowledge base of the prior knowledge of the students.

In considering prior knowledge, science teachers should ask "What ideas do my students have about the topic I will be teaching? Does my lesson allow students to build a scientifically correct understanding of the topic? This question encourages a teacher to consider what his or her students know about the topic, and this provides direction for the structure of the lesson.

Select Appropriate Instructional Strategies

Upon identification of the goals of the lesson and the students' prior knowledge, an early career science teacher can begin to consider different instructional approaches that will support the learning of all students. It may be that students need to make explanations from data, or that students need to consider the limitations of a model. These different instructional approaches have different implications for student learning.

For early career science teachers, identifying appropriate instructional strategies can be a challenge. Often they have a variety of resources to draw upon, but limited experience in modifying the materials for their students. As a result, new science teachers tend to want to create lessons, which is time consuming. By just modifying a lesson through the inclusion of an activity that allows students to interact with the presented science phenomena, students will learn more and new teachers will be able to focus on student learning.

After identifying the instructional goals of the lesson and the prior knowledge of students, a teacher should select a lesson that will support student learning in these areas. For instance, if a lesson will be about speed, a teacher may find a lesson that has students walk and run in order to understand speed. A lesson that provides walking and running directions to the students can easily be modified so that students can decide how fast they will move, and over what distance. Or the lesson can provide data to students and ask them to make explanations from the data. These two modifications allow students to learn how to participate in science, as well as challenge their own ideas about speed.

New teachers can also select different instructional strategies to help differentiate a lesson. For instance, some students may learn best by creating a model of an ecosystem, while others would benefit from an online simulation to investigate nutrient cycling. By providing students with different activities, the learning of each student can be maximized.

Selecting appropriate instructional strategies is at the heart of teaching science. When a new teacher is planning a lesson, the teacher should consider whether there is an existing lesson that can be modified. Then the new teacher should ask "What instructional strategies best support the learning of all of my students, given the goal of the lesson and their prior knowledge?" Answering such a question may result in several different instructional modifications during a lesson.

Consider the Progression of Lessons

To create a coherent storyline between lessons, it is important for a new teacher to consider how the lessons are linked together, and how the sequencing of key ideas and activities relate to the overarching conceptual idea (Roth et al., 2011). Instructional activities should be sequenced to build upon an idea, yet allow students to challenge their existing and emerging knowledge base.

In creating a purposeful progression of lessons, new teachers should also consider the research around the development of student ideas. This research base suggests potential progressions pertaining to student understanding in science, which can guide the selection of instructional activities. For instance, students should know that objects are made of matter before they can understand that solids, liquids, and gases are forms of matter (see Smith, Wiser, Anderson, & Krajcik, 2006). This understanding can be a basis for the organization of lessons and the selection of activities.

In creating a clear progression of lessons, a teacher should ask "What should be the progression of lessons when considering how students come to understand a concept?" In answering this question, a new teacher will consider how the lessons build upon each other, and how they focus on core ideas, the goals of the lesson, and prior knowledge of the students.

IN ACTION: INSTRUCTING

The instructional phase takes place in the classroom and involves the use of selected strategies that support student learning. Within science, there are several EIPs that a science teacher should implement in a classroom. These strategies support students as they encounter scientific phenomena, and they allow a teacher to monitor student understanding. The following sections discuss the essential strategies of collaborative learning, purposeful discourse, and ongoing assessment of student learning. These strategies should be a part of each lesson that is taught.

Collaborative Learning

The power of collaborative learning is well known (Slavin, Hurley, & Chamberlain, 2003), yet teachers often create instructional environments that result in students working individually. By collaborating, students have opportunities to contemplate the scientific information that they encounter, discuss their emerging ideas, and evaluate their own understandings. The back and forth exchange between two or more students can encourage a deeper understanding of the content. In science in particular, collaboration allows students to experience the social component that is inherent in all scientific activities.

In order to support collaborative learning, students need to be physically close to one another so that they can discuss the presented information. This means sitting in small groups, working together during laboratories, or sitting around a table. In addition, students need guidance in learning how to work in a collaborative group. A teacher can present guidelines that support collaborative conversations, which can include: only one student talks at time, an acknowledgement of the ideas presented, or all students need to participate in the conversation. Collaborative guidelines help ensure that all students have an opportunity to participate in the conversation.

When infusing collaboration into instruction, the teacher should ask "How can I make the classroom environment conducive to collaboration during the lesson?" When answering this question, the new science teacher should consider the organization of the classroom and the guidelines that are provided to students to support their interactions with one another.

Purposeful Discourse

The discourse between student and teacher plays an important role in learning. As the teacher engages a student in conversation, she asks questions and elicits explanations. The student gains new knowledge during the exchange, while the teacher gains a deeper understanding about how the student learns.

EIPs can help create such an exchange of information. One strategy involves the use of wait-time. Early research by Rowe (1986) revealed the importance of pausing while talking to students. Specifically, she found that when a teacher asked a question and waited for a response, a potentially more elaborate response came from the student. By providing a student with a small period of time to consider the question that was asked, the student had an opportunity to craft a more complete response.

Another strategy involves the use of questions that challenge students cognitively. Chinn (2006) observed several science lessons in order to determine how the interactions of teachers supported student learning. From this data, Chinn (2006) suggested that an

> …[A]cknowledgement of students' contributions, restatements of students' responses, and, more importantly, her ability to pose subsequent questions that build on students' earlier responses and that stimulate use of various cognitive processes, all appear to promote productive talk activity in students at a level beyond mere recall. (pp. 13–43)

Clearly, purposeful discourse is important in terms of building student understanding and participation in science.

In order to support student learning in science, early career teachers should be aware of and use the practice of wait-time. In addition, new teachers should consider what questions they can ask to build student understanding of science. Questions that can be easily used in a science class have been suggested by Penick, Crow and Bonnstetter (1996) and include: How did you arrive at this answer? How does this finding or result relate to another finding or result? How does this relate to our everyday world? What do you think the results would be? What could you have done differently?

Assessment of Student Learning

The ongoing assessment of student learning during instruction is important. As a teacher interacts with students, the teacher is collecting information about the learning of the student and the effectiveness of the lesson. A new teacher can collect data from a student by asking the student a question, looking at their work, or giving them a problem to solve and considering the answer. These sources of data can guide the teacher in the construction of future lessons or in modifications during instruction.

One of the most important types of assessment is formative or informal assessment, which can occur in the classroom as a teacher is engaged in instruction. Formative or informal assessments indicate what knowledge the student holds about a concept, which can vary across students (Wiliam & Black, 1996). This type of assessment can be explicit and prompted by a teacher, or it can be conducted as students engage in the instructional activity in the classroom. The information gained from formative or informal assessment can impact classroom instruction immediately or it can alter future lessons.

A more common form of assessment is summative or formal assessment. This type of assessment involves collecting data about student learning in a way that has consistency across students (Wiliam & Black, 1996), such as multiple choice tests, true/false questions, or any other type of measure that is administered to all students. This type of assessment often indicates what students have learned during a specific period of time.

A more important function of both formative/informal and summative/formal assessment is the information that is provided about the enacted lessons. The data collected on students indicates if the curriculum and learning environment supported student learning, and what modifications need to be made to future lessons.

During instruction, a new teacher can ask "How do I know that my students are learning the science concepts the lesson was designed to teach? What evidence do I have regarding their learning?"

LINKING TO PLANNING: REFLECTING

The reflection phase involves purposeful examination of the taught lesson. This is as important as the planning of a lesson, since this is when the lesson is evaluated in light of student learning. During the reflection phase, there should be an examination of the documents that the students complete, or records of students' experiences during the instruction of the lesson. These different sources of data allow the new science teacher to evaluate the learning of the students, and the impact of the instructional decisions of the new teacher.

Evaluation of Student Learning

Students' work reveals their involvement and what they learned from the lesson. Indicators of student learning can include: laboratory reports, documents from practical work, students' responses during instruction, or even written assessments completed by the students. These artifacts should be considered in light of the core concepts and goals of the lesson. By looking for evidence of student learning in these areas, the new teacher can determine what the students learned from the lesson.

A rubric represents a simple approach to the examination of student work to determine student learning. Rubrics can be provided to the students at the onset of instruction, or the new teacher can hold on to the rubric and evaluate the collected artifacts. The assessment of student work can be holistic or analytical (Luft, 1997). Holistic rubrics have descriptive levels, while analytical rubrics often have specific descriptions about various levels of performance. Both types of rubrics can provide information about teacher and student learning.

When engaged in this process, the new teacher should ask "What does the evidence suggest about student learning?" and "What does the evidence suggest about my lesson in terms of supporting student learning?"

Evaluation of Instructional Strategies

After a lesson, it is important to consider the effectiveness of the instructional strategies that were used by the new science teacher. Again, students' work reveals the effectiveness of the instruction. By examining the learning of students in light of the instruction and the goals of the lesson, it is possible to evaluate the instructional strategies.

In order to evaluate the instructional strategies that were used, a new teacher should compare the learning of students to the lesson plan. In this way, the teacher can determine which parts of the lesson supported or inhibited student learning. For instance, in a lesson on cell division, which involved students making conclusions from images of cells dividing, it was determined that most students had inaccurate explanations about genetic material. This type of evidence suggests that the instructional strategies should have been changed or modified in order to better support student learning. With an additional analysis of the instruction, a modification or change in instruction could be determined by the new teacher.

When looking at the evidence of student learning, the new teacher should ask "Was this the best way to structure the lesson to support student learning? What should I do next time when I teach this lesson?"

Modification of Future Lessons

Based on the evidence collected by the new teacher, it is possible to determine how the next lesson should be configured. For instance, if only a few students achieved the desired level of understanding, then there is no need to move to another topic. However, if most students have a strong understanding of the concept presented, then additional instruction for the lagging students can be provided within the next lesson.

In addition, the new teacher can determine the best instructional strategy for the upcoming lesson. If students struggled to derive explanations based on the evidence, then the teacher can create an opportunity to continue to build this skill. The teacher may also provide more support for the students to learn how to make an explanation from evidence.

When considering the next lesson, the new teacher should ask "What have I learned that impacts the design of the next cycle of instruction?"

IN SUMMARY

When learning to teach science, it is important for teachers to embrace a cycle of instruction that includes planning, instructing, and reflecting. Most new science teachers engage in planning and instructing, but few purposefully reflect on the effectiveness of the lesson. By focusing on all three components, they can tailor future lessons to meet the learning needs of the students. Table 1 summarizes the cycle of instruction and lists considerations for new science teachers. These considerations support the use of EIPs.

Finally, it is important to reiterate that this process is cyclical. New science teachers should engage in it anew with each lesson. Over time, as they gain experience, they will be able to expedite the process as they hone their teaching skills.

Table 1. Essential instructional practices (EIPs) for science teaching

Cycle of instruction	*Instructional practices*
Planning	Understand the core idea of the lesson Identify instructional goals Determine student prior knowledge Select appropriate instructional strategies Consider the progression of the lessons
Instructing	Collaborative learning Purposeful discourse Assessment of student learning
Reflecting	Evaluation of student learning Evaluation of instructional strategies Modification of future lessons

FURTHER READING

Bennett, R. E. (2011). Formative assessment: A critical review. *Assessment in Education: Principles, Policy & Practice, 18*(1), 5–25.

Mortimer, E. F., Scott, P., & El-Hani, C. N. (2012). The heterogeneity of discourse in science classrooms: The conceptual profile approach. In B. J. Fraser, K. G. Tobin, & C. J. McRobbie (Eds.), *Second international handbook of science education* (pp. 231–246). The Netherlands: Springer.

Osborne, J. (2014). Scientific practices and inquiry in the science classroom. In N. G., Lederman & S. K. Abell (Eds.), *Handbook of research on science education: Volume II* (pp. 579–599). New York, NY: Routledge.

Treagust, D. F., & Tsui, C-Y. (2014). General instructional methods and strategies. In N. G., Lederman & S. K. Abell (Eds.), *Handbook of research on science education: Volume II* (pp. 303–320). New York, NY: Routledge.

Zeichner, K. M., Liston, D. P. (2014). *Reflective teaching: An introduction.* New York, NY: Routledge.

REFERENCES

Abd-El-Khalick, F., Boujaoude, S., Duschl, R., Lederman, N. G., Mamlok-Naaman, R., Hofstein, A., & Tuan, H. L. (2004). Inquiry in science education: International perspectives. *Science Education, 88*(3), 397–419.

Chin, C. (2006). Classroom interaction in science: Teacher questioning and feedback to students' responses. *International Journal of Science Education, 28*(11), 1315–1346.

Davis, E., Petish, D., & Smithey, J. (2006). Challenges new science teachers face. *Review of Educational Research, 76*, 607–651.

Luft, J. A. (1997). Design your own rubric. ***Science Scope*****, *20*(5), 25–27.**

Luft, J. A., Firestone, J., Wong, S., Adams, K., Ortega, I., & Bang, E. J. (2011). Induction programs and beginning science teachers: Beliefs, knowledge and practices during the first two years. *Journal of Research in Science Teaching, 48*(10), 1199–1224.

NGSS Lead States. (2013). *Next generation science standards: For states, by states.* Washington, DC: The National Academies Press.

Penick, J. E., Crow, L. W., & Bonnstetter, R. J. (1996). Questions are the answers. The *Science Teacher, 63*(1), 26–29.

Posner, G. J., Strike, K. A., Hewson, P. W., & Gertzog, W. A. (1982). Accommodation of a scientific conception: Toward a theory of conceptual change. *Science Education, 66*(2), 211–227.

Roth, K. J., Garnier, H. E., Chen, C., Lemmens, M., Schwille, K., & Wickler, N. I. (2011). Videobased lesson analysis: Effective science PD for teacher and student learning. *Journal of Research in Science Teaching, 48*(2), 117–148.

Rowe, M. B. (1986). Wait time: Slowing down may be a way of speeding up! *Journal of Teacher Education, 37*(1), 43–50.

Slavin, R. E., Hurley, E. A., & Chamberlain, A. (2003). Cooperative learning and achievement: Theory and research. In W. Reynolds & G. Miller (Eds.), *Handbook of psychology: Part 3* (pp. 177–198). New York, NY: Wiley.

Smith, C. L., Wiser, M., Anderson, C. W., & Krajcik, J. (2006). FOCUS ARTICLE: Implications of research on children's learning for standards and assessment: A proposed learning progression for matter and the atomic-molecular theory. *Measurement: Interdisciplinary Research & Perspective, 4*(1–2), 1–98.

Wiliam, D., & Black, P. (1996). Meanings and consequences: A basis for distinguishing formative and summative functions of assessment? *British Educational Research Journal, 22*(5), 537–548.

Zembal-Saul, C., Blumenfeld, P., & Krajcik, J. (2000). Influence of guided cycles of planning, teaching, and reflection on prospective elementary teachers' science content representations. *Journal of Research in Science Teaching, 37*(4), 318–339.

F. RIGA, M. WINTERBOTTOM, E. HARRIS AND L. NEWBY

19. INQUIRY-BASED SCIENCE EDUCATION

CHAPTER SUMMARY

By all accounts, science and inquiry should go hand in hand. Whether the same is true of science *education* and inquiry is quite another matter. This chapter charts the evolution of inquiry-based approaches in science education from their first appearance at the beginning of the 20th century to the present time. The fundamental need to teach science in as close a manner as is feasible (within a school context) to the manner in which it is conducted by the scientific community is set out. We examine the way inquiry-based science education (IBSE) is understood and interpreted today, alongside some of the models science educators have used to enact inquiry. Both the challenges facing science teachers when implementing inquiry approaches in their classrooms, and the problems of assessing inquiry-based work, are also discussed. The chapter ends with a look at what the future might hold for IBSE. It is intended that this chapter clarifies what is meant by IBSE and its place in science education, and presents teachers with ideas about how they might incorporate inquiry into their everyday classroom teaching.

REINVENTING SCIENCE EDUCATION

> Science has been taught too much as an accumulation of ready-made material with which students are to be made familiar, not enough as a method of thinking. (Dewey, 1910, pp. 122, 124)

The demand for a new way of conceptualizing the teaching of science – as *inquiry-based* – is not a recent idea. First indications of a *disconnect* between the way science was being taught in schools and how it was practised by the scientific community became apparent as early as the beginning of the 20th century. In England, Armstrong, a forerunner of inquiry-based approaches, advocated the use of *heuristic methods* in teaching school science – "Heuristic methods of teaching are methods which involve our placing students as far as possible in the attitude of the discoverer – methods which involve their *finding out* instead of being merely told about things" (Armstrong, 1910, p. 236) – an approach popularly known at the time as the 'Armstrong Method' (Jenkins, 1979). He argued that "the use of eyes and hands", i.e. scientific method, "cannot be taught by means of the blackboard and chalk or even by experimental lectures and demonstrations alone" (p. 9). Rather,

K. S. Taber & B. Akpan (Eds.), Science Education, 247–261.

it was of "fundamental importance" that children, "from the outset learn to acquire knowledge by their own efforts" (p. 10).

The notion that science should be taught in ways that would kindle students' curiosity and stimulate their thinking was supported by both educators and scientists alike (Armstrong, 1910; Dewey, 1938; Schwab, 1962; Rutherford, 1964). Despite this, school science continued to be routinely taught as the transmission of a series of unchanging facts from teacher to students, with students being required to learn these facts by heart. The 1950s saw a resurgence of ideas that insisted science be taught as an "effective method of inquiry" (Dewey, 1910, p. 124) which included an appreciation of science as a subject whose laws and concepts were continually open to inspection, challenge, and update.

By the 1970s, changes had also become evident in the way psychologists were viewing learning. Behaviourist theories had viewed student learning as an externally imposed obligation by an educator who utilized conditioning (classical or operant) to encourage favourable behaviours and discourage unfavourable ones. The advent of constructivist theories challenging this perception, asserted that individuals *themselves* constructed their own knowledge through experience, and the development of their cognitive structures (Kalat, 2005). Concurrently, educational psychologists such as Ausubel and Bruner began to describe learning in terms of concept formation and the meaningful assimilation of new ideas/concepts (Ausubel et al., 1978; Bruner, 1962, 1986, 1990). According to Ausubel et al. (1978), learning was "the product of an active, integrative interaction between new instructional materials and relevant ideas in the learner's existing structure of knowledge" (p. 40). He laid strong emphasis on *meaningful learning* (as opposed to the learning of meaningful material – usually by rote), proposing that, by definition, meaningful learning involved "the acquisition of new meanings" (p. 67), where new meanings, in turn, "are interactional products of a meaningful learning process, in which new ideas are related to, and interact with, relevant ideas in existing cognitive structure" (p. 72). Therefore, new understandings come about as the product of the interaction of new ideas/concepts *and* elements already present in the individual's cognitive structure.

Such cognitive and constructivist views of learning led to a mass of research literature in the 1980s and 1990s which argued that students did not come to the science classroom as *tabula rasas*, but brought with them their own prior notions of concepts which have been referred to variously as alternative conceptions/frameworks, prior conceptions, preconceptions or misconceptions. Advocates of inquiry-based learning and teaching have suggested that inquiry approaches can potentially provide the sort of environment in which "meaningful science learning can occur" (Asay & Orgill, 2009, p. 57), and that "learning through inquiry accords with modern views of the psychology of learning, which sees learners having an active role in their learning" (Harlen, 2004, p. 7).

> When students are developing their understanding of the natural and made world around them, then, like scientists, they can use inquiry to arrive at ideas

> and theories that help them explain what they observe. Students also have to change their ideas as they encounter new and conflicting evidence. And, like scientists too, they do not begin from a clean slate, but from what they already know and the ideas they have already. (Harlen, 2004, p. 5)

If all the above strands of thought are viewed in concert, it is perhaps no accident that the idea of inquiry-based approaches to science education began to take hold at the time they did, because inquiry was seen by many as a means of reinventing science education to meet the needs of a modern society. Speaking in 2013 to a group of EU science teachers,[1] Bybee stressed the importance of 'scientific literacy' – a term which began to appear in educational literature in the 1950s (Hurd, 1958). Despite the fact that its meaning has not always been consistent amongst its users, 'scientific literacy' is the term generally used to describe "the intentions of science education" (Holbrook & Rannikmae, 2009, p. 275). Bybee contends that if we are to apply science to solving the problems of society in the 21st century, a scientifically literate society is needed – and to achieve this, a *linked-up* approach to teaching and learning science is crucial. It therefore follows that subjects such as biology, physics and chemistry should be studied in an *integrated* way, as opposed to studying these subjects independently through a single-science lens (Bybee, 2013). Developing an inquiry-based approach to science education is seen as a means of achieving scientific literacy.

> "even the casual observer recognises that science with its applications in technology has become the most characteristic feature of modern society" and as such "more than a casual acquaintance with scientific forces and phenomena is essential for effective citizenship." (Hurd, 1958, p. 13)

Although many educators believed the way forward in science education – especially with a view to achieving scientific literacy – was through inquiry approaches, the matter of how precisely IBSE could be *defined* remained something of an enigma, giving rise to innumerable questions. Could the term in fact *be* pinned down? And, supposing it could, how would inquiry be implemented? What might be the implications of an inquiry-based agenda? How effective would such an approach be in delivering a science education which could pave the way for scientists of the future to address 'modern' problems? Moreover, how would IBSE impact on teachers' and/or students' roles? What would teachers' attitudes be towards IBSE? What were the obstacles and challenges? In the subsequent sections of this review, we attempt to ascertain/discern how educators and researchers have attempted to address these, amongst other, questions.

DEFINING IBSE

Despite IBSE having a long history in science education, the term has been used loosely by people to mean different things, making it difficult to arrive at a clear and consistent definition – possibly because the same term has been used in describing "both *teaching* and *doing* science" (Colburn, 2000, p. 42). The publication of the *National Science Education Standards* (NSES) (National Research Council (NRC), 1996) and the *Inquiry and the National Science Education Standards: A Guide for Teaching and Learning* (INSES) (NRC, 2000) went some way towards addressing this "confusion" by attempting to define IBSE and setting out clear guidelines on how inquiry could be recognised.

The NRC's point of departure was to describe *scientific* inquiry as "the diverse ways in which scientists study the natural world and propose explanations based on the evidence derived from their work" (National Research Council, 1996, p. 23), before going on to define IBSE. This publication then portrays inquiry as an approach to teaching science which engages students in the same sorts of activities, practices, and thinking processes that scientists use in their work, i.e. in their pursuit of scientific inquiry. Furthermore, inquiring into "authentic questions" stemming from students' prior experiences is put forward as the main strategy for teaching science (NRC, 1996, p. 31) – a standpoint which also accords with constructivist principles.

According to Anderson (2002), the NRC's (1996) use of the term *inquiry* encompasses three distinct aspects:

1. *Science as inquiry*, i.e. "scientific inquiry refers to the diverse ways in which scientists study the natural world and propose explanations based on the evidence derived from their work" (NRC, 1996, p. 23).
2. *Learning as inquiry*, i.e. students are encouraged to be actively engaged in the learning process – "something that students do, not something that is done to them" (National Research Council, 1996, p. 2). So, "inquiry also refers to the activities of students in which they develop knowledge and understanding of scientific ideas, as well as an understanding of how scientists study the natural world" (NRC, 1996, p. 23).
3. *Teaching as inquiry*, i.e. "inquiry into authentic questions generated from the students' experiences is the central strategy for teaching science" (NRC, 1996, p. 31).

The NRC summarizes the core components of inquiry as follows:

> Inquiry is a multifaceted activity that involves making observations; posing questions; examining books and other sources of information to see what is already known; planning investigations; reviewing what is already known in light of experimental evidence; using tools to gather, analyze, and interpret data; proposing answers, explanations, and predictions; and communicating the results. (National Research Council, 1996, p. 23)

In order to arrive at scientific knowledge and understanding, a broad margin is often applied to the sorts of inquiry methods/techniques employed to achieve this. This suggests that inquiry activities could include (a) those tasks in which students set up their own questions and collect their own data to analyze, *as well as* (b) those in which questions and data are provided to students, who then analyze these to derive their own conclusions. Therefore, inquiry-based learning is thought to refer to "all forms of scholarly exploration and investigation carried out by students as part of their studies or in extra-curricular contexts" (Levy et al., 2011, p. 7). However, as Bell et al. point out, most students would probably require "substantial scaffolding" before they would be able to 'plunge into' developing their own scientific questions and designing data collection procedures to address these questions (2005, p. 30). This may create a dilemma for teachers because, from a Vygotskyan standpoint, *scaffolding* was intended as a resource only to be used up to the point when new knowledge has been 'internalized' (Vygotsky, 1962, 1978). Consequently, educators need to be cautious, and provide appropriate scaffolding only if and when needed, and even then, only up to a point – so as not to jeopardize the 'inquiry'.

While the *Standards* (NRC, 1996) encouraged students to work in groups to design and "conduct investigations that begin with a question and progress toward communicating an answer to the question" (p. 141), the follow-up publication in 2000 identified the five "essential features" of inquiry as follows:

1. Learner engages in scientifically oriented questions
2. Learner gives priority to *evidence* in responding to questions
3. Learner formulates *explanations* from evidence
4. Learner connects explanations to scientific knowledge
5. Learner communicates and justifies explanations (National Research Council (NRC), 2000, p. 29).

How much direction and autonomy teachers give students can vary within each of the above five features, resulting in a spectrum of *levels of inquiry* which can range from highly structured 'recipe-style' activities/tasks at one end to open-ended projects where students enjoy complete autonomy. Nowadays, practitioners frequently use four levels to distinguish between the various levels of support or 'scaffolding' supplied to learners.

1. *Confirmation* – these are traditional 'recipe-style' laboratory activities where students are given step-by-step guidance in order to confirm an already-known principle.
2. *Structured inquiry* – these are activities where the teacher provides the question to be explored, as well as equipment and instructions, but students do not know what the result/solution will be.
3. *Guided inquiry* – in these activities, teachers only provide students with a problem/question, and students design/choose the methods used to collect and analyse data.
4. *Open inquiry* – these are activities where teachers provide a general topic and allow students to generate their own scientific question to investigate. Students have complete autonomy in designing and conducting the investigation.

The purpose of science education is generally held to be the learning of a curriculum of science content so as to achieve scientific literacy. Nevertheless, the learning of content in itself does not typically lead to scientific literacy, and this is where inquiry-based approaches may assist in delivering this purpose of science education. Inquiry should not be perceived in terms of *replacing* the teaching of content – "inquiry is not process versus content; rather it is a way of learning content" (Drayton & Falk, 2001, p. 25).

ENACTING IBSE

Much like defining IBSE, the term 'implementing inquiry' has come to mean different things to different people. On the one hand, there are practitioners who believe implementing inquiry entails using *only* inquiry approaches (meaning 'open inquiry') to teach science content, i.e. students "should themselves find out by inquiring into the world, rather than simply being told what science has found out" (Taber, 2011, p. 258). On the other hand, there are those who believe 'implementing inquiry' means *incorporating* inquiry approaches – when possible and where appropriate – amongst other strategies they use. This latter interpretation is the one generally held by many practitioners in the UK, where teachers tend to plan lessons by first setting out intended learning goals before deciding on the most appropriate approach(es) to achieving them. Bearing in mind that experiencing inquiry science at school is quite different from practicing inquiry as a professional scientist (Kyle, 1980), implementing inquiry at school does not automatically imply that students are practicing science as 'real' scientists do. With this in mind, Colburn (2000, p. 44) recommends teachers "find the right mix of inquiry and non-inquiry methods" to engage their students in learning science.

Gauging whether a task or activity *is* inquiry-based or not, can be difficult unless criteria or standards are specified, and even then, a task/activity may exhibit some (but not all) the said criteria. Bell et al. stipulate two requirements be met for any task/activity to be considered inquiry-based: first, students must have a *research question* to answer, and second, students must undertake *data analysis* to draw conclusions.

Although there may be no blueprint for the optimal inquiry lesson, there are a number of recognizable characteristics that facilitate inquiry-based learning (IBL) which has been described as:

> a cluster of strongly student-centred learning and teaching approaches in which students' inquiry or research drives the learning experience. Students conduct small- or large-scale inquiries that enable them to engage actively with disciplinary or interdisciplinary questions and problems. Learning takes place through an emergent process of exploration and discovery. Guided by subject specialists and those with specialist roles in learning support, students use the scholarly and research practices of their disciplines to move towards autonomy in creating and sharing knowledge. (Levy et al., 2011, p. 6)

Table 1. Information practitioners might offer their students, together with the kinds of questions explored by learners engaged in 'structured', 'guided' or 'open' inquiry activities

	STRUCTURED	*GUIDED*	*OPEN*
Information provided by the teacher	• The question or issue to examine • The resources needed • Instructions presented to students in a step-by-step format	• The question or issue to examine • The resources needed	None provided – learner makes all the decisions about • what to investigate • how to conduct the investigation • why to research this particular question (within a particular area of study)
Questions learners might ask themselves when engaged in inquiry activities	• What observations do I need to make and record? • How should I record my observations? • Can I explain what the observations mean?	• How can I go about answering the question? • What procedure(s) or method(s) can I devise/think up that will enable me to answer the question? • What observations do I need to make and record? • How should I record my observations? • Can I explain what the observations mean? • Can I find out how other people have gone about answering the question?	• What question should I decide to investigate? • How should I phrase the question? • What background research will I need to conduct before proceeding? • How should I go about investigating this question? • What procedure(s) or method(s) can I devise/think up that will enable me to answer the question? • What observations do I need to make and record? • How should I record my observations? • Can I explain what the observations mean? • Can I find out how other people have gone about answering the question? • How can I best present my findings to the class? • Will/How will I be able to defend the decisions I have made throughout the process?

As outlined in the previous section, implementing inquiry could involve anything from highly structured tasks to open-ended project-work, though Bell et al. stress that most learners may require considerable “scaffolding” from teachers “before they are ready to develop scientific questions and design effective data collection procedures to answer these questions” (Bell et al., 2005, p. 30). Table 1 presents the kinds of information teachers could provide to students as well as the kinds of questions learners might ask while engaged in ‘structured’, ‘guided’ or ‘open’ inquiry activities.

MODELS FOR ENACTING IBSE

A number of models have been put forward for enacting IBSE. These models fall into two broad categories:

1. *Instructional models* – aimed at practitioners – are primarily concerned with passing on recipes on ‘how to’ implement inquiry, and these tend to present ‘ideals’;
2. *Academic models* – which provide the theoretical underpinning for instructional models – stem from pedagogic and cognitive theories about how to learn science, and form the theoretical framework moulding together the diverse elements that make up inquiry-based education.

Where academic models explain what inquiry is (or should be) and why it can aid students in understanding scientific concepts, instructional models *interpret* the theory behind inquiry for practitioners to use in their teaching practice. Hence, instructional models help to translate academic models into everyday teaching practice by interpreting how the theoretical basis for IBSE can be implemented/ enacted in learner environments (a term used here to include both *in-school classrooms* and *out-of-school learning situations*).

A number of instructional models have emerged since the first *learning cycle* model (known as *3E* – exploration, invention, discovery) was presented by Atkin and Karplus (1962). Two of the more popular, recent models are: Bybee’s *5E* learning cycle model (engagement, exploration, explanation, elaboration, evaluation) (2002) – with Eisenkraft extending this to a *7E* model by adding elicitation and extension (2003) – and Llewellyn’s ‘six stages’, i.e. inquisition, acquisition, supposition, implementation, summation, and exhibition (2002).

Instructional models (e.g. the learning cycles above) also address the three aspects of inquiry mentioned earlier in this chapter. Activities with elements of ‘invention’ or ‘discovery’ could be seen to address the *science as inquiry* aspect (i.e. how scientists conduct inquiry in their work); those with elements of ‘engagement’, ‘elaboration’, or ‘implementation’ could fulfil the *learning as inquiry* aspect (i.e. engaging students in doing activities that develop their knowledge and understanding of scientific ideas); and, activities with elements of ‘elicitation’, ‘inquisition’, ‘supposition’, ‘exploration’, or ‘implementation’ could address the *teaching as inquiry* aspect (i.e. students undertaking inquiry into authentic questions derived from everyday experiences). Thus, elements from the instructional models (such as those mentioned above) embody the theoretical basis for inquiry outlined in academic literature, albeit conceding that science in schools cannot be practiced precisely as it is by the scientific community.

TEACHERS' VS STUDENTS' ROLES IN IBSE

To enact IBSE successfully, the actions of both teachers and learners are crucial in driving the thinking processes that are expected to bring about meaningful learning. "Inquiry learning results in deep understanding of many aspects of science, as opposed to learning through more prescriptive methods" (Leonard & Penick, 2009, p. 41). Given that the goals of IBSE differ from those of transmission modes of teaching (or direct instruction modes), the traditional role of the learner as a "passive follower" who accumulates information (often by rote), needs to be replaced with roles more attuned to those of an "active designer" (Brickman et al., 2009, p. 16).

Crawford (2000), reporting on a year-long case study where the beliefs and practices of a biology teacher are examined as he applies inquiry approaches in an ecology class of 20 students (all in their final 2 years of high school in the USA), singles out six key characteristics evident in this teacher's classroom: "*situating instruction in authentic problems; grappling with data; collaboration of students and teacher; connection with society; teacher modelling behaviours of a scientist, and fostering student ownership*" (p. 927).

In her study, Crawford isolated a set of *roles* assumed by the teacher. She observed that different tasks required the teacher to assume different roles – frequently undertaking "myriad" roles, some demanding "a high level of expertise" (p. 932). Student roles, too, were many and varied and could encompass roles previously viewed as "reserved" exclusively for teachers. A summary of the roles identified in Crawford's study is as follows:

Teacher's Roles:	motivator, diagnostician, guide, innovator, experimenter, researcher, modeller, mentor, collaborator, learner.
Student's Roles:	*traditional* roles – learner, listener, receiver of information. *new* roles – active collaborator, leader, apprentice, teacher, planner.

CREATING AN INQUIRY-ORIENTED CLASSROOM/INSTRUCTIONAL ENVIRONMENT

Another aspect, key to implementing inquiry (yet often overlooked!), is the physical appearance of the classroom, i.e. the classroom environment. Scientists work in many different modes and the classroom environment should reflect this. The physical space in which inquiry activities take place should not only provide the resources/materials to equip students in pursuing these activities/inquiry tasks (i.e. they should have what they need easily within reach, e.g. computers, internet access, books, journals, chart paper, coloured pens, magnifying glasses, etc.), but should also provide a space which is conducive to eliciting the kind of thinking processes that inquiry requires. Thus, does the classroom environment reflect that *science as inquiry*, *learning through inquiry*, and *teaching as inquiry*, is taking place? How are desks/furniture arranged around the room – and is their position static or changed according to the task/activity students are engaged in? Are inducements to investigation present (e.g. posters on walls, displays of students' own work, displays reflecting students'/the teacher's particular interests) – and are such inducements changed or updated regularly?

> The inquiry-oriented classroom has many tools and instruments around – some in current use, some used a few times during the year ... any of these instruments may serve as an incitement to investigation for the student who happens to notice and wonder about them. (Drayton & Falk, 2001, p. 28)

CHALLENGES TO IMPLEMENTING IBSE

In 1958, Schwab identified four reasons why educators clung to "the rigid literalism of nineteenth-century science" education:

- *Time consuming* – the time required to implement an inquiry-based approach was seen as only being possible "at the expense of coverage" (p. 376), i.e. content would have to be sacrificed to make time for inquiry-based approaches.
- *Confusion due to complexity of inquiry* – presenting students with doubts and an array of alternatives to choose from would only bewilder them.
- *Job requirements* – pressure from industry (and other areas) for particular skills (e.g. engineers) would place restrictions on the curriculum and direct it towards particular fields.
- *Economics* – inquiry-based approaches would be too costly to implement in everyday classrooms.

Although inquiry approaches have generally been encouraged for over 50 years, research suggests that changes within classrooms towards incorporating more inquiry-based approaches have been slow. Classroom lessons are generally still teacher-centred and textbook-based, with Schwab's reasons unfortunately still resonating with us today.

ASSESSMENT OF INQUIRY-BASED WORK

Inquiry-based instruction transcends the transfer of content-knowledge from teacher to learner by seeking to elicit and advance autonomous thinking and critical thinking skills in learners. Taken together with the broader range of learning goals identified when teachers engage in inquiry-based approaches – compared to conventional science practices – the challenge for educators to come up with assessment procedures that will accurately reflect their students' knowledge and understanding seems formidable.

Harlen describes the two main goals of assessment as being "to help students while they are learning" (formative assessment) and "to summarize and report it ... to find out what they have learned at a particular time" (summative assessment) (Harlen, 2013, p. 16).

For assessment to incorporate the broader goals of IBSE, Harlen (2013) contends that *formative assessment* is essential for inquiry-based approaches to be appropriately implemented. She claims that formative assessment's chief strength lies in its ability to develop deeper understandings and better competency skills which are crucial for learning to progress. The "continual formative assessment of student understanding through observation, student questioning, and written assignments" will help teachers to discern when it might be feasible to encourage learners to move towards more open-ended inquiry tasks, and alternatively, when they should "backtrack and scaffold" students' learning instead (Colburn, 2004, p. 66). Formative assessment can also be a useful tool for assessing which of the "multiple levels of inquiry" (i.e. *confirmation*, *structured*, *guided*, or *open* inquiry) a learner belongs to or is advancing towards (Banchi & Bell, 2008).

The following formative assessment strategies that teachers might employ within an inquiry learning environment have been put forward by Harlen (2013):

- teachers' questions and allowing sufficient time for answering
- teachers giving feedback on students' work
- teachers listening to students' feedback on their teaching
- student self-assessment and peer-assessment

The aims of *summative assessment* practices, on the other hand, are often viewed as follows: to ensure students learn the intended subject matter; to track students' learning for both school authorities and parents/guardians; and, to supply information of value to school improvement. Such purposes can influence learning directly, though they might not initially seem to do so (Harlen, 2013). However, concerning IBSE, the gap between "what can be assessed" and "what ought to be assessed" is much larger because its objectives are associated with deepening understanding and developing those competencies used by 'real' scientists (p. 22).

The following methods for applying summative assessment within IBSE have been set out by Harlen (2013):

- tests incorporating knowledge application as well as simply recall;
- questions and tasks that assess science inquiry skills;
- verbal and/or written explanations to justify events,
- data, and/or predictions;

- portfolios of work generated over some time (including accounts, reflections, photos, etc.);
- regularly checking student notebooks and/or electronic postings;
- presentations by groups and/or individuals.

THE FUTURE OF IBSE

We have seen that reinventing science education has not been an instant happening, but a process which began with the notion of making the study of science at school *more like* the way scientists practise their profession in the real world. Over the last 50 years, there have been clear indications from research in science education that inquiry-based approaches might be better suited (than hitherto traditional methods) to giving learners a more holistic and realistic view of science and its methods of operation. Knowing certain scientific concepts, understanding the meaning of these concepts, knowing how they operate and how they relate to everyday life, is what science literacy/education is all about.

From the research literature on IBSE, its future seems to rest to a large degree on:

1. The extent to which IBSE proves to be *effective in advancing scientific literacy* in learners.

   As we have seen, scientific literacy is more than the retention of scientific information well beyond a person's school life – it is the achievement of deeper understandings of scientific concepts. However, establishing scientific literacy is no easy matter. Knowing when scientific literacy has been achieved seems to be inextricably linked to the assessment practices in place (i.e. how assessment is measured) – one can only determine if a person is literate in science by assessing his/her knowledge. But, how *can* one assess scientific knowledge to establish scientific literacy? The answer seems to lie in ascertaining *what* precisely *is* being assessed. If assessment practices are measuring student understanding of concepts and scientific methods, rather than focusing primarily on factual content, then scientific literacy and achievement should go hand in hand. Hence, assessment practices would need to be more explicit in capturing the extent of learners' understandings of scientific concepts and methods.
2. IBSE's ability to *move beyond being 'the new kid on the block'*.

   In the past, much of the research associated with IBSE was concerned with establishing whether or not it represented a more effective way of teaching science. What is now needed is for IBSE to move beyond this into areas such as:
   a. the development of more easy-to-access IBSE *resources* by teachers so as to make their lives easier – with clear objectives and geared towards making students more self-sufficient/autonomous learners;
   b. more *intervention-type research* to establish *how* and *why* IBSE improves science literacy (rather than seeking to establish that it does);

c. the *logistics of changing classroom environments* from 'traditional' to 'inquiry-based' spaces – and the effects of such changes in classroom environment.

A key ingredient in inquiry-based approaches in education has been getting students to think for themselves – both while working together collaboratively in groups and when working alone. How proficient a student becomes in the practice of thinking independently is what will ultimately determine whether s/he will be a 'disciple' or an 'inquirer' (Dewey, 1980).

NOTE

[1] Taken from Prof. Rodger W. Bybee's talk, entitled *Strategies for Developing Scientific Literacy*, given at the 'Best Practices in Inquiry-Based Science Education Summer School' for teachers in Crete, 30 June 2013.

LIST OF RECOMMENDED READINGS

Abd-El-Khalick, F., & Lederman, N. G. (2000). Improving science teachers' conceptions of nature of science: a critical review of the literature. *International Journal of Science Education, 22*(7), 665–701.

Abd-El-Khalick, F., Boujaoude, S., Duschl, R., Lederman, N. G., Mamlok-Naaman, R., Hofstein, A., Niaz, M., Treagust, D., & Tuan, H. L. (2004). Inquiry in science education: International perspectives. *Science Education, 88*(3), 397–419.

Anderson, R. D. (2002). Reforming science teaching: What research says about inquiry. *Journal of Science Teacher Education, 13*(1), 1–12.

Colburn, A. (2008). An inquiry primer. In E. Brunsell (Ed.), *Readings in science methods K–8* (pp. 33–36). Arlington, VA: NSTA Press. Retrieved from http://matkinsscienceeducation20042014.wmwikis.net/file/view/ReadingsinScienceMethodsbook.pdf#page=51

Gibson, H. L., & Chase, C. (2002). Longitudinal impact of an inquiry-based science program on middle school students' attitudes toward science. *Science Education, 86*(5), 693–705.

Keys, C. W., & Bryan, L. A. (2001). Co-constructing inquiry-based science with teachers: Essential research for lasting reform. *Journal of Research in Science Teaching, 38*(6), 631–645.

Lawson, A. E. (2010). *Teaching inquiry science in middle and secondary schools*. Los Angeles, CA: Sage.

Minner, D. D., Levy, A. J., & Century, J. (2010). Inquiry-based science instruction—what is it and does it matter? Results from a research synthesis years 1984 to 2002. *Journal of Research in Science Teaching, 47*(4), 474–496.

Thomas, J. W. (2000). A review of research on project-based learning. Retrieved from http://bie.org/object/document/a_review_of_research_on_project_based_learning; http://www.newtechnetwork.org.590elmp01.blackmesh.com/sites/default/files/dr/pblresearch2.pdf

REFERENCES

Anderson, R. (2002). Reforming science teaching: What research says about inquiry. *Journal of Science Teacher Education, 13*(1), 1–12.

Armstrong, H. E. (1910). *The teaching of scientific method and other papers on education.* London: MacMillan. Retrieved from https://archive.org/stream/sciencemethostea00armsuoft#page/n7/mode/2up

Asay, L. D., & Orgill, M. (2009). Analysis of essential features of inquiry found in articles published in the science teacher, 1998–2007. *Journal of Science Teacher Education, 21*(1), 57–79.

Atkin, J., & Karplus, R. (1962). Discovery of invention. *The Science Teacher, 29*(5), 45–47.

Ausubel, D., Novak, J., & Hanesian, H. (1978). *Educational psychology: A cognitive view* (2nd ed.). New York, NY: Holt, Rinehart & Winston.
Banchi, H., & Bell, R. L. (2008). The many levels of inquiry. *Science and Children, 46*(2), 26–29.
Bell, R. L., Smetana, L., & Binns, I. (2005). Simplifying inquiry instruction. *The Science Teacher, 72*(7), 30–33.
Brickman, P., Gormally, C., Armstrong, N., & Hallar, B. (2009). Effects of inquiry-based learning on students' science literacy skills and confidence. *International Journal for the Scholarship of Teaching and Learning, 3*(2), 1–22.
Bruner, J. S. (1962). *A study of thinking*. New York, NY: Science Editions, Inc.
Bruner, J. S. (1986). *Actual minds, possible worlds*. Cambridge, MA: Harvard University Press.
Bruner, J. S. (1990). *Acts of meaning*. Cambridge, MA: Harvard University Press.
Bybee, R. W. (2002). *BSCS 5E instructional model*. Colorado Springs, CO: Biological Sciences Curriculum Study.
Bybee, R. W. (2013). Keynote talk entitled *Strategies for developing scientific literacy* given at the 'Best Practices in Inquiry-Based Science Education Summer School' in Crete.
Colburn, A. (2000). An inquiry primer. *Science Scope, 23*(6), 42–44.
Colburn, A. (2004). Inquiring scientists want to know. *Association for Supervision and Curriculum Development, 62*(1), 63–66.
Crawford, B. A. (2000). Embracing the essence of inquiry: New roles for science teachers. *Journal of Research in Science Teaching, 37*(9), 916–937.
Dewey, J. (1910). Science as subject-matter and as method. *Science, 31*(787), 121–127.
Dewey, J. (1938). *Logic: The theory of inquiry*. New York, NY: Holt, Rinehart, and Winston.
Dewey, J. (1980). Democracy and education. In J. A. Boydston (Ed.), *John Dewey: The middle works, 1899–1924: Vol. 9. 1916* (pp. 1–370). Carbondale, IL: Southern Illinois University Press.
Drayton, B., & Falk, J. (2001). Tell-tale signs of the inquiry-oriented classroom. *National Association of Secondary Schools Principals (NASSP) Bulletin, 85*(623), 24–34.
Eisenkraft, A. (2003). Expanding the 5E model: A proposed 7E model emphasizes 'transfer of learning' and the importance of eliciting prior understanding. *The Science Teacher, 70*(6), 56–59.
Harlen, W. (2004). *Evaluating inquiry-based science developments*. Retrieved from http://socrates.usfca.edu/xapedoe/ibl12/page1/page20/assets/wharlen_inquiry_mtg_paper.pdf
Harlen, W. (2013). *Assessment & inquiry-based science education: Issues in policy and practice*. Trieste, Italy: Global Network of Science Academies (IAP) Science Education Programme (SEP). Retrieved from http://www.interacademies.net/File.aspx?id=21245
Holbrook, J., & Rannikmae, M. (2009). The meaning of scientific literacy. *International Journal of Environmental & Science Education, 4*(3), 275–288.
Hurd, P. D. (1958). Science literacy: Its meaning for American schools. *Educational Leadership, 16*(1), 13–16, 52.
Jenkins, E. W. (1979). *From Armstrong to Nuffield: Studies in twentieth century science education in England and Wales*. London: John Murray.
Kalat, J. W. (2005). *Introduction to psychology* (7th ed.). Southbank, Vic: Thomson/Wadsworth.
Kyle, W. C. (1980). The distinction between inquiry and scientific inquiry and why high school students should be cognizant of the distinction. *Journal of Research in Science Teaching, 17*(2), 123–130.
Leonard, W. H., & Penick, J. E. (2009). Is the inquiry real? Working definitions of inquiry in the science classroom. *Science Teacher, 76*(5), 40–43.
Levy, P., Little, S., McKinney, P., Nibbs, A., & Wood, J. (2011). *The Sheffield companion to inquiry-based learning*. CILASS, Centre for Inquiry-based Learning in the Arts and Social Sciences, The University of Sheffield, UK.
Llewellyn, D. (2002). *Inquire within: Implementing inquiry-based science standards*. Thousand Oaks, CA: Corwin Press.
National Research Council. (1996). *National science education standards*. Washington, DC: National Academy Press.
National Research Council (NRC). (2000). *Inquiry and the national science education standards: A guide for teaching and learning*. Washington, DC: National Academy Press.

Rutherford, F. J. (1964). The role of inquiry in science teaching. *Journal of Research in Science Teaching, 2*, 80–84.
Schwab, J. J. (1958). The teaching of science as enquiry. *Bulletin of the Atomic Scientists, 14*(9), 374–379.
Schwab, J. J. (1962). The teaching of science as enquiry. In J. J. Schwab & P. F. Brandwein (Eds.), *The teaching of science* (pp. 3–103). Cambridge, MA: Harvard University Press.
Taber, K. S. (2011). Inquiry teaching, constructivist instruction and effective pedagogy. *Teacher Development, 15*(2), 257–264.
Vygotsky, L. (1962/2012). *Thought and language* (E. Hanfmann, G. Vakar, & A. Kozulin, Eds., & Trans). Cambridge, MA: The MIT Press.
Vygotsky, L. S. (1978). *Mind in society: The development of higher psychological processes* (M. Cole, V. John-Steiner, S. Scriber, & E. Souberman, Eds.). Cambridge, MA: Harvard University Press.

KEITH S. TABER

20. MODELS AND MODELLING IN SCIENCE AND SCIENCE EDUCATION

This chapter discusses the nature and roles of models in science, and in science education. It is argued that models and modelling are important in science teaching both because of the need to authentically reflect the importance of modelling in science itself, and because of the pedagogic role of models. It is suggested that effective teaching practice requires teachers to distinguish these two different roles of models in the science classroom. There are extensive literatures relating to the role of models in the practice of science, and to the use of models in science teaching, and the present chapter sets out to introduce readers to some key ideas about this important topic.

WHAT ARE MODELS?

A model can be understood as something that stands for something else, but which provides an affordance that goes beyond a simple representation, thus allowing the model to be a tool for some kind of action. Sometimes that may be a physical action, but often models used in science are primarily thinking tools. In particular, models are used to develop and test scientific explanations (Gilbert, 1998). It is in the nature of models then to be different from what they are modelling. One key feature is that models are often simpler. Many phenomena that scientists study are complex and models can offer carefully selected simplifications. One example would be Daisyworld which was used to explore an idea about the role of feedback cycles in natural ecosystems in maintaining stability despite perturbations. Daisyworld was designed to test an aspect of James Lovelock's Gaia theory which suggested that the natural environment needs to be understood in terms of complex interactions between physical, geological and biological features. Lovelock argued that the evolution of life on earth involved the development of complex interactions that, within certain limits, worked to keep conditions stable.

One problem in understanding the Earth's hospitality for life is how the planet has remained suitable for life despite significant changes in the Sun's energy output (as a result of the gradual shifts in the Sun's composition due to the nuclear reactions that cause the Sun to shine). All other things being equal, the Earth should have got a lot hotter – and so should either have been too cold for complex life when such lifeforms first appeared, or be too hot for complex life now. Yet the fossil record

K. S. Taber & B. Akpan (Eds.), Science Education, 263–278.

shows that the climate must have remained moderately stable over periods when the Sun's output has changed considerably. The geological record shows that there have certainly been many shifts in the Earth's climate but these have never been extreme enough to threaten life. Lovelock suspected there were feedback cycles that maintained conditions within certain bounds.

The biota on the model world, Daisyworld, comprised of just two varieties of daisies (black and white) which suited different conditions. Now such a simple biota would not be viable, and certainly does not reflect the complex range of organisms on earth. However, the idea of Daisyworld was to offer a very simple scenario that could test an idea. In the model the two types of daisy interacted with the environment differently (the black ones absorbing more radiation from the planet's sun, and re-emitting it at wavelengths that would heat the planet; the white ones reflecting more radiation back into space) and thrived in different conditions (the white daisies, less able to warm up by absorbing radiation, thrived better when the planet was warmer). This simple model showed that as the Daisyworld sun's radiation intensity increased, the balance of white and black daisies shifted in response, which changed the albedo of the planet sufficiently to counteract the increased incoming radiation, and so maintain a temperature viable for the daisies to survive. Whilst Daisyworld was far simpler than the real earth, it illustrated that in principle an ecosystem can include negative feedback cycles to maintain constant conditions in response to substantial (if not extreme) perturbations. Daisyworld was actually a simulation programmed into a computer, which allowed the evolution of the system to be speeded up massively compared with the rate at which a star's output actually changes.

Models then are usually simpler than what they model, but they may also be different in other ways that facilitate enquiry that would be difficult to undertake with the real system. So for example, a scale model of an aerofoil, such as an airplane wing, may be placed in a wind tunnel, and subjected to tests to see what happens to the flow of air over the model wing surface under different conditions (wind speed, the wing's 'angle of attack'). Building and testing models is a good deal more resource-effective than building and testing full size wings, and allows problematic designs to be rejected. Another example might be the use of cadavers as models of patients with medical problems requiring surgical treatment. A dead body that has been bequeathed to medical science can sometimes be modified to model a disease condition, allowing surgeons to develop new techniques (or novice surgeons to develop skills) without putting live patients at risk.

One of the most famous examples of a scientific model is the molecular model of the structure of DNA built by Francis Crick and James Watson at the Cavendish Laboratory in the 1950s (Watson, 1968/1980). Students seeing the photographs of that model now may think that it was intended simply as a model *to represent* a structure. However, Crick and Watson did not initially know what the structure was, and used modelling as a way of finding a structure that fitted with the parameters suggested by various laboratory measurements (such as the known ratios between base pairs, and the interpretation of Rosalind Franklin's X-ray diffraction images).

Building models to test against available data (that was difficult to interpret directly in terms of structural features) was a useful complement to the laboratory research.

Crick and Watson's approach was somewhat novel at the time (although it had been used by the Nobel laureate Linus Pauling). Nowadays it is very common for pharmaceutical researchers to model and test potential drug compounds within computer simulations. As synthesis of new drug compounds can be an expensive and slow process, the use of computer simulations allows researchers to explore likely properties of vast numbers of potential structures, to decide which compounds are likely to be most worth synthesising for testing in laboratory work and clinical studies.

Although it is suggested above that a model goes beyond a representation, the distinction may not always be obvious. For example the following equation represents a chemical reaction:

$$P_4 + 6Cl_2 \rightarrow 4PCl_3$$

We would normally think of this as a symbolic *representation* rather than a model, although the equation represents the reaction in such a way as to support calculations about the actual chemical system – such as how much chlorine reacts with a certain amount of phosphorus (see Chapter 24: '*Teaching and Learning Chemistry*'). This suggests it may not be productive to seek to be too definitive about what does or does not count as a model. Whether something is a model depends on how it is understood and used, rather than just its inherent properties.

Metaphors, Similes, and Analogies

Metaphors, similes and analogies are important model-related notions in science and science education. Similes and metaphors are figures of speech that are used to help communicate meaning. A metaphor suggests one thing is *the same as* another (although this is not intended to be literally so) and a simile suggests one thing is *like* another. A metaphor would be that the *cellular nucleus is the brain of the cell.* A cell nucleus is quite different from a brain, but someone using this metaphor would be suggesting that there is some similarity between a brain and a cell's nucleus. A person using this metaphor would not be trying to persuade the listener or reader that a nucleus and a brain *are* the same, but rather that it is helpful to think about brains when considering some aspects of the nature of a cell (for example, that the nucleus has a major role in controlling activity in the cell similar to the role the brain has in controlling bodily activity).

The astute reader may have noticed that terms like '*the* nucleus', '*the* cell', and '*the* brain' have just been used as if they refer to definite entities – particular cells and brains – when clearly the comparison is a quite general one. It is common in science to refer to '*the* cell', '*the* heart', '*the* atom', and so forth when making general statements that refer to classes of objects, for example '*the* [sic] heart pumps blood around *the* [sic] body'. There is a kind of modelling going on here in the way we use a mental construct of a generic example. Members of a general class of objects (hearts) are considered similar enough to be represented in scientific arguments by a generic representation of the class ('the heart…'). In everyday language the phrase "the athlete is strong" (rather than "athletes are strong") would normally be assumed to apply to a specific athlete, but in science when we say that "the kangaroo is a marsupial mammal" we generally mean "the kangaroo" to stand for the general class of all kangaroos. So the statement "the dog is a four-legged animal" refers to the conceptual model of the generic dog and stands even though (due to specific contingencies) there are some particular dogs that do not have four legs.

When metaphors come to be used habitually they can actually take on the meaning that was previously only *implied* by the metaphor. Such metaphors are said to be dead (!) metaphors (so here one metaphor is being used to describe the nature of another metaphor – the figurative power of the metaphor has 'died' as it no longer represents a juxtaposition of two distinct ideas). Examples of dead metaphors that arise in science teaching would be saying that *covalent bonding is electron sharing* or that *the electron has spin*. These ideas have come to be accepted as literately true because within the context of the scientific topic the metaphor has been adopted as part of the informal or formal technical language of the subject. Although chemists realise that atoms cannot share anything, they know what is implied by the commonly – if informally – used notion that electrons are 'shared' in covalent bonds. By contrast, we can say that electrons *do actually* have spin because they have angular momentum, but the meaning of 'spin' here has been formally extended beyond the usual everyday notions of something rotating. Clearly there is scope for such figures of speech and associated shifts in meaning to confuse science learners, and teachers need to be careful not to use such language without ensuring learners know precisely what is implied.

A simile has a very similar role to a metaphor, but is phrased in terms of explicitly referring to the similarity ('the nucleus of the cell *is like* the brain'). The difference between simile and metaphor is therefore in terms of the phrasing. To say that "an enzyme catalyses reactions because it fits substrate molecules *like* a lock and key" is to offer a simile. A lock is designed so that only the intended key will open it, and (in a somewhat similar way) enzymes have evolved so they interact with very particular substrate molecules in ways that catalyse specific reactions.

Metaphors and similes are used to help us think about how one thing is much like, or in some way like, another. We can see here something of the process of modelling (one thing stands for another, to support thinking about some system or other), but we would normally not consider these figures of speech to be fully developed models. That said, they might well act as starting points for modelling. An example here might be the notion of the 'liquid drop' model of the atomic nucleus. This idea was proposed by Lise Meitner and her nephew Otto Robert Frisch. Meitner had left her laboratory in Germany to escape Nazi persecution and had been sent details of experimental results obtained by her colleagues Otto Hahn and Fritz

Strassmann. These results did not make sense in terms of what was then understood about nuclear processes. Meitner and Frisch came up with the idea that if the nucleus was considered to be somewhat like a drop of liquid, then the absorption of a particle (that would initially lead to an increase in nuclear mass and nucleon number) could initiate excitations that might lead to the liquid drop (nucleus) breaking into smaller drops (nuclei). This comparison allowed people to visualise the process and understand how the absorption of a nucleon by a heavy nucleus could actually lead to lighter (less massive) products. Although the idea was initially little more than a metaphor or simile, it was developed by scientists into a sophisticated model. This process is referred to as nuclear fission, by analogy with the process of cellular fission, where one cell divides into two smaller ones.

Scientists often form visualisable mental models that help them simulate processes in their minds, and sometimes to run Gedankenexperiments (thought experiments). Einstein for example was well-known for running thought experiments in his mind in this way. A more contemporary example would be the engineer Temple Grandin who designs systems for humanely treating animals used in farming. She has described how she tests her designs by running simulations in her mind (for example imagining the experience of a cow being led into a slaughterhouse). Grandin, who is autistic (Sacks, 1995), considers visual imagery so important to her work that she sees verbal language as secondary by comparison.

Faraday visualised magnetic fields having field 'lines' as a way of making sense of magnetism. Although the lines of force used to visualise magnetic fields are only imaginary, this proved to be a very useful tool, and modern textbooks still use these kinds of diagrams. Indeed physicists calculate the flux of (imaginary) field lines as a measure of field strength, and explain high energy events on the surface of the Sun in terms of what is happening to these (non-existent) lines. In a similar way, rays of light are used in optics to model the paths of light through prisms, lenses and so forth. A ray of light can be considered as a light beam that is infinitesimally narrow (i.e. a conceptual model formed by abstraction from a real phenomenon). Light rays are – like magnetic field lines – imaginary, but useful, mental tools for modelling real physical systems.

Students also form their own mental models of natural processes in developing their understanding of scientific ideas. These mental models help learners visualise and explain scientific phenomena, in just the same way that scientists themselves use such mental models. Research suggests, however, that learners' mental models may often be inconsistent with scientific models, as they often draw upon alternative conceptions (Taber, 2014). As an example, young children may explain the cycle of day and night in terms of the sun moving behind an obstruction such as a mountain – something based on their experience and observations of real events (Vosniadou & Brewer, 1994). Although the notion is not scientifically accurate, the model can be run in the head as a mental simulation that explains why it is sometimes daylight and sometimes dark. When mental models cannot be readily constructed to test some explanatory idea, it may be possible to fabricate models in the laboratory which do the job.

The Gaia theory referred to above in relation to the Daisyworld model was framed in terms of the analogy between the earth and a living organism. An organism, such as a person, relies for survival on the ability of the system to maintain the conditions needed by the component cells – not too hot or cold, not too acid or alkaline, enough oxygen and sugar, not too high a concentration of toxic waste products, etc. Keeping a wide range of variables close to optimal operating conditions relies on a series of feedback cycles that have evolved such that significant variations from optimal conditions are detected and action taken to counter the change (breathing more deeply, dilating some blood vessels, producing glucose from glycogen,…). Seeing the earth as a supra-organism suggested that the interactions within the ecosystem may also show something analogous to homeostasis, based on feedback cycles that had evolved to be part of the system.

An analogy can be the basis of a model by going beyond mere simile and offering an explicit mapping of the parallels between two systems. Consider the following two equations representing heat flow and current flow respectively:

$$^{\Delta Q}/_{\Delta t} = - K\, A\, ^{\Delta T}/_{x}$$

$$I = - \sigma\, A\, ^{V}/_{x}$$

These can be considered analogous by mapping between the two systems (see Table 1):

Table 1. Comparing two analogous system

Thermal system	*Electrical system*
$^{\Delta Q}/_{\Delta t}$ (rate of heat flow)	I (current – rate of charge flow)
– (heat flows from high to low temperature)	– (current flows from high to low potential)
K (coefficient of thermal conductivity)	σ (coefficient of electrical conductivity)
A (cross sectional area of material)	A (cross sectional area of material)
$^{\Delta T}/_{x}$ (temperature gradient across material)	$^{V}/_{x}$ ([electrical] potential gradient across material)

An analogy has negative features as well as positive features, in the sense that only some aspects of the analogy directly map onto the target system. Consider for example the idea (sometimes found in introductory science texts) that an atom is like a tiny solar system. The atom has been modelled in science through a complex series of models that have been developed over an extended period, but the simple orbital model of the atom (i.e., with electrons in orbits around the nucleus) has been considered to be like a planetary system. We might consider that in some ways a solar system acts as a good analogy to this model of the atom: but not all features of solar systems map across. Table 2 shows how in some ways the two systems are similar; how in other ways they are different; and how some features of the solar system simply do not have anything to map to (Taber, 2001).

Table 2. Mapping an analogy

Feature of analogy (solar system)	*Feature of target (atomic model)*	*Nature of mapping*
The star (sun) is the central body	The nucleus is the central body	positive
Most of the mass of the system is at the centre	Most of the mass of the system is at the centre	positive
A number of planets orbit the sun	A number of electrons orbit the nucleus	positive
A number of comets and asteroids also orbit the sun	[No parallel feature]	neutral – no relevant mapping
Planets are found at different distances from the sun	Electrons can occur in shells – so several are at the same distance from the nucleus	negative
Planets vary in size, composition etc.	Electrons are identical	negative
Planets may have their own satellites (moons)	[No parallel feature]	neutral – no relevant mapping
Centripetal force causes the planets to orbit (rather than leave the system)	Centripetal force causes the electrons to orbit (rather than leave the system)	positive
The centripetal forces are gravitational in nature	The centripetal forces are electrical in nature	negative
The orbiting bodies can interact through forces	The orbiting bodies can interact through forces	positive
Orbiting bodies (planets) attract each other	Orbiting bodies (electrons) repel each other	negative
Orbits may decay in time due to interactions	Orbits are indefinitely stable	negative

Analogical models make use of the analogy between two different systems to allow one system to be used to stand for the other. If exploration of the analogue is used to draw inferences about the target system, it is important to understand the limits of the analogy. It is not unusual for students to mistakenly assume electrons orbit the atomic nucleus because of gravitational attraction, by analogy with the solar system. Teachers using analogical models, as well as similes and metaphors, need to ensure students are clear about the nature of the comparison being made.

TEACHING ABOUT SCIENTIFIC MODELS

An important part of teaching science is teaching students about the nature of science (see Chapter 2: '*Reflecting the nature of science in science education*'). That is, not only should students learn about scientific ideas, but also about the nature of those ideas (as theories, or laws of nature, or models, for example) and how they are derived. Most of what we teach in science is theoretical, and much of it consists of, or heavily relies upon, models of one kind or another.

So when students learn about reaction mechanisms of nucleophilic substitution reactions, for example, they are taught about hypothetical changes at submicroscopic level, based on models of how matter is structured at that level. The S_N1 and S_N2 mechanisms that may be taught in upper secondary school chemistry are models designed to explain the products produced in nucleophilic reactions under different conditions. Students may assume that scientists know precisely what is happening to the molecules during these reactions as we can draw out the reaction mechanisms – but these schemes are inferences from what is necessarily indirect evidence, as no one has ever seen the interactions between the molecules (see Chapter 24: '*Teaching and Learning Chemistry*').

Scientific typologies are a kind of model. An important part of the work of scientists is to describe nature, and offer meaningful classifications of natural phenomena. Some of the typologies that scientists produce reflect features of nature well: for example the different chemical elements. In that case it is now fairly obvious to scientists how to distinguish one element from another, and so where to 'draw the line' between different elements. Historically this was not always the case.

Other classification systems may not reflect such obvious distinctions in nature. For example, classifying elements as metals and non-metals, or into metals, metalloids, and non-metals, requires some judgements about where to best put the boundaries between categories. Any periodic table which shows different groups of elements in these terms is a model that has involved some compromises in considering the different properties of some of the elements (where the same element has a range of properties which individually suggest distinct classifications). Similarly, deciding which acids should be considered strong and which weak is a matter of judgement as dissociation is always technically an equilibrium no matter how nearly completely an acid may be dissociated under some conditions. So in many chemistry laboratories bottles of mineral acids that are considered strong acids are provided as standard bench reagents: often hydrochloric acid, nitric acid (often both as 2M solutions) and sulphuric acid (often as 1M solution). Strong acids are often said to be those that dissociate 'completely' in solution, but nitric acid has also been described as an 'almost' strong acid, suggesting that even though a solution would contain very few undissociated HNO_3 molecules compared with the number of ionised products, dissociation is not 'complete'. Considering nitric acid as one of the strong acids is appropriate for most purposes, but this is based on a model that simplifies the complexity of nature.

Another example would be the use of the species concept. Scientists classify living organisms into types at the levels of kingdom, phylum, class, order, family, genus and species. The principle is that any example of a living organism can be classified according to this system, which is based on the assumption that natural types such as species are discrete. This then is a model of the way organisms fit into distinct categories. This system works well most of the time, but evolutionary theory tells us that over time species change, and sometimes split into separate populations that then evolve into separate species. This tends to be a very slow process so that at any time there is very little ambiguity in a system that assumes discrete species: the vast majority of specimens found can be considered to be clearly members of one species or another. However, there will always be some unclear cases, as the species model simplifies the complexity of the relationship between different organisms.

When it is known that students commonly hold mental models at odds with scientific models in a topic (see Chapter 9: '*The Nature of Student Conceptions in Science*') it is possible to develop activities that ask students to compare different models. One such activity asked students working in groups to explore how well two different models of ionic bonding (the model taught in the curriculum, and a model representing common alternative conceptions) explained a range of phenomena (Taber, 2007).

Student Understanding of Scientific Models

Research suggests that most school age students have quite naive notions of models – often thinking of them as scale replicas (Treagust, Chittleborough, & Mamiala, 2002). Of course in learning science students meet some models that are of that kind – such as scale models of the human torso containing removable organ systems. These are intended however as teaching models (see below) rather than as scientific models. A teacher would be aware of ways in which such a model is not just smaller than what it stands for, but is also a considerable simplification. For example the model does not reflect connective tissues which prevent the real organism from being so easily dismantled, or the fine networks of blood vessels and nerves that permeate through the body. The teacher may assume these omissions are obvious: but that may not be the case to many students.

As scientific models are simplifications, and often abstractions, students can have learning difficulties if they do not realise this. (An interesting question is whether students in many science classes realise that magnetic field lines and rays of light are not real objects.) It was suggested above that "whether something is a model depends on how it is understood and used, rather than just its inherent properties". A corollary of this statement is that when a teacher presents a model of some scientific system, it does not function as a model for learners unless they appreciate how it models the target system.

For example, the orbital atom model referred to above was once useful scientifically, and can still be used to explain some of the science taught in schools (it links to valencies, and patterns in ionisation energies for example), but has largely been superseded by more advanced and sophisticated models of the atom (Justi & Gilbert, 2000). As the orbital model is still taught, it is important that students know it is a model, and therefore a useful thinking tool, but also limited and not a precise description of reality. Where students instead form a realist understanding of the model (that it is a much larger version of what an atom is *actually* like) they may find real difficulties understanding the (incompatible) orbital concept needed for progression in learning chemistry (Taber, 2005).

This should not be seen as simply a limitation of weaker learners. Scientists themselves have been known to suffer learning blocks by putting too much reliance on their models. For example, for many years there was a widely accepted 'central dogma' in molecular biology based on a simple model of the relationship between proteins and nucleic acids. The model can be summarised as in Figure 1.

Figure 1. The so-called 'central dogma' of molecular biology suggested a one way process whereby information stored in DNA determined the structure of RNA and so indirectly the structure of proteins

The model proposed was actually more subtle than shown in Figure 1 (Crick, 1970), but came to be widely understood as suggesting information only flowed from DNA to RNA, and then to protein – with no exceptions. It is now known that some viruses (including the HIV virus associated with the disease condition AIDS) use an enzyme called reverse transcriptase to modify cellular DNA in host cells, so that they become factories (sic, notice the metaphor there) for producing the materials needed for the virus to reproduce. The viral RNA codes for new cellular DNA, so information can pass either way between DNA and RNA (see Figure 2). The central dogma – a model that was often assumed to be realistic – prevented some scientists looking for these kinds of effects for some years.

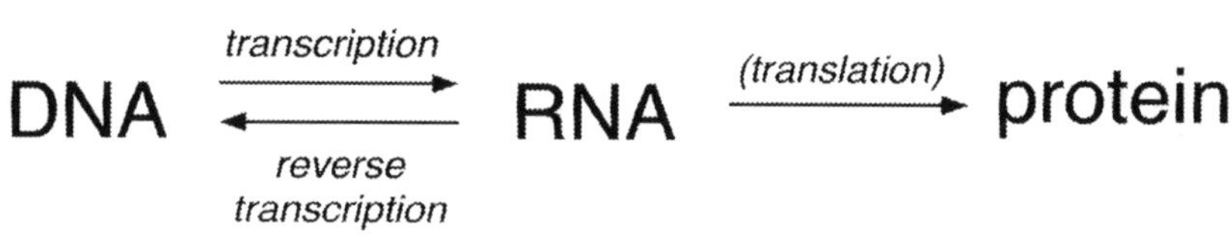

Figure 2. The model of information transfer in molecular biology has been amended now it is known that information in RNA can sometimes be transferred to, and so stored in DNA, before later being used in producing proteins

In is interesting to note in this context that one theory for the development of life on earth posits a time before DNA was produced when all genetic information was stored in RNA. Scientists developing the 'RNA world' theory try to model how simple life might have been based on RNA. If life on earth did pass through an RNA world phase then DNA was adopted at a later stage (as a more stable store of genetic information) at which point the information already represented in RNA must have been transferred to the first DNA molecules.

TEACHING STUDENTS ABOUT THE ROLE OF MODELLING IN SCIENCE

As models and modelling are so important in science, an authentic science education will emphasise models and modelling. This will mean that students will be taught about the status of scientific models *qua* models when they are presented in the curriculum. An authentic scientific education should also include opportunities for students to actually undertake modelling activities.

Curricular Models and Teaching Models

The role of models in teaching science is complicated by the existence of models which do not derive from scientific activity, but have been developed for educational purposes: curricular models that simplify scientific knowledge, and teaching models developed by educators to help teach science. Where scientific models have currency in science (or in some cases are superseded historical models that were once used by scientists) and so have been used as thinking tools to develop scientific explanations, pedagogic models are simplifications designed to help learners find out about the essence of some scientific idea or principle.

So curriculum authorities and designers may set out a simplified account of scientific knowledge as target knowledge considered suitable for learners. This is likely where the learning demand (Leach & Scott, 2002) of the scientific knowledge is considered too great for students – where the gap between students' starting points in terms of knowledge and understanding and the state of current scientific knowledge is considered too large to reasonably expect students to master the full complexity of the canonical scientific understanding. Such a curricular model can be considered authentic as long as it is true to the core of the scientific idea, and offers a suitable basis for later further learning that shows progression towards the full scientific account. Models that are oversimplifications can act as learning impediments (Taber, 2000).

There are also examples of teaching models that are designed to help students make sense of particular teaching points. Like scientific models, these teaching models may be of various kinds such as physical models, computer simulations or mathematical models. It was suggested above that children may often enter classrooms with alternative, scientifically questionable, mental models of the day-night cycle. It is common in teaching to illustrate the scientific model of this cycle (as well as seasonal changes) using a physical model with a globe rotating to show how the pattern of illumination from a light source (representing the Sun) changes over time. The model is a simplification, and not to scale, but offers a means of demonstrating the basic principle.

Another example of a physical model would be the model mine. This is basically a rectangular box, with a hole in the roof at each end, and a candle placed beneath one of the holes. When the candle is alight the smoke from a burning taper or splint will reveal air flow, and show that there is an air current passing through the channel between the two holes. (It is important to point out to students that the smoke is used simply to make the air flow visible.) This model is meant to demonstrate an application of convection, showing how mines were sometimes ventilated by a fire beneath one shaft. The model offers a considerable simplification of a real mine system but can help students visualise the convention process and understand the application of the principle.

Another common model used in teaching is the 'model lungs' composed of two balloons inside an open bottomed large glass jar fitted with a rubber sheet covering the bottom and sealing the apparatus. The rubber sheet acts as a diaphragm which can be manipulated to mimic the way a person's diaphragm moves during breathing. The balloons are attached to tubes passing through the bung sealing the top of the jar, into the air. When the rubber is pulled down, increasing the volume of the air inside the jar and so leading to a decrease in pressure, air flows through the tubes into the balloons due to the pressure difference (as it does into the lungs during inhalation). As the model can be manipulated it supports student visualisation and conceptualisation of aspects of how their own breathing occurs, although structurally it lacks superficial similarity to the actual system being modelled. For example, a negative aspect (cf. Table 2) is the way that even when inflated the balloons only occupy a small portion of the jar and are surrounded by air. A teacher can overcome this drawback by using a range of resources when making this teaching point – the physical model (three dimensional and dynamic, but not anatomically realistic) can be complemented by animations (dynamic, but two dimensional) and anatomically accurate models (lacking the dynamic features). Using a range of models and representations, and being explicit about their relative strengths and weaknesses, can help learners appreciate which features of particular models are (and which are not) meant to reflect the target learning, and also reinforce the nature of models as simplifications that represent only some aspects of the system being modelled.

Students can be asked to explicitly explore the strengths of teaching models used in the curriculum. For example, in an electricity module for lower secondary age classes (11–12 year olds) students were asked to use, and critique three different teaching models for representing current flow in circuits (Taber et al., 2015). The models were pedagogic models rather than scientific models, but students were told that the process of exploring models and testing them in relation to empirical evidence (experienced through making predictions using the models, and then building circuits) was an important part of science. The three models used (vans delivering bread to shops; a rope ring; and a physical simulation where students move around the room as if charge carriers in a circuit) are not used by scientists to model electrical circuits. It was important that students realised that these were simply tools used in teaching to help learners think about what was going on inside circuits. Yet it was also emphasised that the activity asked students to undertake a process (thinking with models) that was a common feature of scientific thinking.

Again, although it may be obvious to the science teacher which models presented in class are authentic scientific models (albeit perhaps simplified) and which are pedagogic tools, it is important not to assume students will always recognise the difference. If students are taught to appreciate the centrality of models and modelling in science, and how many historical scientific models now discarded were once at the cutting edge of science, then they should better appreciate why they are sometimes taught with teaching models that represent but do not match current scientific models.

Students Creating their Own Models

An authentic scientific education should also include opportunities for students to undertake modelling activities of different kinds. It is common in lower secondary science to ask students to build model cells, for example, using different objects (sometimes sweets of different shapes and sizes) to represent components in animal and plant cells. This can be a fun activity that allows students to be creative, whilst – if organised well – focusing their attention on the nature of the structures they are modelling. In this context the generic notion of 'the cell' referred to above may be an unhelpfully overgeneralised abstraction, but the teacher will likely refer to '*the* [sic] animal cell' and '*the* [sic] plant cell' as conceptual representations of two subclasses of the broader class of 'cells'.

However, the level of modelling activity included in science classes should not be limited to the building of scale replicas, when – as suggested above – scientific models tend to be more abstract and schematic. So students should also be given opportunities to build models that – like scientific models – are designed to offer explanatory accounts of patterns in data rather than just represent structures. Science is often seen as a subject which relies on logic and rational thought. This is certainly so, but the creative impulse is also important in science, and indeed the scientific process relies as much on scientists' creativity as their logic (Taber, 2011). Figure 3 offers a schematic suggesting how creative thinking is as important in scientific enquiry as logical thought.

Although creativity is central to science this is not always reflected in learners' experiences in the classroom, especially when most of curriculum time is used to teach students about scientific ideas that have already been tested and established. If students are to experience the excitement of science they need opportunities to be allowed to be creative – to suggest and explore their own ideas. Building,

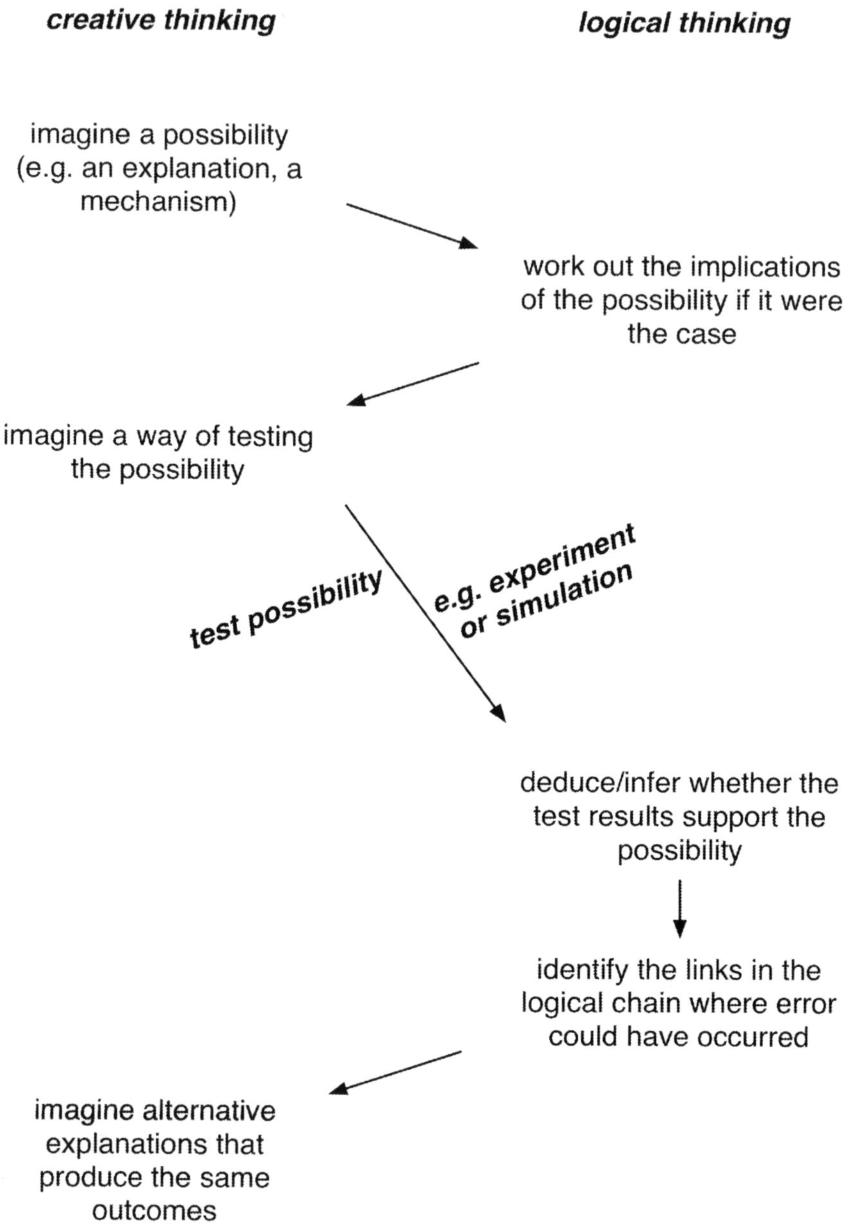

Figure 3. Scientific thinking requires an interplay of logical and creative thought. Whilst scientific enquiry relies upon rational thought, it just as much depends upon the use of imagination

developing, and testing models offers an authentic scientific activity that will engage learners and allow them to see the role of imagination in scientific work.

Teaching activities can be designed with suitable structure to support model building. An introductory activity could be to simply ask students to suggest, and justify, their own metaphors and similes for scientific concepts (Taber, 2016). Then students could be supported in building models based upon the scientific knowledge they are learning in curriculum topics. For example, one activity designed to be challenging for more able secondary students asked them to coordinate information from biology, from chemistry, and from physics, to build a holistic understanding of plant nutrition (Taber, 2007). As students gain experience in such activities and progress in their learning they can be set more advanced modelling tasks. For example, the 'Advancing Physics' course designed by the Institute of Physics in the UK for senior secondary students (16–18 year olds) incorporates software to support students in mathematical modelling (Ogborn, 1999).

CONCLUSIONS

This chapter has suggested that:

a. models and modelling are central to science
b. authentic scientific education should put an emphasis on models and modelling, so

i) that teachers should be explicit about the status of scientific models taught (i.e. that they are models);
ii) teachers and students should as a matter of course explore the strengths, and the limitations, of models met in the curriculum;
iii) science learning should involve opportunities to actively engage in creative modelling activities, not just to passively learn about existing models;
iv) progression in understanding the nature of models and modelling in science should be carefully supported as a long-term goal;

c. teaching science tends to draw heavily on pedagogic models, some – but not all – of which may reflect current or historical scientific models;
d. teaching models offer opportunities to explore the nature and affordances of models, but teachers should make it clear to students when teaching models have scientific currency, and when they are simply being used as pedagogic tools.

FURTHER READING

Clement, J. (2008). The role of explanatory models in teaching for conceptual change. In S. Vosniadou (Ed.), *International handbook of research on conceptual change* (pp. 417–452). New York, NY: Routledge.

Coll, R. K., France, B., & Taylor, I. (2005). The role of models/and analogies in science education: Implications from research. *International Journal of Science Education, 27*(2), 183–198. doi:10.1080/0950069042000276712

Harrison, A. G., & Treagust, D. F. (2000). A typology of school science models. *International Journal of Science Education, 22*(9), 1011–1026.

Passmore, C., Gouvea, J. S., & Giere, R. (2014). Models in science and in learning science: Focusing scientific practice on sense-making. In R. M. Matthews (Ed.), *International handbook of research in history, philosophy and science teaching* (pp. 1171–1202). Dordrecht: Springer Netherlands.

REFERENCES

Crick, F. (1970). Central dogma of molecular biology. *Nature, 227*(5258), 561–563.

Gilbert, J. K. (1998). Explaining with models. In M. Ratcliffe (Ed.), *ASE guide to secondary science education* (pp. 159–166). London: Stanley Thornes.

Justi, R., & Gilbert, J. K. (2000). History and philosophy of science through models: Some challenges in the case of 'the atom'. *International Journal of Science Education, 22*(9), 993–1009.

Leach, J., & Scott, P. (2002). Designing and evaluating science teaching sequences: An approach drawing upon the concept of learning demand and a social constructivist perspective on learning. *Studies in Science Education, 38*, 115–142.

Ogborn, J. (1999). New hope for physics education. *Physics World, 12*(10), 29. Retrieved from http://stacks.iop.org/2058-7058/12/i=10/a=22

Sacks, O. (1995). *An anthropologist on mars*. London: Picador.

Taber, K. S. (2000). Finding the optimum level of simplification: The case of teaching about heat and temperature. *Physics Education, 35*(5), 320–325.

Taber, K. S. (2001). When the analogy breaks down: Modelling the atom on the solar system. *Physics Education, 36*(3), 222–226.

Taber, K. S. (2005). Learning quanta: Barriers to stimulating transitions in student understanding of orbital ideas. *Science Education, 89*(1), 94–116.

Taber, K. S. (2007). *Enriching school science for the gifted learner*. London: Gatsby Science Enhancement Programme.

Taber, K. S. (2011). The natures of scientific thinking: Creativity as the handmaiden to logic in the development of public and personal knowledge. In M. S. Khine (Ed.), *Advances in the nature of science research: Concepts and methodologies* (pp. 51–74). Dordrecht: Springer.

Taber, K. S. (2014). *Student thinking and learning in science: Perspectives on the nature and development of learners' ideas*. New York, NY: Routledge.

Taber, K. S. (2016). 'Chemical reactions are like hell because…': Asking gifted science learners to be creative in a curriculum context that encourages convergent thinking. In M. K. Demetrikopoulos & J. L. Pecore (Eds.), *Interplay of creativity and giftedness in science* (pp. 321–349). Rotterdam: Sense Publishers.

Taber, K. S., Ruthven, K., Howe, C., Mercer, N., Riga, F., Hofmann, R., & Luthman, S. (2015). Developing a research-informed teaching module for learning about electrical circuits at lower secondary school level: Supporting personal learning about science and the nature of science. In E. de Silva (Ed.), *Cases on research-based teaching methods in science education* (pp. 122–156). Hershey, PA: IGI Global.

Treagust, D. F., Chittleborough, G., & Mamiala, T. L. (2002). Students' understanding of the role of scientific models in learning science. *International Journal of Science Education, 24*(4), 357–368. doi:10.1080/09500690110066485

Vosniadou, S., & Brewer, W. F. (1994). Mental models of the day/night cycle. *Cognitive Science, 18*(1), 123–183. doi:10.1016/0364-0213(94)90022-1

Watson, J. D. (1968/1980). The text of the double helix: A personal account of the discovery of the structure of DNA. In G. S. Stent (Ed.), *The double helix: A personal account of the discovery of the structure of DNA* (Norton Critical Edition ed., pp. 1–133). New York, NY: W W Norton and Company.

JACK HOLBROOK AND MIIA RANNIKMAE

21. CONTEXT-BASED TEACHING AND SOCIO-SCIENTIFIC ISSUES

This chapter introduces the concept of context-based teaching and the value of teaching through socio-scientific issues in science teaching. It does this by restating the goals of science education, especially drawing attention to education aspects important for promoting scientific literacy. Related to the idea of context-based teaching is the need to establish relevance of the learning in the eyes of students. A 3-stage model is introduced as the finale of this chapter. This is intended to illustrate the operationalisation of a context-based approach, related to a socio-scientific relevant issue. It is intended to guide the teacher in enabling students to gain the conceptual science background identified from the relevant context and then this taken to meaningful lead, based on newly acquired science ideas, to the inclusion of argumentation to make a reasoned decision as a key learning attribute in science education.

When you have worked through this chapter you should be able to:

- Explain context-based teaching
- Indicate problems with science education
- Specify goals of education
- Give meaning to intrinsic relevance
- Relate relevance to context teaching
- Give meaning to STL
- Explain socio-scientific issues
- Appreciate the importance of argumentation
- Recognise the need for SSI within the context-based approach
- Illustrating a context-based, SSI approach – the 3 stage model

WHAT IS CONTEXT-BASED TEACHING ?

Context-based teaching is when the teacher introduces a topic or a lesson from a real world context and relates this to the learning of conceptual science ideas. The real world context can be included in a number of different ways, for example, a product used in society, a situation described, or an event, which is occurring or has occurred. It is an alternative to initiating the teaching from the science content, derived from a textbook chapter, the specified school or national curriculum, or relating to questions that may occur on an examination paper.

K. S. Taber & B. Akpan (Eds.), Science Education, 279–294.

WHAT ARE SOCIO-SCIENTIFIC ISSUES?

Socio-scientific issues (SSI) are complex, open-ended, often controversial situations, with no definitive answers. In response to socio-scientific dilemmas, valid yet opposing arguments can be constructed from multiple perspectives. Just as scientists employ informal reasoning to gain insights on the natural world, ordinary citizens rely on informal reasoning to bring clarity to the controversial decisions they face. In a democratic society, science and technology are constantly involved in socio-scientific issues, and the processes of informal reasoning allow individuals to address these issues, formulate positions, and provide supporting evidence.

TRADITIONAL FOCUS OF SCIENCE EDUCATION AND ISSUES FACED

The traditional focus for school science lessons has been the content. Science was introduced as a school subject in the 19th century, especially to cater for students entering university to read science subjects (Fensham, 2008). It provided a content background. Since then, change in school science has been traditionally slow, unlike the pace of scientific and technological development within the society, so much so that there was a danger that the changing world made the relevance of current science education and its content-led approach suspect. This was not only in terms of content and related conceptual understanding for a modern society, but also in its approach to developments, changing perceptions of relevant learning and the wider range of skills demanded of the 21st century science teacher.

Of concern is that research indicates that school science teaching with its content approach is out-of-touch with today's world (Holbrook & Rannikmae, 2014) and implications of this are that:

a. science subjects are not popular among students; less students are thinking about careers and further studies in science-related areas;
b. science as taught in schools is not relevant for students. Students do not see its usefulness for their lives and future developments;
c. science content is static in nature, overloaded with facts and theories taken from the past (Rannikmäe, 2001) bearing little relationship with everyday needs;
d. students perceive school science as dominated by content and with too little challenge;
e. science education is isolated from the values components of education. It tends to be portrayed as values free, yet at the same time, the community needs increasingly to address moral and ethical issues and related problems;
f. teaching lacks attention to higher order learning among students, limiting development of problem solving and decision –making skills among school graduates.

It seems there is a need to rethink the rationale for teaching science in schools, the context in which it is put forward and the manner in which science teaching is seen by students to be of value for their future lives. Essential to this is reflecting on an understanding of science itself, the 21st century meaning of science education and the operationalisation of science teaching to enhance its relevance for a changing world (Holbrook & Rannikmae, 2014).

UNDERSTANDING THE MEANING OF SCIENCE AND SCIENCE EDUCATION

Science can be considered as both:

a. a body of knowledge that represents current understanding of natural systems, and
b. a way of thinking associated with how the body of knowledge has been established and continues to be developed, refined, and revised.

The body of knowledge includes scientific facts leading to highly developed and well-tested theories. The theories form a basis for explaining data, predicting experimental outcomes and as a means for further subject development.

With this in mind, it is important to teach science because:

1. science is a significant part of human culture and represents an area of challenge for human thinking capacity;
2. it provides valuable experiences for developing language, logic, and problem-solving skills;
3. as democracy demands that its citizens make personal and society decisions about issues in which scientific endeavours plays a fundamental role, a knowledge of science as well as an understanding of scientific methodology, is needed;
4. for some students, it can become, or support, a lifelong vocation;
5. society is dependent on the technical and scientific abilities of its citizens for its economic competitiveness and development.

GOALS OF EDUCATION

As education in general is intended to develop individuals and lay a foundation for learning throughout life, acquiring a body of knowledge, plus a range of skills and dispositions (attitudes and values) are necessary to function in today's changing world. Education thus needs to enable students to develop attributes, such as:

- Basic Skills for functioning in today's society.
- Lifelong Learning attributes to relate to a changing technological world.
- Interrelate with others and develop a sense of responsibility.
- Acquire self-concepts and gain spiritual development.
- Ensure a positive lifestyle.
- Gain awareness of career and sifting employment patterns.
- Development of responsible citizenship.
- Able to be adaptable and focused in response to changing conditions.

GOALS OF SCIENCE EDUCATION

Although the above attributes are clearly a focus for education as a whole, science education, as a component of education, needs also to relate to these. However, in relating to the learning of science (for which the term science education is used[1]), two aspects need to be regarded as key goals for school science teaching.

- School education should assure a good foundation of scientific literacy for all.
  Looking at the world from a scientific perspective enriches the understanding and interaction with phenomena in nature and technology, enables students (and therefore future adults) to take part in societal discussions and decision-making processes, and gives them an additional element from which to form interests and attitudes. These goals do not only refer to the students' personal and individual development: a culture that is critical but open-minded for science and technology is the necessary basis for raising students' interests in scientific careers.
- Teaching and learning about and from school science must also raise an interest in scientific or science-related studies, careers and employability.
  Whereas many people regard science as important for society and cultural development, they do not regard it as important for their own daily lives or for their own career perspectives. Following this goal of raising interest in science careers, school education must provide students with an authentic view of science-related careers and a fundamental background of competences and attitudes about science that enables further learning in these areas.

Using the aforementioned criteria, we can summarise the goals for science education as the acquiring:

- scientific knowledge;
- scientific methods;
- skills to engage with and resolve social issues;
- personal developmental needs, and
- career awareness.

This integration of scientific knowledge and skills with personal development and social attributes is termed the development of enhancing scientific literacy, or in recognising the strong interaction between science and technology, as enhancing scientific and technological literacy (STL) (Holbrook & Rannikmae, 2007).

WHAT IS ENHANCING STL?

STL is put forward to mean 'developing an ability' to creatively utilise appropriate evidence-based scientific knowledge and skills, particularly with relevance for everyday life and a career, in solving personally challenging yet meaningful scientific problems as well as making, responsible socio-scientific decisions' (Holbrook & Rannikmae, 2009). This is based on acquiring intellectual, attitudinal, communicative, societal and interdisciplinary learning through studies, based on conceptual science.

RELEVANCE IN SCIENCE EDUCATION

A major factor in making science in school more popular, and which can be expected to lead to greater public awareness of science by students in the future, is the relevance of the learning in the eyes of students. Students need to see the relevance of the learning, as it applies to them personally (their own lives, their interests, their career expectations). Making the science education provision relevant to students, illustrating that the provision is helping to determine a career, and showing how it is of importance for them as a responsible member of society, can give the science component more meaning in their education.

The relevance from the students' perspective can be considered as intrinsic relevance (Holbrook, 2008), while relevance, as perceived by the teacher, related to, for example, the curriculum and examinations, can be termed extrinsic relevance. The need to strive for students' intrinsic relevance (Holbrook, 2008) of science education suggests that:

- the manner in which the teaching is approached needs careful consideration;
- the relevance of the subject is more apparent coming from familiarity within society or interests associated with aspects of society;
- the structure of the teaching, initiated from a real life concern, allows the learning of conceptual science to stem from an association with the concern and thus be seen to have a connection with reality rather than be unrelated abstract learning;
- the structure of science lessons should be less about putting forward a series of scientific and technological conceptual topics than relating to science and technology in real life.

Intrinsic relevance can be interpreted as importance, usefulness or meaningfulness to the needs of the students. A more personal interpretation of relevance defines relevance as a student perception of whether the content or instruction satisfied his/her personal needs, personal goals, and career goals. These visions suggest that relevance influences motivation and in particular intrinsic motivation to learn. Furthermore, a number of science educational literature studies have also equated relevance with students' interest (Matthews, 2004). Relevance is seen as the key to raising student interests by making it more useful in the eyes of students. This, of course, begs the question whether science education made interesting (extrinsic motivation) by the teacher can lead to intrinsic relevance. Little research seems to have occurred in this area.

RATIONALE FOR CONTEXT-BASED SCIENCE EDUCATION

The rationale for this approach is that it is more relevant for students, which can stimulate intrinsic motivation to acquire the underlying science. Furthermore, if the context is familiar to students, they can, and need to, be strongly encouraged to use their prior knowledge so the learning builds on existing scientific literacy (Holbrook & Rannikmae, 2014).

Gilbert (2006) identified five major problems in science education for which a context-based approach can be considered advantageous:

1. curriculum overload;
2. curriculum content is too fragmented;
3. student transfer of learning to new situations;
4. learning not relevant to students' lives;
5. confusion as to why learn through science subjects.

Together with Pilot and Bulte, Gilbert (2011) went further to frame the perspectives to be included for context-based approach. These were detailed as:

1. Inclusion of a specific setting – provision of a social, spatial or temporal framework.
2. Behavioural environment setting enabling actions – particularly of importance here being the enabling of student involvement in the initial discussion/
3. Use of specific communication attributes, including language, with respect to the aspects being considered.
4. Enabling linkages to prior and new knowledge.

Context-based learning can thus be identified with an appropriate behavioural environment related to real life, specific communication attributes related to this and also establishing links between the prior and new science literacy learning. While contexts can be used as an application of a concept, this approach is, basically, the reverse of the usual content-led teaching approach, where the application comes first instead of last. Its advantage is that it makes applications of the science familiar to the students from the start and in this sense provides a degree of relevance to students. Its disadvantage is that the content is quickly seen as the major focus and the educational aspects (related to personal and social competences and especially problem solving and decision making) are largely dominated by conceptual subject learning. A similar focus arises if the context is used simply to illustrate the science concepts.

Context can be the starting point from which teaching and learning can emanate in a new science education direction, where students' input from prior learning can be strongly encouraged and where the new science learning to be acquired can be indicated, to a smaller or larger extent, by the students themselves. When such a situation is developed from a context, seen as familiar and thus having personal relevance for students, context-based learning takes on a new perspective, leading to student involvement and meaningful learning. Student involvement is strengthened and the teacher is provided with a base from which to develop the new learning within the science education frame from a relevance standpoint.

Aspects such as topics, modules and themes are frequently used for establishing a relevant personal or societal context. Terms like health, environment and fuels represent broad areas shown to be interesting and relevant as areas of study (Teppo & Rannikmae, 2008). Marks and Eilks (2010), however, challenge context-based chemistry education by claiming it can be superficial, arguing contexts do not automatically motivate students, and suggesting reflection is needed for effective use of these new approaches. There is no doubt the choice of behavioural setting needs to be carefully considered. Terms such as health need to broken down further so that the relevant focus becomes clear in a real life sense.

In terms of the actual approach to teaching, four context-based teaching phases can be identified (Gilbert et al., 2011):

i. the phase of initiation, in which the relevance of the situation is identified and from this links to students' prior-knowledge in a science education sense are made (although care is needed in the direction of the teaching so that discussions do not digress heavily into social experiences or concerns);
ii. the phase of learning (the discussion/interaction) needs to allow students to recognise that their current science background is insufficient to provide an explanatory input, but arouses curiosity enabling students to have the opportunity to raise science-related questions and thereby guiding students to play their part in 'setting the scene' to acquire the curriculum-related science;
iii. the phase of development, where the students become involved in the new science learning (enhancing competences associated with science knowledge and skills), complete meaningful activities to develop their ideas, and finally
iv. the phase of deepening, where the relevance of the learning is appreciated from a science standpoint, incorporated into the science conceptual frame (concept map) and forming a platform for better appreciation of the context situation.

A SOCIO-SCIENTIFIC TEACHING APPROACH FOCUSING ON CONTEXT-BASED LEARNING

Instead of initiating the context-based approach, via a theme or area of science-related interest, a more relevant behavioural approach is proposed to relate to the context to a real life problem or issue. While both a real life problem and real life issue can relate to the learning of science knowledge and both can provide a focus for a context-based approach, only an issue forms? a basis for informal discussion, argumentation and perhaps a contentious decision. A context-based issue relates to a socio-scientific approach, in which the concern or issue has relevance, if not familiarity to the students, and involves both science and wider educational learning.

An issue differs from a problem in that there is no specific, accepted conclusion. Thus, rather than focusing on solving a problem, which may have little relevance for students, the familiar issue, if chosen well, can motivate students to want to learn more, stimulate self-determination (Ryan & Deci, 2000) and even to reflect on societal action.

AN INTRODUCTION TO SOCIO-SCIENTIFIC ISSUES (SSI)

Social issues with conceptual or technological ties to science e.g. cloning, stem cells, global warming and alternative fuels, have become common elements of the national vocabulary, as well as the currency of political debates. Because of the central roles of both social and scientific factors in these dilemmas, they have been termed socio-scientific issues. Several science educators have argued for the inclusion of socio-scientific issues in science classrooms, citing their central role in the development of a responsible citizenry capable of applying scientific knowledge and habits of mind (Driver, Newton, & Osborne, 2000). The socio-scientific issue movement's aim is to focus more specifically on empowering students, in a science education sense, to handle science-based issues that shape the current world and those which will determine the future world.

INFORMAL REASONING & ITS RELATION TO SOCIO-SCIENTIFIC ISSUES

Informal reasoning involves the generation and evaluation of positions on issues that lack clear-cut solutions leading to decision-making. Informal reasoning is thus non-structured and gives opportunities for high order thinking, especially where information is less accessible. It is needed when problems are open-ended and especially when the concern is debatable, complex and ill-structured. It is especially important when an issue is being considered when students need to build an argument to support a claim.

Informal reasoning is involved in formulating and supporting positions for socio-scientific issues. This can be affected by numerous factors. These include the needed educational skills of argumentation, the ability to evaluate information/evidence and the conceptual understanding of the material, all of which underlies the issue.

ARGUMENTATION

In everyday usage, an argument is an unpleasant situation, in which two or more people differ in their opinions and they become heated over this difference. However, in science education, the goal is to approach a consensus, in which differences are supported or refuted. The argumentation can be based, for example, on an underlying agreement of the conceptual science, but disagreement as to the degree of impact of the conceptual science in making a reasoned decision. It involves extracting as much information and understanding from the situation under discussion as possible, not only in a science sense, but in terms of the social attributes such as environmental concerns, economic concerns, ethic and moral aspects, employability and political aspects, etc. In the argumentation, alternative points of view are valued as long as they contribute to the process within the accepted norms of science and logic, but purely social positions are not. [For example, 'Should plastic bags be banned,' where the debate may well be over the health risk or cost of plastic production, or could be related to environmental concerns such as blocking drains – this is actually the case in Bangladesh!]. Because the role, mode, and acceptance of arguments, in its everyday sense, are cultural variables, it is important to teach skills and acceptable modes of scientific argumentation, and this is so for both teachers and students.

Socio-scientific argumentation involves informal reasoning because negotiation and resolution are involved. This makes participating in socio-scientific argumentation more difficult. In a common model (Toulmin, 1969), persuasive argument elements provide useful categories by which an argument may be analysed, such as:

a. A claim – a statement that you are asking the others to accept. This includes information to be accepted as true, or actions you want accepted and to be enacted.
b. Grounds – the basis of real persuasion and is made up of data and facts, plus reasoning behind the claim. It is put forward as the 'truth' on which the claim is based. Grounds may also include proof of expertise and the basic premises on which the rest of the argument is built.
c. A Warrant – this links data and other grounds to a claim, legitimising the claim by showing relevance of the grounds. The warrant may be explicit or unspoken (implicit). It answers the question '*Why* does the data given mean they claim is 'true'?'
d. Backing – this gives additional support to the warrant by answering different questions.
e. A Qualifier – indicates the strength of the link from the data to the warrant and can be an indicator of the limits for which the claim applies. Indicators include terms like 'most', 'usually', 'always' 'often'.
f. A Rebuttal – a counter argument. These counter arguments can be pre-empted during the initial presentation of the argument if the presenter is able to be persuasive.

A CONTROVERSIAL ISSUE (OR REASONABLE DISAGREEMENTS)

A controversial issue is one where there no generally agreed point of view Often this derives by discussants holding different beliefs, when the issue cannot be settled by reference to evidence. This can relate to socio-scientific issues. Decisions are inevitably influenced by feelings and emotions [or having different values] rather than relying on an objective point of view. But the teacher needs to take precautions so that:

1. the scientific evidence is appropriate and accurate;
2. students do not persuade others when there is an emotional resistance to change even when the evidence is compelling.

IMPLICATIONS FOR SCIENCE EDUCATION

Socio-scientific issues can provide a powerful vehicle for teachers to help stimulate the intellectual and social growth of their students. To develop meta-cognition as a component of scientific literacy students need opportunities to engage in informal reasoning, including the contemplation of evidence and data, and express themselves through argumentation. As cited research suggests, socio-scientific issues can provide a context for informal reasoning and argumentation (Driver, Newton, & Osborne, 2000).

It seems the promotion of argumentation skills appears to be a difficult educational goal. Argumentation and the informal reasoning that underlies it are complex processes that require time and practice to develop In fact, it is reasonable to expect that significant improvements (via classroom learning) in argumentation and informal reasoning only occurs following extended learning experiences focused specifically on this goal.

PUTTING CONTEXT-BASED TEACHING AND INCLUSION OF SOCIO-SCIENTIFIC ISSUES TOGETHER – A 3 STAGE MODEL

Figure 1 illustrates a 3-stage model (Holbrook & Rannikmae, 2010), which is based on the recognition that there is a need to initiate science education learning from a *familiar and student relevant socio-scientific issue, thus establishing intrinsic relevance*. The diagram below illustrates how relevance is intended to trigger student's self-motivation to promote self-involvement in the learning. Such motivation is sustained by student involvement but also by extrinsically relevant aspects supplied by the teacher.

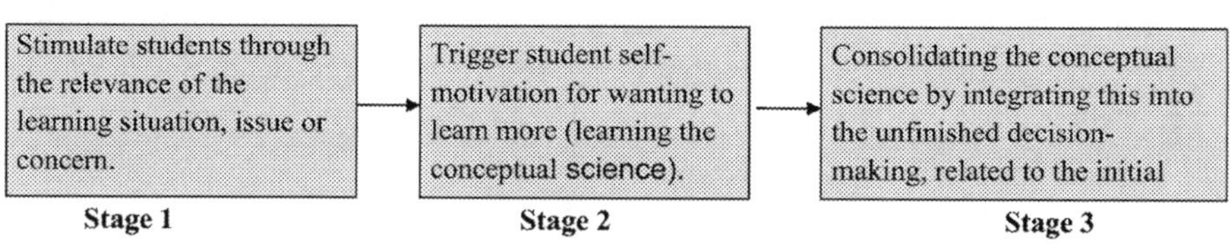

Figure 1. Stages in the 3 stage model

STAGE 1 BEGINNING WITH A STUDENT RELEVANT SITUATION (SCENARIO)

The use of a relevant context-based scenario is important. Not any situation is appropriate. Research shows that students identify with specific words, or expressions and these play an important function in determining whether the scenario chosen is appropriate. So important is the title and the depiction of the issue in a suitable manner that, if this fails to be relevant and motivational for students, the situation should not be used further and the teaching associated with this approach abandoned. This is because relevance is a very useful precursor for developing students' personal interest and a powerful stimulus for science learning. It provides students with a desire to pursue the learning further, going beyond the scenario and into the important science learning component.

The learning approach is thus 'intrinsic relevance first,' leading to science education second. This contrast with the usual suggested approach – make the science itself interesting within the context so that it will then motivate the students (but, alas, in so many cases, it doesn't!). The theoretical construct is that relevance drives students' motivation to learning and once relevance is established, the motivation for involvement can go beyond of the context-based scenario and lead into scenario-related conceptual science learning. Unfortunately, standard approaches, which assume science is inherently interesting for students, if taught well, have been shown to have little appeal to many students at the secondary level (Osborne, Simon, & Collins, 2003).

Once intrinsic relevance is established and the learning parameters defined, further learning is, in fact, the curriculum-based conceptual science ideas, which students acquire as steps towards enhancing their scientific literacy. For the learning to be meaningful, the science learning builds on a familiar, socio-scientific scenario as shown in the flowchart in Figure 2.

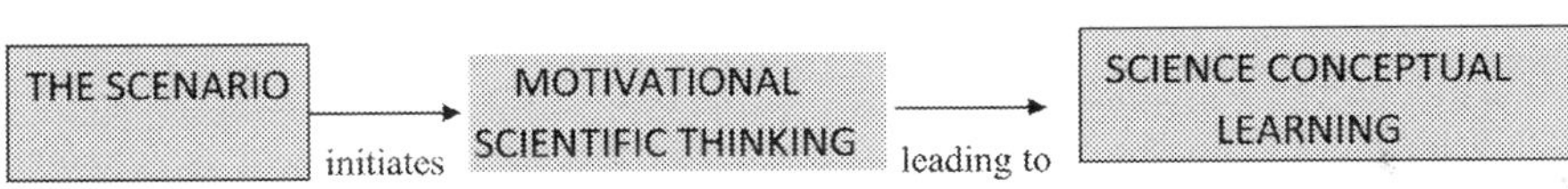

Figure 2. A flowchart showing the role of the scenario in the 3 stage model

The purpose of the scenario is to stimulate students' interest in the learning and to do this from a familiar and student relevant perspective. It is thus importance to persuade teachers to make changes to the scenario, if appropriate, to ensure such an approach. Starting from a carefully worded title (intended to be familiar and of interest to the target students), the teaching progresses, as in Figure 3 below.

EMANATING FROM THE SCENARIO

Once teachers realise the need to *initiate motivational scientific thinking in their students*, the next step is to determine students' prior science knowledge in the area related to the socio-scientific issue depicted in the scenario. In most cases, the teacher needs to expect that the students' prior knowledge is limited and students are unfamiliar with the science ideas associated with the socio-scientific issue, However,

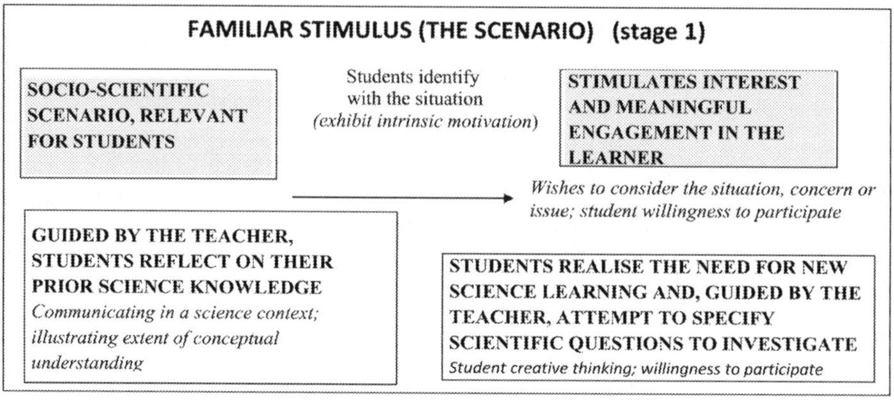

Figure 3. The role of the scenario as a stimulus for future learning

if this is not the case and students really do have a meaningful background in the underlying science, then going further to discuss the scenario *will not involve science learning*. The means the teaching needs to re-focus and the approach abandoned (why study what is already known!).

PREPARING FOR STAGE 2

While stage 1 is initially about establishing relevance for learning science, stage 2 is the important stage for gaining new conceptual science. Experience has shown that teachers need guidance on how to move from stage 1 into stage 2. The expected steps (considered within stage 1) are to:

a. enable students to recognise that they can discuss little about the scenario [OR their discussion of the scenario is limited] without acquiring the underlying science ideas, and then
b. develop the scientific question(s) (by the students if possible, otherwise by the teacher guiding the students – trying hard to not tell) to be answered in stage 2.

Moving from the scenario to developing the scientific question *is heavily dependent on the skill of the teacher*.

UNDERTAKING STAGE 2

This is likely to be the stage where most of the teaching/learning time is spent and where students gain conceptually as well as competences at the personal and social educational levels. The approach here is one of maximising student-constructed learning (with an emphasis on inquiry-based/problem-based learning) and the pace of teaching will depend heavily on the extent to which students' inquiry and process skills have been developed on prior occasions.

If students have prior experience in utilising process skills, then undertaking evidence-gathering learning (*a key element within a scientific approach*) is much facilitated. Inquiry-based learning can be expected to take far less time than in cases where students have not had prior opportunities for *student-centred approaches*. Within this stage, there is a need to stress the importance of the evidence gathering aspects, whether by experimentation, or by other means.

Inquiry-based science education (IBSE) involves, *although not usually seen as process skills:*

- *identifying* the science in a socio-scientific situation;
- *putting forward* scientific questions (questions that can be investigated scientifically);
- if necessary, breaking down questions into sub-questions that can be investigated separately.

Students can also be expected to learn to use *communication skills* to present their conclusions in suitable ways (written, oral, ICT) and, as appropriate, *discuss limitations* associated with the solutions they reach in attempting to solve the problem (that is answer the scientific question). Furthermore, inquiry learning is also very much interrelated with the development of *social skills*, especially interpersonal (student-student and student-teacher) skills and also *personal skills*, associated with aptitudes that support inquiry learning such as initiative, ingenuity, safe-working and perseverance.

Teachers can undertake inquiry learning with their students in different ways. The intended, ultimate goal is to enable students to undertake inquiry learning with no, or minimum, teacher interference (i.e. students undertake project work or 'open' inquiry). For that, teachers need to teach students to construct their thinking for the different stages of inquiry learning. And teachers must realise that *the more practice students have in IBSE,* the more easily they will undertake enquiry and the more capable they will be in undertaking high levels of student-constructed IBSE. Teachers need to recognise that progression to less (direct) teaching involved approaches given is not expected to be linear and teacher scaffolding needs to be ever-present. The type of teacher supported IBSE depends on the module being promoted and student prior experiences.

PREPARING FOR STAGE 3

The solution to the scientific question, carefully detailed and recorded, is expected to be the gateway to stage 3. But first, the conceptual science learning, emanating from the inquiry-based learning needs to be consolidated. This can for example, be enabled by student presentations on their findings and its interpretation, or through the construction of a scientific or socio-scientific concept map.

CREATING CONCEPT MAPS

Stage 2 incorporates conceptual science learning. It brings in new science. To be useful, this science needs to be put into a scientific context and, in particular, interrelated with other science knowledge. Scientific concepts can be interlinked by means of a concept map, centred on a theoretical construct (Novak & Cañas, 2006). Compiling concept maps can be a useful formative assessment exercise in which students can illustrate their learning of scientific patterns – a valuable aspect in developing the science ideas further.

UNDERTAKING STAGE 3

Stage 3 has two major components:

a. to consolidate the science ideas introduced in stage 2. This is achieved by involving students in additional tasks (above and beyond the module) related to the concepts, preferable interlinking with the students' prior concepts which were identified in stage 1. These tasks may be presented in different formats e.g. oral discussions; answering written exercises; jigsaw method, etc.
b. utilise the science ideas gained, [transferred to in the context of?] the original scenario situation, so as to enable students to discuss the scenario situation in more detail, using the newly acquired science. This is an important component of the learning and is expected to achieve two major learning targets (i) being able to transfer scientific ideas to a new, contextual situation, and (ii) participate meaningfully in a decision-making exercise to arrive at a justified decision related to the initial socio-scientific situation outlined in the title of the module.

Part (b) will involve student groups, or whole class interactions, in activities such as debates, role-playing, or discussions. Students are expected to put forward their points of view, while the teacher ensures the new science is incorporated in a meaningful and *appropriately correct* manner. Students are thus involved in aspects of *argumentation*, where the end result is a set of small group decisions, or a consensus decision made by the class as a whole. The actual decision is not, in itself, as important as the justifications put forward, but can be expected to comply with social values accepted by the local society as a whole.

SUMMARY

This chapter provides answers to two fundamental questions:

a. What is context based teaching?
b. What is a socio-scientific issue?

It also introduces STL and the role of science in science education

FURTHER READING

In the 2nd International Handbook of Research on Science Education (2012). (Eds.). Barry Fraser Kenneth Tobin Campbell J McRobbie. Dordrecht, Heidelberg, London New York: Springer.

- *Learning Science through Real World Context* by Donna King and Stephen M. Ritchie (pp. 69–79)
- *Argumentation Evidence Evaluation and Critical Thinking* by Maria Pilar Jimenez-Aleixandre and Blanca Puig (pp. 1001–1015).

NOTE

[1] Stimulate students through the relevance of the learning situation, issue or concern.

REFERENCES

Driver, R., Newton, P., & Osborne, J. (2000). Establishing the norms of scientific argumentation in classrooms. *Science Education, 84*, 287–312.

Fensham, P. (2008). *Science education policy-making.* Paris: UNESCO.

Gilbert, J. K. (2006). On the nature of "context" in chemical education. *International Journal of Science Education, 28*(9), 957–976.

Gilbert, J., Pilot, A., & Bulte, A. (2011). Concept development and transfer in context-based science education. *International Journal of Science Education, 33*(6), 817–837.

Holbrook, J. (2008). Introduction to the special issue of science education international devoted to PARSEL. *Science Education International, 19*(3), 257–266. Retrieved from www.icaseonline.net/seiweb

Holbrook, J., & Rannikmae, M. (2007). Nature of science education for enhancing scientific literacy. *International Journal of Science Education, 29*(11), 1347–1362.

Holbrook, J., & Rannikmae, M. (2009). The meaning of scientific literacy. *International Journal of Environmental and Science Education, 4*(3), 275–288.

Holbrook, J., & Rannikmae, M. (2010). Contextualisation, de-contextualisation, re-contextualisation: A science teaching approach to enhance meaningful learning for scientific literacy. In I. Eilks & B. Ralle (Eds.), *Contemporary science education* (pp. 69–82). Aachen: Shaker Verlag.

Holbrook, J., & Rannikmae, M. (2014). The philosophy and approach on which the PROFILES project is based. *Center for Educational Policy Studies Journal, University of Ljubljana, 4*(1), 9–21.

Marks, R., & Eilks, I. (2010). The development of a chemistry lesson plan on shower gels and musk fragrances following a socio-critical and problem-oriented approach: A project of participatory action research. *Chemistry Education: Research and Practice, 11*(2), 129–141.

Matthews, B. (2004). Promoting emotional literacy, equity and interest in science lessons for 11–14 year olds: The 'Improving Science and Emotional Development' project. *International Journal of Science Education, 26*(3), 281–308.

Novak, J. D., & Cañas, A. J. (2006). The origins of the concept mapping tool and the continuing evolution of the tool. *Information Visualization Journal, 5*(3), 175–184.

Osborne, J., Simon, S., & Collins, S. (2003). Attitudes towards science: A review of the literature and its implications. *International Journal of Science Education, 25*(9), 1049–1079.

Rannikmäe, M. (2001). *Operationalisation of scientific and technological literacy in the teaching of science* (PhD thesis). Tartu University Press, Tartu.

Ryan, R. M., & Deci, E. L. (2000). Self-determination theory and the facilitation of intrinsic motivation, social development, and well-being. *American Psychologist, 55*, 68–78.

Teppo, M., & Rannikmäe, M. (2008). Paradigm shift for teachers: More relevant science teaching. In J. Holbrook, M. Rannikmäe, P. Reiska, & P. Ilsley (Eds.), *The need for a paradigm shift in science education for post-Soviet societies* (pp. 25–46). Germany: Peter Lang Verlag.

Toulmin, S. (1969). *The uses of argument.* Cambridge, England: Cambridge University Press.

PAUL BLACK

22. ASSESSMENT IN SCIENCE EDUCATION

INTRODUCTION

Two functions of assessment, the formative and the summative, are often seen as independent of one another, and even in conflict. This chapter will first present an overview of the roles on assessment in supporting learning and in the broader context of a model of pedagogy. The account of formative assessment will argue that it is a central feature of effective teaching, and will stress the importance of feedback in guiding teachers in the effective implementation of plans to engage pupils in their learning of science. It will also be argued that science education can contribute to the broader development of pupils as effective learners. The links between, and distinct purposes of, informal and formal summative assessment will be explored, stressing the specific role of formal summative assessment in guiding learners in making decisions about future stages in their study of science. Overall, the analysis presents opportunities and challenges to both teachers and their pupils.

In 1998 my colleague Dylan Wiliam and I published an article entitled *Assessment and classroom learning* which reviewed about 250 research papers about the topic and drew some general conclusions from them (Black & Wiliam, 1998a). One overall conclusion, based on results of a diverse set of research studies which focussed on the feedback that teachers gave to their pupils, showed that attention to this aspect of teaching produced a significant increase in their pupils' subsequent test achievements. Over the next sixteen years, this article, and a short booklet for teachers summarising its findings (Black & Wiliam, 1998b), have been cited by many authors, and has been a starting point of further work by Dylan Wiliam and myself, with other colleagues at King's College, and by research workers in many countries. In consequence, our understanding of this aspect of teaching had evolved, both with respect to its theoretical implications and to its practical applications. The latter will serve as the starting points for the presentation in this chapter, but as each practical activity is explained, the ways in which it relates to fundamental implications for theories of learning and of pedagogy will be explained.

The key concept at the heart of this work has been formative assessment. This may be defined in the following way:

> An assessment activity can help learning if it provides information to be used as feedback, by teachers, and by their students, in assessing themselves and each other, to modify the teaching and learning activities in which they are engaged.

K. S. Taber & B. Akpan (Eds.), Science Education, 295–309.

Three features of this definition require emphasis. One is the term 'feedback', which implies some interaction between teacher and pupils.[1] The second is the inclusion of both teachers and pupils in the use of feedback, implying that both teachers and pupils learn from one another. The third is the term 'modify', which implies that information derived in interactive feedback can lead the teacher to alter, even abandon, the original design of the teaching work because the feedback has shown that it assumes an existing understanding by pupils which they have not achieved: the approach taken has to be changed.

A direct exploration of these features is presented in the next section on the ways in which assessment can support learning. The subsequent section explores the question of how assessment fits in to the broader context of an overall model of pedagogy. Two further sections examine implications, in turn for teachers as assessors, and for pupils as learners. A closing section summarises the principles involved and the personal challenges for change that they present.

ASSESSMENT SUPPORTING LEARNING

Feedback in Oral Dialogue

A teacher ought to start any lesson on a new topic with a question designed to explore what the pupils already know and understand about the topic: teachers sometimes find that their pupils know far less than they had anticipated, but at other times or with a different class, that pupils have everyday experiences and ideas about the topic on the basis of which they have built ideas or practices which may appear to contradict the more powerful concepts about which they ought to learn.

Here, for example, is a quotation from a teacher who had taken seriously the idea that she must start by helping pupils to explain the ideas that they already had about a new topic:

- *Questioning*
  My whole teaching style has become more interactive. Instead of showing how to find solutions, a question is asked and pupils given time to explore answers together. My Year 8 [i.e. 12–13 year olds] target class is now well-used to this way of working. I find myself using this method more and more with other groups
- *No hands*
  Unless specifically asked pupils know not to put their hands up if they know the answer to a question. All pupils are expected to be able to answer at any time even if it is an 'I don't know'.
- *Supportive climate*
  Pupils are comfortable with giving a wrong answer. They know that these can be as useful as correct ones. They are happy for other pupils to help explore their wrong answers further. (Black et al., 2003, p. 40)

This teacher found that giving pupils time to explore answers together was essential – research has shown that many teachers expect pupils to answer within less than one second and that if no-one does so, they answer their own questions or ask another one. But if a thoughtful answer is expected, pupils needs time to think and to compose ways of expressing their thoughts. The teacher had also found that where pupils were expected to volunteer an answer, the same pupils would put up their hands each time, whilst many others would not do so through lack of time to think, or through fear of producing a 'wrong answer'. To deal with such problems, she made her class realise that she wanted to know what they thought. So, for example, if a teacher were starting lessons in science about light, instead of telling the class the laws of reflection and refraction, he or she might ask:

> *Which is the odd one out – piece of white paper, mirror, picture, television? Why?*

There is no 'right answer' to this question, but in arguing about alternative answers the pupils will reveal their existing ideas about the nature of light. In this example, it is clear that if pupils are to learn effectively, they have to express and share any naïve conceptions or misconceptions which they already have and to understand how the scientific principles form a more coherent and effective way of dealing with problems. Other examples of such 'open' questions are:

> *If you keep a drink with ice cubes in a thermos flask, do you need to leave room for the ice cubes to melt?*

or

> *What are the similarities, and what are the differences, between combustion and respiration?*

The aim of such questions is to get pupils talking about the subject of the lesson. If such talking is to be encouraged, the teacher has to encourage it: so to ignore a strange response, or to state that it is wrong, is not helpful – a far better response might be 'Why do you think that?', and then to accept any explanation and ask the class 'Does anyone have a different idea?' The teacher's task here is a delicate one, for on the one hand the discussion must not be allowed to wander too far away from the main aim of the lesson, whilst on the other hand pupils must be helped to understand the new ideas with which the lesson is challenging them. A more detailed exploration of classroom dialogue in relation to scientific inquiry has been reported by Ruiz-Primo and Furtak (2006).

Some might object that there are better ways of using the time that such discussions will require. The issue here is about the value of talk in developing learning. Alexander (2008) expresses a clear view on this issue

> Children, we now know, need to talk, and to experience a rich diet of spoken language, in order to think and to learn. Reading, writing and number may be

> acknowledged curriculum 'basics', but talk is arguably the true foundation of learning. (p. 9)

Most teachers will realise, on reflection, that they learn a great deal from talking with colleagues about problems that arise in their own work.

In summary, by encouraging pupils' involvement in classroom dialogue, teachers pursue two aims. One is to match the work to the capacity of the pupils to develop their understanding of the concepts and methods involved. The second, and more general, aim is to develop their pupils' skills in learning, through discussion, to engage in reasoning about their own ideas and about those of fellow-pupils.

Feedback in Written Dialogue

The above account has focussed on oral dialogue. There is also the possibility of dialogue in writing, i.e. of interaction by the exchange of written work. When teachers set pupils a task to write on their own, either as homework or as work in class time, the teacher collects and studies that work and returns it to each pupil with feedback. Such feedback can take the form of a mark (or grade), or of some comments about the work, or of both marks and comments. Research studies have shown that the provision of comments can improve pupils learning, whereas the provision of marks does not do so. The same work has also shown that if both marks and comments are provided, the positive effects of the learning are not produced (Butler, 1988). Pupils regard the marks as a judgment of them and will not think of the feedback as a help to their learning unless they are only given helpful comments. A teacher who took this evidence to heart reported as follows:

> Students do work on targets and corrections more productively if no grades are given. (The researcher) observed on several occasions how little time students spend reading my comments if there were grades given as well. (Black et al., 2003, p. 43)

This teacher was using each pupil's work, with her comments about how to improve it and the subsequent efforts of the pupil to improve the work, as a learning dialogue: this made full use, of the effort invested in setting the work, in the pupil's time spent in doing it, and in the teacher's time in reading it to give feedback. By contrast, simply putting down and recording a mark usually makes little contribution to pupils' learning. Some teachers have kept a record on the comments which they made and of each pupil's responses: they found that such a record gives a far better guide for reporting on each pupil's progress than a set of marks, both for reporting to that pupil's parents and for reporting to the school management.

In summary. feedback in writing adds to the advantages for oral feedback. It is more feasible to ensure that each individual pupil receives guidance about individual problems, and by being at a slower pace it helps pupils to reflect on their own work and to think critically about the quality of their own arguments. However, there is a more fundamental issue involved here.

Feedback can Promote Confidence in Learning

If pupils are given marks repeatedly, whether on written work, or on class tests, or on both, they come to regard these as judgments of themselves. The research by Butler et al. (1988) and more extensive work by Dweck (2000) shows that feedback can develop one or other of two opinions that pupils form about themselves as learners. Some see the feedback as a way to compare themselves with others and for a view that they are either inherently of high intelligence or inherently of poor intelligence, and that there is little that they can do about this. Others see any feedback as a stimulus to improve and believe that they can improve by their own efforts. If they form the first of these two 'mind-sets' they become reluctant to take risks, failures simply damage self-esteem, and they react badly to new challenges, such as those which arise when they change schools or go on to higher education. If they form the second, they become willing to take on new challenges and to learn from failure, and they cope more positively with the challenges presented by new environments or new learning requirements. Such work has also shown that feedback given as marks can make pupils develop the first mind-set, whilst feedback given only as comments to guide improvement develop the second. It is because of such findings that Dweck advises that parents and teachers should never praise a child, but rather that they should praise what the child has achieved, i.e. to say "you are a clever child" encourages the inherent fixed ability view, whereas to say "your answer to this question was very good" encourages the confidence that one can, with effort, overcome difficulties. Teachers, parents, and the pupils find it difficult to accept a change from marks with or without comments, to comments only – but it has been achieved. To report one teacher's experience:

> Students are not good at knowing how much they are learning, often because we as teachers do not tell them in an appropriate way
>
> When asked by a visitor how well she was doing in science, the student clearly stated that the comments in her exercise book and those given verbally provide her with the information she needs. (Black et al., 2003, p. 46)

In summary, to replace the frequent provision of marks, grades and rank-order lists by comments aligned to the need of each individual pupil is important for two reasons. At one level, they can reflect upon their own first attempts and learn how to improve on them. At a deeper level pupils can be moved away from seeing feedback as a judgment on their 'innate ability' to seeing it as guide which encourages the belief they can improve performance by their own thoughtful efforts.

Peer-assessment by Pupils

If it is assumed that the teacher is the only person in the classroom who can help pupils with their learning, then the time that can be given in interactive dialogue with any one pupil is obviously limited. However, pupils can help one another, and often do so. Teachers can encourage such peer interaction. One example was that the teacher looked, after a lesson, at the written work pupils had handed in, but gave their work back to the pupils at the start of the next lesson without writing anything in their books. Pupils were then asked to work in groups of about five, asked to circulate one another's work around the group, and then discuss the strengths and weaknesses of each example. The main aim was that this would involve every pupil in discussing the differences in quality between their own work and that of fellow pupils. One teacher reflected on the value of such work as follows:

> We regularly do peer marking—I find this very helpful indeed. A lot of misconceptions come to the fore and we then discuss these as we are going over the homework. I then go over the peer marking and talk to pupils individually as I go round the room. (Black et al., 2003, p. 50)

This practice helped that teacher to decide where she might best spend her time, whilst ensuring that all pupils would be involved. The involvement of pupils in comparing one another's work has two advantages. One is that any opinion will have to be related to the criteria by which such work should be assessed: as pupils come to think about the meaning of such criteria in relation to the concrete examples from their own and one another's work, their understanding of the aims of their learning work will be improved: this matters because pupils can only enhance the quality of their own work if their efforts are guided by their understanding of the aims of the work and of the criteria by which achievement of such aims may be judged.

A related advantage is that, by seeing how their work compares with that of peers and is evaluated by them, pupils may become better at reflecting on their own work and at estimating its strengths and weaknesses. The importance of this feature was expressed by a psychologist as follows:

> Such encounters are the source of experiences which eventually create the 'inner dialogues' that form the process of mental self-regulation. Viewed in this way, learning is taking place on at least two levels: the child is learning about the task, developing 'local expertise'; and he is also learning how to structure his own learning and reasoning. (Wood, 1998, p. 98)

In summary, peer-assessment can contribute to the development of self-assessment, which is an important ability in the development of pupils as mature and independent learners, able to structure their own learning and reasoning.

Effective Group Work

However, peer-assessment can only be effective if pupils are able to work co-operatively in groups. Studies of group work in schools show that this is not always achieved, and that if the pupils see the group as an arena for competition then the work will not improve learning. Other research has confirmed this finding and has been linked to studies of ways to train pupils to co-operate more effectively (Blatchford et al., 2006; Mercer et al., 2004). Mercer's training was based on requiring pupils to respect four rules: these were that no pupil should dominate the conversation and that no pupils should remain quiet for too long, that groups must work under pressure to achieve consensus and to report that consensus to the whole class, that all contributions must be treated with respect, and that any pupil who states an assertion or a contradiction must be required to give a reason for their statement. This training produce two important effects: one was that after the training such words as 'think', 'should' and 'because' occurred three times more frequently in the group discussion than they had occurred in the same groups before the training. Another was that groups so trained gained higher scores in subsequent test of the topics discussed than comparable groups who had not been trained.

These studies have also shown that the number in a group should be more than two, because when most pupils talk in pairs they tend to agree with one another too readily, but limited to about five to ensure that all members have opportunities to participate. It also helps if there is a range of achievement history within a group so that a range of different levels and types of response will arise, but that this range must not be so large that those at one extreme cannot communicate with those at the other extreme. It has also been found helpful, in mixed gender classes, to have a mix of genders in each group. Finally, pupils should not be left to choose existing friends to form a group: as one pupil put it "I don't argue with my best friend".

In summary, effective group discussions can develop further the benefits of feedback on regular written home-work or class-work. More fundamentally, it can help pupils develop their own 'mental self-regulation' whilst also making them more capable of contributing to, and learning from, group work with their peers.

Formative use of Summative Assessments

In addition to their use in enhancing pupils' learning from their regular written work, group discussions can also be used to help pupils reflect on their performance in informal testing. Any assessment may be used either to serve as a guide to how learning may be improved, or as a summary of the quality of the learning achieved at a given stage. Thus a question asked informally in a classroom may serve as a confirmation that there is no need to repeat an explanation, whereas a formal test set at the end of a year may serve to guide teachers, or parents, or school principals, about the best level of study for a given pupil in the next year. In the first case the purpose is formative, in the second case it is summative. However, it is possible for a test to serve both purposes.

A short test at the end of the teaching of a topic may help teachers to decide that they can move on to the next topic, or it may indicate that there is a need to attend to one gap in understanding because this gap will be an obstacle to the next stage in learning. In fact, an informal end-of-topic test can be treated in the same way as a written homework task – each pupil can be given feedback about how to improve some answers, or the pupils' answers can be explored by peer-assessments in groups. At the same time, each teacher can note the overall mark of each pupil for future summaries of that pupil's work. As for any written task, the work invested in the setting and marking a test might be used more productively if the test is not set when the topic work is due to end, but a short period time before that so that there is time to use the test results to repair faults that the test has revealed. A further shift in perspective is to regard such a test as a guide to learning rather than as a final judgment. One teacher reflected on the effects of using this approach in the following way:

> After each end of term test, the class is grouped now to learn from each other. [The researcher] has interviewed them on this experience and they are very positive about the effects. Some of their comments show that they are starting to value the learning process more highly and they appreciate the fact that misunderstandings are given time to be resolved, either in groups or by me. They feel that the pressure to succeed in tests is being replaced by the need to understand the work that has been covered and the test is just an assessment along the way of what needs more work and what seems to be fine. (Black et al., 2003, p. 56)

That teacher's pupils had come to see that one valuable function of a test is to serve as a review for each of them of achievements at the end of a phase of learning. This change in the attitudes towards, and best use of, summative testing does not of course mean that all tests must be solely formative. It does imply that a test can be helpful to learning in more than one way but it does not imply that the high-stakes summative test have no other valuable function.

In summary, an overall review at the end of work on a topic can be particularly useful to pupils if they can be involved in using it to check for any faults in their understanding of that work and to correct them.

ASSESSMENT IN A MODEL OF PEDAGOGY

To implement each of the learning activities discussed above, teachers will have to insert them into their teaching plans and thereby enrich these plans. So the question to be considered in this section is how such activities should be located within a teaching plan.

It is obvious that dialogue at classroom level is located in the everyday work of classroom teaching. However, interactive dialogue can only work well if it arises out of an activity, often a problematic question, which can both engage the attention of the pupils and present them with a challenge about which they can think and talk. Thus prior to the classroom dialogue, there is the teacher's work of preparing each lesson.

The lesson-planning stage involves thought about both the short term-purpose, of evoking an interactive dialogue, and the longer-term purpose of developing particular aspects of the development of the pupils as effective learners of science. Such development will be a marriage of the understanding of key concepts of a science topic with the aim of improving pupils' capacity to reflect on and be critical of their thinking, and to engage in discussion with others. This implies that the planning stage will interact with a prior stage of formulation of the general aims, for these aims should guide the teaching of science.

The three steps outlined above, of classroom interaction, lesson planning, and formulation of aims, form, in reverse, a natural time sequence, i.e. aims *first*, leading to lessons plans *second*, leading to class room activity *third*. However, the activities of summative testing come after these: a teacher may set an informal summative test as a review of the work in a set of lessons, which will help to consolidate the classroom work and may lead back to more classroom work if the test exposes that some important aim has not been achieved. This testing activity is closely linked to the classroom aims which it san serve to strengthen and complete. It can also be seen as a *fourth* step in the model of pedagogy.

This model cannot be complete without the addition of a *fifth* stage, which is the formal summative test. What distinguishes this stage from the informal testing in the *fourth* stage in that the main aim of the test is to provide evidence which can be used to guide decisions about the future work of the pupil. For a test at the end of one of the sequence of years within a school, such decisions might be about whether or not to continue the study of a particular subject in the next school year, or the choice in assigning each pupil to the most suitable class group in the next year. For a test at the end of the top level in that school, such decisions might be about the best choice of the next school or of future career after education. In these stages, and depending on the state system of education, the tests may be externally set and marked. There are extra dimensions involved here, as aggregated test scores may be used to appraise the work of a particular teacher, or of the school as a whole. So this stage is characterised by decisions and by high- stakes consequences.

The above outline may be summarised as presenting a model of pedagogy in terms of five stages as follows:

a. Decide learning aims, both for subject concepts and for the development of the learner.
b. Select and plan activities to reflect the aims and engage pupils' interest.

c. Implement in classroom interactive dialogue.
d. Review a topic or stage to consolidate and check for gaps using informal summative assessments.
e. Guide decisions about the next stages using formal summative assessment, perhaps high-stakes.

This model was composed to meet the need for a model of pedagogy in which the role of assessment could be made clear. I have argued elsewhere that some of the well-established works about the nature of pedagogy do not treat the role of assessment seriously, i.e. it is seen as a marginal component rather than a central one (Black, 2013). By contrast, the role(s) of assessment stand out clearly in the above. Formative feedback is a main component of stage C and the planning in stage B has to foresee how this can be ensured. The role in stage D might be a dual one insofar as test results will be used formatively to complete the gap-filling function, yet summatively to guide decisions. The role in stage E is clear, but the extent to which this stage is in the control of teachers and schools will vary between school years and between different state systems.

These five stages do not function as a linear sequence in the decisions that teachers make about their pedagogy. An activity designed in stage B might encounter problems on implementation in stage C which might lead to a new design in B for use next time. A deeper problem may arise if, on reflection, what seemed like a successful classroom activity had not helped pupils to think more carefully about the principles, e.g. of control of variables in an experimental investigation. This might be an example of imperfect match of the stage B planning to the stage A aims. In a similar manner, results of tests in stages D or E ought to be scrutinised to see whether they really provide evidence of achievement of the aims of stage A. One common problem here is that it is easy to specify very attractive, and often very vague and general, aims in stage A, but when these are compared with the actual evidence of the final test in stage E, it is evident that the latter only calls for a far more limited set of achievements. Indeed, some have argued for a 'bottom-up' approach (Klenowski & Wyatt-Smith, 2014). They reported the result of using this approach with teachers in the following way

> So basically once you have the assessment firmly in place the pedagogy becomes really clear because your pedagogy has to support that – that sort of quality assessment task … that was a bit of a shift from what's usually done, usually assessment is that thing that you attach on the end of the unit whereas as opposed to sort of being the driver which it has now become. (p. 105)

Thus, any assessment instruments planned for stage E ought to be formulated at the first stage of curriculum planning, to ensure that there a match between these and the aims, or that any mis-match is explored and resolved before going further.

TEACHERS AS ASSESSORS

The view developed in the previous section is very different from the view of assessment held by many teachers. The common view is that assessment is the unpleasant dimension of learning, that there are stages when one has to stop helping pupils to engage in and enjoy their learning and start instead to 'teach to the test' with the implication that such teaching involves rote learning designed to anticipate the types of question which are often asked. This negative view is obviously most strong at times when external tests, set by bodies external to the school, are approaching.

Another cause of the negative views of testing held by many teachers is that study of the concepts and of the different possible methods of assessment is often neglected in teacher training. This is a serious fault because it means that teachers lack both confidence and skill in designing their own summative assessments. Yet for many years of schooling, years in which externally set tests do not operate, the quality of the summative assessments set and marked by teachers for use in their own classes is important in that the results are used to guide pupils – and to assess or guide teachers themselves.

A limited intervention study, lasting for two years, which aimed to help about 15 teachers, of mathematics or of English, from three schools, to survey and improve their own summative assessments, gave evidence of these difficulties (Black et al., 2010, 2011, 2013). Where externally set tests were not imposed, many teachers used sets of questions from past national tests, or from testing companies, for their end-of-year tests, without exploring the quality, or relevance to their teaching, of these instruments. What was needed was to focus their attention on the concept of validity, stressing that the main and over-arching criterion was that a high test mark would justifiably imply, to those who used the results, the degree to which each pupil was good at doing the subject. This debate was found to be valuable, as one teacher of mathematics reflected:

> It all points towards the 'what does it mean to be good at maths' question and how we [get] the students to show this – surely tests in a formal way (if properly constructed) have a role to play in allowing students to demonstrate this – and does also leave scope for teacher assessment – if the teachers are confident in this. (Black et al., 2010, p. 223)

The issue was whether doing well in the assessment was valid evidence of capability in doing the subject, so the teachers had to debate what being good at the subject meant to them. The consensus that they achieved, through their discussions of this topic, led them to realise that their summative assessments were not fit for purpose and needed to be re-designed. This was an example of fruitful interaction between stages A and E outlined in the previous section.

PUPILS AS LEARNERS

The main aim of this chapter has been to emphasise the several ways in which assessment helps pupils to develop as effective learners. Thus:

> *Feedback in oral dialogue* helps pupils' to engage, through discussion, in reasoning about their own ideas and about those of fellow-pupils.
>
> *Feedback in written dialogue* helps all pupils to reflect on their work and to think critically about the quality of their own arguments.
>
> *Feedback can promote confidence in learning* if a focus on specific comments on their work helps pupils to see feedback, not as a judgment on their 'innate ability', but as guide to improvement: this can encourage them to believe that they can improve performance by their own thoughtful efforts.
>
> *Peer-assessment by pupils* can contribute to their development of self-assessment, and thereby help them become more mature and independent learners, able to structure their own learning and reasoning.
>
> *Effective group work* enhances peer-assessment in its development of pupils 'mental self-regulation', whilst also making them more capable of contributing to, and learning from, group work. Finally,
>
> *Formative use of summative assessments* can help pupils to appreciate the value of an overall review of their work – by identifying faults in their understanding of that work and in correcting these before moving to a new topic.

One overall feature of this list is the stress on pupils being helped to take responsibility for, their own learning. The US author, Thomas Groome, has emphasised one dimension of this argument in the following way

> Educators can take over functions that learners should be doing – learning how to learn, making up their own minds, reaching personal decisions. Such imbalance ill serves learners and can be destructive to educators. There is a fine line between empowering learners as their own people and overpowering them– making them too dependent or indebted to teacher or parent. Walking this tightrope is an aspect of the educator's spiritual discipline of a balanced life. (Groome, 2005, p. 348)

One of the bad effects of accountability measures, where these are implemented with formal testing which rewards a limited range of learning behaviours, is that they motivate teachers to 'teach to the test'. Such teaching can make their pupils, as Groome says, too dependent on the teacher, even although there is evidence that teaching with a broader focus on 'empowering learners' can produce better performance, even in such tests (e.g. in mathematics see Boaler, 2002).

Related, but more fundamental, issues are argued in the following extract from Stanley et al. (2009)

> ...the teacher is increasingly being seen as the primary assessor in the most important aspects of assessment. The broadening of assessment is based on a view that there are aspects of learning that are important but cannot be adequately assessed by formal external tests. These aspects require human judgment to integrate the many elements of performance behaviors that are required in dealing with authentic assessment tasks. (p. 31)

This highlights two issues. Science education can make a strong contribution in helping pupils, through 'Authentic assessment tasks', to be more capable in dealing with a wide range of academic and work-place problems. To make this contribution, science teachers should aim to help pupils understand how scientists work, which means that they should engage them in open-ended inquiry. Such inquiry involves setting pupils a problem about a phenomenon which they have observed, or can be guided to observe, and asking them to find out what they can about the factors which give rise to, and/or about any interventions which can change, what is observed. Where the example is well chosen, and the pupils are given a minimum of guidance, they have to exercise both reasoned judgment, initiative and skills in collaboration, to make progress, and so to achieve the aim 'to integrate the many elements of performance behaviors'. This potential contribution of science education to the broader development of young people has been spelt out in detail by work within the European Community and the innovations that they have supported have been shown to achieve the aims set out in the quotation above (see, for example, Fibonacci, 2010).

A second issue is that formal and externally set tests cannot support or validly assess such activities, that work has to be done by the teachers themselves. It follows both that teachers need more help in developing their skills and confidence in their own assessments, and that national systems must support and endorse their judgments.

CHANGING ASSESSMENT: PRINCIPLES AND PERSONAL CHANGE

In 1998, the Swiss researcher, Perrenoud, wrote a profound critique of the 1998 Black and Wiliam review: his main point may be illustrated by the following extract:

> This [feedback] no longer seems to me, however, to be the central issue. It would seem more important to concentrate on the theoretical models of learning and its regulation and their implementation. These constitute the real systems of thought and action, in which feedback is only one element. (Perrenoud, 1998, p. 86)

The point of this critique, the need to locate formative assessment in a broader view of pedagogy, has been addressed, both at greater length elsewhere (Black & Wiliam, 2009), but also in this chapter in the discussion of a model of pedagogy.

A different aim in this chapter has been to develop ways, that assessment can enhance learning, which are directly applicable in practice. These have implications for the day-to-day work of both teachers and pupils. Many teachers who have attempted to develop dialogue with and between pupils in their classrooms have said 'It's pretty scary': the way they perceived and implemented their role as teacher was changing, and there was fear that they might be losing control. However, teachers will only discover how such development can empower both themselves and their pupils if they are willing to hand over to their pupils more of the control of the work. As one teacher expressed it:

> I was focusing on the girls' understanding and not on their behaviour. I often found that once the understanding was there, the behaviour followed. (Black et al., 2003, p. 96)

Many have also found that pupils themselves have, at first, resisted the change in their roles that the new emphases on oral and written dialogue require, so that any changes in practice required sensitive handling. As Perrenoud put it:

> … a number of pupils … are content to 'get by'… Every teacher who wants to practice formative assessment *must reconstruct the habits acquired by his pupils*. (Perrenound, 1991, p. 92, Emphasis in the original)

Thus, whilst this chapter is in part a theoretical analysis, it is also a challenge to practice.

NOTE

[1] Throughout this chapter, the terms 'pupils' and 'students' will be regarded as equivalent. The former term will be used in the main text, but where the latter was used in the author's original, it will be retained in any quotation.

FURTHER READING

Alexander, R. (2008). *Towards dialogic thinking: Rethinking classroom talk* (4th ed.). York, UK: Dialogos.

Black, P. (2016). The role of assessment in pedagogy – and why validity matters. In D. Wyse, L. Hayward, & J. Pandya (Eds.), *The Sage handbook of curriculum, pedagogy and assessment* (Ch. 45; pp. 725–739). London, UK: Sage.

Black, P., & Atkin, M. (2014) The central role of assessment in pedagogy. In N. G. Lederman & S. K. Abell (Eds.), *Handbook of research on science education: Volume II* (Chapter. 38, pp. 775–790). Abingdon, UK: Routledge.

Black, P., Harrison, C., Hodgen, J., Marshall, B., & Serret, N. (2013) *Inside the Black box of assessment: Assessment of learning by teachers and schools.* London: GL Assessment.

Klenowski, V., & Wyatt-Smith, C. (2014). *Assessment for education: Standards, judgment and moderation.* London: Sage.

Mercer, N., Dawes, L., Wegerif, R., & Sams, C. (2004). Reasoning as a scientist: Ways of helping children to use language to learn science. *British Educational Research Journal, 30*(3), 359–377.

Wiliam, D. (2011). *Embedded formative assessment.* Bloomington, IN: Solution Tree.

REFERENCES

Alexander, R. (2008). *Towards dialogic thinking: Rethinking classroom talk* (4th ed.). York, England: Dialogos.

Baines, E., Blatchford, P., & Kutnick, P. (2009). *Promoting effective group work in the primary classroom.* London, UK: Routledge.

Black, P. (2013). Pedagogy in theory and in practice: Formative and summative assessments in classrooms and in systems. In D. Corrigan, R. Gunstone, & A. Jones (Eds.), *Valuing assessment in science education: Pedagogy, curriculum, policy* (pp. 207–229). Dordrecht: Springer.

Black, P., & Wiliam, D. (1998a). Assessment and classroom learning. *Assessment in Education, 5*(1), 7–74.

Black, P., & Wiliam, D. (1998b). *Inside the Black Box.* London: GL Assessment.

Black, P., Harrison, C., Lee, C., Marshall, B., & Wiliam, D. (2003). *Assessment for learning: Putting it into practice.* Buckingham: Open University Press.

Black, P., Harrison, C., Hodgen, J., Marshall, B., & Serret, N. (2010). Validity in teachers' summative assessments. *Assessment in Education, 17*(2), 215–232. ISSN 0969-594X

Black, P., Harrison, C., Hodgen, J., Marshall, M., & Serret, N. (2011). Can teachers' summative assessments produce dependable results and also enhance classroom learning? *Assessment in Education, 18*(4), 451–469.

Blatchford, P., Baines, E., Rubie-Davies, C., Bassett, P., & Chowne, A. (2006). The effect of a new approach to group-work on pupil-pupil and teacher-pupil interactions. *Journal of Educational Psychology, 98*(4), 750–765.

Boaler, J. (2002). *Experiencing school mathematics: Traditional and reform approaches to teaching and their impact on student learning* (Revised and Expanded ed.). Mahwah, NJ: Lawrence Erlbaum Association.

Butler, R. (1988). Enhancing and undermining intrinsic motivation: The effects of task-involving and ego-involving evaluation on interest and performance. *British Journal of Educational Psychology, 58*(1), 1–14.

Dweck, C. S. (2000). *Self-theories: their role in motivation, personality and development.* Philadelphia, PA: Psychology Press.

Fibonacci Consortium. (2010). *The Fibonacci Project: Disseminating inquiry-based science and mathematics education in Europe.* Retrieved French Academy of sciences [website] January 20, 2011, from http://www.fibonacci-project.eu

Groome, T. H. (2005). *Educating for life.* New York, NY: Crossroad.

Perrenoud, P. (1991). Towards a pragmatic approach to formative evaluation. In P. Weston (Ed.), *Assessment of pupils' achievement: Motivation and school success* (pp. 79–101). Amsterdam: Swets and Zeitlinger.

Perrenoud, P. (1998). From formative evaluation to a controlled regulation of learning processes. Towards a wider conceptual field. *Assessment in Education: Principles, Policy and Practice,* (1), 85–102.

Ruiz-Primo, M. A., & Furtak, E. M. (2006). Informal formative assessment and scientific inquiry: Exploring teachers' practices and student learning. *Educational Assessment, 11*(3&4), 205–235.

Stanley, G., MacCann, R., Gardner, J., Reynolds, L., & Wild, I. (2009). *Review of teacher assessment: What works best and issues for development.* Oxford, England: Oxford University Centre for Educational Development; Report commissioned by the QCA

Wood, D. (1998). *How children think and learn.* Oxford, UK: Blackwell.

JAMES DE WINTER

23. TEACHING AND LEARNING PHYSICS

INTRODUCTION, *WHAT THIS CHAPTER IS AND WHAT IT IS NOT*

The transition from one side of the desk to the other, from a learner of physics to a teacher of physics can present a number of challenges. Perhaps the greatest is what Margaret Donaldson would call the *ability to decentre* (1986); stepping outside your own knowledge and understanding of a subject you know fluently and intimately, and re-imagining it from the perspective of someone who may struggle with ideas that are so innate or internalised to you that you require no effort to understand. For example, by the time you start a degree in physics, an action such as the understanding of Ohm's law, the ability to rearrange the equation into any form and draw an appropriate graph, may be so fluent that it is hard to appreciate how this may feel impossible to some students, and may take a long time even for those familiar with it.

The aim this chapter is to explore some of the challenges that teachers and students face in the physics classroom to help support and promote professional dialogue, as well as to make suggestions for further reading. The focus will be on areas that, in most cases are relevant across much of the specific subject content that you might be expected to teach students aged 11–19. It does not promise to be a 'how to' guide, or to suggest the best way to teach forces or any other topic. Whilst much work has been done in physics education research and though our understanding has developed, it is almost impossible to suggest the 'best' way to teach any topic, particularly without specific reference and consideration of the students and context. What I do hope to provide is a closer look at some of the key areas that beginning physics teachers should consider in their teaching, which may help support them as the step into the classroom or lab in the early part of their career. Specific content will not be overlooked though, as each section will contain specific examples from common physics courses to illustrate the points that I wish to make.

The areas of focus in this chapter will be

- *The Language of Physics and the Language of Everyday Life:* How common use and physics specific of use of language can cause challenges.
- *Ways of Representing and Solving Problems:* Seeing and solving questions in more than one way.
- *Maths and Physics:* The mathematical requirements of teaching and learning physics.
- *Seeing the Unseen:* Using models to visualise concepts and the challenge of microscopic and macroscopic views.

K. S. Taber & B. Akpan (Eds.), Science Education, 311–323.

The Language of Physics and the Language of Everyday Life: How common use and physics specific of use of language can cause challenges.

If a student says "Miss, I have no energy to do this work" then it's highly likely that you have a clear idea what they mean. They are tired; don't particularly want to do whatever it is that you have asked. The term 'energy' in the example above, while used in a rather vague and ill defined way, has a meaning that can be understood and responded to. By contrast, a very precise meaning for words is often a necessity in physics; it avoids ambiguity and potential confusion and provides clarity. Permittivity and permeability may look and sound similar, appear in equations that have a similar form and have a joint connection to the speed of light, but they mean different things and to confuse, combine and interchange them would cause some significant problems for an engineer. The advantage of these two technical terms is that most students are unlikely to come into contact with them until they are introduced in a physics classroom, at which point their definitions and differences can be clarified and reinforced. In this case the lack of common usage of these terms may help, as from the point of introduction students will be aware that these are two similar terms and caution is needed. As well as such technical terms, the vocabulary and usage of normal English in a science context can cause problems for students (Cassels & Johnstone, 1985). Further work has been undertaken to detail these challenges and offer suggestions for teachers of physics (Arons, 1996; Farrell & Ventura, 1998). Unfortunately, in physics teaching there also seems to be a middle ground between the technical and normal English, whereby a number of technical terms used in physics have a common, everyday usage that may be related to the 'correct' physics definition but have a less defined and thus less helpful usage, causing confusion for students. Some particular examples relevant to physics teaching are addressed in Table 1.

The language or shorthand notation that teachers use when writing text down for students to describe connected and related events and sequences can sometimes leave out or imply details that may be obvious to some but not for others. Considering the following text that may be written on a whiteboard for students when studying Boyle's law:

Pressure Increases → Volume Decreases

In one sense this is correct, but there are some omitted details that just as important to the relationship under consideration, namely that the mass of the gas needs to remain constant as does the temperature. When teaching DC circuits and Ohmic conductors a teacher might be inclined to say "when the voltage increases, so does the current" because saying "if the resistance of the component remains constant and the voltage across the component is increased then the current flowing through the component will increase" takes longer and seems more complex. Caution should be exercised here because the relationships mentioned above are between three or four variables and it is important to clarify and articulate the hidden information. One

Table 1. Words and collections of words that can create problems in physics teaching

Word/Words	*Notes/Comment*
Energy	The study of energy is, perhaps more than any other, one that causes disagreement and opposing views on the right way to teach the topic. Terms such as transfer, transform, convert and change are often used in descriptions relating to energy and words such as heat, temperature and work are known to cause problems (Jewett, 2008). Common usage can often lead to the idea that energy can be used up or run out, leading to potential misconceptions.
Temperature and Heat	Following on from potential confusion with the term Energy, these two terms are often muddled, although the extent to which this is as a result of common (mis)usage or the challenging nature of the physics concepts is unclear. The level of complexity needed for a meaningful discussion is a barrier here. Temperature might be described as *The degree of hotness or coldness of an object* and to some extent this will be understood, but it does not really develop a student's understanding of the ideas. A more 'correct' answer may involve discussion of kinetic energy of particles, or even an average of them. Considering the idea of quantities as related to some kind of average (temperature) or a sum (heat) may help to allow students to consider the nature of what is being measured and the differences between them. Parallels with speed and distance in the study of motion could be made here. For some, a better starting point might be simply "temperature is the reading from the thermometer" developing an understanding from experiences that students already have through use of multiple examples (Millar, 2011).
Current and Voltage	The term 'electricity' is often used as an overarching term to describe the study of current, voltage and resistance and the relationships between these quantities. Careful and precise use of the individual terms and the word electricity itself can help. Some teachers prefer always to use the term *potential difference* rather than voltage in a classroom context to reinforce the nature of the quantity that is being measured and the fact that measurement at two points is required.
Speed/Velocity/ Acceleration/ Momentum	Each of these words has a very precise meaning but students often merge and overlap them into an all-describing-term – 'movement' (Knight, 2002). The consequence of this can be that students can give descriptions and explanations of situations that do not adequately separate them, leading to errors. The fact that there are vector and scalar versions of related quantities (distance/displacement, speed/velocity) can add to the difficulties that students face.
Force	As with Energy, this is another word with common usage often used interchangeably with words such as power and energy. Providing a clear and unambiguous definition of what a force is can be difficult, particularly when working with younger students as the definition used might include the uncommon (to them) term *interaction* as well as requiring a secure understanding of 'motion' (see above).

way of addressing this is to use a set of defined intermediate terms that separate out connections between stages, as shown below

Table 2. Levels of intermediate terms to support causal linear reasoning (Viennot, 2014, p. 132)

Level	*Term*
Logical	hence
Intermediate	then
Chronological	next

Using this approach, and taking an example from DC electricity, the following description of the relationship between current and voltage for a filament lamp might develop from

Voltage Increases → Current Increases → Lamp Heats → Resistance Increases

to

The potential difference across the lamp is increased and hence the current through it increases. Then, as a result of this increased current flowing through it, the temperature of the filament increases. In metals the resistivity rises with temperature hence the resistance of the lamp increases

There is an obvious tension for the classroom teacher when dealing with the challenges of the language between absolute precision and a consideration for the realities of a classroom filled with students who are not physics experts. It might be argued that one should always use the correct physics term with a precise and unambiguous definition but on a practical level this is not always possible, particularly when some definitions require an understanding of underlying or foundation concepts which cannot be assumed. Inevitably some compromise needs to be made between clarity and supporting the understanding of students. Teachers may prefer not to use the examples above, considering them too lengthy and providing too much detail for students to process and remember. When teaching DC current, some might find the use of the term 'flow of current' incorrect, since the term current already implies a flow of charges. However as a teacher one may wish allow or even encouraging the addition of *flow* to the description in the hope of reinforcing, rather than undermining, student understanding.

Ways of Representing and Solving Problems: Seeing and solving questions in more than one way

As well as being a specific body of knowledge, Physics is often seen a process or way of looking at the world. One of its further strengths is that it can offer multiple of representing a particular problem and thus approach and solve it. One framework suggests four main ways a physics problem could be presented: by words; pictorially; physically and mathematically, often with students working through all four step-by-step (Van Heuvelen, 1991). An example based on the original author's work is given in Figure 1 below although commonly students will be presented with only one of these representations in a single question.

A Gymnast Falls on a Crash Mat: Multiple Representations of a physics problem

a. *Words*: A gymnast of mass 60kg falls from a trapeze down onto a crash mat. Just before they hit the mat, the gymnast is travelling at $15ms^{-1}$ and they come to rest 2 seconds after first touching the mat. Calculate the average force acting upon the gymnast as they come to a halt.
b. *Pictorial Representation*:

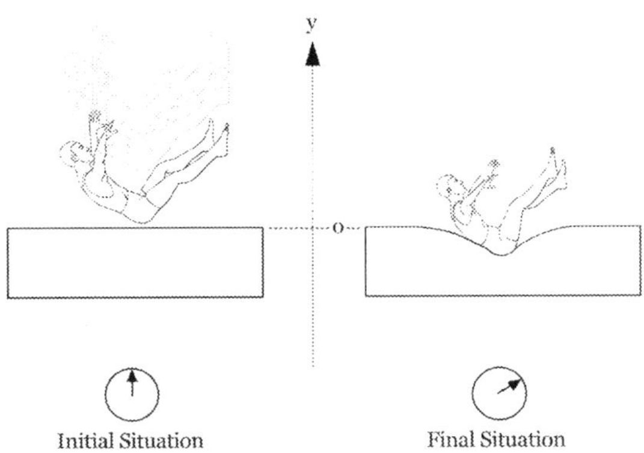

$v_0 = 15m/s$ $\quad$ $v_f = 0m/s$

$t_0 = 0s$ $\quad$ $t = 2s$

c. *Physical Representation* (free-body diagram)

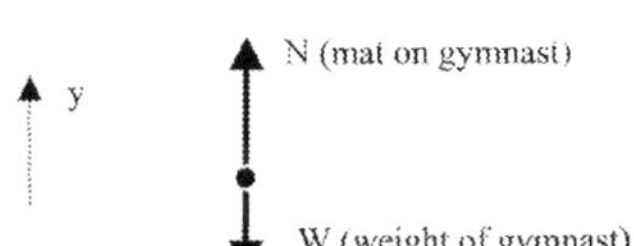

d. *Mathematical Representation*

$$\sum \vec{F} = m\left(\frac{\vec{v}_f - \vec{v}_0}{\Delta t}\right) = \frac{m\vec{v}_f - m\vec{v}_0}{\Delta t}$$

Figure 1. Multiple representations of a physics problem. (after Van Heuvelen, 1991, p. 892)

Activity: Eve and Alice are swimming to the end of a 25m long pool and back. This is shown in the diagram below. At $t = o$ seconds Eve starts and swims at 2 m/s all of the way to the end, turns and returns to the starting point. Alice was not ready at $t = o$ and did not start swimming until $t = 4$s, she then swims at 1.5m/s to the end of the pool and back. Ignore the time taken for swimmers to turn.

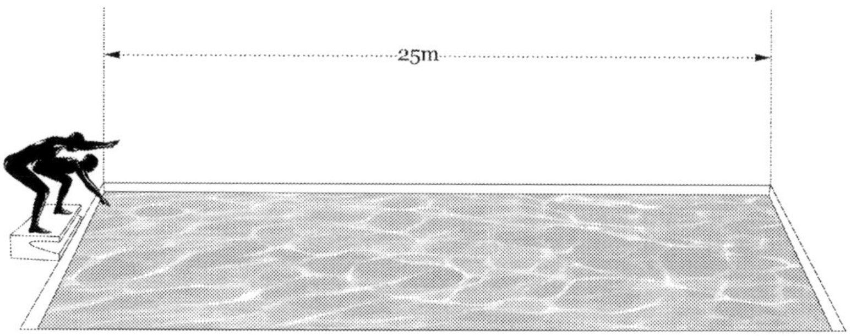

Question 1: Draw a strobe diagram showing the positions of Eve and Alice every second during the race. (*A strobe diagram shows the positions of Eve and Alice every second as if a set of photographs were taken producing a second-by-second record of the positions*). Using this diagram, work out answers to the following questions

a. How far away from the starting point is Alice when Eve passes her on her way back to the start?
b. At what time this will happen?

Question 2: Using *only* the equations of motion and algebra, work out answers to the following questions

a. How far away from the starting point is Alice when Eve passes her on her way back to the start?
b. At what time this will happen?

Question 3: Draw position vs time graphs for Eve and Alice, begin both at the time when Eve starts swimming. Using these graphs, work out answers to the following questions

a. How far away from the starting point is Alice when Eve passes her on her way back to the start?
b. At what time this will happen?
c. Do your answers agree with the ones you have worked out from Questions 1 and 2?

Reflection: Of the three methods used in this activity…

a. … which is easiest to use to get the answers you wanted?
b. … which gives you the most information about the motion?
c. … which would you like to learn better how to use?
d. … which would you recommend others use to answer these types of questions?

Figure 2. Example question encouraging students to solve the same physics question in more than one way and reflect upon the relative strengths and limitations of each approach (after Leonard, Dufresne, & Gerace, 1999)

On occasion a single problem could be presented in some or all of the different ways listed above. In practice, when teachers provide a model or exemplar answer they often only provide one solution based on one representation (Kibble, 1999) creating a missed opportunity to develop thinking. One physics education research-informed curriculum development project structured some of its resources to make students solve the same problem by using multiple methods (Leonard, Dufresne, & Gerace, 1999). Figure 2 gives an example of this type of question.

As well as leading the students to consider and solve the same problem in multiple ways, note the reflective step at the end to encourage the students to evaluate the process of the problem solving. This encourages meta-cognitive thinking intended to enhance learning.

Physics problems students encounter are commonly of the mathematical type. Considering the intimate links between maths and physics addressed elsewhere, this may seem a sensible approach. However, in a synthesis of many years of work, Lillian McDermott suggests that on its own, the usual quantitative questions may not be enough. She suggests that "Facility in solving standard quantitative problems is not an adequate criterion for functional understanding. Questions that require qualitative reasoning and verbal explanation are essential for assessing student learning and are an effective strategy for helping students learn" (McDermott, 2001, p. 1133). Paul Hewitt's Next Time questions (http://www.arborsci.com/next-time-questions) provide many examples that are designed not to be solved by a traditional quantitative method, an approach also taken by others (Mazur, 2013). The example shown in Figure 3 requires students to show a fluency with the relationships between current, voltage and resistance without any numerical calculations.

If we accept that there may never be a 'right' way to pose or answer a question or indeed a single 'right' way to answer one then incorporation of multiple representations can provide a framework for teaching that follows McDermott's advice and moves beyond using solely standard qualitative problems.

Maths and Physics: The mathematical requirements of teaching and learning physics

> It is impossible to explain honestly the beauties of the laws of nature in a way that people can feel, without their having some deep understanding of mathematics. (Feynman, 1992, p. 39)

Feynman's position is not a new one, since as far back as Aristotle and Archimedes and probably before that, the relationship between maths and physics has been explored and philosophers continue to try and unpick it. Regardless of this, it is well documented that the mathematical skills and competences that are required to teach and learn physics can cause a challenge for some students (Lenton & Stevens, 2000). Debate continues over whether a version of physics without maths

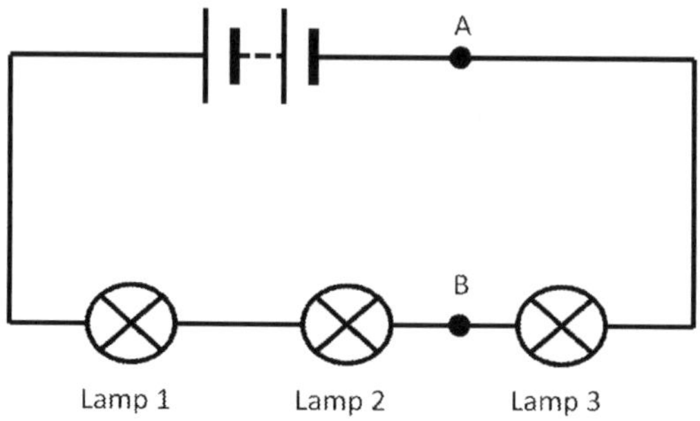

In the circuit above, three identical lamps are all glowing with equal brightness. A wire is connected between points A and B.

Q1) Describe and explain what happens to the brightness of lamp 3
Q2) Describe and explain what happens to the current leaving the battery.
Q3) Does the potential difference across lamp 2 increase, decrease or stay the same?
Q4) Compared to the circuit before the wire was connected between points A and B, will the time taken for the battery to run down increase, decrease or stay the same?
Q5) Explain your answer to Q5.

Figure 3. A conceptual question not requiring mathematical substitution. (after Paul Hewitt)

might be more appealing and successful (Taber, 2009). Others consider whether a purely qualitative teaching approach would be more appropriate (Uhden, Karam, Pietrocola, & Pospiech, 2011).

On a pragmatic level it could be argued that there are two key stages for a physics teacher to plan when considering these mathematical challenges. Initially one must identify the mathematical skills, techniques and processes required by the course, and then how their development is to be supported in the classroom. In many situations students studying physics will also be studying maths at the same level – but this is not always the case and so caution needs to be applied in any assumptions made here. Early on in his guide, *Teaching Introductory Physics*, Arnold Arons identifies what he calls the ‘underpinnings’ in a short section listing what he suggests are some foundations for study of physics which, if not addressed early and thoroughly in the study of physics, may cause significant problems later on (1996, pp. 1–20).

- Area, Volume
- Ratios and Division
- Arithmetical reasoning involving division
- Graphs and arithmetical reasoning
- Scaling and Ratio

- Elementary Trigonometry
- Horizontal, Vertical, North, South, Noon, Midnight (measurement and reference)
- Interpreting and algebraic statements

The list itself is interesting and enlightening but more broadly it raises questions for physics teachers: what is the 'toolkit' needed for physics; when and if students need to be equipped with it; how this is to be done; and what might be the consequence of not addressing it directly. There may be disagreement over any list between colleagues, and this will doubtless evolve as courses and specifications do. However the list itself represents the challenge faced. There is a whole body of work on the pedagogies of mathematic teaching that will not be explored here but there are suggestions in the further reading below.

One issue emerging from some research in this area is over the value, or not, in distinguishing between teaching mathematical skills and conceptual understanding (Lenton & Stevens, 2000). Others suggest a possible classification that considers the algorithmic uses of mathematics that can exit separate from the physics concept, known as the *technical role* and one that is a more conceptual, where the maths penetrates into the construction of the physics concept, known as the *structural role* (Uhden et al., 2011). Using acceleration as an example, one might consider that the physical concepts here are of primary importance, and therefore the teacher must focus on helping students develop a sense of a moving object and how the measurement of that movement can be quantified. However, acceleration is a rate of a rate (the rate of change in velocity, which also is rate of change in displacement) and so the idea of rate as a mathematical construct is hard to separate out from the physics concept. This distinction seems of value to the teacher as it provides a more nuanced way to look at 'maths and physics' and consider how we prepare our teaching to approach the maths in an embedded or a separate fashion. This continues to be an area of scrutiny with no clear resolution (Kurt & Pehlivan, 2013). Where this work and that of Aarons seem to align is in constructing a narrative that moves from a set of technical competences or skills to the use of them in a context, to enrich and empower the study of physics. There may be some disagreement across the literature but the need for teachers to have a considered position on the role of mathematics in physics education seems widely accepted and justified.

Seeing the Unseen: Using models to visualise concepts and the challenge of microscopic and macroscopic views

The use of modelling is a common theme across science teaching, although I would suggest that physics is a special case simply because at school level there are so many abstract ideas and concepts – forces, magnetic fields, energy, electric current – requiring some kind of model to help students develop their own understanding. Significant research has been done over the years in this field (e.g. Gilbert & Boulter, 2012) and so here I will only touch on some key themes and present some suggestions for classroom consideration. If one takes a model, whether mathematical, diagrammatic, verbal or graphical, as a way to help students connect the 'real' world to what is inside their heads, with explicit links between the representations, then for teachers it is worth considering how we can help students understand the process of modelling as well as using it to enhance and enrich their learning. Once they have observed or thought about a phenomenon, it is likely that the students will develop or refine some kind of mental model of it, and the teaching should aim to provide some kind of experience, dialogue or explanation that helps this connect this to the 'preferred' or 'correct' model that we wish them to have.

The abstract nature of some of the physics content presents a challenge in itself and thus the process of constructing and using models for teachers and students is not an easy one. McDermott notes "Connections among concepts, formal representations, and the real world are often lacking after traditional instruction. Students need repeated practice in interpreting physics formalism and relating it to the real world" (2001, p. 1133). What may be obvious to teachers may not be to students and so thought should be given to the choice and use of models and analogies. She also suggests that, rather than being given models, the students need to be active participants in constructing the qualitative models themselves, and use them to predict and explain real-world phenomena (McDermott, 2001). A simple taxonomy of models to be shared with students allowing them to interrogate and challenge their own models, may help. This process could generate discussions about identifying the nature, strengths and limitations of a model which in themselves are valuable and worthwhile activities to support learning. In starting this process, students could consider whether the models they create are *descriptive* and/or *explanatory* and/or *predictive* (Etkina, Warren, & Gentile, 2006). This could lead to examining the limitations of the models or comparing their relative benefits and values. Etkina et al. (2006) develop this approach further, providing a framework for teachers and students to help them scrutinise the nature of the models themselves, as shown below.

- Models of objects *(e.g. a car becoming a single point of mass)*
- Models of interactions *(e.g. force between two electrostatic charges)*
- Models of systems *(e.g the behaviour of an ideal gas)*
- Models of processes (qualitative/ quantitative) *(e.g. consideration of the behaviour of a ideal gas inside a piston)*

A conceptual challenge in the classroom often arises when teachers and students have to move from the large to the small and back again within descriptions and explanations. The behaviour of gases as described by the gas laws can considered on a macroscopic level with examination of the relationships between pressure, volume, mass and temperature as bulk properties. It can also be explored on a microscopic level, in the behaviour of individual particles of an ideal gas from which kinetic theory can be derived. This is a challenge across a significant number of topics covered in a physics course. Many lessons will move from the macro to the micro and back again on multiple occasions, expecting students to keep up – from the universe to a subatomic particle and all points in-between. Taking an example from the teaching of thermal conduction, a collection of rods of different metals with drawing pins held on with wax at one end have the other end placed in a flame and the class have to describe what they see and then try and explain why this happens. The *what* is relatively straightforward – the drawing pins fall as the wax melts and this happens at different times. This might be developed to deduce something like '*different metals conduct heat at different rates*'. However, they *why* usually involves some kind of discussion of atomic or electron behaviour and this should be connected with the macroscopic behaviour for a coherent explanation. Another example of this micro-macro challenge comes from work examining the reasoning of students studying electrostatics and electrodynamics. In this case, to support learning, a three stage model of student reasoning was developed, shown below (Eylon & Ganiel, 1990, p. 79).

- *quantitative relationships*, which are defined by algebraic expressions between circuit parameters
- *functional relationships*, which involve qualitative considerations, and lead to a correct description of the interplay between circuit variables; and
- *processes involving macro-micro relationships*, where the macroscopic circuit parameters are tied with microscopic models and rules.

All three steps were suggested as requirements of a solid understanding, but the third and last one was identified as the area where the complexity of the thinking has the potential to be too challenging for some students, even when the context is a relatively simple one, such as in basic electrical circuits. As with any framework there will be some limitations and universal application many not be possible. In the thermal conduction example above, explaining the falling drawing pins in terms of differing thermal conductivities would be an example of a functional explanation; however quantitative relationships, though they may exist, may not be appropriate for the age group. As with the taxonomy of models presented earlier, this does not necessarily provide an 'answer' helping students develop a secure conceptual understanding of any particular topic, but it does help provide a structure for their written and discursive work when using the models and ideas in the topic. The aim is to facilitate an active, engaged process to support learning.

CONCLUDING THOUGHTS

My focus on these broad areas is just a starting point for anyone who wishes to begin to engage with what is known about some of the challenges faced in a physics classroom. My intention is that the examples, frameworks and taxonomies given here will prompt thought, reflection and further reading (see below). However, these only form part of the picture and the best way to get a fuller, more rounded experience of what this all really means is to spend some time in a classroom or laboratory with students engaged in the teaching and learning of physics. The academic perspectives above have the greatest value when they inform what happens in that room where the teacher will be well placed to develop and refine them further.

FURTHER READING

General Physics Education: Redish, E. F. (2003). *Teaching physics: With the physics suite*. Hoboken, NJ: John Wiley & Sons.

General Physics Education: Duit, R., Schecker, H., Höttecke, D., & Niedderer, H. (2014). Teaching physics. In N. G. Lederman & S. K. Abell (Eds.), *Handbook of research on science education* (2nd ed., pp. 434–456). New York, NY: Routledge.

Language: Wellington, J., & Osborne, J. (2001). *Language and literacy in science education* (1st ed.). Buckingham & Philadelphia, PA: Open University Press.

Maths: Watson, A., Jones, K., & Pratt, D. (2013). *Key ideas in teaching mathematics: Research-based guidance for ages 9–19* (1st ed.). Oxford, United Kingdom: Oxford University Press.

The three works listed below by Arons, Knight and Viennot are well worth reading in detail.

REFERENCES

Arons, A. B. (1996). *Teaching introductory physics*. New York, NY: John Wiley & Sons.

Cassels, J., & Johnstone, A. (1985). *Words that matter in science*. London: Royal Society of Chemistry.

Donaldson, M. (1986). *Children's minds* (New edition). London: HarperCollins.

Etkina, E., Warren, A., & Gentile, M. (2006). The role of models in physics instruction. *The Physics Teacher*, *44*(1), 34–39. Retrieved from http://doi.org/10.1119/1.2150757

Eylon, B.-S., & Ganiel, U. (1990). Macro-micro relationships: The missing link between electrostatics and electrodynamics in students' reasoning. *International Journal of Science Education*, *12*(1), 79–94. Retrieved from http://doi.org/10.1080/0950069900120107

Farrell, M. P., & Ventura, F. (1998). Words and understanding in physics. *Language and Education*, *12*(4), 243–253. Retrieved from http://doi.org/10.1080/09500789808666752

Feynman, R. P. (1992). *The character of physical law* (New edition). London: Penguin.

Gilbert, J. K., & Boulter, C. (2012). *Developing models in science education*. Dordrecht, The Netherlands: Springer Science & Business Media.

Jewett, J. W. (2008). Energy and the confused student III: Language. *The Physics Teacher*, *46*(3), 149–153. Retrieved from http://doi.org/10.1119/1.2840978

Kibble, B. (1999). How do you approach a physics problem? *Physics Education*, *34*(1), 16. Retrieved from http://doi.org/10.1088/0031-9120/34/1/014

Knight, R. D. (2002). *Five easy lessons: Strategies for successful physics teaching* (1st ed.). San Franciso, CA: Addison Wesley.

Kurt, K., & Pehlivan, M. (2013). Integrated programs for science and mathematics: Review of related literature. *International Journal of Education in Mathematics, Science and Technology, 1*(2), 116–121. Retrieved from http://eric.ed.gov/?id=ED543277

Lenton, G., & Stevens, B. (2000). Numeracy in science: Understanding the misunderstandings. In J. Sears & P. Sorenson (Eds.), *Issues in science teaching* (pp. 80–88). London & New York, NY: Routledge.

Leonard, W., Dufresne, R., & Gerace, W. (1999). *Minds on physics: Motion, activities and reader: 1*. Dubuque, IA: Kendall Hunt Pub Co.

Mazur, E. (2013). *Peer instruction: A user's manual* (1st ed.). Upper Saddle River, NJ: Pearson.

McDermott, L. C. (2001). Oersted Medal Lecture 2001: 'Physics Education Research—The Key to Student Learning'. *American Journal of Physics*, *69*(11), 1127–1137. Retrieved from http://doi.org/10.1119/1.1389280

Millar, R. (2011). Energy. In D. Sang (Ed.), *Teaching secondary physics* (2nd ed., pp. 13–17). London: Hodder Education.

Taber, K. S. (2009). Maths should be the last thing we teach. *Physics Education*, *44*(4), 336. Retrieved from http://doi.org/10.1088/0031-9120/44/4/F01

Uhden, O., Karam, R., Pietrocola, M., & Pospiech, G. (2011). Modelling mathematical reasoning in physics education. *Science & Education*, *21*(4), 485–506. Retrieved from http://doi.org/10.1007/s11191-011-9396-6

Van Heuvelen, A. (1991). Learning to think like a physicist: A review of research-based instructional strategies. *American Journal of Physics*, *59*(10), 891–897. Retrieved from http://doi.org/10.1119/1.16667

Viennot, L. (2014). *Thinking in physics*. Dordrecht: Springer Netherlands. Retrieved from http://link.springer.com/10.1007/978-94-017-8666-9

KEITH S. TABER

24. TEACHING AND LEARNING CHEMISTRY

This chapter is about teaching chemistry as a science subject, and teaching chemistry topics within more general science courses. Given the very real differences in scientific practices across different disciplines, a school science that does not reflect the major disciplinary differences offers a poor reflection of the diversity within the sciences (Jenkins, 2007). Teaching chemistry has much in common with teaching other science subjects, but chemistry education is also recognised as a field of research and scholarship in its own right – having its own specific journals and conferences. There is a great deal of specific advice that could be given on teaching chemistry topics in secondary school and much of this can be found in the recommended readings at the end of the chapter. The present chapter seeks to highlight some of the key issues and challenges fundamental to teaching the subject. In particular, the chapter will use one commonly met chemical reaction (the combustion of methane) as an example to discuss some of the complexity of core chemical ideas when first met by learners.

THE NATURE OF CHEMISTRY AS A SCIENCE

Chemistry is sometimes referred to as the central science because it can be considered to be 'between' biology and physics in terms of its subject matter. Actually there is considerable overlap between chemistry and its disciplinary neighbours. Organic chemistry has strong links with biology through sciences such as biochemistry and pharmacology. Some topics studied in physical chemistry are also studied in physics (where they are collectively referred to as chemical physics). Chemistry also informs sciences such as geology, where there is again a linked specialised subject of geochemistry, and the interdisciplinary area of environmental science. Chemistry is then a broad subject with much variety. Whilst the traditional division at upper secondary and tertiary level into inorganic, organic and physical chemistry is no longer as widely followed as it once was, the study of chemistry involves topics of diverse nature, and requires a wide range of skills. Chemists may specialise in such different areas as organic synthesis, inorganic quantitative analysis, statistical thermodynamics, or quantum chemistry – and school chemistry needs to offer a flavour of this range.

K. S. Taber & B. Akpan (Eds.), Science Education, 325–341.

THE CHALLENGE OF THE SUBJECT MATTER OF CHEMISTRY

Chemistry as a science is the study of substances and their properties, and especially their interactions with each other. This simple statement belies one of the challenges of teaching and learning chemistry in that it is concerned with substances, but the idea of a substance (in the scientific sense) is not something familiar to students from the use of the word in everyday discourse. What are familiar from everyday experience are materials, such as steel, wood, various plastics and the like, and mixtures such as orange juice and air. Some common materials are in effect substances (ice and some plastics would be examples) but conceptually 'substance' is a very special category of things, distinct from 'material'. Chemistry is then inherently a subject about abstractions, because its deals with the simplified, ideal cases of pure samples of substances that are only rarely met outside of the laboratory.

Teaching students about chemical substances can either involve starting with familiar materials and later considering how they are chemically constituted, or teaching through the more traditional chemical demonstrations involving reagents that are generally unfamiliar to learners. The latter approach is typically appreciated by younger secondary students (who often enjoy working in a real laboratory, with 'chemicals', and in particular Bunsen burners!), but may make it difficult to impress upon students the relevance of chemistry to everyday life and the ubiquity of chemistry in the environment. A key challenge for the teacher of chemistry is to attempt to help students see the wide relevance of chemistry as a subject (Eilks & Hofstein, 2015), whilst retaining some of the awe and wonder that many students experience when first allowed to do chemistry practical work in the school laboratory.

THE CHEMISTRY TRIPLET

One of the key ideas in chemistry education is what is sometimes known as the chemistry triplet, which is the idea that discourse about chemistry works at three levels. This idea was first popularised by Johnstone (1982) who suggested this principle applied, in somewhat different ways, in biology, chemistry and physics. The idea has been seen as particularly useful and important in teaching chemistry, although it has been developed in a number of ways (Taber, 2013b; Talanquer, 2011).

The key idea is that learning chemistry involves:

- discussing phenomena at the level of what can be seen and handled;
- using explanatory models that invoke conjectured entities at a scale much too small to be visible (such as electrons, ions, and molecules);
- novel forms of representation that are part of the specialist language of the subject.

However, it is not simply that chemistry students need to be able to handle these three features of chemistry classroom discourse, but rather that teachers will often draw upon, and move between, them within a single segment of teacher talk.

There are several features of this challenge that each contribute to the difficulty of learning chemistry. To some extent these are problems inherent in teaching and learning chemistry and it is not suggested teachers can or should seek to avoid them. However being aware of how unfamiliar and complex the subject matter can seem from the learners' perspective allows the teacher to scaffold students' learning by (a) not introducing too many complications at once, and (b) being explicit about the levels being referred to and any shifts in level that are made.

RE-DESCRIBING PHENOMENA AT TWO THEORETICAL LEVELS

The first aspect of learning chemistry that can be difficult for students is how chemists talk about chemical phenomena such as something burning – for example when igniting the natural gas supply to use a Bunsen burner. The phenomenon – what the student experiences – is the light and heat produced by the flame. The chemist describes this process at two levels (see Figure 1). Firstly, it is conceptualised in terms of technical categories and concepts at the macroscopic (bench) scale – as a type of chemical reaction between two substances referred to as combustion and which is a subcategory of a wider class of reactions known as oxidation or redox reactions. This already poses quite a learning demand on the student not familiar with these ideas. Then it is common to re-describe what is seen at the macroscopic scale in terms of submicroscopic molecular level models – bonds breaking in reactant molecules and the formation of new molecules of the products. So here students are expected to think with, and develop accounts of chemical changes using, a whole realm of unfamiliar and abstract entities that have been invented by scientists. To say invented rather than discovered is not to suggest that molecules and the like do not really exist, but rather that these submicroscopic models are inventions in the sense

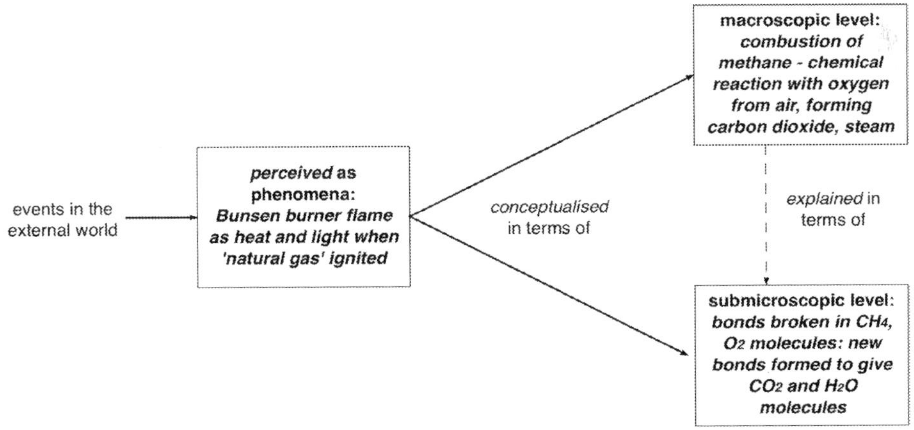

Figure 1. Chemists re-describe observable phenomena in theoretical terms at two levels (after Taber, 2013b, fig. 2)

of ideas thought-up to make best sense of empirical data that provides evidence of the nature of matter at an extremely small scale (see Chapter 2: '*Reflecting the nature of science in science education*').

THE CHALLENGE OF ADOPTING THE CHEMISTS' PARTICLE MODELS

Scientists consider that the observable behaviour of the material world arises as emergent properties of vast numbers of molecules and ions interacting together. That is, the unfamiliar (and sometimes counter-intuitive) properties of particles so small that quantum effects become significant give rise to quite different properties of matter on the scale that can be directly observed. For example, atoms, ions and molecules do not have definite surfaces, but rather have indeterminate volumes (with their electron 'clouds' becoming less dense the further from the nuclear cores). Yet when enormous numbers of molecules or ions clump together to give visible particles of matter these have (at observable scales) definite surfaces, and volumes that can be precisely measured.

Learners however commonly misunderstand the logic of the particle model as emergent properties of a complex system, and instead assume that materials have the properties they do *because* they are made up of particles with those particular properties (so they may assume that butter is made of soft particles; glass is made of transparent particles; copper is made of conducting particles, and so on). This is just one area of chemistry where students form alternative conceptions that are inconsistent with scientific thinking (see Chapter 9: '*The Nature of Student Conceptions in Science*').

BRIDGING THE TWO THEORETICAL LEVELS USING AMBIGUOUS REPRESENTATIONS

Another complication occurs at the symbolic level, where particular specialist forms of representation are adopted in chemistry (Gilbert & Treagust, 2009). One aspect, explored further below, is the range of different forms of representation used in the subject. A particular issue, however, is how some core forms of representation are ambiguous in terms of their reference to either the macroscopic or submicroscopic scale. So, for example,

$$CH_4 + 2O_2 \rightarrow CO_2 + 2H_2O$$

or

$$CH_{4(g)} + 2O_{2(g)} \rightarrow CO_{2(g)} + 2H_2O_{(g)}$$

is an example of a very common form of representation of chemical reactions (chemical change): the chemical 'equation'. Yet this same representation could be referring to the 'macroscopic' bench scale (where two moles of oxygen are required to fully react with each mole of methane), and where the equation implies 16g of methane react with 64g of oxygen to produce 44g of carbon dioxide and 36g of water (see below); or to the explanatory model at the ('submicroscopic') molecular level where the reaction is understood in terms that each molecule of methane interacts with two molecules of oxygen. (This is a conceptual model simplifying what is actually likely to happen in the chaos of a real flame with various molecular fragments moving around at high speeds.)

As the same equation can refer to either of these different conceptualisations, such representations can act as a kind of conceptual bridge (see Figure 2) that allows the teacher to shift the explanation between these two levels, and so connect the chemical description at the macroscopic, bench level with the explanation in terms of theoretical particle models. That is very useful, but clearly can potentially confuse learners unless the teacher is very careful to be explicit about when referring to each level, and when they are making shifts between levels.

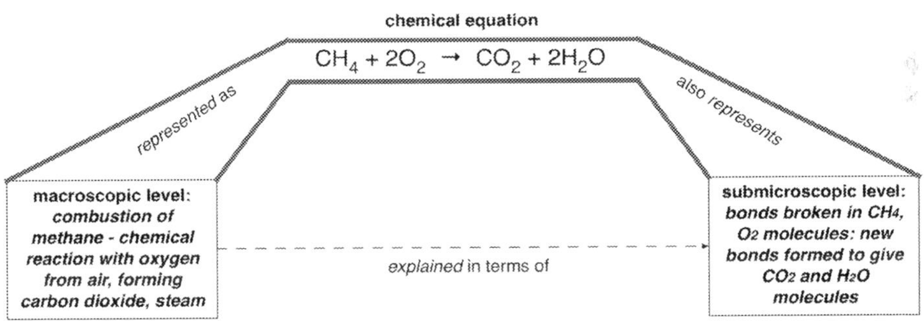

Figure 2. Using symbolic representation to bridge the two theoretical levels of chemistry

OVERSIMPLIFICATION MAY ACT AS AN IMPEDIMENT TO CONCEPTUAL DEVELOPMENT

Another challenge for the chemistry teacher when introducing students to abstract and unfamiliar ideas is that students will often adopt what are intended as models, general rules, heuristics, approximations, and the like as definite and absolute (see Chapter 20: '*Models and modelling in science and science education*'). Students often see science as a subject providing factual descriptions of the world, whereas scientists are often offering theoretical accounts. Students may take models (both scientific models, and teaching models used to simplify complex material) to be precise representations of what is being modelled, and may consider analogies and metaphors used to help make the unfamiliar familiar to be intended as precise and literal accounts (Treagust, Chittleborough, & Mamiala, 2002).

A spiral curriculum (Bruner, 1960) enables teachers to revisit ideas at greater levels of sophistication to build up student knowledge over time. This is important, as unfamiliar material needs to be introduced in manageable learning quanta, and such new learning is often initially quite labile and tenuous until it is subsequently regularly reinforced. Only once well consolidated is such learning robust enough to act as foundations for further learning. So complex ideas may need to be revisited in diverse contexts, and then developed over extended periods of time. However, the teacher needs to be aware that teaching sequence can have unintended effects. For example, it is common for students to develop alternative conceptions of ionic bonding and to assume ionic substances contain molecule-like groups of ions. It seems likely that learning about covalent bonding in simple molecules sets up a bias, or expectation, through which teaching about ionic structures is interpreted. (The common use of formulae representing the simplest ratio of ions, e.g. NaCl for sodium chloride, can also mislead students into assuming the basic unit is Na^+-Cl^-.)

When chemical reactions are first met and represented by equations the examples used tend to be of reactions that (in effect) 'go to completion', that is where the reactants are changed (virtually) completely to products (something represented by the → symbol). Usually students have some years of working with such chemical equations before meeting the idea of chemical equilibria, by which time they may well find it difficult to modify their conception of a chemical reaction. Arguably then, the possibility of reactions that do not go to completion, illustrated with a few common examples, should be introduced when students first meet the formalism of chemical reactions so they will not form a fixed conception that is difficult to modify later (when the teacher prefers them to think of all reactions as potentially equilibria, and those that can be considered to go to completion under particular conditions as the special cases).

Some other examples of the potential for introductory teaching itself to act as 'pedagogic learning impediments' (Taber, 2014) to later teaching are:

- seeing the metal to non-metal dimension as a dichotomy (i.e. an element is clearly either a metal or non-metal, rather than falling somewhere on a scale of electronegativity);
- seeing covalent and ionic bonding as a dichotomy (so, for example, when polar bonds are met they are assumed to be a type of covalent bond);
- learning a model of atoms based on electron shells without appreciating its nature as a model (interfering with later learning about orbital models);
- learning about neutralisation based only on examples of strong acids reacting with strong alkalis (so developing an alternative conception that neutralisation always leads to neutral products);
- learning about pH using standard indicator paper (so assuming strong acids have pH1, and that this is the minimum of the pH scale).

Teachers do need to simplify complex ideas and focus on uncomplicated examples when first teaching abstract material – but also need to acknowledge the existence of complications from the start so that learners do not adopt overly rigid notions that can be difficult to shift once established. Arguably teachers should (usually!) make it a habit to qualify general statements with 'generally', 'normally', 'usually', 'often', 'typically', and similar terms. If students are told, and learn that – for example – combustion is *usually* a reaction involving oxygen, they are better prepared to make sense of reactions such as the combustion of iron in chlorine. If however they develop a conception that combustion *always* involves oxygen, then they will find it difficult to later learn about combustion in chlorine without misinterpreting the teaching (for example, assuming that iron and chlorine are both reacting with oxygen). Research on how students remember science that does not fit their expectations suggests that even when students seem to accept what they are taught at the time, their later recollections can be distorted. So even if learners seem to accept in the lesson that they have seen a combustion reaction that was a binary synthesis between the two elements iron and chlorine, they will likely later remember this as being a reaction involving oxygen if they already 'know' that combustion is burning in oxygen.

Metaphors, similes and analogies are important teaching tools to help make the unfamiliar familiar to learners, but need to be used as conceptual bridges that can only support progression in learning when they are quickly passed over and soon left behind. The common teaching analogy that the atom is like a tiny solar system may have some potential value (when learners are familiar with the structure of the solar system, but not of atoms), but can easily lead to students adopting negative aspects of the analogy as part of their model of the atom – assuming that gravitational forces act to keep electrons in the atom for example (Taber, 2013c).

Another example is the common use of social metaphors to talk about the behaviour of atoms and molecules. Teachers may use anthropomorphism to put molecular activity into a narrative that is more accessible to students: but students commonly adopt and retain this way of talking and thinking, and this contributes to the very common alternative conceptual framework (Taber, 2013a) that explains chemical processes in terms of what atoms want (i.e., the atom needs a full shell of electrons), rather than in terms of physical models such as bond strengths, energy levels, and electrical forces. In all these different cases sensible pedagogic choices can have unintended and unhelpful outcomes. General advice to the teacher of chemistry is:

- regularly emphasise the nature of scientific models and be explicit about the nature of typologies, models, general rules, etc met in chemistry as useful thinking tools but not absolute accounts of nature;
- be explicit about the use of teaching models, analogies, metaphors and the like: make sure students are aware of their limitations as well as their value, and only use them to bridge to the concepts being introduced (e.g., if later students use them as if scientific accounts, reflect back their comments by rewording their points, modelling the scientific language);
- focus on simple examples when introducing abstract ideas, but be clear when the examples used are special cases that do not fully reflect the concept being taught (as with neutralisation to give neutral products, or chemical reactions that go to completion, or combustion in oxygen).

USING CHEMICAL FORMULAE AS THINKING TOOLS

There are actually a great many kinds of representations used in chemistry, beyond the ubiquitous use of formulae (CH_4, O_2, NaCl, etc.) and equations (see later in the chapter). However, even these core representations may be extended in various ways such as when undertaking chemical calculations (see Figure 3), or when exploring changes in oxidation state to demonstrate that a redox reaction has a balance of oxidation and reduction steps (see Figure 4).

$$CH_4 + 2O_2 \rightarrow CO_2 + 2H_2O$$

$$1 : 2 \qquad\qquad 1 : 2$$

$$12 + 4(1) : 2(2(16)) \qquad\qquad 12 + 2(16) : 2(2(1)+16)$$

$$16g : 64g \qquad\qquad 44g : 36g$$

$$(80g \rightarrow 80g)$$

$$4g : 16g \qquad\qquad 11g : 9g$$

Figure 3. Representing mass relations in the combustion of methane

Figure 3 shows annotation of the basic chemical equation with a series of lines of numbers. Immediately beneath the balanced equation is a representation of the number of molecules of each reactant and product species (one molecule of methane reacts with two molecules of oxygen to form one molecule of carbon dioxide and two molecules of water). As the equation can bridge between the particle model and the macroscopic description, this also represents the mole ratios in the reaction: one mole of methane reacts with two moles of oxygen. Therefore it is possible to work out the ratio of reacting masses using the relative atomic masses to calculate molecular masses, and so molar masses – again making use of the ability to scale between molecules (one molecule of methane has a mass of 16 relative atomic mass units, so one mole of methane has a mass of 16g). As the equation is balanced (the same number of atoms of each element – carbon, hydrogen, oxygen – appear in the products as in the reactants), mass must be conserved – which it is: 80 grammes of reactants gives 80 grammes of products. Finally, a ratio is conserved by scaling each component, so we can find the simplest representation for the ratio using integer values of mass (4:16::11:9).

Figure 3 is readily understood by the chemist or experienced chemistry teacher. This kind of representation is useful because it allows us to represent a good deal of chemical thinking, and relates our theoretical model of what is going on in terms of molecules with actual laboratory operations. Some pupils however may tend to panic as soon as the teacher writes a chemical equation, and others may despair at the first sight of mathematics being applied in the subject – so these powerful formalisms need to be carefully introduced. The representations support shifts in thinking between the molecular and the macroscopic – but something that is a bridge between levels for the expert can become a barrier to comprehension for the novice. Powerful representations such as these are important in teaching, but need to be introduced in stages, and with a good deal of verbal scaffolding, if students are to understand the formalisms and appreciate the power of the representational tool. Most students will need opportunities to work through examples and become confident working at each stage (e.g. identifying mole ratios, before considering masses) so as not to overload working memory with too much new information at once.

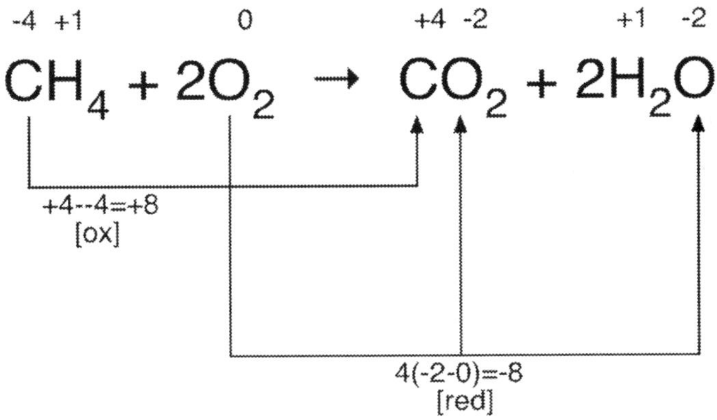

Figure 4. Representing a redox process

Figure 4 considers the same equation, but annotated in a different way. This time oxidation states for each element in each substance are designated, and this is used to identify changes in oxidation state (i.e., final oxidation number minus initial oxidation number) and so where oxidation occurs (the oxidation state of carbon increases) and where reduction occurs (the oxidation state of oxygen decreases). The formalism used here also demonstrates that (as must be the case in a balanced equation) the total number of oxidation steps is balanced by the total number of reduction steps – eight in each case.

This links to the need to conserve charge in chemical changes – although in this reaction the reactants and products are all comprised of neutral species. This form of representation then not only draws upon mathematics (albeit very basic arithmetic) but also relates to an additional abstraction from what students can observe – beyond both seeing the flame of burning laboratory gas as a reaction between chemical substances, and then seeing those substances (invisible gases) as composed of a multitude of submicroscopic molecules. Oxidation states here reflect treating covalent substances, as if they were ionic. Oxygen is more electronegative than carbon so in the carbon dioxide molecule the charge distribution is more akin to $C^{4+}O^{2-}_2$ than to $C^{4-}O^{2+}_2$ (although actually it reflects neither of these extremes). The logic of oxidation numbers is to consider how a molecule might most readily ionise (i.e. a hypothetical, mental operation). Most chemical compounds are somewhat between the ideal models of covalent and ionic bonding, and so can be considered as a resonance of different canonical forms (see Figure 5). The ionic form would not exist under normal chemical conditions as the highly charged carbon cation would be too unstable – although students may not realise this as one of the most common alternative conceptions in chemistry is that any species with an octet of valence electrons or a full outer shell is stable (Taber, 1998).

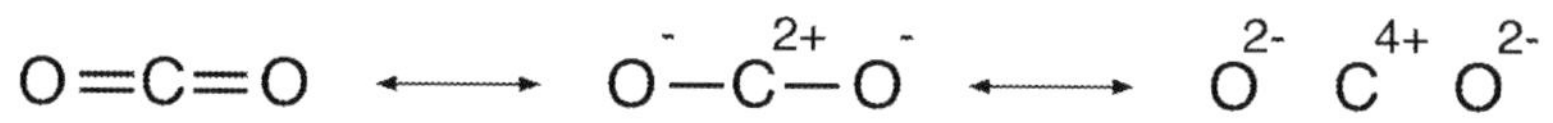

Figure 5. Representing the carbon dioxide molecule as ions

Figure 5 then represents a particular conventional formalism used in chemistry to represent a theoretical mental operation carried out on conjectured submicroscopic entities (molecules) as part of an abstract explanatory scheme. The links between such a representation and the phenomena students can actually observe are – to say the least – indirect.

ELEMENT IS A MANIFOLD CONCEPT

Students here also have to cope with how chemists use one of the most fundamental concepts in the subject – element – in multiple ways. An element is a pure substance that cannot be broken down into anything more simple or basic by chemical means, and is considered to be comprised from only one type of atom. This basic concept is inherently challenging when first met. For one thing it is not obvious from observation what makes one substance more fundamental than another. For example, heating different substances in air might led to decomposition (to something chemically more simple), or to combustion (which if what is heated is an element leads to something chemically more complex), or to melting, and evaporation or boiling; or sublimation (changes of state without any chemical change and so no change in chemical complexity). Without already knowing whether the substance being heated is an element or not, there is no obvious sign of whether the process produces something chemically more simple. Solids are inherently more ordered than liquids, or gases, which have component particles with more 'degrees of freedom'. So the physical changes produced by heating do reflect changes in levels of complexity – but not in a chemical sense. This distinction is not something that a student can infer from simply observing a substance being heated.

Describing the substance in particle terms is more straightforward – methane contains two types of atom, and oxygen only one. So oxygen is an element. However, oxygen does not comprise of atoms, but molecules where the constituent parts of two atoms have been reconfigured into a molecule (which is not understood in chemistry as just two atoms 'stuck' together, but rather the result of an interaction which leads to chemical bonding, conceptualised in terms of energy states, or the formation of molecular orbitals – ideas not readily understood by novices recently introduced to the idea of atoms). Moreover, the 'same' kind of atom, for this purpose, potentially incorporates different isotopes – as chemically the proton number is critical, but the neutron number is generally of little, chemical, consequence. Again this seems familiar and obvious to the chemist or science teacher, but is another complication to be made sense of by students when first meeting these ideas.

Yet there is an even more challenging aspect to the chemist's thinking here. A common alternative conception that many students develop is the common-sense assumption that a compound should have a combination of the properties of its constituent elements. The chemistry teacher has to emphasise that compounds are not mixtures of elements but unitary substances in their own right, and that when new substances form in chemical reactions they have their own individual properties. No sane person would want to add a mixture of a very reactive inflammable metal and a choking gas to their meals, yet sodium chloride – table salt – has traditionally been used for just that purpose. The compound has very different properties than those of the elements sodium and chlorine, and whilst too much salt added to food can increase blood pressure, NaCl is not generally seen as a serious chemical hazard. Indeed, a fair proportion of the species on the planet live out their lives immersed in sea water, an impure solution of salt. So a key teaching point in the subject is that the *elements are not present in their compounds*, as a sample of element ceases to exist when it reacts to become a new substance.

Yet Figure 4 represents how, despite this core principle, chemists feel justified in assigning oxidation states to the 'elements in' compounds even though no elements are actually present. The element concept then has several related but distinct meanings, and *there is a sense* of an element as a kind of essence that survives into compounds and can be isolated again. The element is 'in there' because it can be retrieved. The particle model suggests that what is actually unchanged is atomic cores – as chemical change does not influence the nuclei, or usually inner electron shells, present. We can recover the element through various chemical (or electrochemical) processes as at a particle level what determines the element is the nucleus (with its particular proton or atomic number) which is unperturbed by chemical changes going on around it which reconfigure arrangements of atomic cores and valence electrons. So the numbers written above the element symbols in the compounds in the chemical equation in Figure 4 – the oxidation states – refer to the oxidation state of the element$_{\text{[essence]}}$ in its combined form where we are using the term element in a distinct way (i.e. not the element$_{\text{[substance]}}$ that we usually mean in chemistry). If the teacher is not aware of, and careful about, how they talk about 'elements' in these different ways there is clearly considerable scope for confusing and frustrating learners.

OTHER FORMS OF REPRESENTATION USED IN CHEMISTRY CLASSES

Other forms of representation commonly used in chemistry include graphs, such as those recording changes of temperature, pressure or volume during a reaction. Graph-like schematics such as reaction profiles (e.g. Figure 6) are also commonly used. Here what is plotted on the y-axis is the abstract notion of energy (variously calculated as free energy, enthalpy changes, bond energies etc.), whilst the x-axis reflects the change in configuration of the molecular components as first bond fission, then bond formation, occurs. Although schematic in nature these types of representations can be developed in relation to aspects of reaction kinetics, the effects of catalysts, and reaction mechanisms (which are mental simulations of the conjectured processes of molecular interaction, deconstruction and reconfiguration during the reaction) i.e. showing transition states (the most unstable conjectured configurations during the process) and the formation of intermediates that can under some circumstances be isolated. Again then, learners are presented with formalised representations based on imagined operations on the conjectured submicroscopic entities used to develop theoretical explanations of chemical phenomena. These are very indirect abstractions from what learners can actually observe during a reaction.

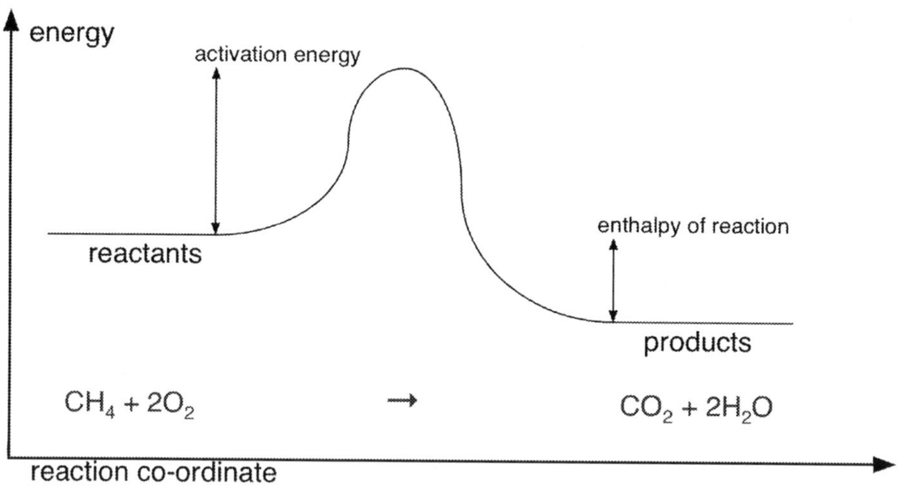

Figure 6. Representing the 'reaction profile' as a reaction proceeds

Various ways of representing the particles at the molecular scale are also widely used in chemistry (Taber, 2009). Atoms may be shown as circles, or concentric spheres, or with scatter-type figures showing electron density, or with lobes representing orbitals and so forth. Bonds may be shown by lines or dots (electron pairs) or contour lines or density of grey or dot scatter. Atoms (or strictly atomic cores) in molecules may be shown by their elemental symbols, or their nuclear charge. Complex organic molecular structures are often shown in a skeleton form with C-C bonds as default and hydrogen atoms bonded to carbon centres excluded (such as in the second image in Figure 7). The chemist or science teacher becomes so used to this range of representations that they learn to 'see' past the specific formalism to what is represented. However, for the novice, these variations may be salient and seem very significant (Taber, 1994).

Some molecular representations involve further abstraction or complications – such as circles inside hexagons for aromatic groups (see Figure 7), and the use of element-like symbols for radical groups (Me for $\bullet CH_3$, Et for $\bullet C_2H_5$, Ph for $\bullet C_6H_5$, etc.), or symbolism to show three dimensional structure (for example in stereoisomers). As one example, Figure 7 presents three different ways of representing a molecule of one substance: phenol. Unless learners are carefully supported in learning such formalisms they will construct their own meanings for the symbolism involved – for example, students have assumed that the circle in a hexagon representing the aromaticity of a benzene ring is meant to show a container for 'spare' electrons in the structure.

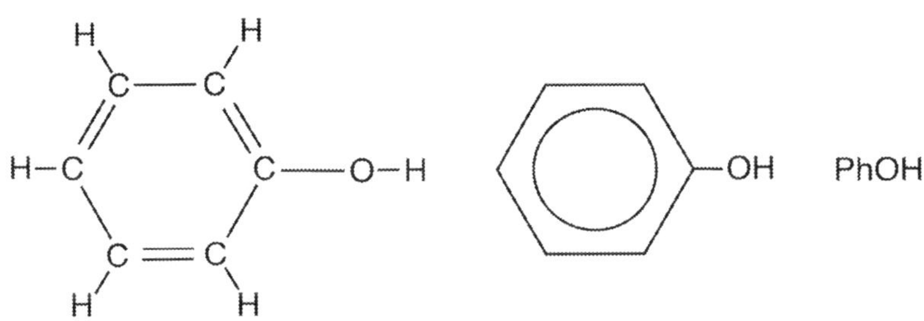

Figure 7. Three representations of the phenol molecule structure

These are all forms of representation that can appear on the page or screen, but chemistry also uses various kinds of structural models to represent molecules and lattices. Again there are different formalisms used (such as touching spheres in some crystal models, or representing bonds between atomic centres with springs or straws or sticks). All of these different forms of representation have been developed to help communicate specific aspects of the subject matter of chemistry – yet each new form of symbolism or representation adds to the material to be learnt and mastered by the student. This does not just mean recognising a suitable interpretation, but being able to select the most appropriate form of representation from among a number of possible options when communicating ideas. Students therefore need to not only be introduced to the forms of representation themselves, but also to their particular affordances and ranges of application. Again this is an area where the teacher, often being very familiar with a wide range of representational tools for discussing chemistry, needs to be careful to scaffold student learning. As in most areas of learning, students need to meet ideas in manageable learning quanta, and then to have opportunities to apply the ideas in a range of contexts, with regular review to consolidate learning.

This could seem very negative, but is not all bad news for the teacher. For one thing, it is believed that learning is often more effective when it is multi-modal – so for example when verbal descriptions are accompanied by diagrams and models (Jewitt, Kress, Ogborn, & Tsatsarelis, 2001). Chemistry naturally lends itself to this kind of multi-modal teaching. Effective teaching needs to link together the phenomena seen in the laboratory with the descriptions at both macroscopic and submicroscopic scales and various relevant forms of representation – and this draws upon verbal description, gesture, diagrams, models, simulations etc.

OPPORTUNITIES TO CHALLENGE THE MOST ABLE (GIFTED) STUDENT

This chapter has deliberately focused on some of the fundamental challenges of teaching and learning chemistry, but concludes by suggesting that these very complications make chemistry an excellent subject to challenge the most able learners and to teach about the nature of science. Effective classroom teaching is educative because (a) students are faced with challenges that force them to shift out of their 'comfort zones' (where they can work algorithmically to undertake tasks that are in effect exercises), and rather experience genuine problems that require them to develop their thinking, but (b) tasks are scaffolded by teachers so that problems are soluble with the amount of support provided to particular students. Good teaching then offers the right balance between challenge and support. For many learners first meeting the ideas behind chemistry, the emphasis needs to be on scaffolding support so that students are eased into new ways of thinking. For the most able students, however, who (at least in some national contexts) seldom find school science lessons genuinely stretch them, there are opportunities for the teacher to present significant levels of challenge (Taber, 2010).

As one hypothetical example, where most learners might initially struggle to understand the ideas behind figure 4, a gifted learner in a class might point out that carbon is only somewhat more electronegative than hydrogen, so it would be possible to offer an alternative version of figure 4 (see figure 8) where the carbon is considered to be oxidised by fewer steps, and where the oxidation state of hydrogen increases as well. This could be supported by arguing that although the methane molecule is, theoretically, somewhat less likely to form $C^{4+}H^{-}_{4}$ rather than $C^{4-}H^{+}_{4}$ (i.e. by considering resonances, cf. figure 5) the similarity in electronegativity between carbon and hydrogen makes any ionic form of limited relevance to the structure. A teacher should be impressed by a student offering such a suggestion as it would show they understood and were actively engaging with the ideas met in class. The gifted student may also point out that alternative notions of oxidation (oxygen is added to hydrogen as well as carbon) as well as the relative degrees of polarity in the bonds in CH_4 and H_2O suggests that figure 8, with hydrogen also being oxidised, could be the preferred analysis. This kind of discussion (about abstract theoretical models and formalisms) is likely to be highly motivating for a gifted student who has quickly acquired robust understanding, but in most classes is best carried out privately as it is likely to only confuse and mystify those classmates with a more fragile grasp of the basic ideas requiring consolidation before they can be used as the foundations for further learning.

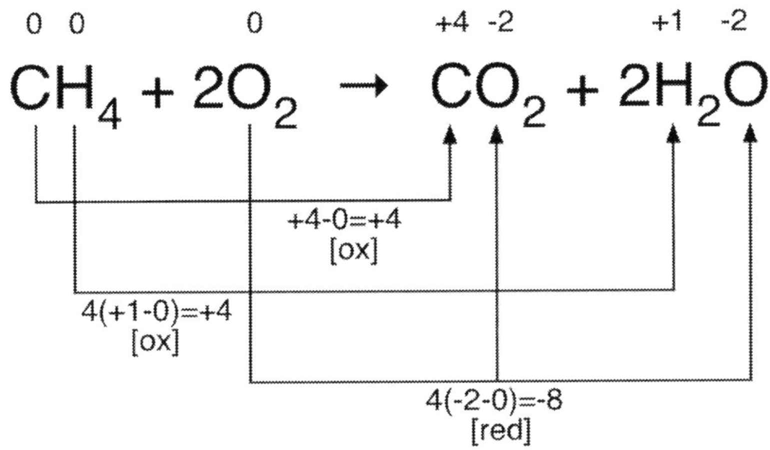

Figure 8. Representing an alternative conceptualisation of a redox process

OPPORTUNITIES TO TEACH ABOUT THE NATURE OF SCIENCE IN CHEMISTRY

Part of the reason students often find school science becomes dull once past the initial excitement of being allowed to work in laboratories is that it can seem an endless list of things to be learnt – things previously discovered and so already known. The various models, and formalisms, used in chemistry could easily fall into this category. There is also an inherent issue of relevance to students' lives in chemistry. Chemistry is a widely applicable and relevant science, but usually the applications are at least one step removed from the ideas met in school science (again because chemistry works with substances and so is a level of abstraction away from the materials science of more direct relevance).

Something to aspire to here then is what has been termed epistemic relevance (Taber, 2015). This means that rather than teaching students a wide range of chemical ideas and models – especially when research suggests students often have limited appreciation of the nature and roles of models in science – and then try to show how these have applications, it is better to present students with chemical phenomena that can generally fire their imagination and use this as a starting point for showing students how science involves an interplay between empirical evidence and theory, and so how theory taught in chemistry is motivated by the desire to explain actual observations.

It is not suggested that students will easily arrive at the concepts, category systems, models, and formalisms used by professional chemists (Driver, 1983). These will inevitably need to be introduced: but only once students have been set the task of making sense of phenomena and given an authentic experience of what it is to try to develop new ideas and representations to explain, report, and predict chemical phenomena. By contrast much current practice involves teaching students about other people's solutions to *problems they have never had*. A more authentic chemistry education may need to limit its scope in terms of the topics covered, but will give a much better flavour of chemistry as science rather than just a catalogue of facts and strange categories and formalisms.

Some readers may wonder if it is counterproductive to ask students to develop their own representations given that these will inevitably need to be put aside in favour of taught conventions. Yet such activity can support effective learning (Tytler, Prain, Hubber, & Waldrip, 2013). If groups work on developing their own representations and formalisms, and have to argue for the logic of their approach, they will come to appreciate both (i) that many of the representations used in chemistry are historically contingent and not in some way inevitable (as different groups will devise different alternatives), and (ii) how science comes to adopt such conventions through argumentation and community agreement. They will also better appreciate that the processes of science are creative and call upon imagination – and that scientific ideas are often only widely accepted after being refined by extensive work (see Chapter 2: '*Reflecting the nature of science in science education*').

There are widespread calls to involve students in more genuine enquiry in science lessons (Osborne, 2014). Enquiry processes encompass all stages of scientific work (see Chapter 19: '*Inquiry-Based Science Education*'). There are limits to the kinds of student-initiated chemical investigations that can be safely carried out in school laboratory conditions (in those countries where these are available) but the teacher can use standard chemical demonstrations and class practicals as starting points for exploring the other side of enquiry – the scientific work of finding ways to construct understanding of the world.

FURTHER READING

Chemistry Education Research and Practice: The top ranking research journal on chemistry education is provided free to access for all readers by the publisher (The Royal Society of Chemistry) – at pubs.rsc.org/en/journals/journalissues/rp#

Taber, K. S. (Ed.). (2012). *Teaching secondary chemistry* (2nd ed.). London: Hodder Education.

Eilks, I., & Hofstein, A. (Eds.). (2013). *Teaching chemistry – A studybook: A practical guide and textbook for student teachers, teacher trainees and teachers*. Rotterdam, The Netherlands: Sense Publishers.

Taber, K. S. (2002). *Chemical misconceptions: Prevention, diagnosis and cure*. London: Royal Society of Chemistry.

de Jong, O., & Taber, K. S. (2014). The many faces of high school chemistry. In N. Lederman & S. K. Abell (Eds.), *Handbook of research in science education* (Vol. 2, pp. 457–480). New York, NY: Routledge.

Kind, V. (2004). *Beyond appearances: Students' misconceptions about basic chemical ideas* (2nd ed.). London: Royal Society of Chemistry.

Gilbert, J. K., de Jong, O., Justi, R., Treagust, D. F., & Van Driel, J. H. (Eds.). (2002). *Chemical education: Research-based practice*. Dordrecht, The Netherlands: Kluwer Academic Publishers BV.

Garcia-Martinez, J., & Serrano-Torregrosa, E. (Eds.). (2015). *Chemistry education: Best practices, opportunities and trends*. Weinhein, Germany: Wiley.

REFERENCES

Bruner, J. S. (1960). *The process of education*. New York, NY: Vintage Books.

Driver, R. (1983). *The pupil as scientist?* Milton Keynes: Open University Press.

Eilks, I., & Hofstein, A. (Eds.). (2015). *Relevant chemistry education: From theory to practice*. Rotterdam, The Netherlands: Sense Publishers.

Gilbert, J. K., & Treagust, D. F. (Eds.). (2009). *Multiple representations in chemical education*. Dordrecht: Springer.

Jenkins, E. W. (2007). School science: A questionable construct? *Journal of Curriculum Studies, 39*(3), 265–282.

Jewitt, C., Kress, G., Ogborn, J., & Tsatsarelis, C. (2001). Exploring learning through visual, actional and linguistic communication: The multimodal environment of a science classroom. *Educational Review, 53*(1), 5–18. doi:10.1080/00131910123753

Johnstone, A. H. (1982). Macro- and microchemsitry. *School Science Review, 64*(227), 377–379.

Osborne, J. (2014). Scientific practices and inquiry in the science classroom. In N. G. Lederman & S. K. Abell (Eds.), *Handbook of research on science education* (Vol. 2, pp. 579–599). New York, NY: Routledge.

Taber, K. S. (1994). *Can Kelly's triads be used to elicit aspects of chemistry students' conceptual frameworks?* Paper presented at the British Educational Research Association Annual Conference, Oxford. Retrieved from http://www.leeds.ac.uk/educol/documents/00001482.htm

Taber, K. S. (1998). An alternative conceptual framework from chemistry education. *International Journal of Science Education, 20*(5), 597–608.

Taber, K. S. (2009). Learning at the symbolic level. In J. K. Gilbert & D. F. Treagust (Eds.), *Multiple representations in chemical education* (pp. 75–108). Dordrecht: Springer.

Taber, K. S. (2010). Challenging gifted learners: General principles for science educators; and exemplification in the context of teaching chemistry. *Science Education International, 21*(1), 5–30. Retrieved from http://www.icaseonline.net/sei/2010march.html; http://www.icaseonline.net/sei/march2010/p2.pdf

Taber, K. S. (2013a). A common core to chemical conceptions: Learners' conceptions of chemical stability, change and bonding. In G. Tsaparlis & H. Sevian (Eds.), *Concepts of matter in science education* (pp. 391–418). Dordrecht, The Netherlands: Springer.

Taber, K. S. (2013b). Revisiting the chemistry triplet: Drawing upon the nature of chemical knowledge and the psychology of learning to inform chemistry education. *Chemistry Education Research and Practice, 14*(2), 156–168. doi:10.1039/C3RP00012E

Taber, K. S. (2013c). Upper secondary students' understanding of the basic physical interactions in analogous atomic and solar systems. *Research in Science Education, 43*(4), 1377–1406. doi:10.1007/s11165-012-9312-3

Taber, K. S. (2014). *Student thinking and learning in science: Perspectives on the nature and development of learners' ideas*. New York, NY: Routledge.

Taber, K. S. (2015). Epistemic relevance and learning chemistry in an academic context. In I. Eilks & A. Hofstein (Eds.), *Relevant chemistry education: From theory to practice* (pp. 79–100). Rotterdam, The Netherlands: Sense Publishers.

Talanquer, V. (2011). Macro, submicro, and symbolic: The many faces of the chemistry "triplet". *International Journal of Science Education, 33*(2), 179–195. doi:10.1080/09500690903386435

Treagust, D. F., Chittleborough, G., & Mamiala, T. L. (2002). Students' understanding of the role of scientific models in learning science. *International Journal of Science Education, 24*(4), 357–368. doi:10.1080/09500690110066485

Tytler, R., Prain, V., Hubber, P., & Waldrip, B. G. (Eds.). (2013). *Constructing representations to learn in science*. Rotterdam, The Netherlands: Sense Publishers.

MARK WINTERBOTTOM

25. TEACHING AND LEARNING BIOLOGY

INTRODUCTION

Biology is a popular subject. It is popular because many students already have pre-conceived ideas about how biological systems work, which give them confidence about their ability to understand biology. Indeed, they have made sense of their immediate biological environment through their own informal processes of inquiry. By observing, generating hypotheses, observing again, and refuting or refining their ideas (albeit in an informal way), many of the ideas students have about biology have been 'worked out' by themselves. Such 'working out' of biological ideas is a nice way to think about the teaching and learning of biology. It has long been recognised that learners are not simply empty vessels into which to pour knowledge, but that they actively construct their knowledge, sometimes alone and frequently with others. As early as 1909, John Dewey (1910) suggested that science teaching had become a presentation of ready-made knowledge, with no attempt to communicate to students that science is a method of inquiry. By placing the onus onto the learner to inquire and to build their knowledge themselves, albeit with appropriate scaffolding from the teacher, they can start to build ideas themselves, and to understand and assimilate such an inquiry process. Indeed, such an approach helps learners to take on new concepts through "[interacting] with relevant ideas in [their] existing cognitive structure" (Ausubel et al., 1978, p. 67). Hence, new understandings come about through the product of the new ideas with elements already present in the individual's cognitive structure. By giving learners a meaningful role in their learning through constructivist and inquiry based approaches (Harlen, 2004), they can learn biology more effectively.

This chapter offers perspectives on teaching and learning about biology illustrated through a discussion of some key topics to exemplify how the research-informed teacher might approach planning and teaching biology. Immediately below, we examine three areas in which it is beneficial to engage with learners' starting points, and provide them with the best scaffold to 'working out ideas for themselves'. We then look at two areas which should facilitate children's learning of biology, outlining contemporary ideas about the utility of practical work in learning biology, and about the utility of fieldwork in learning biology.

K. S. Taber & B. Akpan (Eds.), Science Education, 343–353.

TEACHING PHOTOSYNTHESIS. IS IT TIME TO DITCH THE STARCH TEST?

With photosynthesis, the clue is in the name. 'Light-making' provides a useful starting point. And from that point, many teachers will present the equation for photosynthesis below.

> Carbon dioxide + Water → Glucose + Oxygen (in the presence of light and chlorophyll)

There then follows a series of practicals which involve testing leaves for starch. Table 1 shows the various treatments and expected results.

Table 1. Conditions and outcomes in a traditional practical about photosynthesis

	Carbon dioxide	*Water*	*Light*	*Chlorophyll*	*Starch present?*
Normal plant	Yes	Yes	Yes	Yes	Yes
Normal plant in bell jar with conc. Sodium hydroxide	No	Yes	Yes	Yes	No
Normal plant with stencils over part of each leaf.	Yes	Yes	Only on exposed parts of leaf	Yes	Only in exposed parts of leaf
Variegated plant	Yes	Yes	Yes	Only in green parts of leaf	Only in green parts of leaf

If we did not need to think about children's prior ideas, there would be no problem. So what are the problems? Remember that learning comes from the product of the interaction of new ideas with pre-existing ideas in the student's cognitive structure. These practical activities can often simply reinforce what the equation has already told them, rather than engaging with students' prior ideas in order to help build the scientific model. There are other things wrong too:

1. Starch is not the primary product of photosynthesis but formed by polymerisation of glucose which is, in fact, produced in photosynthesis. Asking students to accept that starch presence indicates the production of glucose is not trivial, and requires explicit attention in class.
2. Each practical, with one exception, confirms the importance of a variable (light, chlorophyll) with a positive starch test. To confirm the importance of carbon dioxide requires a negative starch test. Unless they understand the logic of the tests completely, students frequently refer to the leaf deprived of carbon dioxide as showing carbon dioxide is NOT required for photosynthesis, which is wrong.

In order to enable children to learn appropriately about photosynthesis, they need to encounter new ideas in a way that recognises their pre-existing ideas and problems they may encounter. For example, many students think that plants get food from the soil. Many find it implausible that a gas (carbon dioxide) can contribute to plant growth, because they don't understand that gases have mass. A team of researchers at the Centre for Studies in Science and Mathematics Education at the University of Leeds (Hind et al., 2002) recommended a sequence which explicitly addressed such ideas in order to teach plant nutrition. This included the following steps:

1. Asking students to think about (a) what factors affect plant growth, (b) what food is, (c) what food is needed for, and (d) how animals and plants get their food. The aim is to open up pupils' ideas about food, shedding light on their pre-existing ideas.
2. Present the scientific model of plant nutrition in terms of producing sugar from a chemical reaction involving carbon dioxide and water. It is helpful to present students with statements about the implausibility of this model: (a) a gas and liquid can react to form a solid, (b) carbon dioxide has mass, (c) carbon dioxide gas and water can react to form sugar. Ask students to respond to these statements, and to devise ways to test them.
3. Students test the implausibilities of the model: (a) bubble carbon dioxide through limewater and centrifuge the resulting suspension, (b) weigh a balloon containing carbon dioxide and compare with the mass of an empty balloon, (c) using cardboard atoms, rearrange the atoms in six molecules of water and carbon dioxide to form a sugar molecule (almost like a jigsaw).
4. Return to students' ideas about the importance of light. Remind them of the involvement of energy in chemical reactions; identify light as the source of energy here. Then look at what is left over from the rearrangement of atoms in step 3. Six molecules of oxygen should be left. You can then demonstrate that pond weed produces oxygen 'over water'.

Having run through those steps, bicarbonate indicator provides a reagent which enables students to test predictions on the basis of their new knowledge. Bicarbonate indicator can be used to detect the presence or absence of carbon dioxide. Students can make predictions about the effect of varying light intensity or carbon dioxide concentration based on the scientific model.

By using students' starting points, and building knowledge from that point, they learn more effectively about how photosynthesis works, rather than simply churning through some conventional practical activities, which may be fun, but whose utility to learning is dubious.

TEACHING GENETICS. IS IT TIME TO DITCH MENDEL AND GENETIC ANALYSIS?

Inquiry based instruction can often follow the discoveries made by scientists in history, simply because following the historical steps in discovery can mirror the questions which students themselves are likely to ask, and can demonstrate to students that underpinning scientific principles have always emerged from evidence. Many curricula include the work of Gregor Mendel, whose experiments with peas are commonly used to help students learn about monohybrid and dihybrid genetic crosses. So what is wrong with that?

Well, students who study the work of Mendel to help them undertake genetic analysis learn to do genetic crosses very effectively. However, many students do not understand the location of the alleles depicted in a genetic diagram, and fail to understand the difference between somatic and sex cells, or the role of meiosis in transferring genetic information (Lewis & Wood Robinson, 2000). It is possible to learn genetic analysis without really understanding the mechanism underpinning it. Indeed, students can often be confused about the relationship between genes and chromosomes, knowing that a zygote requires a full set of chromosomes, but failing to understand the mechanisms which bring this about (Lewis & Wood Robinson, 2000; Mills Shaw et al., 2008). Many students also think that alleles are intrinsically dominant or recessive, a belief which inhibits their understanding of the mechanism of dominance when they meet it later in further or higher education. The T, t depiction of alleles may propagate this belief further (Dougherty, 2009). In fact, dominance is a relationship between alleles, so an allele may be dominant to one particular allele, but recessive to another one. Likewise, because we start with dominant and recessive, many students see this type of monogenic inheritance as the norm. In fact, such a deterministic view of genes is rarely correct; most phenotypic variation is not caused by 'classical' dominance at all, and variation most frequently has a polygenic cause, rather than one gene being responsible for one phenotype (Dougherty, 2009). Indeed, a Mendelian approach reinforces such a deterministic view, rather than seeing multiple genes acting within an environmental context, the interaction between both determining phenotype. Such a belief can make it difficult for students to engage with the theory of evolution by natural selection (Dougherty, 2009), and can negatively influence individual behaviour; attributing a health problem to genetics may make individuals fail to heed doctors' advice to alter their diet or behaviour (Mills Shaw et al., 2008).

If we continue to use Mendel as the start of an inquiry-based teaching approach, then we should follow – step by step – the historical development of ideas about inheritance, the gene, the chromosome and the Central Dogma of the relationship between DNA, RNA and protein. Engaging students in the knowledge building process from Mendel, step-by-step through to the development of the Central Dogma and beyond, can encourage them to see that science is not simple truth, but also allow them to develop ideas, and to build upon them and revise them, as they progress through their learning, using such understanding to eventually engage with contemporary genetics in an era where genetic testing is now commonplace. However, such an approach may still allow the above alternative conceptions to 'bed in' and become resilient in students' minds.

As we have seen above, if students fail to understand the genetics knowledge taught in the classroom, they may therefore be unable to apply such knowledge to their daily lives (Mills Shaw et al., 2008). Prescriptive curricula can force teachers to adopt piecemeal approaches to teaching inheritance and genetics, with meiosis appearing separately to genetic analysis which appears separately to reproduction. In reviewing our teaching of inheritance and genetics, perhaps we should consider approaches which are more revolutionary.

For example, should our starting point for genetics teaching be protein? Lewis and Wood-Robinson (2000) noted that the majority of students did not see protein as the link between genotype and phenotype. Because proteins are the functional molecules, should a study of genetics start with proteins as biologically active agents, and work outwards from them, creating a strong link between genotype and phenotype in students' minds, and drawing on mechanisms of inheritance to help students understand the process, rather than 'summary' genetic analysis? Such a focus on the role of protein may provide a firmer basis for understanding genes, genotypes, phenotypes and inheritance (Dougherty, 2009).

Finally, should we spend more time thinking about genomics. By comparing genomes across organisms, we can begin to understand how organisms share related processes, how they behave and how species adapt to different environments. Likewise, students will increasingly need to engage with personal genomics, given that identifying the effects of single genes is not necessarily helpful to students in making decisions which affect their health and lifestyle. We should consider how students will need to apply their genetic knowledge now and in the future and build the curriculum from there, with the aim of creating genetically-literate citizens. Such an approach contributed to the development of the science curriculum in England following the Beyond 2000 report, where the authors looked at the success and failure of current provision, and asked what science education was actually needed by students to make informed life choices (Millar & Osborne, 1998).

Discussion of the structure of genetics curricula is long-standing, with many authors realising the need for change, but generating such change requires tangible examples and resources to trial. In the further reading section, you can look at Michael Dougherty's (2009) ideas to invert the genetics curriculum to avoid some of the difficulties expressed above.

TEACHING OSMOSIS: TAKING STUDENTS FROM PARTICLE TO TISSUE

Diffusion is the net movement of particles from a region of high concentration to a region of low concentration. Up until the age of 16, we tend to define osmosis as the diffusion of water particles from a region of high water concentration (low solute concentration) to a region of low water concentration (high solute concentration) across a partially permeable membrane. Learning about osmosis is hard, because (a) it builds upon often poorly understood foundations about diffusion, (b) it is hard to link the process to the reality students see in plants, and (c) we usually only talk about solutes diffusing, rather than water diffusing as a solvent (Winterbottom, 2011).

Students' ideas about solvents and solutes can also be poorly formed (Barker, 2000). They can think that solutes 'disappear' when dissolved in solvents. They can attribute decision making powers to particles, suggesting that particles 'want' to go from a region of high concentration to low concentration, and hence that all particles move away from a region of high concentration, rather than understanding the random nature of particle movement. Indeed, they can think that particles stop moving when equilibrium is reached, a misconception which is responsible for students failing to understand the idea of net movement (Odom, 1995).

So how do you enable students to build their learning, whilst overcoming the problems above? Stephen Tomkins, formerly of the Faculty of Education, University of Cambridge, suggests a four-step approach (Winterbottom, 2011).

Step 1. Reinforce particle and diffusion theory and challenge misconceptions. Stress that water can diffuse. It is sensible to start with everyday examples of diffusion, asking students to observe and explain their observations (such as potassium manganate (VII) diffusing, or smells diffusing from the front to the back of the lab), enabling you to identify any misconceptions. Using beans or balls to provide models of diffusion, using ICT simulations of diffusion, or using role play can help students to visualise their ideas more tangibly.

Step 2. Having understood that water and solutes can diffuse, the next step is for children to work out that water, like a solute, can have a concentration gradient. As a first step, you need to demonstrate that a particular volume of strong sugar solution contains fewer water particles than the same volume of pure water. Present two large measuring cylinders. Label them A & B. Fill both with warm water to the top graduation mark. Ask the class what will happen to the level in A if some sugar is poured in (Likely answer: it will go up). Add a 150ml beaker of sugar. Ask the class what actually happened (It did go up!).

Shake the cylinder until the sugar dissolves (the level does not go down). Then pour off the difference into a small beaker until the levels are the same again. Ask the class which cylinder has the most sugar in it (Likely answer A). Ask the class which cylinder has the most water in it (Likely Answer B, although some may say they are the same). Challenge those who get it wrong by pointing out the water in the little beaker. Ask students what would happen if the solutions in A and B were brought into contact with each other, steering the conversation to help them realise that water can diffuse from one solution to another from a high water concentration to a low water concentration.

Step 3. Use models to reinforce and apply the ideas of water diffusing across a partially-permeable membrane (the cell membrane and other intra-cellular membranes). The important thing here is that the solute cannot diffuse across the membrane, and that a concentration gradient of water exists across the membrane.

Some of the ideas below may be helpful in allowing students to see that osmosis takes place, to visualise the mechanism, and to see the effects of osmosis.

- Use Visking tubing filled with black treacle and submerged in pure water so you can see that water diffuses in, both with Visking tubing enlarging in size, and with the colour being diluted.
- Use a model to help students visualise how the membrane prevents diffusion of the solute. You can use different sized balls in a tray, with only some being small enough to pass through holes in a divider across the middle.
- ICT simulation of osmosis. Try to find a simulation with variable speed control, or which moves slowly so students can follow individual particles, enabling them to see the random and continuous movement of particles, allowing you to discuss the idea of net movement.
- Use a balloon (cell membrane) inside a 'paper box' (cell wall) model to look at the effect of osmosis on turgidity and plasmolysis.

Step 4. Apply and test the ideas with living plant tissue. Having built the ideas in models, use students' understanding to predict and explain what will happen when osmosis happens in living tissue.

- Measure (length or mass), bathe in pure water, and remeasure a row of ten sultanas.
- Irrigate a slide of red onion cells with (a) water and (b) concentrated sugar solution. Look at the effects on turgor and plasmolysis.
- Measure (length or mass), bathe, and measure ten potato chips in different concentrations of sugar solution.
- Test children's understanding. Take a potato half, hollowed out with a pile of sugar inside the hollow. After 2–3 hours, this fills with water that has diffused out of the potato cells. Ask students to explain the effect, by talking about osmosis.

UTILITY OF PRACTICAL WORK IN LEARNING ABOUT BIOLOGY

Practical work is considered to be integral to becoming a biologist, and teachers see practical work as an essential part of learning science; hence, use of practical work is common across school laboratories and could be considered almost routine (Millar, 2002). Such routine use is further reinforced by teachers' views that practical work is motivating to students (Wellington, 1998). If practical work becomes routine, there is a risk that teachers will fail to critically analyse the benefits of such work to students' learning (Abrahams & Reiss, 2012). Indeed, use of the term practical work suggests limited focus on how a piece of practical work supports learning. It may include 'recipe-following' work, demonstrations by the teacher, or a full open-ended investigation. Teachers frequently favour shorter tasks, as they fit with curriculum objectives more easily, whereas more open ended investigatory work may be perceived as lacking efficiency in securing assessed curriculum objectives. However, the utility even of shorter tasks is equivocal, with reviews suggesting that practical work gives no advantage to students' learning over non-practical approaches (Lazarowitz & Tamir, 1994). So how do teachers assess the potential utility of practical work for students' learning? Abrahams and Millar (2008) devised a model for evaluating a practical task. It is very much based on examining whether students did and learnt what they were intended to do and learn (see also Chapter 29: *'Minds-on practical work for effective science learning'*). If the practical work is effective at level 1, they do and see what they were intended to do and see. If it is effective at level 2, they learn what they were intended to learn. Abrahams and Reiss (2012) took the examination of effectiveness slightly further, distinguishing the two levels between two domains of knowledge: the domain of ideas, and the domain of observable objects and events. We can use these two classifications to develop an effectiveness matrix, which can help biology teachers to examine the utility of a particular practical exercise.

Take a simple practical activity, in which students exercise and measure their heart rate before and after increasing amounts of exercise. Let's start with the domain of observables.

a. To be effective at level 1, students need to do and see what they were intended to do and see. So the practical is effective if students can find their pulse, and if they can count their heart beat before and after exercise and calculate their heart rate per minute.
b. To be effective at level 2, students need to be able to design and carry out an investigation to find out how different intensities of exercise affect heart rate.

Then look at the domain of ideas.

a. To be effective at level 1, students need to do and see what they were intended to do and see. So the practical is effective if students can talk about the difference in heart rate before and after exercise, and how heart rate increases with increased duration of exercise.
b. To be effective at level 2, students need to talk about how more exercise requires more energy and so increased respiration, how increased respiration requires more oxygen, and because oxygen is carried in the blood, supplying more oxygen requires an increased heart rate.

If you now think about the practical activity, without considerable input and scaffolding from the teacher, it is unlikely that the practical will achieve effectiveness at level 2 in the domain of ideas. There is some evidence that this need for scaffolding can be forgotten. It is straightforward for a teacher to be kept busy, ensuring students are busy doing what they are intended to do, and seeing what they are intended to see, without giving enough consideration to ensuring students learn what they are supposed to learn. By using this framework for analysis prior to running a practical activity, it is possible to ensure that the practical becomes effective for students' learning, rather than just a motivational time filler with limited effect on learning.

UTILITY OF FIELDWORK IN LEARNING ABOUT BIOLOGY

Many biology teachers cite the importance of fieldwork in learning about biology, although often their assessment of such utility is based on professional experience and judgement, rather than being based upon research evidence. Without evidence, it is difficult to argue against the challenges which fieldwork faces in schools. Schools are sensitive to perceived health and safety risks during fieldwork, and any associated litigation. Teachers may be short of time, resources and support. The perception that fieldwork constrains curriculum time is often prevalent, and availability of staff and course timetabling can be incompatible with provision of fieldwork (Rickinson et al., 2004).

To argue against such constraints, it is important to understand the benefits of fieldwork, as well as the factors which can improve its efficacy. Nundy (2001) focused on three benefits associated with fieldwork, namely (a) a positive impact on long-term memory due to the memorable fieldwork setting, (b) affective benefits, such as individual growth and improvement in social skills, and (c) reinforcement between affective and cognitive benefits. Rickinson et al. (2004) reviewed the evidence for cognitive benefits and affective benefits. A number of authors found gains in knowledge and attitudes attributable to fieldwork (e.g. Bogner, 1999; Manzanal et al., 1999), with some demonstrating benefits in both compared to conventional classroom learning (Eaton, 2000). In terms of affective benefits, Nundy (1999, 2001) noted that collaborative tasks during fieldwork had a positive impact on students' cooperative skills, leadership qualities, perseverance, initiative and motivation.

Positive outcomes of fieldwork can only be achieved by planning well for students' learning. The structure, duration and pedagogical approach in fieldwork are all important (Rickinson et al., 2004). There is ample evidence that longer programmes are more effective for students' learning than shorter ones. Bogner (1999) compared a one day and five day version of an environmental education course, finding students' learning to be greater in the five day version. Preparatory work has also been found to influence the success of a field experience; so, for example, even simply involving students in setting a behaviour code for a field visit (Aleixandre & Rodriguez, 2011) was found to be beneficial to students' eventual learning outcomes. Likewise, effective follow up work is essential. Orion and Hofstein (1994) thought that the fieldwork should be placed early in a sequence of lessons, so the field visit itself 'drives' the curriculum in subsequent lessons. Designing curriculum related learning opportunities is important, as is implementing familiar routines and structure to establish trust and ensure discipline (New Economics Foundation, 2004). By making the structure and format of the learning activity closely aligned with the goals of the activity, students are more likely to learn effectively.

FURTHER READING

Abrahams, I., & Reiss, M. (2016). *Enhancing learning with effective practical science 11–16*. London: Bloomsbury.

Association for Science Education. (2010). *Outdoor science. A coordinated approach to high-quality teaching and learning in fieldwork for science education*. Hertford: Association for Science Education. Retrieved from https://www.field-studies-council.org/media/154119/2010_outdoor_science.pdf

Dougherty, M. (2009). Closing the gap: Inverting the genetics curriculum to ensure an informed public. *American Journal of Human Genetics, 85*(1), 6–12. Retrieved from http://www.ncbi.nlm.nih.gov/pmc/articles/PMC2706960/?tool=pubmed

Reiss, M. (Ed.). *Teaching secondary biology*. London: Hodder Education.

Winterbottom, M. (1999). *Teaching biology to Key Stage 4*. London: Hodder Education.

REFERENCES

Abrahams, I., & Millar, R. (2008). Does practical work really work? A study of the effectiveness of practical work as a teaching and learning method in school science. *International Journal of Science Education, 30*(14), 1945–1969.

Abrahams, I., & Reiss, M. (2012). Practical work: Its effectiveness in primary and secondary schools in England. *Journal of Research in Science Teaching, 49*(8), 1035–1055.

Aleixandre, M. P. J., & Rodriguez, L. R. (2001). Designing a field code: Environmental values in primary schools. *Environmental Education, 7*(1), 5–22.

Ausubel, D., Novak, J., & Hanesian, H. (1978). *Educational psychology: A cognitive view* (2nd ed.). New York, NY: Holt, Rinehart & Winston.

Barker, V. (2000). *Beyond appearances: Students' misconceptions about basic chemical ideas*. London: Royal Society of Chemistry.

Bogner, F. X. (1999). Empirical evaluation of an educational conservation programme introduced in Swiss secondary schools. *International Journal of Science Education, 21*(11), 1169–1185.

Dewey, J. (1910). *How we think*. London: D.C. Heath and Company.

Dougherty, M. (2009). Closing the gap: Inverting the genetics curriculum to ensure an informed public. *American Journal of Human Genetics, 85*(1), 6–12.

Odom, A. L. (1995). Secondary and college biology students' misconceptions about diffusion and osmosis. *The American Biology Teacher, 57*(7), 409–415.

Harlen, W. (2004). *Evaluating inquiry-based science developments*. Retrieved from https://www.google.co.uk/url?sa=t&rct=j&q=&esrc=s&source=web&cd=1&cad=rja&uact=8&ved=0ahUKEwidoL_uta7MAhWEtxQKHYE8Dy0QFggfMAA&url=http%3A%2F%2Fstem.gstboces.org%2FShared%2520Documents%2FSTEM%2520DEPLOYMENT%2520PROJECT%2520RESEARCH%2FNAS_paper_eval_inquiry_science.pdf&usg=AFQjCNGUMTodIhVnm-qEmGHKML0O94ySaQ&bvm=bv.120551593,d.d24

Hind, A., Lewis, J., Leach, J., & Scott, P. (2002). *Teaching science for understanding*. Centre for Studies in Science and Mathematics, University of Leeds. Retrieved from http://www.education.leeds.ac.uk/assets/files/research/cssme/PlantNutrScheme.pdf

Lazarowitz, R., & Tamir, P. (1994). Research on using laboratory instruction in science. In D. L. Gabel (Ed.), *Handbook of research on science teaching and learning*. New York, NY: Macmillan.

Lewis, J., & Wood-Robinson, C. (2000). Genes, chromosomes, cell division and inheritance – do students see any relationship. *International Journal of Science Education, 22*(2), 177–195.

Manzanal, R. F., Barreiro, L. M. R., & Jimenez, M. C. (1999). Relationship between ecology fieldwork and student attitudes toward environmental protection. *Journal of Research in Science Teaching, 36*(4), 431–453.

Millar, R. (2002). Thinking about practical work. In S. Amos & R. Booka (Eds.), *Aspects of teaching secondary science: Perspectives on practice*. London: Routledge Falmer.

Mills Shaw, K. R., Van Horne, K., Zhang, H., & Boughman, J. (2008). Essay contest reveals misconceptions of high school students in genetics content. *Genetics, 178*(3), 1157–1168.

New Economics Foundation. (2004). *Forest school evaluation project. A study in Wales: April to November 2003.* London: NEF.

Nundy, S. (1999). The fieldwork effect: the role and impact of fieldwork in the upper primary school. *International Research in Geographical and Environmental Education, 8*(2), 190–198.

Nundy, S. (2001). *Raising achievement through the environment: The case for fieldwork and field centres.* Doncaster: National Association of Field Studies Officers.

Orion, N., & Hofstein, A. (1994). Factors that influence learning during a scientific field trip in a natural environment. *Journal of Research in Science Teaching, 31*(10), 1097–1119.

Rickinson, M., Dillon, J., Teamey, K., Morris, M., Choi, M. Y., Sanders, D., & Benefield, P. (2004). *A review of research on outdoor learning.* London: National Foundation for Educational Research/ King's College.

Wellington, J. (1998). *Practical work in school science. Which way now?* London: Routledge.

Winterbottom, M. (2011). Transport in organisms. In M. Reiss (Ed.), *Teaching secondary biology.* London: Hodder Education.

SECTION V

RESOURCES FOR SCIENCE TEACHING

AVI HOFSTEIN

26. THE ROLE OF LABORATORY IN SCIENCE TEACHING AND LEARNING

INTRODUCTION: THE HISTORY OF THE LABORATORY IN SCIENCE EDUCATION.

Throughout the chapter I use the terms practical work, which is common in the UK and Germany context, and laboratory work, which is common in USA, interchangeably. A precise definition is difficult as these in school practice embrace an array of activities, but generally they refer to experiences in school settings in which students interact with equipment and materials or secondary sources of data to observe and understand the natural world (Hegarty-Hazel, 1990).

Laboratory activities have long had a distinctive and central role in science curriculum as a means of making sense of the natural world. Since the ninetieth century, when schools began to teach science systematically, the laboratory has become a distinctive feature of science education

After the first-world-war, and with the rapid increase of scientific knowledge, the laboratory was used mainly as a means for confirming and illustrating information previously learnt in a lecture or from textbooks. With the reform in science education in the 1960s in many countries, the ideal became to engage students with investigations, discoveries, inquiry, and problem solving activities. In other words, the laboratory became the core of the science learning process (Shulman & Tamir, 1973).

The *National Science Education Standards* (*NSES*, 1996, p. 23) defines such learning activities (e.g. inquiry) as: "the diverse ways in which scientists study the natural world and propose explanations based on the evidence derived from their work. Scientific inquiry also refers to the activities through which students develop knowledge and understanding of scientific ideas, as well as an understanding of how scientists study the natural world."

For many years science educators have suggested that many benefits accrue from engaging students in science laboratory activities (Tobin, 1990; Hofstein & Lunetta, 2004). Tobin (1990) for example wrote that: "Laboratory activities appeal as a way of allowing students to learn with understanding and at the same time engage in the process of constructing knowledge by doing science" (p. 405).

In curricular-type -projects developed during the 1960s the laboratory was intended to be a place for inquiring, developing, and the testing of theories as well as providing students with the opportunity to 'practice being a scientist'. Many research

K. S. Taber & B. Akpan (Eds.), Science Education, 357–368.

studies (summarized for example by Bates, 1978; Hofstein & Lunetta, 1982) were conducted with the goal in mind to explore the effectiveness of the laboratory for attaining the many objectives (both cognitive as well as affective) that had been suggested in the science education literature.

The traditional list of objectives includes:

- Understanding of scientific concepts
- Interest and motivation
- Attitude towards science
- Scientific practical skills and problem solving abilities
- Scientific habits of mind
- Understanding the nature of science (NOS)
- The opportunity to *do* science

Over the years, hundreds of papers and essays were published with the goal being to explore and investigate the uniqueness of the science laboratory in general and its educational effectiveness in particular. In addition, it has been widely believed that the laboratory provides the only place in school where certain kinds of skills, abilities, and understanding can be developed (Lazarowitz & Tamir, 1994). In other words, as they suggested, the laboratory provides a unique mode of instruction, learning, and assessment.

Precisely what kind of objectives and aims will be attained in the laboratory dependent on a wide range of factors? It is suggested that, amongst others, these will include the teacher's goals, expectations, subject and pedagogical content knowledge as well as the degree of relevance to the topic, the students' abilities and interests, and many other logistical and economic considerations related to the school settings and facilities (see Table 1).

Table 1. Suggested goals for laboratory activity (after Bennett, 2003)

Goals for Laboratory Activity
to encourage accurate observation and description
to make scientific phenomena more real
to enhance understanding of scientific ideas
to arouse and maintain interest (particularly in younger pupils)
to promote a scientific method of thought

It should be noted that some of these goals, such as "*enhancing learning of scientific concepts*" coincide with the broad goals for science education that are not necessarily laboratory based. The teacher should be in the position to judge whether the laboratory is the most effective learning environment for attaining a certain objective while teaching a certain topic. Teachers should be aware that there has been a great deal of discussion and numerous research studies about which goals are in fact better achieved through laboratory instruction than through other instructional (pedagogical) approaches (Hofstein & Lunetta, 1982; Hofstein & Lunetta, 2004). The many research studies and essays that were cited in Hofstein and Lunetta reviews criticized the tradition of conducting experiments without clear purposes and goals. In addition, they revealed a significant mismatch between teachers' goals for learning in the science laboratory and those that were originally defined by curriculum developers and the science education milieu.

RESEARCH BASED IDEAS RELATED TO: LEARNING IN FROM THE SCIENCE LABORATORY

The main goal of this chapter is to argue and demonstrate that the laboratory in science education is a unique learning environment (Lazarowitz & Tamir, 1994; Hofstein, 2004; Lunetta, Hofstein, & Clough, 2007) so that if designed in an articulated and purposeful manner with clear goals in mind has potential to enhance some of the more important learning skills such as learning by inquiry, metacognition, and argumentation.

Laboratory activities have long had a distinctive and central role in the science curriculum, and science educators have suggested that many benefits accrue from engaging students in science laboratory activities (Dori, Sasson, Kaberman, & Herscovitz, 2004; Hofstein & Lunetta, 1982; Tobin, 1990; Lazarowitz & Tamir, 1994; Lunetta, 1998; Hofstein & Lunetta, 2004; Lunetta, Hofstein, & Clough, 2007). More specifically, they have suggested that, when properly developed, inquiry-centered laboratories have the potential to enhance students' meaningful learning, conceptual understanding, and their understanding of the nature of science. Inquiry-type laboratories are central to learning science, since students are involved in the process of conceiving problems and scientific questions, formulating hypotheses, designing experiments, gathering and analyzing data, and drawing conclusions about scientific problems or phenomena.

At the beginning of the 21st century we are entering a new era of reform in science education. Both the content and pedagogy of science learning and teaching are being scrutinized, and new standards intended to shape and rejuvenate science education are emerging (National Research Council, 2005; AAAS, 1990; Millar & Osborne, 1998; Bybee, 2000). In general, one of the characteristics of this reform is the change in the goals articulated for science teaching and learning namely, that science education should be targeted to all students (attaining scientific literacy for all students) and should be extended beyond the preparation of science oriented students for academic careers in the sciences This is in fact a call for also rethink the goals for the learning in and from laboratory work. There are several buzz words that characterize current reform. Among these are student-centered learning, learning by the inquiry method, and development of high learning skills such as argumentation, metacognition and asking relevant questions (relevant to the experimental situation). Inquiry in the context of science learning in general and inquiry in the science laboratory in particular are amongst the important components of this reform (Bybee, 2000; Lunetta, 1998; Hofstein & Lunetta, 2004; Sere, 2002). Bybee (2000) wrote that inquiry in terms of skills and abilities includes the following components identifying and posing scientifically oriented questions, forming hypotheses, designing and conducting scientific investigations, formulating and revising scientific explanations, and communicating and defending scientific arguments. It is suggested that many of these abilities and skills are in alignment with those that characterize inquiry-type laboratory work (practical work to include project-based learning), an activity that puts the student in the center of the learning process (see also Tobin, 1990; Hofstein, Shore, & Kipnis, 2004; Hofstein, Mamlok-Naaman, Navon, & Kipnis, 2005; Dori & Sasson, 2007). To attain this goal he suggested that students should be provided in the laboratory, with opportunities to reflect on findings, clarify understandings and misunderstandings with peers, and consult a range of resources that include teachers, books, websites, and other learning materials. His review reported that such opportunities rarely exist since teachers, in the laboratory, are so often preoccupied with technical and managerial activities. Similarly, Hodson (1993) suggested that although teachers generally professed belief in the value of student-driven, open, practical investigation, in general, their teaching practices in the laboratory failed to support this claim. He also argued that the research literature failed to provide evidence that standard school laboratory activities encouraged knowledge construction. He was critical of the research literature:

> Despite the very obvious differences among, for example, practical exercises designed to develop manipulative skills or to measure 'physical constraints', demonstration-type experiments to illustrate certain key-concepts, and inquiries that enable children to conduct their own investigations, there is a tendency for researchers to lump them all together under the same umbrella title of practical work. (p. 97)

Tobin wrote that teachers' interpretation of practical activity should be elaborated and made a part of the research design since a laboratory session could be open-ended inquiry in one classroom and more didactic and confirmatory in another teacher's classroom.

Based on their review of the literature regarding the laboratory Lazarowitz and Tamir (1994) joined the long list of writers claiming that the potential of the laboratory as a medium for teaching and learning science is enormous. They wrote that the laboratory is the only place in school where certain kinds of skills and understanding can be developed. They are among those who have suggested that one of the complicating factors associated with research on the effectiveness of the school laboratory is that often the goals articulated for learning in the laboratory have been experiences in the laboratory and related assessment practices have remained relatively unchanged. In addition, almost synonymous with those articulated for learning science more generally. Hart et al. (2000) claimed that much practical work is purposeless and often the explicit objectives of the practical work do not coincide with the purpose of practical experiences. They also claim that many practical tasks have too many different teaching/learning objectives to focus on during instruction. Similarly, Sere' (2002) in France, reporting on a long-term project (Lab-Work in Science Education) conducted in seven European nations wrote that:

> The intention [of the study] was to address the problem of the effectiveness of lab-work, which in most countries is recognized as being essential to experimental sciences, but which turns out to be expensive and less effective than wished. (p. 624)

The project focused mainly on the effectiveness of lab-work conducted in the context of science learning in the upper secondary schools. Information on practice was gathered through 23 case studies, surveys, and a tool that helps to map and describe the domain of laboratory work. Sere reported that the objectives typically articulated for laboratory work (i.e., understanding theories, concepts, and laws; conducting various experiments; learning processes and approaches; and applying knowledge to new situations) were too numerous and comprehensive for teachers to address successfully in individual laboratory sessions. In response, she suggested that the scope of the objectives for specific laboratory activities should be limited. Science curriculum developers and science teachers should make conscious choices among specific learning objectives for specific lab activities and clearly articulate the specific objectives for their students.

Gunstone (1991), wrote that helping students' develop scientific ideas from practical experiences is a very complex process and that students generally did not have sufficient time or encouragement to express their interpretations and beliefs and to reflect on central ideas in the laboratory. Research on learning in the school laboratory makes clear that to understand their laboratory experiences, students must manipulate *ideas* as well as materials in the school laboratory and they must be helped to contrast (and align) their findings and ideas with the concepts of the contemporary scientific community. Manipulating materials in the laboratory is *not* sufficient for learning contemporary scientific concepts. This accounts for the failure of "cookbook" laboratory activities and relatively "unguided" discovery activities to promote desired scientific understanding. Several studies suggested that while laboratory investigations offer excellent settings in which students can make sense of phenomena and in which teachers can better understand their students' thinking, laboratory inquiry alone is *not* sufficient to enable students to construct the complex conceptual understandings of the contemporary scientific community (Lunetta, 1998). In the laboratory, students should be encouraged to articulate and share their ideas, to help them perceive discrepancies among their ideas, those of their classmates, and those of the scientific community.

At the end of the twentieth century there was increasing understanding from the cognitive sciences that learning is contextualized and that learners construct knowledge by solving genuine, meaningful problems (Wenger, 1998). The school science laboratory can offer students opportunities to have some control of their activities, enhance their perception of sense of *ownership* and *motivation* (Johnstone & Al-Shuaili, 2001). It can be an environment particularly well suited for providing a meaningful context for learning, determining and challenging students' deeply held ideas about natural phenomena, and constructing and reconstructing their ideas. Though a complex process, meaningful learning in the laboratory *can* occur if students are given sufficient time and opportunities to interact, reflect, explain, and modify their ideas. Engaging in *metacognitive* behaviors of this kind enables students to elaborate and to apply their ideas; the process can promote conceptual understanding as well as the development of problem-solving skills. The challenge is to help learners take control of their own learning in the search for understanding while providing opportunities that encourage them to ask questions, suggest hypotheses, and design investigations, "minds-on as well as hands-on" (Gunstone, 1991).

NEW ERA-NEW GOALS: GOALS FOR LEARNING IN AND FROM THE SCIENCE LABORATORY IN THE 21ST CENTURY

Introduction

We are operating in an era in which high-order learning skills are seen as important as the content of science (Hofstein & Kind, 2012). Higher order thinking/learning skills and activities in the context of learning science are considered to be complex, non-algorithmic, and involve applications of multiple criteria instead of memorizing facts. These activities include asking research questions, solving authentic problems, argumentation, metacognitive skills, drawing conclusions, making comparisons, dealing with controversies, and taking a stand (White & Mitchell, 1994). Gunstone (1991) claimed that meaningful learning in the laboratory occurs when students are given ample opportunities for interaction and reflection in order to initiate discussion. It is suggested that some of these skills could be developed as part of inquiry-based science laboratories. Many of these abilities and skills are in alignment with those that characterize inquiry-based chemistry laboratory work, an activity that places the student at the center of the learning process (Sandoval, 2005); Researchers claim that learning in the laboratory might provide a constructivist environment that fosters higher order thinking, metacognitive and argumentative skills (Kind, 2003). In this chapter I shall elaborate on two of these variables namely the development of argumentative skills and the ability to ask high level and relevant questions, these in the context of the science laboratory.

Argumentation in the Science Laboratory

Several researchers who focused on the issue of argumentation suggested that the inquiry-type laboratory in science education can provide opportunities for students to develop argumentation skills (See a detailed discussion in Hofstein and Kind (2012)). However, only few research studies were conducted with the goal in mind of accepting or rejecting this assumption. For example, Rickey and Stacy (2000) found that students who participated in guided inquiry-type laboratories were better at evaluating evidence obtained from their research.

Two recent studies reported in the literature regarding the nature of the experiments as a platform for evoking argumentation both in quantitatively (number of arguments) and qualitatively (level of arguments). Kind et al. (2011) in the UK investigated the quality of argumentation among 12 to13 year old students in the UK in the context of secondary school physical science program. Their study explored the development of argumentation among students who undertook three different designs of laboratory-based tasks. The tasks described in their paper involved the students in the following: collecting and making sense of data, collecting data for addressing conflicting hypothesis, and paper-based discussions in the pre-collected data phase about an experiment. Their finding showed that the paper-based task (the 3rd one in the above tasks list) generated larger number of arguments in a period of time compared to the two other tasks. In addition, they found that in order to encourage the development of high-level and authentic argumentation there is a need to change the practice that generally exists in the science laboratories in England. They suggested that more rigorous and longitudinal research is needed in order to explore the potential of the science laboratory as a platform for development of students' ability to argue effectively and in an articulated way.

The second study was conducted in Israel in the context of 12 years of research and development of inquiry-type laboratories in the context of upper secondary school in grades 10–12 (for more details about the philosophy and rationale of the project see, Hofstein, Shore and Kipnis (2004). The implementation and effectiveness of this project were researched intensively and comprehensively and were reported in a series of manuscripts (e.g. Hofstein et al., 2004; Kipnis & Hofstein, 2008). The research study conducted by Katchevich et al. (2013, 2014b) focused on the process in which students constructed arguments in the chemistry laboratory while conducting different types of experiments. It was found that, *inquiry-type* experiments have the potential to serve as an effective platform for formulating arguments, owing to the special features of this learning environment. The discourse conducted during inquiry-type experiments was found to be rich in arguments, whereas that during *confirmatory-type* experiments was found to be sparse in arguments. In addition, it was found that the arguments, which were developed during the discourse of an inquiry-type experiment, were generated during the following stages of the inquiry process: hypothesis-building analysis of the results, and drawing appropriate conclusions. On the other hand confirmatory-type experiment revealed small number of arguments. In addition, the arguments that were posed in the confirmatory-type experiments were of law-level in their characteristic.

On the basis of a detailed analysis of the discourse in the chemistry laboratory (conducted by Katchevitch et al., 2013), one can conclude the open-ended inquiry experiments stimulate and encourage the construction of arguments, especially the stages of hypotheses definition, analysis of the results and the drawing of conclusions. Some arguments arise from individuals and some from the group. Both types of arguments consist of explanations and scientific evidence, which link the claims to the evidence. Therefore, it is suggested, that the learning environment of open-ended inquiry experiments, is a platform for raising arguments. In this study the researchers wanted to highlight the main factors that stimulate raising argumentation in open-ended inquiry experiments, as well as to characterize situations in which argumentation develops a significant discourse.

Asking Questions in the Science Laboratory

In attempt to develop scientific literacy among students, teachers must create effective learning environments in which students are given opportunities to ask relevant and scientifically sound questions. Usually, questions asked during a lesson are those initiated by the teacher and only rarely by the students, and those questions do not emerge spontaneously from students; rather, they have to be encouraged. The content of a question can indicate the level of thinking of the person who raised it. It should be noted that in general the cognitive level of a certain question is determined by the type of answer that it requires (Yarden et al., 2001). Several studies noted the importance (and value) of questioning skills. For example, Zoller (1987) in the context of learning high school chemistry, suggested that questioning is an important component in a real world, involving problem-solving and decision-making processes. Cuccio-Schirripa and Steiner (2000) suggested that:

> Questioning is one of the thinking processing skills which is structurally embedded in the thinking operation of critical thinking, creative thinking, and problem solving. (p. 210)

This quote is in alignment with the results of a study conducted in chemistry by Dori and Herscovitz (1999) who found, that fostering 10th grade students' capabilities to pose questions improved their problem-solving ability. Hofstein et al. (2005), conducted a research study that focuses on the ability of high-school (11th and 12th grade – [give student ages here for an international readership]) chemistry students, who learn chemistry through the inquiry approach (see Hofstein, Shore, & Kipnis, 2004) to ask meaningful and scientifically sound questions. Two aspects were investigated in this study: (a) the ability of students to ask questions related to their observations and findings in an inquiry-type experiment (a practical test), and (b) the ability of students to ask questions after critically reading a scientific article. The student population consisted of two groups: an inquiry-laboratory group (experimental group) and a traditional laboratory-type group (control group). Three common features were researched: (1) the number of questions that were asked by each of the students, (2) the cognitive level of the questions, and (3) the nature of the questions that were chosen by the students, for the purpose of further investigation. Importantly, it was found that students in the inquiry group who had experience in asking questions in the chemistry laboratory outperformed the control group in their ability to ask more and better questions. The activity of asking inquiry questions (that are, by definition, high-level questions) is one of the operations that the students are required to do during every full inquiry experiment. In contrast, the students of the control group, who had learned the traditional-type program, which does not contain the inquiry experiments, did not have any opportunity to practice the activity of asking questions and specifically asking inquiry questions, which are higher-level questions, and therefore their skills in asking questions, as was indicated by the test, were lower.

HOW ARE SCIENCE LABORATORIES USED?-DIFFERENT FORMS OF PRACTICAL WORK

Teachers' and Students' Practice in Science Laboratories

The question to be asked in the first theme is to what degree the use of practical work has changed at schools. We will be looking at research describing how laboratories are used by teachers and students, as well as the nature of laboratory activities and facilities.

Hofstein and Lunetta (2004) in their review wrote that although many biology teachers' articulated philosophies appeared to support a hands-on investigative approach with authentic learning experiences, the classroom practice of those teachers did not generally appear to be consistent with their stated philosophies. Several studies have reported that very often teachers involve students principally in relatively low-level, routine activities in laboratories and that teacher-student interactions focused principally on low-level procedural questions and answers. Marx et al. (1998) reported that science teachers often have difficulty helping students ask thoughtful questions, design investigations, and draw conclusions from data. More recently, Abrahams and Millar (2008) in the UK investigated the effectiveness of practical work by analysing a sample of 25 "typical" science lessons involving practical work in English secondary schools. They conclude that the teachers' focus in these lessons was predominantly on making students manipulate physical objects and equipment. Hardly any teachers focused on the cognitive challenge(s) of linking observations and experiences to conceptual ideas. Neither was there any focus on developing students' understanding of scientific inquiry procedures.

These are findings that echo the situation at any time in the history of school science. Basic elements of teachers' implementation of practical work seem not to have changed over the last century: students still carry out recipe-type activities which are supposed to reflect science procedures and teach science knowledge, but which in general fails on both. This is not to say everything is the same. Science education has moved forward in the last decades and improved teachers' professional knowledge and classroom practice, but this improvement has not sufficiently caught up with the challenges of using laboratory work in an efficient and appropriate way. Teachers still do not perceive what is required to make that laboratory activities serve as a principal means of enabling students to construct meaningful knowledge of science, and they do not engage students in laboratory activities in ways that are likely to promote the development of science concepts. In addition, many teachers do not perceive that helping students understand how scientific knowledge is developed and used in a scientific community is an especially important goal of laboratory activities for their students.

Aligned with this situation for teachers we find a matching picture in students' experiences and laboratory teaching material. Attempts have been made to develop protocols for analyzing laboratory activities (Lunetta & Tamir, 1979). This tool was adopted in Australia by Fisher et al, (1999). Domin (1998) in the USA analysed laboratory guides and found that students are seldom given opportunities to use higher-level cognitive skills or to discuss substantive scientific knowledge associated with the investigation, and many of the tasks presented to them continue to follow a 'cookbook' approach concentrating on the development of lower level-type skills and abilities.

The reviews discussed earlier in this chapter reported a mismatch between the goals articulated for the school science laboratory and what students regularly do in those experiences. Ensuring that students' experiences in the laboratory are aligned with stated goals for learning demands that teachers explicitly link decisions regarding laboratory topics, activities, materials and teaching strategies to desired outcomes for students' learning. The body of past research suggests that far more attention to the crucial roles of the teacher and other sources of guidance during laboratory activities is required, and researchers must also be diligent in examining the many variables that interact to influence the learning that occurs in the complex classroom laboratory.

SUMMARY

The biggest challenge for practical work, historically and today, is to change the practice of "manipulating equipment not ideas". The typical laboratory experience in school science is a "hands-on" but not a "minds-on" activity. This problem is related to teachers' fear for losing control in the classroom and to give students more responsibility for their learning. Also to be blamed for the current situation is an assessment practice which does not pay enough attention to higher order thinking and a long tradition of developing fool-proof laboratory tasks which guide students through activities without requiring any deeper reflection. The review in this chapter has demonstrated a relationship between these problems in practical work and the "common sense" ideas about science inquiry as a step-wise method.

It has taken science education research a long time to reveal this practice, analyze its underlying rationales and presented alternatives. The development has required a move away from quantitative research methods, which were not sensitive to students' learning in the laboratory, towards more authentic ways of studying what actually goes on in the laboratory. It has also required a thorough analysis of the nature of science inquiry and what makes someone good at doing it. The alternatives which are prominent today combine socio-cultural perspectives on science and learning, but also link to new aims for school science as an important provider of skills and knowledge for citizenship.

At the turn of the century we may claim science education is at a better position than ever before for developing a meaningful and appropriate practice for laboratory work. The situation is most promising because of the results and knowledge which were accumulated and achieved. There are many places to start for new development of laboratory teaching strategies and professional development of teachers. These and other tasks calls for science education researchers still to engage with practical work and help develop this area further.

FURTHER READING

Hegarty-Hazel, E. (1990). The student laboratory and the science curriculum: An overview. In E. Hegarty-Hazel (Ed.), *The student laboratory and the science curriculum* (pp. 3–26). London: Routledge.

Hofstein, A., & Kind, P. (2012). Learning in and from science laboratories. In B. Fraser, K. Tobin, & K. McRobbie (Eds.), *International handbook on science education* (pp. 189–207). Dordrecht: Springer.

Lazarowitz, R., & Tamir, P. (1994). Research on using laboratory instruction in science. In D. L. Gabel (Ed.), *Handbook of research on science teaching and learning* (pp. 94–130). New York, NY: Macmillan.

Lunetta, V. N. (1998). The school science laboratory: Historical perspectives and centers for contemporary teaching. In B. J. Fraser & K. G. Tobin (Eds.), *International handbook of science education*. Dordrecht, The Netherlands: Kluwer.

REFERENCES

Abrahams, I., & Millar, R. (2008). Does practical work really work? A study of the effectiveness of practical work as a teaching and learning method in school science. *International Journal of Science Education, 30*, 1945–1969.

Bennett, J. (2003). *Teaching and learning science: A guide to recent research and its application*. London: Continuum.

Bybee, R. (2000). Teaching science as inquiry. In J. Minstrel & E. H. Van Zee (Eds.), *Inquiring into inquiry learning and teaching in science* (pp. 20–46). Washington, DC: American Association for the Advancement of Science (AAAS).

Cuccio-Schirripa, S., & Steiner, H. E. (2000). Enhancement and analysis of science question level for middle school students. *Journal of Research in Science Teaching, 37*, 210–224.

Domin, D. S. (1998). A content analysis of general chemistry laboratory manuals for evidence of high-order cognitive tasks. *Journal of Chemical Education, 76*, 109–111.

Dori, Y. J., Sasson, I., Kaberman, Z., & Herscovitz, O. (2004). Integrating case-based computerized laboratories into high school chemistry. *The Chemical Educator, 9*, 4–8. Retrieved September 26, 2006, from http://chemeducator.org/bibs/0009001/910004yd.htm

Fisher, D., Harrison, A., Henderson, D., & Hofstein, A. (1999). Laboratory learning environments and practical tasks in senior secondary science classes. *Research in Science Education, 28*, 353–363.

Gunstone, R. F. (1991). Reconstructing theory from practical experience. In B. E. Woolnough (Ed.), *Practical science* (pp. 67–77). Milton Keynes: Open University Press.

Hegarty-Hazel, E. (1990). The student laboratory and the science curriculum: An overview. In E. Hegarty-Hazel (Ed.), *The student laboratory and the science curriculum* (pp. 3–26). London: Routledge.

Hofstein, A. (2004). The laboratory in chemistry education: Thirty years of experience with developments, implementation, and research. *Chemistry Education Research and Practice, 5*, 247–264.

Hofstein, A., & Kind, P. (2012). Learning in and from science laboraories. In B. Fraser, K. Tobin, & K. McRobbie (Eds.), *International handbook on science education* (pp. 189–207). Dordrecht, The Netherlands: Springer.

Hofstein, A., & Lunetta, V. N. (1982). The role of the laboratory in science teaching: Neglected aspects of research. *Review of Educational Research, 52*(2), 201–217.

Johnstone, A. H., & Al-Shuaili, A. (2001). Learning in the laboratory: Some thoughts from the literature. *The Higher Education Chemistry (RSC), 5*(2), 42–51.

Katchevitch, D., Hofstein, A., & Mamlok-Naaman, R. (2013). Argumentation in the chemistry laboratory: Inquiry and confirmatory experiments. *Research in Science Education, 43*(1), 317–345.

Kind, P. M. (2003). TIMSS puts England first on scientific enquiry, but does pride come before a fall? *School Science Review, 85*, 83–90.

Kind, P. M., Kind, V., Hofstein, A., & Wilson, J. (2011). Peer argumentation in the school science laboratory-Exploring effects of task features. *International Journal of Science Education, 33*, 2527–2558.

Kipnis, M., & Hofstein, A. (2008). The inquiry laboratory as a source for development of metacognitive skills. *International Journal of Science and Mathematics Education, 6*, 601–627.

Lunetta, V. N., & Tamir, P. (1979). Matching lab activities with teaching goals. *The Science Teacher, 46*, 22–24.

Marx, R. W., Freeman, J. G., Krajcik, J. S., & Blumenfeld, P. C. (1998). Professional development of science teachers. In B. J. Fraser & K. G. Tobin (Eds.), *International handbook of science education* (pp. 667–680). Dordrecht, The Netherlands: Kluwer.

National Research Council. (2005). *National science education standards*. Retrieved May 29, 2006, from http://www.nap.edu/readingroom/books/nses/html/index.html

Rickey, D., & Stacy, A. M. (2000). The role of metacognition in learning chemistry. *Journal of Chemical Education, 77*, 915–920.

Sandoval, W. A. (2005). Understanding students' practical epistemologies and their influence on learning through inquiry. *Science Education, 89*(4), 634–665.

Sere, G. M. (2002). Towards renewed research questions from outcomes of the European project lab-work in science education. *Science Education, 86*, 624–644.

Tobin, K. G. (1990). Research on science laboratory activities: In pursuit of better questions and answers to improve learning. *School Science and Mathematics, 90*, 403–418.

Wenger, E. (1998). *Communities of practice: Learning, meaning and identity*. Cambridge: Cambridge University Press.

Yarden, A., Brill, G., & Falk, H. (2001). Primary literature as a basis for a high-school biology curriculum. *Journal of Biology Education, 35*, 190–195.

Zoller, U. (1987). The fostering of question asking capability. *Journal of Chemical Education, 64*, 510–512.

JOSEPH MAKOKHA

27. EMERGING TECHNOLOGIES AND SCIENCE TEACHING

INTRODUCTION

There is general agreement among educators that given society's current pace of transformation, modern teaching and learning should also evolve rapidly to meet 21st century needs. The referenced basic skills include critical thinking, creativity, collaboration, communication, innovation and problem solving. While these skills and their corresponding assessment methods rarely fit with the mandated content standards in many states and countries, it is possible to utilize many affordances of emerging technologies like low cost sensors, web and computing technologies, rapid prototyping platforms, and others in STEM education. On the other hand, incorporating these technologies in classrooms raises new challenges especially with regards to assessment of learning, given that these new skills require different methods of measuring learning outcomes beyond tests, the current default. This problem has been characterized as an overreliance on the product as opposed to the processes undertaken by students during learning activities. Viewed this way, taking a multiple-choice test at the end of the course to demonstrate learning gains seems to focus on the product rather than the process through which the learner gained new knowledge and skills, but incorporating learning technologies promises some alternatives.

Among recent developments that might transform our ability to assess learning is the capacity to quickly collect and analyze massive data. Based on examples highlighted in this chapter, documenting the learning process gives rise to massive data in different media formats. Without proper tools, such data tends to hinder rather than enhance learning given that educators need simple, real time feedback in order to adjust their methods in response to evolving learner needs. Technologies and methods of handling, processing and presenting big data may help solve this problem, giving us a tool to assess the learning process together with the other 21st century skills mentioned earlier.

We may also consider emerging technologies as tools that enable us to extend the human mind. One way this happens is through connecting minds, whose sum total together is greater than the sum of individual minds. Evidence of this abounds from examining the ways in which participatory capabilities have exponentially risen to match new capacities demonstrated by Web 2.0 technologies like Wikipedia, open source software and resources, video streaming on YouTube, social networks,

K. S. Taber & B. Akpan (Eds.), Science Education, 369–383.

podcasts and blogs among others. This rise represents different tools for participatory learning compared to the past when such engagement required organized, face to face, mostly offline activities. While in-person classroom interaction still thrives, it is augmented by these technologies leading to a view of this phenomenon as an approach to learning that is based on conversation and interaction, on sharing, creation and participation, on learning not as a separate activity, but rather, as embedded in meaningful activities such as games or workflows (Downes, 2010). This approach takes focus away from the finished product and emphasizes the process, a key factor in shifting our assessment model to reflect the new set of skills expected of learners.

The role of community as a pivotal component of learning is well documented. Educators often seek ways to foster this aspect, and based on the emerging technology tools, opportunities for creating new learning communities continue to expand (Dede, 2004). Like many learning activities, Web 2.0 technologies combine multiple aspects of games, workflows and curriculum among others, fostering learning through conversation and interaction, creating and sharing artifacts and participation. Learning is therefore embedded in other meaningful activities as opposed to being a separate activity. It is this dimension of learning that takes advantage of the many affordances associated with emerging technologies. Also notable is the fact that as a result of the networked nature of society, collaboration and opportunities for creating new learning communities continue to rise.

Wearable gadgets, especially computers, are coming in many forms. Their ability to effortlessly collect and transmit data presents new tools for learners and teachers. For instance, students may be able to collect project data away from school using their wearable devices, and then send it to a shared space to continue working when they return to class. The diminishing cost of cloud computing as well as the enhanced capabilities of devices, networks and connectivity all contribute to an environment where wearable devices increasingly help collect and transmit data. This data may then be aggregated to provide patterns, trends and contribute to predicting outcomes. Considering the unexpected use of mobile phones to determine traffic conditions in major cities around the world, think of the possibilities presented to educators by wearables embedded with multiple sensors. Another important feature of these wearable devices arises from the premise that by incorporating bio information, geo-location and environmental data, as well as sensors that collect real information, students are likely to experience authenticity from such learning. This may be achieved by combining lessons with meaningful data collection directly pertaining to learners.

Educators have frequently used gaming as a learning tool but we now have rich, engaging media on a variety of platforms including smartphones and consoles that many learners already use. There also exist different gaming applications for desktop and mobile devices, providing greater opportunities for teachers to experiment with educational games. Closely related to gaming is Augmented Reality (AR). We are now able to bring data to real, physical environments such as science labs – for instance information on the inner workings of a body part may be revealed when a learner points a smartphone at a biology lab model. Similar uses can be seen at leading museums where patrons get additional facts by pointing their hand held devices to artifacts. AR apps exist for activities ranging from animations for fun, to others that provide detailed, on-time instructions for technical personnel in a lab or workshop, even for doctors performing complex surgery. This is another example of using technology to extend human thinking and abilities.

Data/Info manipulation has evolved exponentially over the past few years, partly because of improved capacity to collect data as well as lower storage costs. Availability of applications and fast processors as well as new data manipulation methods have made it easier to handle massive data sets and to present the patterns in useful, accessible forms. As for Maker Tools the "just on time" learning demanded by many tasks like designing, measuring, shaping, assembling; the ability to demonstrate skills and concepts through making; as well as learning transferable skills like information technology, equipment/machine operation, material science and collaboration, all make a compelling case for this approach.

The Role of Emerging Technology (ET)

Papert's (1980) vision of ways that computers might affect learning captures the essence of ET's role in general, and narrowly as it pertains to learning. He postulates two distinct ways: by enhancing thinking and by changing patterns of access to knowledge. One can find evidence of these two broad goals in features presented by current technologies like simulations, digital modeling, and making tools as well as information manipulation, communication and processing techniques. By combining these technologies in new ways, it is possible to present mental models and concepts through multiple modes, and to access many learning channels.

ET's place in learning is also articulated by Hopkins' (1991) summary of the 5 general categories of the roles assigned to tools. While the specific examples have evolved, the basic functions still hold, hence emerging technology can be used as a tool for individualized learning (including skills building, referencing …); or for group interaction (e.g., a technology called Discourse enables the teacher or presenter to view learner responses to questions or problem situations); for managing and coordinating learning (the student's personal growth plan, schedule building, portfolio construction, e-mail writing), for self expression (writing across the curriculum, video and multimedia, telecommunication networks); and for knowledge production (HyperCard, Hyperstudio and Lego/Logo projects, for portfolios, MIDI music and new tools for art).

MOOCs

Massive Open Online Classes (MOOCs) were viewed as a vehicle of delivering lessons to multitudes irrespective of their physical location – as long as they had Internet access. They were seen as a viable solution for providing individualized learning and seemed poised to disrupt traditional educational establishments. While many of the initial projections failed to hold, some aspects of the online course dissemination gained acceptance and are common even within traditional schools and colleges. The blended style has become popular in certain disciplines, with up to 30% of Computer Science classes at a leading Silicon Valley university now offered online, or with a mix of online and in-class presence (see Table 1).

Table 1. A comparison of In-Person with Online Classes in 4 Schools/Departments

Department/School	*Total number of classes*	*Online/Blended*
Computer Science	86	29
Business School	127	2
Education	92	5
Engineering	506	28

Also glimpsed from the table is the relatively higher number in this one field compared to others like Education, Business and even the rest of the Engineering courses (note that Computer Science falls under the Engineering School at this university, but numbers were separated to highlight this difference). Examining the course content across the schools reveals that Computer Science classes lend themselves more to several affordances (see Table 2) of online platforms, hence the larger representation in number of online/blended classes.

Table 2. Examples of technology affordances

Some technology affordances for educational purposes	
File Sharing	Website Hosting
Screen Casting	Web Chat
Content Streaming Over Internet	Video Annotation
Interactive screens	Social Media Plug-ins
Cloud Computing	Mobile Web Platforms
Mobile Audio/Video Devices	Visualization Tools
Content Management Systems	Bluetooth/Wifi
Cameras/microphones in Wearables	Network Security/Authentication
Solar Powered Devices	Digital Content/Books
Mobile Applications	Interconnected Devices

Another outcome of the MOOCs has been the rise of educational-centric applications for chatting and corresponding, whether this is within institutional course management platforms (such as CourseWork or BlackBoard) or Google hangouts and online study boards. Some are able to document student activity and progress. In addition to collaboration efforts between universities like the OpenEdX platform, other examples of successful online, certificate-issuing institution that arose from the MOOCs are Cousera and Udacity, which now offer a variety of courses and disciplines.

Several technologies are making online learning increasingly viable. Among them are recent improvements in basic features such as file sharing, which has been boosted by Bluetooth; screen casting and mirroring of mobile device screens onto larger displays; and others. If we add cloud computing, which enables users to run processes on remote servers while handling the least amount of load on local computers (these are increasingly mobile devices such as tablets and smartphones), we achieve an online environment that delivers functionality required for most online learning. Cloud computing by itself offers a new range of possibilities. Apart from availing affordable processing capacity, this computing style makes it possible to start off with a small network and then scale with increasing needs, something that was quite complicated in the past – this was very costly in environments where organizations hosted their own servers. A commentator recently wondered: "Who, as the last century drew to a close, could have taught about Bluetooth, WAP, Jini, or XML, to cite but a few examples of significant new technologies? Furthermore, new applications of these technologies, and new business models enabled by them, are also introduced at bewildering rates" (Fedorowicz & Gogan, 2002, p. 10). Yet this was before current mobile devices, apps and interactive technologies that are fueling new ways of thinking and learning.

There have been many reviews and commentaries on MOOCs over the past three years, including one by a trio that has been actively exploring the future of education (Thille, Mitchell, & Stevens, 2015) in the digital age. In their evaluation, Thille et al. (2015), point out the successes and failures of MOOCs, including areas that previously seemed to be a great fit but turned out to be ineffective – such as delivering quality education to masses of students worldwide. A clear verdict from the MOOCs experiments confirms the traditional educational institutions' enduring existence. It is unlikely that schools and colleges as we know them, will disappear any time soon. Given this position, it is worth seeking ways to evolve these institutions to support new ways of thinking and learning. Davidson et al. (2010) have investigated this issue and think that when considering the future of learning institutions in a digital age, it is important to look at the ways that "digitality" works to cross the boundaries within and across traditional learning institutions. In this regard, the blended models, as well as programs in which students take in-person classes supplemented by online content for credit, are viable emerging trends.

Wearables

There has been a recent spike in wearable devices ranging from watches and fitness trackers, head-mounted glasses and cameras, to tiny chips embedded in clothes. One benefit of this abundance of wearable sensors is ease of massive digital data collection, which previously required a combination of equipment and software applications. Wearable technology, such as the Google Glass promises to make self data collection (think of the quantified self movement) quite easy. The concept that one can capture almost every detail of their hourly/daily life presents new possibilities for learning beyond the collected data. For instance it was observed that students in a university health class consisting mostly of sports men/women now have a new tool (wearables) to collect and aggregate their biometric data, physical activity, sleep and other information that enables them to evaluate their individual nutrition and health needs. Prior to wearable health trackers, it took several gadgets and considerable time to capture, record, store and share such data, which is now managed by a single device connected to others via internet.

An example of a multi-functional wearable device is Google Glass. It captures pictures and video, streams internet content from connected mobile devices using Bluetooth, and can perform tasks from voice prompts among other features.

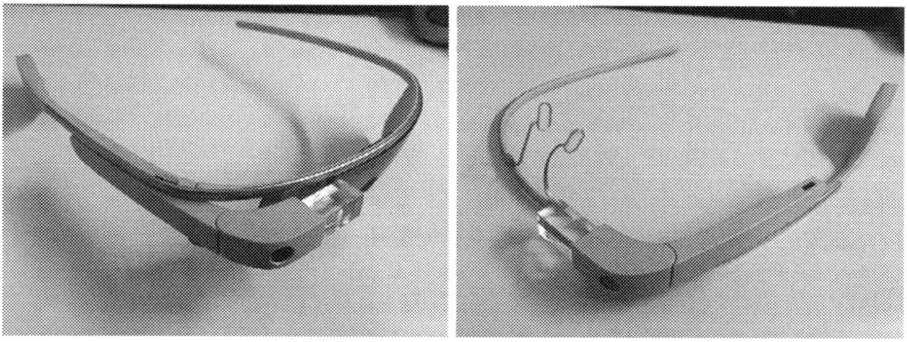

Figure 1. Google glass

Other examples are sensor-embedded watches that collect information such as physical activity and real-time physiological data like heart rate and blood pressure. This kind of information may be used to help students appreciate health concepts by applying their own, real data in completing science projects. An important aspect of such data collection is the simplicity with which young learners, including primary school students can operate the devices. Modifying existing lessons to incorporate digital data gathering creates new versions involving technology tools and resources. What changes in such cases are tools used to gather data, rather than the essence of the lessons.

Wearables have some shortcomings such as the limited number of applications designed to run on the devices, which may cause delays in integrating them into mainstream learning environments where curriculum is mostly scripted. Few lessons based on these technologies exist, and educators willing to try them in class are faced with the problem of reinventing the wheel in terms of lesson planning. Battery life and lack of connectivity across devices from different vendors also pose impediments to using wearables.

Sensing

Sensors are increasingly embedded in wearable devices in addition to stand-alone versions. They are quickly becoming standard feature in conventional devices like phones and home appliances. The result is a rapid collection of massive data, which then presents the problem of analyzing and deriving meaning from such vast datasets. New sensing and data mining technologies including the availability of super-fast processors, software and analytics tools, as well as shared computing (such as cloud computing) make it possible to imagine scenarios where STEM learning activities may be easily captured and analyzed. A number of ongoing research projects are developing methods for educators to use in technology-enhanced learning environments. One research group at a leading university in Silicon Valley is exploring multimodal process-based student assessment by using biosensing, signal- and image processing, text-mining, and machine language, tools previously reserved for exclusive and expensive projects, but now accessible to many small research labs. The rate at which sensors are evolving is evident by comparing the relatively short span between soil sensor technology used in the 2009 Wireless Sensor Networks For Soil Science (Terzis et al., 2009) projects and the current, cheaper but relatively advanced gadgets. At this university lab, school-age students work on long term, evolving projects that provide opportunities to use sensors at different stages, but in the process, a lot of data is also collected including video and audio covering the entire project. This is where the image processing and other digital tools come in handy in analyzing the data.

An interesting case of modern technology involves the intersection of interactive media with sewing machine. Within this context of converging media technologies, the concept of mobile media embedded in wearable material has been introduced. Wearable Computing, Fashionable Technology, and Smart Textile are being developed at the intersection of media, art, design, computer science, and engineering (Reimann, 2011). This pushes the limits on personal data collection as better, more efficient devices are developed. The connected nature of these devices simplifies the gathering process while presenting the previously mentioned challenge of data manipulation and presentation.

A common limitation of these devices is that since most wearable sensors run on battery power, replenishing of energy is a common problem especially in extended use that may be necessary in some projects. Compatibility is also limited, so that users are constrained by proprietary platforms, and their devices may not interact with each other as desired.

Gaming

Playing and learning have always gone hand in hand during the early learning years, and the advent of digital technologies elevated the prospect of getting better learning games in the hands of older students. The combination of mobile apps for handheld devices, developments in wearable gadgets and in augmented reality all led to opportunities to explore this new found capacity to engage with learners through games. Thousands of mobile learning games have been developed in the past five years, and educational researchers have tested their usefulness at different levels of learning. Some widely used games include Motion Math, an app that teaches basic math concepts to young learners through a series of gesture-based interactions. Such interactions became possible with the introduction of accelerometers and other sensors into mobile devices.

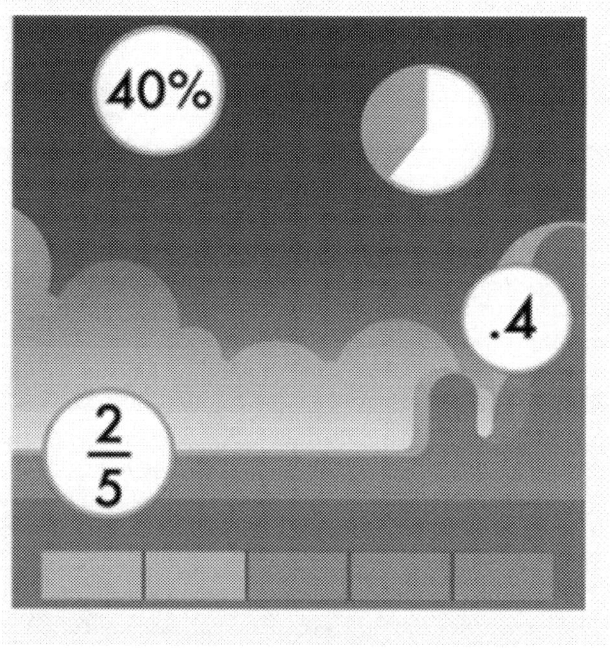

Figure 2. Motion math mobile game screenshot

The effectiveness of simulators as a training tool has been demonstrated in several fields like aviation, and the rise of digital interfaces and their corresponding affordances are fueling a rise in applications for other subjects like physics, chemistry and others. Many of these applications are showing up as learning games that allow students to experiment with a larger set of combinations relative to the real world scenario.

From exploring the composition of elements to mastering the periodic table, learning through games is embodied in apps designed to reinforce science skills. In one chemical reactions app, learners can evade destructive explosions by combining the correct set of compounds. An added advantage of these gaming applications is the customization feature that allows different age and skill levels to use them.

Simulation seems quite accessible to learners, as can be observed from a young child playing with new objects. Discussing this topic, Landriscina (2013) points out that simulation is all around us – that most objects around us have been carefully simulated before they were produced. He notes however, that despite accolades from experts and learners, simulation as a learning tool still lags behind in adoption due to several factors including gaps and weaknesses in the body of research on the use of simulations and games for science learning.

Researchers have also noted that social games are becoming increasingly acceptable in education (Wallneret et al., 2013). This is likely a result of several factors working in concert. Among them are the falling cost of digital devices on which games are run, the "Bring Your Own Device" model adopted by some school districts, familiarity (even active participation) of educators with mobile games among others. Games involving multiple players in creating projects are of interest to educators due to the different learning channels they access – audio, visual, tactile as well as requiring some level of collaboration or engagement with others.

Some common challenges came to fore in the case of the iPads for Learning initiative at a leading university, where 75 iPads were offered to four different instructors and classrooms for part of the school year. While most of the applications used in the classes tended to be pedagogical in the traditional sense (such as enumerating facts), some gaming and social media applications were also used. The diversity of classes and content presented a challenge with the choice of default applications to be loaded on the devices. The process also involved security imaging, resolving circulation policies, determining who pays and owns the apps, finding credible app reviewers, collaborating with instructors, students and teaching assistants in an ongoing manner among other details. One could easily summarize the challenges exactly as the educational researcher Murray (2011) posits in his assessment of a different experiment that while each of the four projects had a different focus, generic objectives were around improvements in student learning, changes in teacher pedagogy, planning for and implementing innovative curriculum and classroom practices, and managing the technical problems. Also, issues highlighted during the implementation were teacher confidence and training, wireless access, device synchronization, copyright and intellectual property.

Augmented Reality (AR)

Augmented reality is a feature that supplements real environments with contextually relevant, mostly computer-generated information. In a way, AR promises to bring additional information on anything that we look at. The user is able to sense extra details (mostly visual for now, though other senses can be invoked) while experiencing a real environment. A simple example is a smartphone user pointing the camera at a site and getting an additional layer of information on the screen in overlaying the actual image viewed through the camera lens. This may be contrasted with Virtual Reality (VR) where the participant is completely immersed in a synthetic environment and cannot see the real world (Kiper & Rampolla, 2013). Two major categories of AR are the forms based on physical markers or objects and those based on locational sensing. In the first category, a camera scans a marker to initiate some action such as displaying a video or animation. The smartphone app Aurasma has utilized advanced technology to use any image instead of the default QR (quick response) codes, presenting opportunities for customizing the experience to match learning tasks and environments. The second category uses Global Positioning System (GPS) on smartphones or tablet computers to overlay additional information on physical places. Simply put, one method uses markers while the other uses GPS on the device to initiate the information overlay.

In their paper on Augmented Reality in learning, FitzGerald et al. (2013) suggest a taxonomy of terms for a variety of AR used in a selection of mobile learning projects. It reveals several aspects that may help to determine settings where AR might be used including mobility, modes of interaction, media types, and the nature of technology. In a highlighted project, learners in two separate settings – a lab and in the field – collaboratively participate in inquiry-based learning through data sharing, hypothesis development and web access to information. Since inquiry-based approach fits many STEM topics, AR can be employed in enhancing student learning in addition to providing a new set of tools and affordances. The main shortcoming with AR is the lack of infrastructure to effectively implement it on a large scale. It needs internet connection in order to receive the real-time data, and physical environments require tags or markers in order to initiate the additional data request. Otherwise, AR presents some compelling use cases for teaching and learning.

Data/Info Manipulation

Information has changed rapidly over the past few years. Modes of learning (Davidson & Goldberg, 2010) have evolved dramatically over the past two decades—our sources of information, the ways we exchange and interact with information, how information informs and shapes us have all evolved rapidly. But our schools—how we teach, where we teach, whom we teach, who teaches, who administers, and who services—have changed mostly around the edges.

Searching aka Googling has revolutionized access to information, so that what students previously found by seeking reference librarians, checking dictionaries, thesaurus and so on, is now readily available by entering search terms in a web browser. Primary school kids now have access to information at a rate and amount previously unimagined even for a doctoral student just 16 years ago when Google was founded. Both scale and speed continue to be important drivers, but it is the data mining, contextualizing and analysis tools that promise to take searching to a new level, including an impact on STEM education.

Speed. With processing power increasing per device, as well as distributed computing – think cloud computing, bit torrents and other shared computing tools; applications can be built to perform functions requiring a lot of processing capacity.

Scale of data. The availability of vast data, the ability to collect a lot of personal data, and the improved tools that enable us to link this data is another new development likely to impact STEM education. Given the reliance of the scientific method on data, it makes the ubiquity of data an added advantage of modern technology.

Better access to information may reduce the time required to conduct research, say on the topic of "polynomials" and lend a better understanding of the subject to the learner. This eliminates ambiguity and then leaves the bulk of the time to be dedicated to studying the concepts, then practicing the applications. One of the common hindrances I encountered as a Science and Math teacher came from the language barrier from terms and practices in these subjects – some of them historical. Think of the Pythagorean Theorem – just the pronunciation of the name may hinder a student from exploring the concept. This is both a language component as well as a technological one, considering that access to the online dictionary, for instance, makes it fast and easy to find relevant information.

Mobile applications such as the Apple's initiative Researchkit are helping collect and securely manage medical data for research at universities and medical centers. An advantage of such medical apps is their relatively low cost given the number of users that a researcher can access. Traditional data gathering techniques would be prohibitive if one tried to reach as diverse and large group of participants as we can now do with mobile apps.

Maker Tools

Maker tools consist of equipment used to design, measure, cut, mould, print, assemble or otherwise create physical artifacts. A recent journal article (Gaines et al., 1996) asserts that technology's new tools are seen as empowering, productive, motivational, and that they make learning fun; but most importantly, they let the user both access and create new realms of knowing and doing. Given the examples seen at a university Fab Lab, it is true that these technologies are permeating new areas, enriching music, art and industrial/vocational education. As stated in the same article, "it's not just the number of tools we make available, the number of new features makes a difference, too. Newer technologies are functionally different than a decade ago". Maker tools are powerful in the sense that they let us move from thinking to doing, to modifying, to creating. As a result of this, knowledge and information are made more accessible to both learners and teachers. One of the greatest maker tools in terms of combining digital and physical domains has been the 3-D printer.

What happens when students are excited, engaged and actively participate in a learning activity? My experience is that meaningful learning happens better under such an environment than one devoid of these components. Watching students in an ongoing pilot project at leading Silicon Valley university learn science and math through making things using 3-d technologies – including printers, scanners and software, one quickly senses the overwhelming excitement and engagement around them. While there are many benefits of this kind of learning compared to the lecture style methods, the students point out the immediate need for knowledge and information (on time information) arising from solving an immediate problem (such as the best shape for longest paper plane flight, or optimal size for a specific gear) as the driving force to seeking new knowledge. This information and knowledge matter to the students because they can see a real and immediate application in their circumstances. We may consider the ability to create things within the classroom as an integral component of the future learning environment. Part of the appeal for creating things comes from the cross-content nature of such projects, such as a group of students that chose to tackle environmental protection by lowering the time spent in the shower in urban homes. While the goals of the project centered around the environment, they conducted research involving taking measurements at their own homes, looking up statistics from websites, interviewing family and friends (and even strangers), then creating a coherent narrative to support their position and rally others behind their cause. This, together with the actual fabrication of the prototype – which involved physics, material science, math and others, provided the opportunity to learn many concepts, facts and applications, and gave the satisfaction of creating a useful gadget. They felt like inventors.

Collaboration

Educators often strive to foster collaboration among their students and research points to the importance of social factors in learning. For instance, Khine and Hung (2006) posit that the goal of formal education should be meaningful learning and emphasize that "meaningful learning is necessarily social, collaborative, intentional, authentic, and active. The result of meaningful, learning lies in its cognitive residue, the learner's mental model". Collaboration helps students tap into each other's strengths and knowledge while building their own. Amongst technologies that enhance collaboration are web tools and applications; physical devices; learning spaces together with equipment such as interactive whiteboards and maker tools. In some sense, many of the technologies highlighted above serve multiple functions including fostering collaboration. It may be noted however, that despite their variety in form and function, some promising technologies have previously disappointed educators, but a few of the current tools – both web-based and tangible, seem like a better fit within the evolving learning environments and are experiencing faster adoption by both students and educators. The Personal Computer (PC) offers a great case study of the former category of technologies that fell short of expectations.

Was it a failed promise? When the PC gained widespread use in the 1990s, educators were enthusiastic about the possibilities presented by this new technology in the classroom. It promised to engage learners in new ways, provide access to multimedia, improve learning assessment and provide powerful applications for learning, modeling and even creating artifacts. While the PC brought many improvements in the classroom, some of the key projections of its impact failed to manifest because of challenges ranging from logistics to the non-personal nature of the computers in classrooms. The later may account for the renewed interest in mobile devices – the smartphones and tablets, which are quite personal. This aspect enables the user (student) to switch between "serious" work like uploading collected data to a class thread, and a personal task such as text messaging a friend about lunch. This difference between the PC and the mobiles is so profound in itself that as one commentator said, it may be considered to be "the" difference. More can be said of the smaller size, lower power consumption and battery life in comparison to earlier personal computing devices. In terms of collaboration, the recent developments attempt to support learning communities better than the PC did back then. This is happening through the features embedded in the devices and applications, which allow for multiple, synchronous interaction, as well as convenient resource sharing.

Finally, convergence of media – text, photo, audio and video, together with tools to manipulate them has given rise to new possibilities. Students can make clearer connections between digital artifacts from patterns detected using data applications that employ complex algorithms – these would have required infinite human cognition to accomplish without computers. This multimedia convergence, together with powerful processors in smartphones has sparked the development of applications that perform multiple functions such as collecting and sharing data, accessing private networks and managing files among others. It is also evident that collaboration between human beings as well as with machines in networked, open environments presents new opportunities to extend the human mind and abilities. However, as noted by researchers, some challenges still remain, especially relating to team formation, role allocation, synchronization of beliefs, communication trade-offs, and information sharing (Scerri et al., 2006). Despite the advances made in improving the communication channels such as wireless networks, Bluetooth, near-field communications and even with regard to machine learning, these challenges remain an impediment to achieving the desired level of collaboration.

Several concerns arise from using digital technologies and tools in education, most commonly pertaining to data privacy of users. This is especially important when working with minors. We have seen a rise in educational applications and platforms developed by private entities like startup companies and corporations. While they promise a level of privacy for those using their applications, doubts remain as to how well app developers and vendors take this into account. In summary, opportunities for educators to explore different ways of motivating, challenging, supporting and exposing new thinking to learners continue to expand, and we expect that many of the current impediments will be resolved as new designs, tools and applications emerge.

FURTHER READING

http://en.wikipedia.org/wiki/Big_data
http://search.proquest.com/docview/222791008?accountid=14026
http://site.ebrary.com/lib/stanford/Doc?id=10367819&ppg=32
https://tltl.stanford.edu/projects/multimodal-learning-analytics
http://www.apple.com/researchkit/
http://www.aurasma.com
https://www.coursera.org
http://www.nmc.org/nmc-horizon/
https://www.udacity.com
online.stanford.edu/opened

REFERENCES

Davidson, C. N., & Goldberg, D. T. (2010). *Future of thinking: Learning institutions in a digital age*. Cambridge, MA: MIT Press. Retrieved from http://site.ebrary.com/lib/stanford/Doc?id=10367819&ppg=31

Dede, C. (2004, September). Enabling distributed-learning communities via emerging technologies. In *Proceedings of the 2004 conference of the Society for Information Technology in Teacher Education (SITE)* (pp. 3–12). Charlottesville, VA: American Association for Computers in Education.

Downes, S. (2010). Learning networks and connective knowledge. In H. Yang & S. Yuen (Eds.), *Collective intelligence and e-learning 2.0: Implications of web-based communities and networking* (pp. 1–26). Hershey, PA: Information Science. doi:10.4018/978-1-60566-729-4.ch001

Fedorowicz, J., & Gogan, J. L. (2002). Teaching that keeps pace with technology. *Journal of SMET Education: Innovations and Research, 3*(3), 10. (Web, 2015)

FitzGerald, E., Ferguson, R., Adams, A., Gaved, M., Mor, Y., & Thomas, R. (2013). Augmented reality and mobile learning: The state of the art. *International Journal of Mobile and Blended Learning (IJMBL), 5*(4), 43–58. doi:10.4018/ijmbl.2013100103

Gaines, C. L., Johnson, W., & King, D. T. (1996). Achieving technological equity and equal access to the learning tools of the 21st century. *T H E Journal, 23*(11), 74–77.

Hopkins, J. M. (1991, April). Technology as tools for transforming learning environments. *The Computing Teacher.*

Khine, M. S., & Hung, D. (Eds.). (2006). *Engaged learning with emerging technologies* (p. 1). Berlin/Heidelberg, DEU: Springer. Retrieved from http://site.ebrary.com/lib/stanford/Doc?id=10134296&ppg=15

Kipper, G., & Rampolla, J. (2013). *Augmented reality: An emerging technologies guide to ar*. Retrieved from [Books 24×7 version] http://common.books24x7.com.ezproxy.stanford.edu/toc.aspx?bookid=47311

Landriscina, F. (Ed.). (2013). An introduction to simulation for learning. In *Simulation and learning: A model-centered approach* (pp. 1–12). New York, NY: Springer. Retrieved from http://dx.doi.org/10.1007/978-1-4614-1954-9_1

Murray, C. (2011). Imagine mobile learning in your pocket. In W. Ng (Ed.), *Mobile technologies and handheld devices for ubiquitous learning: Research and pedagogy* (pp. 209–236). Hershey, PA: IGI Global Publishing. doi:10.4018/978-1-61692-849-0.ch012

Reimann, D. (2011). Shaping interactive media with the sewing machine: Smart textile as an artistic context to engage girls in technology and engineering education. *International Journal of Art, Culture and Design Technologies (IJACDT), 1*(1), 12–21. doi:10.4018/ijacdt.2011010102

Scerri, P., Vincent, R., & Mailler, R. (2006). *Coordination of large-scale multiagent systems* (p. 17). New York, NY: Springer. Retrieved from http://site.ebrary.com/lib/stanford/Doc?id=10129630&ppg=216

Terzis, A., Musaloiu-E, R., Cogan, J., Szlavecz, K., Szalay, A., Gray, J., Ozer, S., Liang, C-J. M., Gupchup, J., & Burns, R. (2009). Wireless sensor networks for soil science. *IJSNet (International Journal of Sensor Networks)*. Retrieved from http://www.msr-waypoint.com/pubs/157594/LUYFJournal.pdf

Thille, C., Mitchell, J., & Stevens, M. (2015, September 22). What we've learned from MOOCs. *Inside Higher Ed*. Retrieved from https://www.insidehighered.com/views/2015/09/22/moocs-are-no-panacea-they-can-help-improve-learning-essay

Wallner, G., Kriglstein, S., & Biba, J. (2013). Evaluating games in classrooms: A case study with DOGeometry. In Y. Baek & N. Whitton (Eds.), *Cases on digital game-based learning: Methods, models, and strategies* (pp. 475–494). Hershey, PA: IGI Global. doi:10.4018/978-1-4666-2848-9.ch024

JACK HOLBROOK

28. 21ST CENTURY SKILLS AND SCIENCE LEARNING ENVIRONMENTS

This chapter promotes a democratic approach to education and emphasises personal and social democratic values as well as scientific knowledge and skills. It considers the goals of science education and especially those sometimes called 21st century skills. Also included is a consideration of the attention needed to the classroom learning environment for the teaching of science, drawing attention to the teaching focus, the classroom atmosphere and the teaching approach.

LEARNING OUTCOMES

When you have worked through this chapter you should be able to:

With regard to what to teach

- Appreciate the need for science education
- Identify and recognise the importance of 21st century skills
- Associate these skills with student needs for responsible citizenship and employability
- Identify the term 'education through science' and determine its meaning

With regard to facilitating learning

- Recognise the importance of establishing a conducive learning environment
- Understand the role of the teacher in developing the learning environment
- Understand the role of support materials (modules) in establishing the learning environment
- Identify and critique classroom learning approaches

TEACHING SCIENCE FOR DEMOCRATIC SOCIETY DEVELOPMENT

A democratic picture of education sees science education being portrayed as important for all students. Less attention is paid to building a widespread knowledge background, but importance is attached to promoting a wider range of competences through science teaching and thereby enabling students to function meaningfully and responsibly in a democratic society. While thinking abilities are still recognised as important, to make sense of the science encountered in everyday life, the focus

K. S. Taber & B. Akpan (Eds.), Science Education, 385–401.

is also towards capabilities in terms of personal self-determination, especially with respect to problem-solving and social consensus making, geared to students developing skills to make reasoned decisions.

In such a focus, science content can derive from relevant contexts, related to the students' world and seeking to establish the importance of science in the day-to-day functioning of society. Thus, for example, scientific concepts behind modern materials, such as polymers, foodstuffs, medicines and different forms of energy providers, all part of everyday life, are promoted, enabling understanding from a safety, environmental or health risk point of view, and hence have a potential impact on raising the quality of life. This impact on the way of life can be perceived as being immediate – discarding versus recycling of waste, or the use of paper, rather than plastic bags – or related to the future, associated with risk awareness, pollution, global warming, derisive impacts on eating habits, poverty alleviation and environmental protection or sustainability versus non-sustainability.

Such an approach inevitably has extensive implications for the type of teaching emphasis. Learning in a societal frame is expected to build on the familiar and hence lend itself to a strong constructivist approach. Inquiry-based teaching can stimulate explorations in a student-centred manner, using materials found locally, while the conceptual areas are carefully selected to take account of the local availability of materials and familiar processes, or concerns. And, of course, student involvement is highlighted by the students playing a major role in seeking and using indigenous materials, the setting up of relevant investigations and seeking the intended explanations.

THE GOALS OF SCIENCE EDUCATION

Within a democratic orientation, a commonly stated science education goal is the achievement of scientific literacy, although there are different interpretations of its meaning. In recognition that a democratic focus for science education needs to relate to society, there is a need to interrelate science and technology. Thus, Holbrook and Rannikmae (2007) put forward scientific and technological literacy (STL) as a more appropriate focus for the goals of science education.

The scientific thrust of STL has its focus on conceptualisations of need-to-know scientific knowledge, but STL also relates to an interaction of the science with society and an awareness of the need for expert opinions, thereby introducing understanding that ordinary citizens do not need to possess. STL teaching further strives to enable students to make decisions in a democratic society, where science-based technology is playing a greater and greater role and an appreciation that while the advantages of technological developments can be great for some, it can be a major disadvantage for others. Furthermore, side-effects related to health, the sustainability of the environment, or economic concerns can become key factors in choosing the most appropriate science–driven technology. STL is seen as developing student capabilities to consider and reflect on all of these.

It is not surprising that a single, simple definition of STL is always likely to be extremely problematic. A definition intended to involve an appreciation of the nature of science, the educational development of the person and functionality in a social domain, while also stressing socio-scientific-making abilities, is (Holbrook & Rannikmae, 1997):

> *developing the ability to creatively utilise sound science knowledge in everyday life, or in a career, to solve problems, make decisions and hence improve the quality of life.*

While STL is rather nebulous, it clearly draws attention to the need for education to embrace the nature of science education (NSE) (Holbrook & Rannikmae, 2007), not simply the nature of science and hence favours the democratic approach to education. Figure 1 illustrates that the nature of science education (NSE) can be taken to be the inclusion of three major educational needs – the nature of the science, personal development and social development in an educational frame. It recognises, for example, that abilities in a range of educational goals, including socio-scientific decision-making and scientific problem-solving, are more important for enhancing true scientific literacy (Shamos, 1995), or multi-dimensional scientific literacy (Bybee, 1997) than a systematic, basic understanding of poorly related, fundamental, content knowledge.

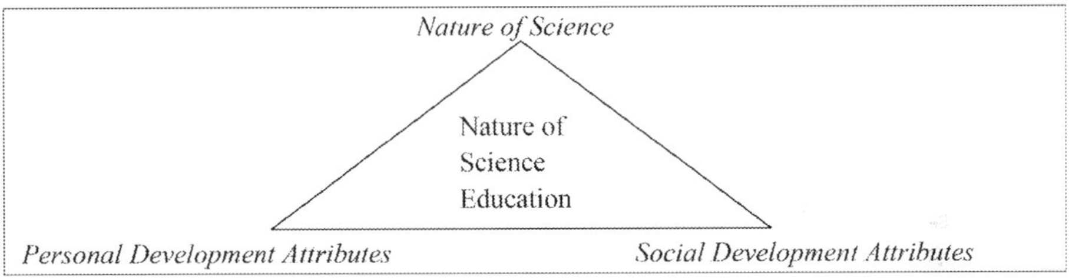

Figure 1. The three domains comprising the Nature of Science Education

EDUCATION THROUGH SCIENCE

While the nature of science education portrayed in Figure 1 recognises three key learning areas, it does not address directly the degree to which science teaching, within a specific educational focus such as for democratic development, relates to the science itself. Hence, the emphasis to be placed on promoting the nature of science, and science academic learning, versus a wider educational emphasis promoting personal and social attributes is not determined. This emphasis very much relates to the teaching focus and whether stress is placed on the scientific or the educational aspects.

It is important to stress that these directions for science education focus on the degree of emphasis rather than the exclusion, of any of the following:

- cognitive learning;
- appreciation of the subject (the nature of science);
- the development of the person to be capable of functioning in a meaningful and responsible manner;
- the development of the person, especially in terms of social values.

With the above in mind, the use of education through science emphasizes the educational learning over the science content. This contrasts with a science through education approach in which the science is the dominant aspect.

21ST CENTURY SKILLS

An emerging body of research suggests that gaining a set of broad citizenship and employability skills is an important focus in the 21st century. Such skills are seen as valuable towards becoming more responsible within society and enabling individuals to be better placed to secure a wide range of jobs in the national economy. Some business and education groups have advocated infusing specifically-identified 21st century skills into the school curriculum, and developments have taken place in this direction (Partnership for 21st Century Skills, 2008; NRC, 2010). The NRC [define for readers] put forward five broad categories or attributes – Adaptability; Complex communication/social skills; Non-routine problem- solving skills; Self-management/self-development; and System thinking. Clearly, these apply across the curriculum and not simply to a single subject area such as school science, but they tend to complement rather than oppose the 'education through science' approach introduced in the previous paragraph.

The earlier arguments have gradually focused on the expectation that gaining science knowledge and skills is not enough. There is a need also to be able to make use of the learning in science lessons, especially within society and in the workplace. This clearly favours the democratic approach to science teaching and the interrelating of the learning within science lessons with the learning taking place in other subject disciplines. It promotes the not-ion that students need to appreciate that the teaching of science in school is not only valuable in terms of forming a scientific knowledge based, but also in promoting competences to enable citizens to function in the 21st century, especially in allowing them to increase their marketability, employability and readiness for citizenship.

Nevertheless, in striving to identify the directions for success in learning, the development of students as individuals, as well as future citizens, the Melbourne Declaration on Education Goals for Young Australians (MCEETYA, 2008) identified more fully the learning that can relate to students. Linking this to 21st century needs, whether expressed by the NRC (table 1, column 2) or by teaching aspects (table 1, column 3), poses a challenge as one focuses on the personal needs and social values while the other sees employability as of much importance. Table 1 attempts to interrelate these two directions towards a new curriculum and teaching focus.

Table 1. Interrelating educationally perceived Student Needs with the promotion of 21st Century Skills

Education Goals (Melbourne Goals Declaration)	*21st Century learning skills (NRC)*	*Teaching to develop 21st Century skills) (Anderman & Sinatra, 2009)*
(a) Successful learners are those who:	• Possible related statements	• Possible related statements
(i) Develop their *capacity to learn* and play an active role in their own learning.	4.2. Develop an ability to work autonomously. 3.3. Foster creativity to generate new and innovative solutions.	1. Foster a productive learning environment. 4. Capitalizing on progressions in learning by revisiting earlier content in more depth. 6. Using assessment strategies that focus on higher order learning;
(ii) Have *essential skills* in literacy and numeracy, are creative and productive users of technology (especially ICT), as a foundation for success in all learning areas.	3.6. Acquire knowledge of how the information is linked conceptually in problem solving situations. 3.3. Foster creativity to generate new and innovative solutions. 2.5. Gain service orientation communication.	3. Developing requisite knowledge, skills, and dispositions necessary for science literacy and to support nascent science career choices. 4. Capitalizing on progressions in learning by revisiting earlier content in more depth.
(iii) Are able to *think deeply* and logically, and obtain and evaluate evidence in a disciplined way as the result of studying fundamental disciplines.	3.1. Ability to narrow the information to reach a diagnosis of the problem to be solved. 3.2. Ability to reflect on whether a problem-solving strategy is working and switch to another strategy if the current strategy is not working. 3.3. Foster creativity to generate new and innovative solutions.	3. Developing requisite knowledge, skills, and dispositions necessary for science literacy and to support nascent science career choices. 5. Promoting an inquiry and problem-based learning approach to science instruction. 6. Using assessment strategies that focus on higher order learning.

(Continued)

Table 1. (Continued)

Education Goals (Melbourne Goals Declaration)	*21st Century learning skills (NRC)*	*Teaching to develop 21st Century skills) (Anderman & Sinatra, 2009)*
(iv) Are creative, innovative and resourceful, and are able to **solve problems** in ways that draw upon a range of learning areas and disciplines.	3.1. Ability to narrow the information to reach a diagnosis of the problem to be solved. 3.2. Ability to reflect on whether a problem-solving strategy is working and switch to another strategy if the current strategy is not working. 3.3. Foster creativity to generate new and innovative solutions. 3.6. Acquire knowledge of how the information is linked conceptually in problem solving situations.	1. Fostering productive learning environments. 6. Using assessment strategies that focus on higher order learning.
(v) Are able to plan activities independently, collaborate, work in teams and communicate ideas.	4.1. Ability to work remotely, in virtual teams. 4.4. Self-monitoring.	1. Fostering productive learning environments. 6. Using assessment strategies that focus on higher order learning.
(vi) Are able to make sense of their world and think about how things have become the way they are.	1.1. Social perceptiveness, 3.4. Integrating seemingly unrelated information. 3.5. Recognize patterns not noticed by novices.	2. Promoting active engagement, based on connections to students' personal interests and career goals. 3. Developing requisite knowledge, skills, and dispositions necessary for science literacy and to support nascent science career choices. 6. Using assessment strategies that focus on higher order learning.

Education Goals (Melbourne Goals Declaration)	*21st Century learning skills (NRC)*	*Teaching to develop 21st Century skills) (Anderman & Sinatra, 2009)*
(vii) Are on a pathway towards continued success in further education, training or employment, and acquire the skills to make informed learning and employment decisions throughout their lives.	4.1. Systems analysis considering things as scientists do. 5.2. Systems decision-making.	2. Promoting active engagement, based on connections to students' personal interests and career goals. 3. Developing requisite knowledge, skills, and dispositions necessary for science literacy and to support nascent science career choices.
(viii) Are motivated to reach their full potential.	4.3. Self-motivation.	6. Using assessment strategies that focus on higher order learning.
(b) confident and creative individuals are those who:		
(i) Have a sense of self-worth, self-awareness and personal identity that enables them to manage their emotional, mental, spiritual and physical well-being – have a sense of optimism about their lives and the future.	1.2. Adaptable in handling work stress. 1.2. Ability and willingness to cope with uncertainty.	2. Promoting active engagement, based on connections to students' personal interests and career goals.
(ii) Are enterprising, show initiative and use their creative abilities.	1.3. Communicating through Instructing, 3.3. Foster creativity to generate new and innovative solutions. 4.1. Ability to work remotely, in virtual teams.	6. Using assessment strategies that focus on higher order learning.
(iii) Develop personal values and attributes such as honesty, resilience, empathy and respect for others.		

(Continued)

Table 1. (Continued)

Education Goals (Melbourne Goals Declaration)	*21st Century learning skills (NRC)*	*Teaching to develop 21st Century skills) (Anderman & Sinatra, 2009)*
(iv) Have the knowledge, skills, understanding and values to establish and maintain healthy, satisfying lives.	3.3. Foster creativity to generate new and innovative solutions.	3. Developing requisite knowledge, skills, and dispositions necessary for science literacy and to support nascent science career choices.
(v) Have the confidence and capability to pursue university or post-secondary vocational qualifications leading to rewarding and productive employment.	1.1. Ability and willingness to cope with uncertain, new, and rapidly changing conditions on the job.	2. Promoting active engagement, based on connections to students' personal interests and career goals.
(vi) Relate well to others and form and maintain healthy relationships.		
(vii) Are well prepared for their potential life roles as family, community and workforce members.	1.1. Ability and willingness to cope with uncertain, new, and rapidly changing conditions on the job. 4.5. Willingness and ability to acquire new information related to work. 4.6. Willingness and ability to acquire new skills related to work.	
(viii) Embrace opportunities, make rational and informed decisions about their own lives and accept responsibility for their own actions.	1.4. Able to be persuasive and participate in negotiation.	

Education Goals (Melbourne Goals Declaration)	*21st Century learning skills (NRC)*	*Teaching to develop 21st Century skills) (Anderman & Sinatra, 2009)*
(c) active and informed citizens are those who:		
(i) Act with moral and ethical integrity.		
(ii) Appreciate existing social, cultural, linguistic and religious diversity, and have an understanding of a country's system of government, history and culture.	2.1. Social perceptiveness.	
(iii) Understand and acknowledge the value of indigenous cultures and possess the knowledge, skills and understanding to contribute to, and benefit from, reconciliation between indigenous and non-indigenous peoples.	1.3. Adapting to different personalities, communication styles, and cultures. 2.1. Social perceptiveness.	
(iv) Are committed to national values of democracy, equity and justice, and participate in the country's civic life.	2.1. Social perceptiveness.	
(v) Are able to relate to and communicate across cultures, especially the cultures and countries of Asia.	1.3. Adapting to different personalities, communication styles, and cultures. 2.1. Social perceptiveness.	
(vi) Work for the common good, in particular sustaining and improving natural and social environments;.	2.1. Social perceptiveness, Service orientation communication.	2. Promoting active engagement, based on connections to students' personal interests and career goals.
(vii) Are responsible global and local citizens.	1.4. Physical adaptability to various indoor or outdoor work environments.	

N.B. Numbers relate to the labelling given to skills in the indicated references

LEARNING ENVIRONMENT

It is proposed that the learning environment needs to be conceived from three major perspectives:

a. The teaching focus,
b. The classroom atmosphere, and
c. The teaching approach.

a. The teaching focus is related to the degree of emphasis on the different aspects of the nature of science education. This relates to the goals of education and the perception of STL. The teaching focus thus pays attention to the emphasis placed on science conceptual learning, alongside the development of the person geared to responsible citizenship and social development. The focus is particularly geared to employability and reasoned, decision making. In turn, the focus relates to whether the democratic orientation is adopted and whether education through science is the preferred orientation.
b. Classroom atmosphere relates to the manner in which the teacher interacts with the students. The role of the teacher is thus crucial in setting the scene for the learning (the classroom atmosphere is controlled by the teacher). Where this role is effective:
   1. the classroom atmosphere promotes student motivation;
   2. student motivation is likely triggered by the teaching perceived by students as being relevant and meaningful;
   3. motivation is sustained through constructivist teaching principles and the use of student-centred learning techniques;
   4. motivation is aided by an 'education through science' approach to learning.

The aspects in the previous paragraph suggest the manner in which the teacher interacts with different students is important. There is a need for consistency and fairness so that all students feel they are being treated equally. This is not actually treating all students in the same manner, as students have different needs and hence appreciate different levels of support. Perhaps consistency and fairness is more appropriately seen in terms of supporting students with their zone of proximal development (what students are capable of achieving with the guidance and support of others, such as the teachers and/or other students). Flexibility is also seen as important in inducing student abilities. One way to give students greater flexibility is to allow them choices in portions of the course content, topics for papers, and questions for class discussion. Flexibility leads to a sense of control, which can contribute to a student's expectation of success.

As a final note on classroom atmosphere, it is much easier for students to learn when something makes sense to them and is related to one's life, interests or aspirations. Teaching at the students' pace, with multiple opportunities for student involvement clearly makes sense. Putting forward science as isolated facts, or unrelated theories can be expected to lead to memorisation and the development of poorly conceptualised learning, or even misconceptions. This does little to promote 21st century skills and enhance employability.

c. A key factor related to both classroom atmosphere and teaching approach is that the teaching of science needs to address its lack of popularity among students. This seems to be better addressed at students grade 7 and above. Top of the list in this respect is relevance of the teaching in the eyes of students.

The literature suggests that for the teaching of science subjects to be more relevant for students, there is a need for:

- student participation in the choice of social context for science learning;
- an increase in student activities and with this greater opportunities for student self-learning; related here is the need for more potential diagnostic measures of the effectiveness of the teacher;
- maximising student involvement and the important move away from teacher-centred approaches.

Table 2 summarises connections between the teaching focus, the classroom atmosphere and teaching approach.

CRITERIA FOR AN EXEMPLARY STL LEARNING ENVIRONMENT

The following criteria are put forward for STL learning environment needs. (Holbrook, 1997; Holbrook & Rannikmae, 1997).

Involve Demanding (Higher Order) Thinking Skills

Undertaking the activity is an appropriate learning exercise for the learner i.e. it provides an intellectual challenge at an appropriate level for the students. It utilizes constructivist principles – moving from information and understanding already in the possession of students, to the new learning situation. It involves analytical or judgemental thought.

Include a Communication Skill Component

Due consideration is given to enhancing a wide range of communication skills appropriate for the dissemination of scientific ideas and social values. This will involve oral (group discussion, debate, role playing), graphical, tabular, symbolic, pictorial as well as written forms.

Table 2. Interrelating the teaching focus, the classroom atmosphere and teaching approach

Teaching focus	*Classroom atmosphere*	*Teaching approach*
Selected features of science teaching	Skills/knowledge needed by teacher	
1. Teacher identifies the curriculum focus related to the intended teaching and learning.	Requires an appreciating of relevance from the students' perception and the degree of familiarity expressed by the students. Establishing prior learning, especially in a social sense.	Motivational: building on the relevance and familiarity of the initial situation. Student centred: involve student in putting forward their current learning as a base for identifying the need for further learning The planning and evidence gathering and interpretation as components of inquiry-based science education (IBSE).
2. Teacher solicits students' initial conceptions of focal phenomena; guides students to represent what they know, then links this to further instruction (new learning) based on expressed currently understanding.	Establishes prior learning of science ideas (concepts and process skills). Requires teacher's ability to construct questions or tasks that are "rich"—i.e. have potential to reveal multiple facets of student thinking about target idea. Requires *analysis of student responses* and comparison against target understanding, to make *principled judgments* about how to design further instruction.	Continues to be motivational: building on the relevance and familiarity of the initial situation. Constructivist approach building on students' identified prior knowledge and skills.

Teaching focus	*Classroom atmosphere*	*Teaching approach*
3. Teacher guides/co-constructs with students to identify deficiencies in students' scientific conceptualization, awareness, skills and other issues of relevance. Focus on deriving the need for key science learning that forms the focus for further curriculum-related teaching. Identifying the scientific question to be addressed.	Requires strategies for guiding student inputs towards relevance as a base for the further learning anticipated by the teacher Requires vision of what type of scientific question is complex enough to be meaningful, can be at least in part put forward by student and can sustain motivation related to the inquiry scientific learning to be introduced.	Continues to be motivational: building on the relevance and familiarity of the initial situation. Critical thinking by students in establishing their lack of needed knowledge and skills to conceptualise the situation from a science perspective. Self-efficacy of students to engage in putting forward the scientific question to investigate and gain new science conceptual knowledge.
4. Teacher guides students to hypothesis, plans investigations and interpret evidence as well as providing resources, scaffolding experiences relevant to answering essential student questions. Resources related to investigation could cover readings (books/ internet), technology, other tools, hands-on experimentation. Also supports *students* in deciding what other kinds of resources and experiences needed.	Requires understanding of how to help students hypothesise, plan, investigate and interpret evidence as well as undertake collaborative group work ideas and construct complex forms of meaning *across* representations (problem solving and meta-representational competence). Requires understanding of how experiments and other forms of testing providing data can be used as evidence to support solutions or explanations answering big questions.	Continues to be motivational: building on the relevance of the investigation and related science conceptualisation. Collaborative learning within peer groups. Developing or enhancing problem solving skills including safety, error identification and limitation, and recording skills. Conceptualisation of new science learning. Development of meta-cognition and meta-representation skills.

(Continued)

Table 2. (Continued)

Teaching focus	*Classroom atmosphere*	*Teaching approach*
5. Teacher supports (scaffolds and assesses) students in monitoring their own progress toward defined goals.	Requires understanding of modelling and how to foster metacognition, self-regulation in students. Requires teacher to possess a broad repertoire of formative assessments. Understands in what contexts they should be used, how they can provide both teacher and student with targeted feedback. Requires specialized discourses around questions such as: "What additional information do I need?" "How do I know we've solved the problem?" "What evidence will count as supporting an explanation?" "How do we address alternative hypotheses?"	Continues to be motivational: building on the relevance of the investigation and related science conceptualisation. Metacognition and self-regulation to ensure learning at the students' pace. Teacher and peer group scaffolding to enabling meeting more demanding challenges.
6. Teacher monitors and evaluates student understanding of science ideas and engagement in authentic scientific conceptualisations and problem solving.	Requires broad repertoire of formative assessments. Understands in what contexts they should be used, how they can provide both teacher and student with targeted feedback.	Continues to be motivational: building on the relevance of the investigation and related science conceptualisation. Metacognition and self-regulation to ensure learning at the students' pace. Teacher and peer group scaffolding to enabling meeting more demanding challenges.

Teaching focus	*Classroom atmosphere*	*Teaching approach*
7. Teacher guides students to interrelate new knowledge with existing conceptualization via techniques such as constructing concept maps. Teacher presses students to compare and integrate ideas across different representations/ disciplines, use secondary data and primary data as evidence to support explanatory models and arguments relevant to essential questions.	Requires understanding of how to weigh different forms of evidence, coordinating evidence and explanations, differentiating between theory and evidence. Requires orchestrating productive discourse by students in collaboratively evaluating solutions to problems or explanatory models. Requires understanding of the rhetorical practices of authentic science.	Continues to be motivational: building on the relevance of the investigation and related science conceptualisation. Metacognition and self-regulation to ensure learning at the students' pace. Teacher and peer group scaffolding to enabling meeting more demanding challenges.
8. Teacher guides students to interrelate the new science learning with the initial situation Interrelate the science with other social factors and determine the relative importance of the science to the familiar situation through argumentation and decision making. Teacher asks students to review the intellectual work of others in ways consistent with scientific practice and in ways that advance the thinking of others.	Requires ability to manage a "community of practice" in classroom. Needs to model discursive interactions over ideas that are appropriately challenging, based on evidence, and civil. Needs to model how one learns from feedback, how one re-considers ideas in light of input from others.	Continues to be motivational: building on the relevance of the investigation and relating the science to the initial societal situation. Metacognition and self-regulation to ensure involvement in development of argumentation skills and decision-making. Teacher and peer group scaffolding to enabling meeting more demanding challenges and reaching consensus among peers.

Include a Comprehensive Teacher's Guide

As problems, issues and concerns coming from society are often interdisciplinary in character, with the science ideas unfamiliar to the teacher, full explanations are needed to help teachers make use of the materials in a meaningful and interesting manner. The teacher's guide also needs to highlight the link between activities put forward and the outcomes expected of the teaching in terms of educational objectives. The teacher's guide needs to detail a suggested teaching strategy by which the student-centred approach is exemplified.

SUMMARY

This chapter is based on the democratic view of education, catering for all students. It sets out to introduce 21st century skills and explain appropriate science learning environments from a teaching focus, classroom atmosphere and teaching approach perspective.

FURTHER READING

Fraser, B. (2012). Classroom learning environments: Retrospect, context and prospect. In B. Fraser, K. Tobin, & C. J. McRobbie (Eds.), *2nd international handbook of research on science education* (pp. 1191–1239). Dordrecht, Heidelberg, London & New York, NY: Springer.

National Research Council. (2009). *Exploring the intersection of science education and 21st century skills: A workshop summary*. Retrieved from http://www.nap.edu/catalog.php?record_id=12771

National Research Council. (2012). *Education for life and work: Developing transferable knowledge and skills in the 21st century***.** Retrieved from http://www.nap.edu/catalog/13398/education-for-life-and-work-developing-transferable-knowledge-and-skills

National Research Council. (2016). *Science teachers' learning: Enhancing opportunities, creating supportive contexts.* Retrieved from http://www.nap.edu/catalog/21836/science-teachers-learning-enhancing-opportunities-creating-supportive-contexts

REFERENCES

Anderman, E. M., & Sinatra, G. M. (2009). *The challenges of teaching and learning about science in the 21st century: Exploring the abilities and constraints of adolescent learners.* Paper prepared for the Workshop on Exploring the Intersection of Science Education and the Development of 21st Century Skills, National Research Council. Retrieved from http://www7.nationalacademies.org/bose/AndermanSinatra.pdf

Bybee, R. W. (1997). Towards an understanding of scientific literacy. In W. Gräber & C. Bolte (Eds.), *Scientific literacy. An international symposium*. Kiel, Germany: IPN.

Holbrook, J. (1997). Resource materials for science education reform. In KEDI: *Globalization of Science Education – moving towards worldwide science education standards* (pp. 205–213). Korea: Proceedings of the International Conference on Science Education.

Holbrook, J., & Rannikmae, M. (2007). Nature of science education for enhancing scientific literacy. *International Journal of Science Education, 29*(11), 1347–1362.

MCEETYA Secretariat. (2008). *Melbourne declaration on educational goals for young Australians.* Australia, Carlton South, Vic. Retrieved from www.mceetya.edu.au

National Research Council (NRC). (2010). *Exploring the intersection of science education and 21st century skills: A workshop summary.* (Margaret Hilton, Rapporteur. Board on Science Education, Center for Education, Division of Behavioral and Social Sciences and Education.) Washington, DC: The National Academies Press.
Shamos, M. (1995). *The myth of scientific literacy.* New Brunswick, NJ: Rutgers University Press.

IAN ABRAHAMS

29. MINDS-ON PRACTICAL WORK FOR EFFECTIVE SCIENCE LEARNING

INTRODUCTION

> The time is surely past when science teachers must plead the case for school laboratories. It is now widely recognized that science is a process and an activity as much as it is an organized body of knowledge and that, therefore, it cannot be learned in any deep and meaningful way by reading and discussion alone. (NSTA, 1970, p. 3)

> Although practical work is commonly considered to be invaluable in science teaching, research shows that it is not necessarily so valuable in science *learning*. The evidence points to the uncomfortable conclusion that much laboratory work has been of little benefit in helping pupils and students understand concepts. Its main justification seems to have been moderate success in the teaching of measuring techniques, and in improving manual dexterity; skills which it might be more appropriate for pupils to acquire through craft based activities. (Clackson & Wright, 1992, p. 40)

This chapter, based on a longer and more detailed account of practical work (Abrahams & Millar 2008), considers the role of practical work in the teaching of scientific knowledge. What the two quotes above, separated by just over twenty years, illustrate is how opinion-based views regarding the role of laboratory-based practical work, however self-evidently true those views might appear, are being challenged by the findings from educational research that has increasingly questioned the role and effectiveness of laboratory-based practical work. The term practical work is used here to refer to any work in which pupils are involved in manipulating and/or observing real (as opposed to virtual) objects and materials.

Such a move, towards a more evidenced-based approach, is important given that a distinctive feature of science education in many countries, not only England, is the frequent and widespread use of practical work most of which takes place in purpose built laboratories (Millar, 1987). Indeed, pupils spend an appreciable proportion of the time allocated to science teaching undertaking laboratory-based practical work with Bennett (2003) reporting that in secondary schools in England the time pupils currently spend on practical work ranges from one third of the total time allocated to science education during their 'A' level (age 17–18) study rising to one half of the total science teaching time for pupils within the 11–13 age range.

K. S. Taber & B. Akpan (Eds.), Science Education, 403–413.

One explanation for the frequent and widespread use of laboratory-based practical work being that many science teachers see its regular use as a *sine qua non* of what it means to be 'a good science teacher'. Indeed, as practical work "seems the 'natural' and 'right' thing to do" (Millar, 2002 p. 53), many teachers see its use as the basic *modus operandi* in their teaching of science. Yet with such frequent and widespread use comes the risk that teachers cease to assess critically whether its use is always the most appropriate way of achieving a specific learning outcome. Indeed, many science teachers see laboratory-based practical work unquestioningly as an essential feature of their everyday work and believe that it leads to more effective learning – pupils are, they believe, far more likely to understand and remember things they have learnt through 'hands-on' activity than things they have just been told by the teacher or read about in a text book. Yet despite this we know, both from experience and research, that pupils frequently do not learn from a practical task the things we wanted them to learn. Furthermore, in the medium to long term many pupils can only recall specific surface details of the practical task they undertook and are unable to say what they learned from it, or the reason that they undertook it.

Research findings regarding the effectiveness of practical work in enhancing the development of conceptual understanding remains ambiguous. Hewson and Hewson (1983) report a significant enhancement of pupils' conceptual understanding amongst that half of their study group (pupils aged 13–20) who had received a primarily practical-based instruction compared to the other half of the study group that had received a traditional non-practical instruction. However, in other similar studies such findings have not been duplicated. Indeed Mulopo and Fowler (1987), in a study of 120 grade 11 pupils studying chemistry, reported no significant difference in the level of conceptual understanding amongst pupils irrespective of whether they had been taught using either practical or traditional non-practical methods. In contrast they report that the most appreciable factor in determining the extent of conceptual development was not the method of instruction but rather the pupil's level of intellectual development.

Indeed what the research literature (e.g. Lazarowitz and Tamir, 1994; Hofstein and Lunetta, 1982) relating specifically to practical work has repeatedly found is that, when outcomes are measured using pen and paper tests, the use of practical work offers no significant advantage in the development of pupils' scientific conceptual understanding.

Although Hofstein and Lunetta (1982) observe that as with a glass that can optimistically be said to be half full and pessimistically half empty, the same is true regarding the effectiveness of practical work in so far as "Many of these studies have reported nonsignificant results, meaning that the laboratory medium was at least as effective in promoting pupil growth on the variable measured as were more conventional modes of instruction" (p. 212). However, given the central role of the laboratory in the teaching of Science in many countries, coupled with its high financial cost, these non-significant findings are, at best, disappointing.

Indeed Yager et al. (1969) have suggested that in some situations some academically able pupils may perceive laboratory work to be wasteful of their time, serving only to delay their pursuit of new theories and concepts. Connell (1971) suggests that even if this were the case, a point he argues remains unclear, this would more than likely only be indicative of a mismatch between the practical work and the pupils' academic ability. Similarly Van den Berg and Giddings (1992) argue that such beliefs, if held by the pupils, would be a criticism of the form of specific practical tasks rather than constituting a criticism of practical work *per se*.

However, these arguments seem, generally speaking, to further reinforce Ausubel's (1968) assertion that "In dividing the labour of scientific instruction, the laboratory typically carries the burden of conveying the method and the spirit of science whereas the textbook and teachers assume the burden of transmitting subject matter and content" (p. 346). However, it is important to note that Ausubel goes on to make a distinction in this context between different forms of laboratory work and states that "Laboratory work in this context refers to inductive or hypothetico-deductive discovery experiences and should not be confused with [teacher] demonstrations" (p. 346).

Hodson (1992) has claimed that it is necessary to introduce the pupils to the relevant scientific concepts *prior* to their undertaking any practical work if the task is to be effective as a means of enhancing the development of their conceptual understanding. More recently Millar (1998) has questioned whether the observation of specific phenomena within the context of a practical task can, unaided, lead to the development of' conceptual understanding. In this context it has been proposed (Millar et al., 1999) that the function of practical work might be better understood in terms of a link, or bridge, between previously taught scientific concepts and subsequent observations.

One possible reason for the lack of research evidence to support the use of practical work as an effective means for developing pupils' conceptual knowledge is that, in contrast to teacher demonstration, its use can generate cognitive overload. Cognitive overload occurs as a consequence of simultaneous demands made of the pupils by practical work in that they need to apply intellectual and practical skills as well as prior knowledge.

Therefore, despite the high hopes and expectations of those who advocated a central role for practical work in the teaching of science, research has consistently found that it is no more successful in achieving most of these generic aims than other non-practical methods of teaching. Indeed, Hodson (1991) claims that: "as practised in many countries, it is ill-conceived, confused and unproductive" (p. 176) whilst Osborne (1998) suggests that practical work 'only has a strictly limited role to play in learning science and that much of it is of little educational value' (p. 156). Perhaps the key phrase in Hodson's claim is 'as practised'. Whilst few would doubt that practical work is an essential part of science education the question we need to consider is whether we use it effectively and, if so, how. In order to answer that question we need first to consider what we mean by 'effectiveness'.

EFFECTIVENESS

To think about the effectiveness of a teaching/learning task of any kind it is useful to consider the steps in both developing such an activity and monitoring what happens when it is used. To do this a model of the processes involved in designing and evaluating a practical task, developed by Millar et al. (1999, p. 37), has been used and this is shown below in Figure 1.

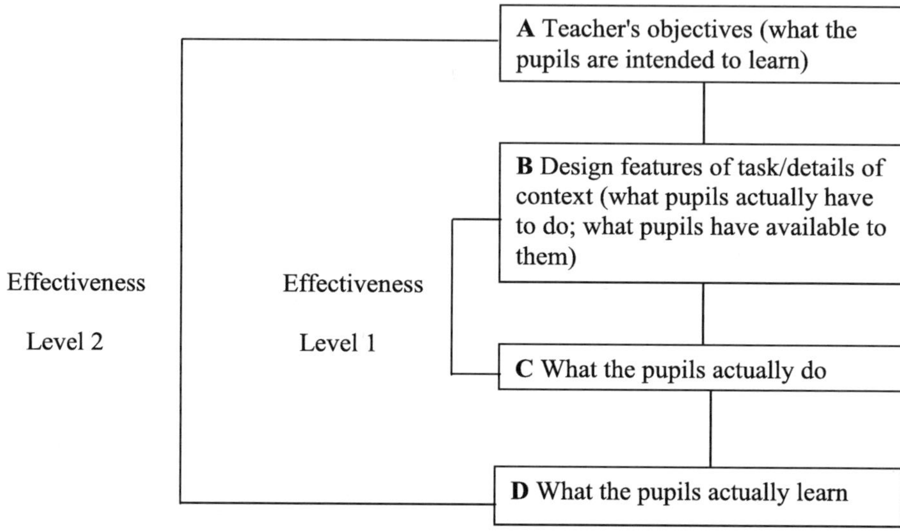

Figure 1. A model of the process of design and evaluation of a practical task (after Millar et al., 1999, p. 37)

Given that the aim of this model is to consider the effectiveness of a specific task *relative* to the aims and intentions of the teacher the starting point (Box A) is an evaluation of the teacher's learning objectives in terms of what it is they want the pupils to learn. Once the teacher has decided what they want the pupils to learn the next stage (Box B) is for them to design a specific practical task (or use an existing one from a scheme of work) that they consider has the potential to enable their pupils to achieve the desired learning objectives.

However, because the pupils might not do exactly as the teacher intended when they designed the task, the next stage in the model (Box C) considers what it is that the pupils actually do as they undertake the task. There are various reasons as to why, and the extent to which, what the pupils actually do might differ from what was intended by the teacher. Pupils might, for example, not understand the instructions or, even when they do and adhere to them meticulously, faulty apparatus could prevent them from doing what the teacher intended. Alternatively even if the task is carried out as intended by the teacher and all of the apparatus functions as intended the pupils still might not engage mentally with the task using the ideas that the teacher had intended them to use. The last stage in the model (Box D) is therefore concerned with the question of what it is that the pupils actually learn as a consequence of undertaking the task.

The use of this theoretical model allows the question of the effectiveness of a specific practical task to be considered at two separate levels. The first of these two effectiveness levels relates to the issue of what pupils *do* relative to what the teacher intended them to do. This level of effectiveness, referred to as 'level 1 effectiveness', is about the relationship between boxes B and C in the above model. The second level of effectiveness considers what the pupils *learn* relative to what the teacher intended them to learn. This second level of effectiveness, referred to as 'level 2 effectiveness', is about the relationship between boxes A and D in the model. This model can therefore be used to clarify what is meant by the 'effectiveness' of a specific practical in terms of:

- Does the task enable the pupils to do the things the teacher actually wanted them to do when they chose to use that specific practical task?
- Does the task enable the pupils to learn what the teacher actually wanted them to learn when they chose to use that specific practical task?

By combining this model of effectiveness with Tiberghien's (2000) model of knowledge in which there are two distinct domains: the domain of observable objects and events (o) and the domain of ideas (i) it is then possible to consider each of the two levels of effectiveness in terms of these two distinct domains. The effectiveness of any practical task can now be analysed and discussed in terms of two levels with each level being further divided into two domains. In terms of task effectiveness these levels are defined in the following way:

- A task is effective at level 1:o if the pupils *do* with the objects and/or materials the things that the teacher intended them to do and, as a consequence, they see the intended outcome.
- A task is effective at level 1:i if the pupils *think* about the task using the ideas that the teacher intended them to use.

At level 2:o and 2:i the issue of effectiveness relates to whether or not the task enables the pupils to *learn* the things intended by the teacher.

- At level 2:o a task is effective if the pupils *learn* and can recollect details about the objects/materials/events that they have observed and/or handled.
- At level 2:i a task is effective if the pupils *learn* and can recollect the scientific ideas that provide an explanation about the objects/materials/events that they have observed and/or handled.

These two levels of effectiveness, each of which can be considered with respect to the two distinct domains of knowledge, can be represented (Figure 2) using a 2x2 effectiveness matrix:

Intended outcomes…	…in the domain of observables (Domain o)	…in the domain of ideas (Domain i)
…at level 1 (what pupils do)	Set up the equipment and operate it in such a manner as to see what the teacher intended.	Think about the task using the ideas and scientific terminology intended by the teacher.
…at level 2 (what pupils learn)	To set up and operate similar equipment. Discover patterns within their observations/ data.	To understand their observations /data by being able to link them, using the ideas and scientific terminology intended by the teacher, with the correct scientific theory.

Figure 2. A 2×2 effectiveness matrix

Effectiveness at level 2:i is therefore a necessary requirement if, as has been suggested (Abrahams & Millar, 2008; Millar et al., 1999), an important function of practical work is to provide a link between the domain of observable objects and/or events and the domain of ideas.

If, as seems reasonable to assume, a task can only be effective at level 2:i if it were also effective at both level 1:o and 1:i, then it appears that effectiveness across level 1 is a necessary pre-requisite for a successful link to be created between the two distinct domains of knowledge.

DOING WITH OBJECTS AND MATERIALS

Despite the fact that closed 'recipe' style tasks might potentially be perceived by pupils as being dull and/or demoralising, and might also not be perceived of being necessarily either meaningful or engaging, many practical tasks in secondary school science are at, or close to, the closed 'recipe' end of the continuum (Abrahams & Millar, 2008). Some teachers, despite stating their preference for open investigations, see the use of such closed tasks as a necessary means of ensuring that most pupils within a given class, irrespective of their academic ability, are able to set-up and produce a particular phenomenon and analyse the results within the relatively short period of time available in one lesson. The fact that this approach is considered by teachers as very successful (Abrahams & Millar, 2008) strongly suggests that teachers see the effective generation of a particular phenomenon, and/or set of results, by the majority of their pupils as being their main priority.

DOING WITH IDEAS

Practical tasks, as Millar et al. (1999) have pointed out "do not [or should not] only involve observation and/or manipulation of objects and materials. They also involve the pupils in using, applying, and perhaps extending their ideas" (p. 44). Whilst 'doing' with objects and materials is relatively self-explanatory, 'doing' with ideas is less obvious and needs some clarification. The theoretical 2×2 matrix representation of practical work, discussed previously, distinguishes in the horizontal dimension between *doing* and *learning* and in vertical dimension between *observables* and *ideas*. In this context the two quadrants on the right-hand side of the matrix refer only to ideas that, in contrast to observables, cannot be directly measured or observed. Doing with ideas therefore refers to the process of 'thinking about' objects, materials and phenomena in terms of theoretical entities that are not directly observable. Clearly not all thinking is synonymous with 'doing with ideas' – far from it. For example, whilst a pupil can observe, and think about, the readings on a voltmeter in terms of observables – in this case the numbers on a dial or scale – thinking about those readings in terms of their being a measure of the voltage – a non-observable property of batteries and other circuit components – is, as the following example illustrates, what constitutes 'doing' with ideas.

Researcher: What type of circuit is this?
Pupil SK18: It's a series circuit.
Researcher: So what's the voltmeter measuring?
Pupil SK22: How much energy is going in and how much energy is coming out.
Researcher: And what will that tell you?
Pupil SK22: How much energy is lost.

Having clarified what 'doing with ideas' entails it is important to remember that task effectiveness, in the context used here, is a measure of what pupils do with ideas relative to what the teacher intended them to do with them. In this respect whilst the pupils can think about a task in any way that they wish, the task only has the potential to be effective (or ineffective) at level 2 if the teacher actually *intends* the pupils to think about the observables using specific ideas.

What has been found (Abrahams & Millar, 2008) is that there is a significant difference between the effectiveness of practical work in the domain of observables and that of the domain of ideas. Whilst pupils are frequently able to make the observations that the teacher intends they rarely talk to each other, or to the teacher, using the ideas that underpin the observable features of the task and whose use would enable them to make scientific sense of their actions and observations.

Indeed many teachers appear (either tacitly or explicitly) to maintain an inductive 'discovery-based' view of learning and, as such, to expect that the ideas that they intend pupils to learn will simply emerge of their own accord from the observations or measurements, provided only that the pupils are able to produce them successfully. The underlying epistemological flaw in this viewpoint, and the practical problems to which it leads, have long been recognised (see, for example, Driver, 1975).

Since science involves interplay between ideas and observation an important role of practical work is to help pupils develop links between observations and ideas.

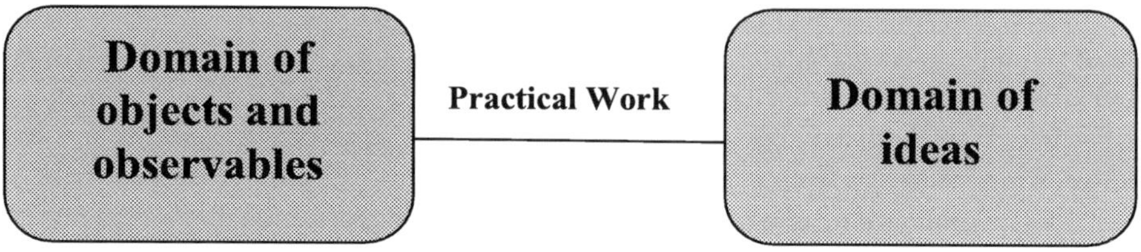

Figure 3. Practical work: Linking two domains of knowledge (after Millar et al., 1999, p. 40)

However, not only do these ideas have to be introduced but it may be important that they are used during the practical activity itself, rather than after it (possibly in a subsequent lesson), in order to account for what is being observed. Solomon (1999) discusses the valuable role that 'envisionment' can play in practical work, in terms of helping pupils to imagine what might be going on 'beneath the observable surface' as they do things with objects and materials and make their observations. Yet the evidence (Abrahams & Millar, 2008) is that few practical lessons are designed to stimulate interplay between observations and ideas *during* the practical activity. Even if these links are developed in subsequent lessons, the fact that the ideas are not available to help pupils make sense of the activity (to see its purpose) or their observations (to interpret these in the light of the theoretical framework of ideas and models) must, arguably, reduce the effectiveness of the practical task as a learning activity.

In terms of implications for practice the suggestion here is that the two domains model, outlined in this chapter, is a useful tool for teachers when thinking about practical work. First, it draws attention to the two domains of knowledge involved, and their separateness – that one does not simply 'emerge' from the other. Second, it provides a means of assessing the 'learning demand' (Leach & Scott, 1995) of a particular task and recognising there is a substantial difference in the learning demand of tasks in which the primary aim is simply for pupils to see an event or phenomenon or become able to manipulate a piece of equipment, and those tasks where the aim is for pupils to develop an understanding of certain theoretical ideas or models that might account for what is observed. If teachers can be helped to distinguish more effectively between tasks of relatively low learning demand and those where the learning demand is much higher, this would then allow them to identify those tasks where pupils might require greater levels of support in order that the intended learning might occur.

To achieve this aim it is necessary to ensure that task design more clearly reflects the fact that 'doing' things with objects, materials and phenomena will not lead to pupils 'learning' scientific ideas and concepts unless they are provided with a 'scaffold'. The process of scaffolding provides, as the following example shows, the initial means by which pupils are helped to 'see' the phenomena in the same 'scientific way' that the teacher 'sees' it (Ogborn et al., 1996).

Dr Starbeck:	[Points to the animated character moving around a stylised circuit on the whiteboard.] Right so we've got something moving around a circuit, a person moves around the circuit. What's moving around a real circuit?
Pupil SK4:	Electrons.
Dr Starbeck:	Ok, electrons, electric charges. So the person [points to character on screen] stands for?
Pupil SK5:	Charge.
Dr Starbeck:	Electrons, charges. People stand for charges. [points to animated character walking around circuit]. What do charges do? They move around the circuit. What do we call a movement of charge? [This had been taught in a previous lesson].
Pupil SK14:	Current.
Dr Starbeck:	Current. Current, right. So we've got charges moving around a circuit, people carrying boxes, charges carrying energy around a circuit. Ok?

Indeed, Lunetta (1998) has argued, "laboratory inquiry alone is not sufficient to enable pupils to construct the complex conceptual understandings of the contemporary scientific community. If pupils' understandings are to be changed towards those of accepted science, then intervention and negotiation with an authority, usually a teacher, is essential" (p. 252). The issue then is the extent to which this intervention, and negotiation with the teacher, is acknowledged and built into the practical task by the teacher.

Given the clear importance of trying to ensure that pupils do what the teacher intends with objects and materials in the limited time available, 'recipe' style tasks are likely to continue to play a prominent role in science practical work. However, if the scale of the cognitive challenge that pupils can sometimes face in linking their actions and observations to a framework of scientific ideas was better recognised, teachers might start to divide the time in a practical lesson more equitably between 'doing' and 'learning'. Such a separation would see a greater proportion of the time being devoted to helping pupils use ideas and concepts associated with the phenomenon that they want them to produce, rather than simply seeing the production of the phenomenon as a successful end in itself.

FURTHER READING

Abrahams, I. (2011). *Practical work in school science: A minds-on approach*. London: Continuum.
Millar, R., Leach, J., & Osborne, J. (Eds) (2000). *Improving science education: The contribution of research*. Buckingham: Open University Press.
Osborne, J., & Dillon, J. (2010). *Good practice in science teaching: What research has to say*. Maidenhead: Open University Press.

REFERENCES

Abrahams, I., & Millar, R. (2008). Does practical work really work? A study of the effectiveness of practical work as a teaching and learning method in school science. *International Journal of Science Education, 30*(14), 1945–1969.
Bennett, J. (2003). *Teaching and learning science: A guide to recent research and its applications*. London: Continuum.
Clackson, S. G., & Wright, D. K. (1992). An appraisal of practical work in science education. *School Science Review, 74*(266), 39–42.
Connell, L. (1971). Demonstration and individual practical work in science teaching: A review of opinions. *School Science Review, 52*, 692–702.
Driver, R. (1975). The name of the game. *School Science Review, 56*(197), 800–805.
Hewson, M., & Hewson, P. (1983). Effect of instruction using student prior knowledge and conceptual change strategies on science learning. *Journal of Research in Science Teaching, 20*(8), 731–743.
Hodson, D. (1991). Practical work in science: Time for a reappraisal. *Studies in Science Education, 19*, 175–184.
Hodson, D. (1992). Redefining and reorientating practical work in school science. *School Science Review, 73*(264), 65–78.
Hofstein, A., & Lunetta, V. N. (1982). The role of the laboratory in science teaching: Neglected aspects of research. *Review of Educational Research, 52*, 201–218.
Lazarowitz, R., & Tamir, P. (1994). Research on using laboratory instruction in science. In D. L. Gabel (Ed.), *Handbook of research on science teaching and learning*. New York, NY: Macmillan.
Leach, J., & Scott, P. (1995). The demands of learning science concepts: Issues of theory and practice. *School Science Review, 76*(277), 47–52.
Lunetta, V. N. (1998). The school science laboratory: Historical perspectives and contexts for contemporary teaching. In K. Tobin & B. Fraser (Eds.), *International handbook of science education*. Dordrecht: Kluwer.
Millar, R. (1987). Towards a role for experiment in the science teaching laboratory. *Studies in Science Education, 14*, 109–118.
Millar, R. (1998). Rhetoric and reality: What practical work in science education is really for. In J. Wellington (Ed.), *Practical work in school science: Which way now?* London: Routledge.
Millar, R. (2002). Thinking about practical work. In S. Amos & R. Booka (Eds.), *Aspects of teaching secondary science: Perspectives on practice*. London: Routledge Falmer.
Millar, R., Le Maréchal, J-F., & Tiberghien, A. (1999). 'Mapping' the domain: Varieties of practical work. In J. Leach & A. Paulsen (Eds), *Practical work in science education*. Dordrecht: Kluwer.
Mulopo, M. M., & Fowler, H. S. (1987). Effects of traditional and discovery instructional approaches on learning outcomes for learners of different intellectual development: A study of chemistry students in Zambia. *Journal of Research in Science Teaching, 24*(3), 217–227.
NSTA Commission on Professional Standards and Practices. (1970). *Conditions for good science teaching in secondary schools*. Washington, DC: National Science Teachers Association.
Ogborn, J., Kress, G., Martins, I., & McGillicuddy, K. (1996). *Explaining science in the classroom*. Buckingham: Open University Press.
Osborne, J. (1998). Science education without a laboratory? In J. J. Wellington (Ed.), *Practical work in school science. Which way now?* (pp. 156–173). London: Routledge.

Solomon, J. (1999). Envisionment in practical work. Helping pupils to imagine concepts while carrying out experiments. In J. Leach & A. Paulsen (Eds.), *Practical work in science education: Recent research studies* (pp. 60–74). Roskilde/Dordrecht, The Netherlands: Roskilde University Press/ Kluwer.

Tiberghien, A. (2000). Designing teaching situations in the secondary school. In R. Millar, J. Leach, & J. Osborne (Eds.), *Improving science education: The contribution of research*. Buckingham: Open University Press.

Van den Berg, E., & Giddings, G. (1992). *Laboratory practical work: An alternative view of laboratory teaching* (Monograph). Western Australia: Curtin University, Science and Mathematics Education Centre.

Yager, R., Engen, H., & Snider, B. (1969). Effects of the laboratory and demonstration methods upon the outcomes of instruction in secondary biology. *Journal of Research in Science Teaching, 6*, 76–86.

SECTION VI

INFORMAL SCIENCE EDUCATION

ANWAR B. RUMJAUN

30. EDUCATIONAL VISITS AND SCIENCE EDUCATION

INTRODUCTION

Students learn science by doing science and inquiry is central to learning science by doing science. Inquiry learning starts by putting key questions where students would have to look for information to answer those questions. The information could either come from documents (books, newspapers…), from the internet, from experimentation conducted in class, from survey by questioning and interviewing people. How do all these searches for information in an inquiry perspective fit with learning science concepts through educational visits? This is what will be discussed in this chapter. The chapter will also include the nature of, role of educational visit, the learning experiences of students engaged in a science class, some concrete examples to show case the steps involved, the pedagogy and the learning theories which support educational visits in science and finally some current challenges which educators and students face in conducting educational visits and the ways and means to address those challenges.

EDUCATIONAL VISIT

School students spend most of their waking time out of formal schooling contexts and yet science educators tend to ignore the crucial influences that experiences of out of school context can have on students' beliefs, attitudes and motivation to learn. They often see "out of school context" as a potential source of misconceptions among students.

WHAT IS AN EDUCATIONAL VISIT?

An educational visit or tour is a journey by a group of people to a place away from their normal environment. When organized by a school for its students, it is also known as a school trip or a field trip (Braund & Reiss, 2006). Educational visits enable students to interact with the settings, displays and exhibits to gain an experiential connection to the idea, concepts or subject under study.

The aim of an educational visit is usually observation for non-experimental research or to provide students with experiences outside their everyday activities. It also aims at observing the subject under study in its natural state and possibly

K. S. Taber & B. Akpan (Eds.), Science Education, 417–429.

collect samples. Thus, educational visits provide first hand experiences, stimulate interest and motivation in science, add relevance and interrelationships, strengthen perceptions and observation skills, and promote personal and social development among students. Educational visits provide students with rich informal learning experiences which complement their formal learning experiences in classrooms.

NATURE OF EDUCATIONAL VISIT

Educational visits could be of different kinds. They could be a tour to a place, a visit to a site (indoor or outdoor), attending a talk out of school. Educational visits could be either formal visits or informal visits. Formal visits are planned, well-orchestrated visits where students follow a documented format. Such types of visits are provided by science museums, research centers, hospitals where students follow the instructions laid down by the staff of these institutions. Thus, one student's experience is essentially the same as another student's experience. By contrast, informal types of educational visits are less structured and offer the students some control and choice on their activities at the sites. Table 1 below summarizes the potential sites for educational visits in science.

Table 1. Potential sites for educational visits in science

Sites	*Subject/Topics/Concepts*
School Garden	Living and Non Living Things – Classification
Outside school- Nature Park	Ecosystems, Biodiversity, Environment, Forces, Energy, Materials, Pollution
Hospitals	Diseases, Treatment, Health, biochemical techniques for disease diagnosis
Zoos and Public Botanical gardens	Biodiversity, Classification, Conservation
Research Centers	Scientific Techniques, Electrophoresis, Electron Microscopy, Mass Spectrophotometer
Science Centers	"Big" science activities, Astronomy (planetarium)

There are various potential sites for educational visits to support teaching and learning of science. An educational visit could be in the form of an investigation of a small area on the school compound. This is the simplest and the less expensive form of educational tour or visit and could be referred to as a simple field work. In such type of visit, students can learn abstracts concepts such as food chains and interrelationships among living organisms and non-living organisms. It could also allow students to look up to the sky and to ask questions such as what are there up in the sky. In this chapter educational visits will be considered as moving students out of their school environment to go to other places of scientific interest.

Students usually go for educational visits out of school and sometimes out of their town and villages. These types of visits entail more planning and preparation and they bear a cost for the transportation. Today we can also have virtual educational visits without the students going out of the classroom (*https://en.wikipedia.org/wiki/Virtual_field_trip*). Virtual educational visit is a guided and narrated tour of Web sites that have been selected by experts and arranged in a "thread" that students can follow from site to site with just the click of a single button. Virtual educational visits are proposed for places which are difficult to access, such as inside the volcanoes or under the sea. There are similarities between virtual and real visits: both include group activities whereby students interact with each other. Both are active learning experiences and both take students mentally out of the classroom. There are differences between the two and the major one is that virtual educational visits lack the sensory experiences which exist in real educational visit.

ROLE OF EDUCATIONAL VISIT IN SCIENCE

The aim of science education today is twofold; first it aims at enabling students to gain an understanding of the established body of scientific knowledge and second it aims at developing students' understanding of how this knowledge has been gained (Millar, 2004). The second aim is crucial as it allows students to understand not only about the scientific facts and knowledge but also they will understand how scientist make claims by the forms of argumentation for scientific knowledge using scientific inquiry. Moreover, as part of the first aim, students in this modern society need to gain an understanding of some scientific knowledge so that they can evaluate those claims which may affect their decisions in their everyday lives, such as about health, diet, energy and resource used and to reach out views on matters of public policy such as genetic therapies or electricity production. It is in these perspectives that educational visits in science find its place. These visits enable students to learn abstract concepts in science and make learning of science meaningful and contextual. Students who go on educational visits as part of their educational experiences show statistically significant learning of the subject content compared to those students who learn the same science concepts in classroom situations (Lisowski & Disinger, 1991). Students exhibit more knowledge about a subject if they learn the subject during educational visits than learning the same subject in a classroom (Wendling & Wuensch, 1985).

Several studies have shown that students enjoy learning during educational visits (Michie, 1998). Braund and Reiss (2006) suggest that students not only find learning fun during educational visits but they enjoy activities during educational visits more than the lessons they are taught on the same subject in class. Educational visits in science motivate students in learning science though interest and curiosity. They also promote student-student and student-teacher social interaction. More importantly, learning of science out of classrooms provides opportunities for students to formulate questions, develop their scientific observation, record information in a systematic way and relate abstract science concepts to their respective contexts. Thus, they help students to be engaged in the concepts- contexts interface.

STEPS INVOLVED IN AN EDUCATIONAL VISIT

Educational visits involve three steps: preparation, on site activity and follow-up/ post visit activity.

1. Preparation

Preparation applies to both the students and the teacher. Teachers should take the time to learn about the destination and the subject before the trip. The preparation phase includes 2 major parts; one is administrative preparation and the other one is academic preparation. In the administrative part, teachers should first of all inform and get the approval of the school management. This step will then lead to making request to and getting the consent from parents of each and every student. These consent forms should include the purpose of the visit, the place of visit, the time of departure from school and the time of return to school, the necessary requirements in terms of clothing, shoes, lunch and drinks any cost required, all these information should be provided to parents to enable them to give their consent. The precise forms of these processes may vary from country to country or from school to school. It is important for teachers to follow the proper policies and procedures in place where they are employed. The preparation phase also includes the arrangements made with the contact person of the site identified and sometimes it entails a prior visit of the teacher to the site to enable him or her to get acquainted and familiarized with the site. For example, the teacher can visit the site to assess the potential dangers and risks for students. This prior visit is also crucial for the teacher as he/she may need to prepare activities to be conducted by the students on the site.

The academic part of the preparation phase includes identifying potential sites which could support the teaching and learning of the science topic under study. The teacher needs to discuss with students about the aim of the visit to the site and he/ she should work out, using a participatory approach, with the pupils the tasks and the relevant activities which would be conducted on the site. The idea is to spend sufficient time on the preparation of activities so as not to waste time once at the site. Worksheets such as questionnaires, observation checklists, collection of samples sheets, as and where applicable, would need to be developed during this pedagogic part of this preparatory phase.

2. On site activities

Once on the site, the students are briefed again on what they are supposed to do and they start to tackle their tasks. These tasks could be observations of displays and demonstrations in science museums or in the outside environment, questioning people on the site to get information using a questionnaire, collecting relevant samples using appropriate tools and in appropriate containers, drawing the site so as to map out the area under study and in some cases, the tasks could also be hands-on activities where students manipulate and interact with objects and apparatus. Attending talks and presentations by staff or viewing videos at the visit site are also common activities organized at science centers.

3. Follow up activity

A follow up activity should be organized in the classroom after the visit usually on the following day or during the session where the teacher next meets the students. The follow up activity usually starts by the teacher brainstorming the students on what they have seen or what they have enjoyed during the visit. The teacher asks the students to look at the information or samples collected and they are asked to analyze them. Teacher can put questions to guide the students in their analysis. This follow up activity can be a very rich and intense session for learning.

LEARNING THEORIES SUPPORTING EDUCATIONAL VISIT IN SCIENCE TEACHING

Inquiry is central to teaching and learning in science. Educational visits in science involve inquiry processes where students interact with a context by exploring, asking questions, discussing and confronting ideas and formulating good questions for inquiry. These questions will lead the students to engage in investigation with the idea of looking for information to answer these questions. It is in this perspective that educational visits should be conducted to allow students to engage with the relevant resources to obtain the required information which they would analyze and discuss in class in the follow up activity session. Several learning theories support these learning processes and learning experiences in science during educational visits. These are socio-constructivism, situated learning and experiential learning.

Socio-constructivism. (Vygotsky, 1896–1934)

This theory of learning supports the idea that knowledge is constructed. Students do have the raw materials from their everyday experiences and the learning situations in which they are engaged, that can enable them to process the raw materials to construct new ideas and new knowledge. However, this can lead to the development of alterative conceptions, unless the construction process is carefully scaffolded by a teacher, education officer at the visit site, or someone else with the necessary knowledge and skills. Students confront and challenge each other's ideas and discuss them in order to co-construct knowledge with respect to the science topic under study. Educational visits provide students with the appropriate platforms to enable them to engage in such learning experiences. The context, the types of interaction, the emotions and the daily life experiences all together help students to make sense and meaning of they are learning. For example, students are more likely to appreciate the causes, the consequences and come up with solutions to environmental problems if they are put and engaged in that particular situation.

Situated learning. (Lave & Wenger, 1991)

Situated learning was proposed as a model of learning in a community of practice. Situated learning is learning that takes place in the same context in which it is applied. Lave and Wenger (1991) argued that:

> Learning should not be viewed as simply the transmission of abstract and decontextualized knowledge from one individual to another, but a social process whereby knowledge is co-constructed; they suggest that such learning is situated in a specific context and embedded within a particular social and physical environment. (Lave & Wenger, 1991)

Situated learning aims at bridging the gap between the theoretical learning in the formal instruction of the classroom setting and the application of that knowledge in the out of classroom environment (Herrington & Oliver, 1995). A situated learning environment provides an authentic context that reflects the way knowledge is used in real life experiences. It thus provides a learning environment whereby the activities are ill-defined. In such circumstances, students *find* as well as *solve* problems in the context. Learning experiences which are offered to students in educational visit enable them to bridge the gap between learning of abstract science concepts and their applications in the environment. Thus they make sense to what they learn and develop science process skills to attain the learning objectives.

Experiential learning. (Kolb, 1984)

Experiential learning is defined as learning through doing, observing and reflecting. Experiential learning focuses on the learning process of the students. According to Kolb (1984), learning is continuously gained through personal and concrete environmental experiences. In order to gain meaningful knowledge from an experience, the students must have four abilities:

- The learner must be willing to be actively involved in the experience;
- The learner must be able to reflect on the experience;
- The learner must possess and use analytical skills to conceptualize the experience; and
- The learner must possess decision making and problem solving skills in order to use the new ideas gained from the experience (Kolb, 1984).

Educational visits in science support the students to develop the four abilities. They are actively involved in the field work on the site, they should be able to use their systematic observation and analytical skills at the site and during the follow up activity, they should decide how to present their information collected and what could be done to address the issue under study at the selected site. All these abilities would certainly be developed if the educational visit is placed in a problem solving perspective.

CONCRETE EXAMPLES OF CONDUCTING AN EDUCATIONAL VISIT FOR TEACHING AND LEARNING OF SCIENCE

It would be very interesting to illustrate an educational visit in the teaching and learning of science for a class at lower secondary level using concrete examples. At this education level, students are of 11–12 years of age (these students' ages can vary with varying education system). Two examples are selected and they aim at providing an overview of ideas and processes involved during an educational visits in two different places. These two examples are as follows:

- Case one: a visit to a typical science center which makes provision for students to observe exhibits and displays of science concepts and processes.
- Case two: a visit to the bank of a river in the open environment.

The above two examples are described in box one and two respectively.

Box 1. Case one: Learning science at a science center

PREPARATION PHASE

First of all the teacher identifies the concepts which should be addressed. The idea which the teacher would have in his/her mind should be to use the science center as resource center to maximize learning of abstract concepts and processes in science. It would not generally be cost effective to move students out of school to learn one concept or one science process or phenomenon. It would be more appropriate and cost effective to conduct a series of lessons on a given topic or topics and inform students that they would see more of these in visual forms and/or in action at the science center. Potential science topics taught at lower secondary school level for which an educational visit would be needed are: clean energy production, recycling of and management of waste.

The teacher should then discuss these issues with pupils. For example, the forms of energy we make use of in our everyday lives, the sources from which we obtain these energy, their impacts on our lives and on the environment and the alternatives options for producing energy we want and the reasons thereof. The same discussion would apply to the waste issue. For example, what are the types of waste we produce, the consequences of generating so much of waste and how to minimize waste in our surroundings. These discussion would be in the form of a brainstorming session whereby the teacher would help students to reflect on these issues and would lead them to ask questions about how and what one can do to address these issues.

The teacher should then initiate actions to obtain approval from the relevant authorities about the educational visit to move students out of school. The teacher should also obtain approval of access to the science center from the management of the said center. Thus, the teacher or the school manager should write to the science center and provides the following details:

- The purpose of the visit,
- The level of schooling of the students,
- The number of students and
- The specific learning objectives related to the science concepts.

In some cases, the teacher will make a first visit to get an insight about the facilities available at the site. This is often recommended to teachers. Finally, the teacher sends consent forms to parents to obtain their approval.

ON SITE ACTIVITY

The teacher allows the students to listen to the guide, to observe and watch the demonstration of science processes, for example, recycling of paper. Students will listen and will take note of the explanation provided by the guide on the displays under study.

FOLLOW UP ACTIVITY

The follow up activity is usually carried out in class. In some cases, this can already start in the bus while driving back to school. The teacher leads the discussion and relates to the observations made at the center with what has been discussed in the class prior to the visit. The teacher can ask students to write a report on their visit by providing a template which includes the aim of the visit, the observations made, the important learning points they came across. The learning difficulties which students meet, are also addressed in this follow up session. In a more structured follow up activity, the teacher can start with an oral debriefing session with the students. He or she captures all the learning points on the board and he or she may set a written assignment to the students. The follow up activity could also take the form of a whole class discussion to consolidate learning at the site. It could also include a task whereby students would need to produce something such as a presentation to the whole class.

Now let us look at example two.

Concrete example two: a visit to a bank of a river

This educational visit to a river bank fits in an environmental education perspective where the teacher exploits this visit to integrate science concepts with concepts from other curriculum areas, namely history and geography, languages and arts.

PREPARATION PHASE

The teacher opens the discussion on the topic of pollution in our environment. Through a brainstorming session, this leads to identifying types of pollution and teacher leads the discussion to water pollution in our rivers. He/she then asks questions such as:

- What are the causes of water pollution?
- What are the consequences to human beings? To the other life forms? To the environment at large?
- What can we do to reduce water pollution?

These three questions are the potential key questions for a contextualized problem in the students' immediate environment. Students will then formulate possible answers. This stage will lead to providing an inquiry learning process. Teacher together with the students identify a potential river site for the educational visit. The teacher can here map out the site of the visit and earmark the potential areas to carry out the investigation. At this stage, he/she relates to concepts/ issues which will be used as a lens for the investigation. Students, under the guidance of the teacher, then prepare appropriate and relevant worksheets including observation checklist, questionnaires to survey people living nearby the riverbank. They also identify tools and equipment to collect specimens for further investigation and experiment in class after the visit. The teacher then follows the steps to get the administrative clearance as already elaborated in example one above.

ACTIVITY ON THE BANK OF THE RIVER

The activity on the bank of the river usually takes the form of a group work activity. Teacher puts one group of students to each of the sites of the river bank identified in preparation phase. Students will then use their appropriate worksheets and will start gathering the required information as follows:

Group A will use the prepared questionnaire to interview people on the site about the past status of the river and its surroundings bank. How to explain the problems/issues which observed on the river bank and its surroundings?

Who is/are responsible for this situation? What has/have been done to address these issues/problems and why these problems/issues are still present? All this information would provide a historical perspective of the issues under study at the river.

Group B will look at the quality of the water flowing in the river. They will look at the turbidity of the water, the aquatic plants and the animals living in the water, the flow of the water in terms of quantity and quality, the smell and what are they?

Group C will draw examples of lived specimens present. They will not collect them as such as the teacher will explain the importance of these living organisms at the site to maintain biodiversity.

Finally, group D will be engaged in testing the quality of water or of the soil there. This group can also collect samples of them to bring in the class for further analysis.

The on-site activity need not take too much time if the educational visit has been well planned, organized and all worksheets are elaborated accordingly with the participation of the students. Getting the students involved in the elaboration of the worksheets provides a sense of ownership of these sheets by the students and these sheets become user friendly for the students.

FOLLOW UP ACTIVITY IN CLASS

The follow up activity takes place usually the day following the visit or the next class with the teacher. To start with, the teacher captures the students' experiences on the educational visit. This is done through a debriefing session where the teacher allows the students to speak freely on their experiences. Then the teacher asks each group (A, B, C and D) to report on their work.

The teacher captures every point reported on the board in the form of a concept map. A concept map on the board enables the students to have a holistic picture of the learning experiences in which they have been engaged. The students will put their ideas together, thereby creating confrontation, argumentation, questioning and debating thus lead to knowledge construction related to the topic under study. The role of the teacher here is to ensure learning of accurate concepts.

This session is very intensive for both the students and the teacher. The latter needs to demonstrate the relevant attitudes and skills to lead such type of active learning activity.

Another way to consolidate the educational visit in the follow up activity would be to ask the students to write a report of their visit and submit to the teacher. The report can be an essay or students could be given a set of questions related to the visit. The students will then work out these questions and return it to the teacher for marking.

The teacher could also find out from the students about their learning experiences at the site of visit. This could take the form of a whole class discussion or it could be a questionnaire administered by the teacher to the students to capture their views and opinions.

An important advantage from this follow up activity is that it allows the teacher to integrate learning of biological and physico-chemical concepts. From a case of a polluted river, issues related to biodiversity and classification of living organisms can be addressed. The session could also be linked up to relationships among the living organisms through concepts such as food chain and food web. The students can also be engaged in further investigation by conducting laboratory tests with the samples collected (if they have not them yet). Thus, the students would be involved in determining the physico-chemical properties of the water and soil collected, such as pH, turbidity test, biochemical oxygen demand (BOD), fecal presence and observing for living organisms using a microscope.

CHALLENGES RELATED TO EDUCATIONAL VISITS IN SCIENCE

Educational visits in science are also associated with many benefits as discussed above. We should also recognize that educational visits can also be associated with challenges. Firstly, in science centers, sometimes the exhibits and displays do not reflect the real science and thus lead to insignificant learning potential. These types of displays do not reflect the scientific thoughts and processes which scientists have been engaged in and thus lead to presenting science as easy and unproblematic (Rennie, 2007). In some cases, the same displays and exhibits are used to teach science concepts to students of all educational levels. Michie (1998) reported seven barriers to successful educational visits namely:

- Transportation,
- Teacher training and experiences,
- School schedule and teachers' abilities to prepare,
- Lack of school administrative support for educational visits,
- Lack of flexibility in the school curriculum,
- Poor students attitudes and behavior and
- Lack of awareness of teachers for potential sites.

Resistance of institutions to open up their venues for educational visits for school students is also a common barrier. The cost entailing the organization of educational visits is also another barrier. Schools do not make provision in the budget to promote learning of science concepts through educational visits.

Tal and Morag (2007) reported that in natural museums, the main visitation pattern consists of guide-centered and task-oriented activity. Their analysis of questions asked by museum guides revealed that most of these questions required mainly lower-order thinking skills. A common questioning pattern was to ask rhetorical questions as a means of carrying on the lecture. Detailed analysis of the scientific vocabulary used by the guides indicates that they used much scientific jargon, with limited explanation. There was only limited social mediation provided by teachers and museum guides. A minority of teachers were involved in the activities or in helping the guide to clarify or in helping the students to understand the explanations. The overall data indicate limited opportunities for meaningful learning, suggesting that the museums should shift from the traditional knowledge-transmission model of teaching to a more socio-culturally contextualized model.

To address the above challenges, the teacher needs to plan for their educational visits well in advance, for example at the beginning of the school year. The teacher should understand the importance of a well-organized and planned visit. They should be able to optimize the visit by providing opportunities for learning experiences related to several concepts and processes in science. They can also use the same visit to integrate issues and ideas from other curriculum areas. Educational visits should be part of the school culture. Schools can resort to virtual educational visits as well although these do not reflect the authentic experience of going on a visit.

CONCLUSION

The outcome of a learning experience depends on the learning context of the student, his/her interest, motivation, prior knowledge and experience. Educational visits in science offer rich learning experiences which help students to appreciate and understand science concepts. At the same time these rich learning experiences increase their knowledge and promote further learning related to other curriculum areas (Behrendt & Franklin, 2014).

The success and the efficacy of an educational visit in science require teacher preparation and interaction. The barriers and challenges discussed need to be addressed by all stakeholders and virtual educational visits should be encouraged and introduced in our schools.

FURTHER READINGS

Fraser, B. J., Tobin, K. G., & McRobbie, C. J. (Eds.) (2012). *Second international handbook of science education*. Dordrecht: Springer.

Schwan, S., Grajal, A., & Lewalter, D. (2014). Understanding and engagement in places of science experience: Science museums, science centers, zoos and aquariums. *Educational Psychologist*, *49*(2), 70–85. (Routledge: Taylor & Francis Group.)

REFERENCES

Behrendt, M., & Franklin, T. (2014). A review of research on school field trips and their value in education. *International Journal of Environmental and Science Education, 9*, 235–245.

Braund, M., & Reiss, M. (2006). Validity and worth in the science curriculum: Learning school science outside laboratory. *The Curriculum Journal, 17*(3), 213–228.

Herrington, J., & Oliver, R. (1995, December 3–7). *Critical characteristics of situated learning: Implications for the instructional design of multimedia* (pp. 253–262). ASCILITE 1995 Conference, University of Melbourne, Melbourne.

Kolb, D. (1984). *Experiential learning as the science of learning and development.* Englewood Cliffs, NJ: Prentice Hall.

Lave, J., & Wenger, E. (1990). *Situated learning: Legitimate peripheral participation.* Cambridge, UK: Cambridge University Press.

Lisowski, M., & Disinger, J. K. (1991). The effect of field-based instruction on students' understanding of ecological concepts. *Journal of Environmental Education, 23*(11), 19–23.

Michie, M. (1998). Factors influencing secondary science teachers to organize and conduct field trips. *Australian Science Teacher's Journal, 44*(3), 44–50.

Millar, R. (2004). *High school science laboratories: Role and vision.* Washington, DC: National Academy of Sciences.

Tal, T., & Morag, O. (2007). School visits to natural history museums: Teaching or enriching? *Journal of Research in Science Teaching, 44*(5), 747–769.

Wendling, R. C., & Wuensch, K. L. (1985). A fifth grade education program: Expectations and effects. *Journal of Interpretation, 10*(1), 11–20.

SHIHO MIYAKE

31. LEARNING SCIENCE IN INFORMAL CONTEXTS

In this chapter, I give an overview of learning science in informal contexts, focused around the following five points.

1. What are informal contexts of learning? Definitions of learning contexts.
2. Where and how does informal learning occur? Examples of research on where and how informal learning occurs.
3. Why is informal learning now in focus? Social trends relating to the demand for this topic.
4. Is there an antagonistic (mutually exclusive) or cooperative (overlapping) relationship between formal and informal learning?
5. Why is it necessary to discuss informal learning in science education?

WHAT ARE INFORMAL CONTEXTS OF LEARNING? DEFINITIONS OF FORMAL LEARNING, NON-FORMAL LEARNING, AND INFORMAL LEARNING

The world educational policy making leaders of the United Nations Educational, Scientific and Cultural Organization (UNESCO), and of the Organization for Economic Co-operation and Development (OECD), define formal learning, informal learning, and non-formal learning as shown in Table 1. Within each of these definitions, there appear to be four defining components: aim, outcome, place, and approach (Figure 1). While "formal learning" is clearly delineated on each of these components, "non-formal" and "informal" have some shared features. For example, on the "place" component, though non-formal learning and informal learning both occur outside school, the former takes place in community-based settings, and the latter takes place on an individual basis. "Aim" and "outcome" are similar in both non-formal and informal contexts, but "approach" is slightly different between the two. As another example, UNESCO states "it [lifelong learning] extends beyond formal education to non-formal and informal learning for out-of-school youth and adult citizens" (p. 1). In this chapter, the phrase "informal contexts" refers both to informal and non-formal learning contexts.

K. S. Taber & B. Akpan (Eds.), Science Education, 431–442.

Table 1. Definition of formal, non-formal and informal learning

[UNESCO (2012)]*	*[OECD (2015)]*
Formal learning takes place in education and training institutions, is recognised by relevant national authorities and leads to diplomas and qualifications. Formal learning is structured according to educational arrangements such as curricula, qualifications and teaching-learning requirements.	Formal learning is always organised and structured, and has learning objectives. From the learner's standpoint, it is always intentional: i.e. the learner's explicit objective is to gain knowledge, skills and/or competences. Typical examples are learning that takes place within the initial education and training system or workplace training arranged by the employer. One can also speak about formal education and/or training or, more accurately speaking, education and/or training in a formal setting. This definition is rather consensual.
Informal learning is learning that occurs in daily life, in the family, in the workplace, in communities and through interests and activities of individuals. Through the recognition, validation and accreditation process, competences gained in informal learning can be made visible, and can contribute to qualifications and other recognitions. In some cases, the term experiential learning is used to refer to informal learning that focuses on learning from experience.	Informal learning is never organised, has no set objective in terms of learning outcomes and is never intentional from the learner's standpoint. Often it is referred to as learning by experience or just as experience. The idea is that the simple fact of existing constantly exposes the individual to learning situations, at work, at home or during leisure time for instance. This definition, with a few exceptions (see Werquin, 2007 (/edu/skills-beyondschool/41834711.pdf)) also meets with a fair degree of consensus.
Non-formal learning is learning that has been acquired in addition or alternatively to formal learning. In some cases, it is also structured according to educational and training arrangements, but more flexible. It usually takes place in community-based settings, the workplace and through the activities of civil society organisations. Through the recognition, validation and accreditation process, non-formal learning can also lead to qualifications and other recognitions.	Mid-way between the first two, non-formal learning is the concept on which there is the least consensus, which is not to say that there is consensus on the other two, simply that the wide variety of approaches in this case makes consensus even more difficult. Nevertheless, for the majority of authors, it seems clear that non-formal learning is rather organised and can have learning objectives. The advantage of the intermediate concept lies in the fact that such learning may occur at the initiative of the individual but also happens as a by-product of more organised activities, whether or not the activities themselves have learning objectives. In some countries, the entire sector of adult learning falls under non-formal learning; in others, most adult learning is formal. Nonformal learning therefore gives some flexibility between formal and informal learning, which must be strictly defined to be operational, by being mutually exclusive, and avoid overlap.

	Formal	Non-formal	Informal
Aim	Learning objective for knowledge, skills and/or competences	Interests, Experience, Leisure	
Outcome	Diplomas and qualifications	Never intentional	
Place	Education and training institutions (e.g. schools)	Outside schools settings Community-based	Individuals
Approach	Structured curricula, qualifications, teaching-learning requirements	Rather organized than flexible	Various and flexible

Figure 1. Outline structure of formal, non-formal and informal learning

WHERE AND HOW DOES INFORMAL LEARNING OCCUR?

Several research findings suggest that knowledge acquisition occurs in everyday life in society, as well as in school settings (e.g., Osborn & Dillon, 2007; Fenichel & Schweingruber, 2010; Yager, 2008; Falk, 2001).

Learning Science Outside of School (Rennie, 2014) reviewed science learning in informal contexts, including learning in and from museums, on trips outside the classroom, and from media. Rennie said, "There may be differences in what is learning about science outside of school compared to inside school, but the processes of learning are similar" (p. 121).

Falk (2001) explained the concept of free-choice learning. He said "seating children in a museum auditorium and requiring them to hear a lecture is somehow different from seating children in a school auditorium to hear a lecture.... The vast majority of learning that occurs outside of school involves free-choice learning – learning that is primarily driven by the unique intrinsic needs and interests of the learner" (p. 7). This means that learning does not take place only by putting a showpiece into view. Learning occurs when people intentionally face the object with the active objective of acquiring information. In museums and zoos, such an active attitude may be indicated by observing a showpiece intentionally, reading descriptions, touching, and feeling. However, the purpose of carrying out these activities and the manner in which they are carried out vary between individuals, so that Falk (2001) called the active attitude "free-choice learning."

Learning Science Outside the Classroom (Braund & Reiss, 2004) also discusses popular informal learning settings outside of school, such as natural habitats, outdoor field centers, botanic gardens, zoos and farms, industrial sites, newspapers, and ICT tools. Patrick, Mathews and Tunnicliffe (2013) discussed how pre-service teachers develop their abilities through visiting zoos. Sanders (2007) argued for the possibility of botanic gardens as educational facilities.

In addition to place, research on learners' motivation and attention also explores one of the fundamental components of teaching science (Wittrock, 1994). Bultitude and Sardo (2012) described how informal science events had impacts for participants. Their work suggests factors that contribute to the perceived success of science communication activities occurring within leisure spaces.

Community-based science learning, especially that which occurs in museums, is reviewed by Rennie and Stocklemayer (2003). They outline how or why people behave in these contexts according to four patterns: (1) visiting exhibitions of various kinds, (2) pursuing interests and hobbies, (3) needing information to interpret one's circumstances, and (4) participating in community education programs.

As the above examples show, informal learning is often depicted as mutually exclusive from formal learning. It is also commonly accepted that informal education has no systematic framework of policies, contents, settings, and target persons (Roger, 2004, p. 99), and therefore that educational activities are developed by the balance of supply and demand among actors.

WHY IS INFORMAL LEARNING A FOCUS OF RESEARCH?

Human Capital

According to Drori (2004), "The assumption regarding the link between science education and economic growth was well-established as a common wisdom, much earlier than any evaluation of such a link. Moreover, in spite of the wide acceptance of science education as an important contributor to establishing economic development, few are the empirical tests of this 'truth' " (p. 27). This comment suggests that there is a close link between science education and the economic growth of our society.

The OECD report on "Human Capital" offered evidence for the impact of science education and learning on economic growth. Human capital refers to "the knowledge, skills, competences, and attributes embodied in individuals, which facilitate the creation of personal, social, and economic well-being" (Kelley, 2007; p. 29). The report also analyzed why people learn. One suggestion is that "to choose well, they [people] will need to keep up to date with a host of changes and trends. And then there's learning for sheer pleasure – something people will have more time for as lifespans continue to lengthen" (p. 80). Another suggestion is, "to stay relevant at work, people will need to go on continually upgrading their education" (p. 81). This means that, under the category of human capital, people learn for sheer pleasure; in informal learning contexts, people learn for their interest and leisure. These two explanations provide quite similar motivations for why people learn.

In sum, that which allows people to feel pleasure in their daily lives, including in their work, is essential for promoting economic growth. Informal education, as a part of science education, contributes to this growth.

Sustainable Development: A Global Consensus on Educational Purpose

The United Nations Decade of Education for Sustainable Development (2005–2014) (UNDESD) promoted alternative education principles and practices on a global scale.

> Sustainable development needs to be described for each of these dimensions—various critical challenges which appear unable to balance the needs of people and planet in the pursuit of peace and prosperity—with regards to their interrelationships in time (past-present-future) and in space (near-far). Sustainable social development (people) is aimed at the development of people and their social organization, in which the realization of social cohesion, equity, justice, and wellbeing plays an important role. A sustainable environmental development (planet) refers to the development of natural ecosystems in ways that maintain the carrying capacity of the Earth and respect the non-human world. Sustainable economic development (prosperity) focuses on the development of the economic infrastructure, in which the efficient management of our natural and human resources is important. It is the finding of balanced ways to integrate these dimensions in everyday living and working that poses, perhaps, the greatest challenge of our time as this requires alternative ways of thinking, valuing, and acting. (UNESCO, 2009, p. 6)

This point of view gives all people, at all stages of life and in all possible learning contexts, opportunities for the continual acquisition of knowledge and skills, acknowledging that peoples' educational needs change over a lifetime (UNESCO, 2014). Since its commencement, science education in Education for Sustainable Development [ESD] has manifested a new energy (Fensham, 2008). An interesting point in the UNESCO's ESD 2014 final report is that Cyprus, a member state of UNDESD, says "Taking into consideration that non-formal and formal education are closely connected, the programs implemented in the communities through the Environmental Education Centers provide the opportunity for students and teachers to investigate an issue outside of the school and consequently expand upon it further within the classroom" (p. 134). Although there were explicit differences (a mutually exclusive model) between formal and informal/non-formal learning in the history of their definition, this comment about Cyprus brushes the difference aside and suggests an extension-model of learning opportunities. That is, people can continue to learn across differently characterised modes of learning settings both inside and outside of school.

IS THE RELATIONSHIP BETWEEN FORMAL AND INFORMAL LEARNING MUTUALLY EXCLUSIVE OR OVERLAPPING?

Practice and Policy

School curriculums adapt to use outside-school settings in school science lessons. For example, the National Curriculum in England and Wales (Department for Education, 2013) mentioned a programme of study of "uses of everyday materials." The Next Generation Science Standards (2013) in the USA pointed out that building home-school connections was important for the academic success of non-dominant student groups, although in practice this was rarely done in an effective manner (NGSS Lead States, 2013). Wellcome Trust (2012) clearly states that "the links between informal and formal science learning can be enhanced for mutual benefit… It is also important to remember that the decisions on whether to take advantage of informal science learning opportunities are taken by individual teachers and school leaders" (p. 51). These curriculum and education standards suggest that informal learning outside school settings provides students an alternative manner in which to grow from school learners into social citizens. Teachers and school leaders play an important role in leading informal learning, and successfully linking activities that take place inside and outside of school. These examples explained that there is a cooperative or overlapping relationship between formal and informal learning.

The Japanese Law of Education distinguishes formal and informal education systems (Figure 2). There is a multi-layered education act under the Constitutions of Japan. The Basic Act on Education (Act No. 120 of 2006) provides the principle on which the Social Education Act (Act No. 207 of 1949) and the School Education Act (Act No. 26 of 1947) lie. The Social Education Act provides the principles on which the Library Act (Act No. 118 of 1950) and the Museum Act (Act No. 285 of 1951) lie (Ministry of Education, Culture, Sports, Science and Technology, n.d.). The existence of these several laws of education means that the concept of education, including determining who is in charge or what should be learned, has been interpreted separately in the history of Japan.

Article 2 of the Social Education Act defines social education as "systematic educational activities performed mainly by youths and adults including physical activities and recreation, however, those educational activities performed as curriculum studies based on the School Education Act are excluded." As a result of this, social educational activities and school education were mutually exclusive for the parties who deliver science activities. For example, a museum staff member, who was in charge of museum education, argued for denying direct support for school curriculum-based activities in museums (Miyake, 2002). It is a pity that the system worked consequently to limit the benefits of science learning for a half-century.

In 1998, however, the course of study in science was reformed and included recommendations to use museums and outdoor field centers for primary and

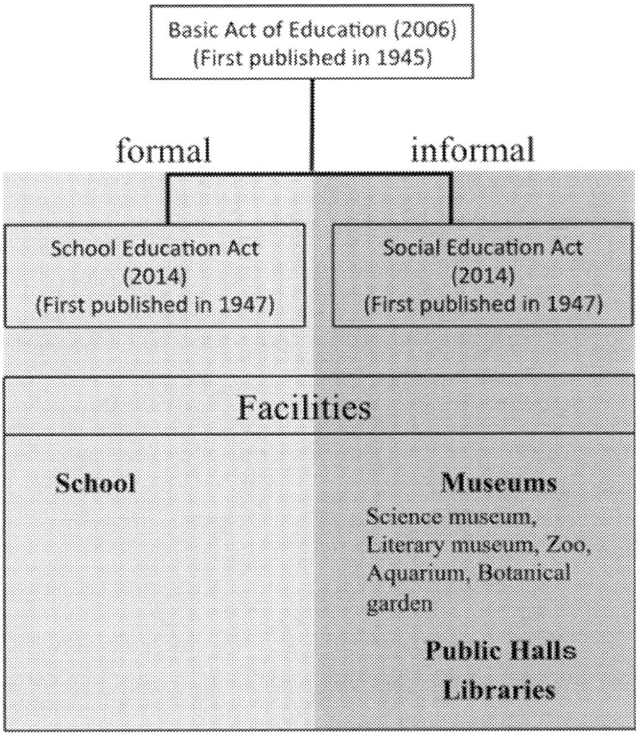

Figure 2. Educational act systems in Japan

secondary science. After this curriculum reform, a new slogan of "Haku-Gaku Renkei [cooperation between social education and school education]" and "Gaku-Sya Yugou [blend social education and school education into one]" was also held in social education contexts (Takada, 2010). Since then, teachers, museum staff members, and science education researchers began to maintain cooperative relationships between schools and museums, and they started to develop programs for students that took place between the two (e.g. Nakayama et al., 2006; Miyake et al., 2011).

An Actual Example of How People Learn through Informal Contexts

> Here, I describe an example of a Japanese science teacher, who wrote an article in Science Education Monthly, a commercial magazine published by an academic association of science teaching in Japan called the Society of Japan Science Teaching. As a teacher, Ms. Satooka one day faced the need to use a natural history museum. (Miyake and Satooka, 2013, translated into English by the author)

> After a couple of years of my teaching experience, I had my new assignment to Kumanoe junior high school in April, 2000. Beautiful untouched nature surroundings such as protected plant habitats and tidal wetlands remained at the periphery of the school. I thought there were a lot of resources to investigate and study with students. However, I had no idea of how to investigate such nature, because my major as a science teacher was chemistry. I looked for a person whom I could ask. But unfortunately, I was the only one science teacher in the school and there was no senior teacher who had the time to solve my problem. I was totally at a loss. A couple days later, I found the Miyazaki Prefectural Museum of Nature and History, which stood nearby my school. A museum staff member and a professor in Miyazaki University supported me in developing a one-year science project entitled "Where is the line between the River Kumanoe and the sea?" (see Satooka et al., 2004)
>
> A couple years later, I was transferred to another junior high school, where there was also the great natural resource of Mt. Kirishima. A crater lake, deep forests, and wild birds were in the surroundings; moreover, the beautiful striped pattern of the volcanic-ash layers were outcropped in the road. Again, I asked for a staff member at the Miyazaki Prefectural Museum of Nature and History to support me in developing a science project for my students. I investigated and studied the outcropping of the volcanic ash with my class. I had an opportunity to deliver a lecture for parents and neighbors, as well as my students, at a community class. We observed volcanic ash and learned about the nature of the area. (see Nakayama et al., 2006)
>
> During these two projects, I continuously told the museum staff members about students' learning progress and sometimes I e-mailed them with questions. I visited the museum on the weekends, and tried to find a showpiece as a resource to support knowledge building in my project. Then, I came to think that the meaning of studying at museum as a teacher was to obtain alternative 'eyes' to see museum showpieces as teaching materials.

From Ms. Satooka's experience above, we can understand a teacher's growth as a learner. She tried to find useful resources to support her work and was always thinking about how to deliver better lessons for her students. She was at a loss when she moved to the Kumanoe junior high school; however, she found the support necessary to study natural resources in the region in the natural history museum. She learned how to use the museum exhibitions and staff members as resources. This experience shows the collaborative relationship that was formed from mutual support between the schoolteacher and museum staff members and/or education researchers. Furthermore, this collaborative project succeeded without professional problem-solving actions on the part of a teacher. This experience is a typical example of informal learning on the part of a teacher, who can choose what and how to learn in a museum freely, based on her genuine form of professional problem solving.

WHY IS IT NECESSARY TO DISCUSS INFORMAL LEARNING IN SCIENCE EDUCATION?

Layton (1973) suggested a dilemma in science education:

> We turn now to examine a quite basic objection to the inclusion of science as a central component of general education, the contention that it fails to contribute to the social and moral development of learners (p. 179). … As long as education was conceived as a process of initiation into a stable and enduring value system, the contribution of science to social and moral development of learners was a restricted one. (p. 181)

His dilemma is that we tend to miss learners' social or moral development in science education. The demand of creating a "moral framework" has been discussed over the decades. In particular, from the perspective of environmental science such as climate change (Stoll-Kleemann, O'Riordan, & Burns, 2003), citizen participation is a most challenging issue, but is necessary in order to change or inform environmental policies (Gough et al., 2003). The Advisory Council on Fostering Science Literacy at the National Museum of Nature and Science, Tokyo (2010) developed a framework to foster science literacy at science museums that included the goals of "cultivation of sensitivity" and "development of the ability to properly respond to circumstances in society." These points of view suggest that the support of informal learning should extend from individual knowledge and skills, to issues of morality.

In considering informal learning, it is important to recognise that learning is occurring all the time and everywhere for everyone. This overview of science learning in informal contexts teaches us what we, science education researchers and practitioners, should do for whom, and how. Ogawa (2013), on the other hand, suggested that we should reconsider our belief that 'all' people have access to targeted science-oriented activities, since there are people who have no interest in the activities. According to his classification of science communication activities (p. 13), there is the 'indifferent public', who is not involved in both driving and target actors within science education efforts. Even if it is a fact, however, it is also true that it is difficult to identify what, how, and why people take part in informal learning activities. Since informal learning is occurring all the time and everywhere for everyone, we should continually observe and conduct research on the phenomenon.

ACKNOWLEDGEMENT

A part of this study is supported by the Grant-in-Aid for Scientific Research (C) No. 15K00998 from the Ministry of Education, Culture, Sports and Technology in Japan.

FURTHER READING

[Books & Handbooks]

Cheng, D., Claessens, M., Gascoigne, T., Metcalfe, J., Schiele, B., & Shi, S. (Eds.). (2008). *Communicating science in social contexts: New models, new practices*. New York, NY: Springer.
Falk, J. H., & Dirking, L. D. (2000/1948). *Learning from museums: Visitor experiences and the making of meaning*. Walnut Creek, CA: Alta Mira Press.
Falk, J. H., Dirking, L. D., & Foutz, S. (Eds.). (2007). *In principle, in practice: Museums as learning institutions*. Lanham, MD: Altamira Press.
Falk, J. H. (Ed.). (2001). *Free-choice science education how we learn science outside of school*. New York, NY: Teachers College Press.
Fenichel, M., & Schweingruber, H. A. (2010). *Surrounded by science: Learning science in informal environments, national research council*. Washington, DC: National Academies Press.
Fraser, B. J., Tobin, K. G., & MacRobbie, C. J. (Eds.). (2012). *Second international handbook of science education*. Netherlands: Springer.
Gilbert, J. K., & Stocklemayer, S. (Eds.). (2013). *Communication and engagement with science and technology*. New York, NY: Routledge.
Kasemire, B., Jäger, J. C., & Gardner, M. T. (Eds.). *Public participation in sustainability science: A handbook* (pp. 37–61). Cambridge: Cambridge University Press.
Lederman, N. G., & Abell, S. K. (Eds.). (2014). *Handbook of research on science education Volume II*. New York, NY: Routledge.
National Research Council. (2009). *Learning science in informal environments: People, places, and pursuits*. Washington, DC: The National Academic Press.
Patrick, P. G., & Tunnicliffe, S. D. (2013). *Zoo talk*. Dordrecht, The Netherlands: Springer.

[Journals]

Public Understanding of Science, Sage Journals.
Science Communication, Sage Journals.
Research in Science Education, Springer.
International Journal of Science Education Part B, Taylor & Francis.

REFERENCES

Advisory Council on Fostering Science Literacy National Museum of Nature and Science, Tokyo. (2010). *Development of a continuous education; Program framework to foster science literacy: For development of the programs focusing on each generation: Summary of the final report*. Tokyo: Social Education in Japan. Retrieved from http://www.mext.go.jp/component/a_menu/education/detail/__icsFiles/afieldfile/2010/06/01/1285289_1.pdf
Braund, M., & Reiss, M. (Eds.). (2004). *Learning science outside the classroom*. London: RoutledgeFalmer.
Bultitude, K., & Sardo, A. M. (2012). Leisure and pleasure: Science events in unusual locations. *International Journal of Science Education, 34*(18), 2775–2795.
Burns, T. W., O'Connor, D. J., & Stoclemayer, S. M. (2003). Science communication: A contemporary definition. *Public Understanding of Science, 12*(2), 183–202.
Department for Education. (2013). *Science programmes of study: Key stages 1 and 2, National Curriculum in England*. London: DfE. Retrieved from https://www.gov.uk/government/uploads/system/uploads/attachment_data/file/425618/PRIMARY_national_curriculum_-_Science.pdf
Drori, G. S. (2004). Science education and economic development. In J. Gilbert (Ed.), *RoutledgeFalmer reader in science education* (pp. 22–38). London: RoutledgeFalmer. (Reprinted from Studies in Science Education, *35*, 27–58, 2000)

Falk, J. H. (Ed.). (2001). *Free-choice science education how we learn science outside of school.* New York, NY: Teachers College Press.
Falk, J. H., Dirking, L. D., & Foutz, S. (Eds.). (2007). *In principle, in practice: Museums as learning institutions.* Lanham, MD: Altamira Press.
Fenichel, M., & Schweingruber, H. A. (2010). *Surrounded by science: Learning science in informal environments.* Washington, DC: National Academies Press, National Research Council.
Gilbert, J. K., & Stocklemayer, S. (Eds.). (2013). *Communication and engagement with science and technology.* New York, NY: Routledge.
Gough, C., Darier, É., Marchi, B. D., Funtowicz, S., Grove-White, R., Pereira, Â. G., Shackley, S., & Wynne, B. (2003). Contexts of citizen participation. In B. Kasemire, C. C. Jäger, & M. T. Gardner (Eds.), *Public participation in sustainability science: A handbook* (pp. 37–61). Cambridge: Cambridge University Press.
Kelley, B. (2007). *Human capital how what you know shapes your life.* Paris: OECD Publications.
Lederman, N. G., & Abell, S. K. (Eds.). (2014). *Handbook of research on science education* (Vol. II). New York, NY: Routledge.
Ministry of Education, Culture, Sports, Science and Technology. (n.d). *Social education in Japan.* Retrieved from http://www.mext.go.jp/component/a_menu/education/detail/__icsFiles/afieldfile/2010/06/01/1285289_1.pdf
Miyake, S. (2002). Collaboration between museums and schools: Case Study of the Museum of Nature and Human Activities, Hyogo. *Research Report of the Japan Society for Science Education, 17*(2), 1–6. [in Japanese]
Miyake, S., & Satooka, A. (2013). 'Rika' of collaboration with the region: With a double mind of a science teacher and a museum staff. *Science Education Monthly, 62*(12). (Society of Japan Science Teaching. [in Japanese])
Miyake, S., Yamada, C., & Nogami, T. (2011). A case study of a veteran elementary teacher's competence by the practical use of a botanical garden for a school science program. *Journal of Research in Science Education, 52*(1), 143–157. (Japan Society for Science Education. [in Japanese])
Nakayama, H., Yamaguchi, E., Satooka, Y., Kushima, Matsuda, K., & Yamamoto, Y. (2006). A case study of in-service training a science teacher as a science communicator at a museum. *Journal of Science Education in Japan, 30*(5), 316–331. [in Japanese]
NGSS Lead States. (2013). Science, technology, society, and the environment. *Next Generation Science Standards for States, by States, 2*, 108–112. (Washington, DC: National Academic Press)
Osborne, J., & Dillon, J. (2007). Research on learning in informal contexts: Advancing the field? *International Journal of Science Education, 29*(12), 1441–1445.
Organisation for Economic Co-operation and Development. (n.d.). *Recognition of non-formal and informal learning: Home.* Retrieved April 5, 2015, from http://www.oecd.org/edu/skills-beyond-school/recognitionofnon- for malandinformallearning-home.htm
Ogawa, M. (2013). Towards a 'Design Approach' to science communication. In J. K. Gilbert & S. Stocklemayer (Eds.), *Communication and engagement with science and technology* (pp. 3–18). New York, NY: Routledge.
Patrick, P., Mathews, C., & Tunnicliffe, S. D. (2013). Using a field trip inventory to determine if listening to elementary school students' conversations, while on a zoo field trip, enhances preservice teachers' abilities to plan zoo field trips. *International Journal of Science Education, 35*(15), 2645–2669.
Rennie, L., & Stocklmayer, S. M. (2003). The communication of science and technology: Past, present and future agendas. *International Journal of Science Education, 25*(6), 759–773.
Rennie, L. (2014). Learning science outside of school. In N. G. Lederman & S. K. Abell (Eds.), *Handbook of research on science education* (Vol. II, pp. 120–144). New York, NY: Routledge.
Roger, A. (2004). *Non-formal education: Flexible schooling or participatory education?* Hong Kong, China: Comparative Education Research Centre the University of Hong Kong & Kluwer Academic Publishers.
Sanders, D. L. (2007). Making public the private life of plants: The contribution of informal learning environments. *International Journal of Science Education, 29*(10), 1209–1228.

Satooka, A., Nakayama, H., Yamaguchi, E., Ito, Y., Kushima, N., Sueyoshi, T., & Nagai, H. (2004). Lower secondary school science learning on tidal flats supported by Miyazaki prefectural museum of nature and history. *Journal of Science Education in Japan, 28*(2), 122–131. (Japan Society for Science Education. [in Japanese])

Stoll-Kleemann, S., O'Riordan, T., & Burns,T. R. (2003). Linking the citizen to governance for sustainable climate futures. In B. Kasemire, J. Jäger, C. C. Jaeger, & M. T. Gardner (Eds.), *Public participation in sustainability science a handbook* (pp. 239–248). Cambridge: Cambridge University Press.

Takada, K. (2010). Expectation and Issues to the application of new curator training course, museum studies. *Japanese Association of Museums, 45*(12), 18–20. [in Japanese]

United Nations Education, Scientific and Cultural Organization. (2009). *Review of contexts and structure for education for sustainable development, 2009*. Paris: UNESCO.

United Nations Education, Scientific and Cultural Organization. (2012). *UNESCO guidelines for recognition, validation and accreditation of outcomes of non-formal and informal learning*. Germany: UNESCO Institute for Lifelong Learning.

United Nations Education, Scientific and Cultural Organization. (2014). *Shaping the future we want UN decade of education for sustainable development (2005–2014) (Final Report)*. Paris: UNESCO.

Wellcome Trust. (2012). *Review of informal science learning*. Retrieved from http://www.wellcome.ac.uk/stellent/groups/corporatesite/@msh_peda/documents/web_document/wtp040862.pdf

Wittrock, M. C. (1994). Generative science teaching. In P. Fensiam, R. Gunston, & R. White (Eds.), *The content of science a constructivism approach to its teaching and learning* (pp. 29–38). London: RoutledgeFalmer.

Yager, R. E. (2008). Using the national science education standards for improving science education in nonschool settings. In R. E. Yager & J. Falk (Eds.), *Exemplary science in informal education settings* (pp. ix–xv). Arlington, VA: National Science Teachers Association Press.

Yager, R. E., & Falk, J. (Eds.). (2008). *Exemplary science in informal education settings*. Arlington, VA: National Science Teachers Association Press.

AMY SEAKINS AND MARIE HOBSON

32. PUBLIC UNDERSTANDING OF SCIENCE

INTRODUCTION

The phrase 'public understanding of science' has developed a dual meaning, as both public attitudes and understanding of scientific concepts and developments, and also the field of research and pedagogical approaches relating to those attitudes and understandings. In this chapter we address both. Firstly, we discuss the engagement and science communication approaches within the public understanding of science field, providing an overview of the key developments and turning points leading to how public engagement is carried out today. We highlight key scientific controversies with importance to public understanding of science from the UK and US. Secondly, we discuss the public understanding of scientific content relating to a particular area of science: climate science. Through this focus area we illustrate how public understanding of science, and attitudes towards science, can be influenced by misconceptions, how levels of understanding and attitudes might be studied, and describe the effect initiatives such as museum exhibitions might have on developing public understanding and awareness of science.

MAIN DEVELOPMENTS IN PUBLIC UNDERSTANDING OF SCIENCE

Historical Context

A report by the Royal Society in 1985 entitled *Public Understanding of Science*, also known as the Bodmer Report, set the tone for the first era of science communication. It warned of a public deficit in knowledge about science and recommended that a better educated citizenship would develop more positive and trusting views about science and scientists. The science communication that followed took a transmission approach – scientific experts were called upon to lecture and present to publics, transmitting their scientific knowledge and 'filling up' the public audience, who were thought of as 'empty vessels'.

Between 1985 and 2000 the deficit model largely dominated the public understanding of science field. However, there were some groups who began to question the effectiveness of such a direct and one-way approach to science communication, and ponder whether the public really had nothing to contribute to a conversation about science (e.g. Wynne, 1992).

K. S. Taber & B. Akpan (Eds.), Science Education, 443–452.

The 1990s and early 2000s saw a difficult period for science in terms of public levels of trust in scientists and scientific research. Controversial socio-scientific issues such as genetic modification (see Box 1), Bovine spongiform encephalopathy (or BSE, known as mad cow disease) in the UK, and stem cell research in the US (see Box 2) shocked the public, and media portrayal led to feelings of mistrust in science and a breakdown in relationships between science and the public.

Box 1. Public understanding of scientific controversies: Genetically modified foods in the UK

When genetically modified (GM) organisms such as Montana's GM soya first appeared in the European food market in the mid-1990s they were met with great unease from the UK population. Amidst the BSE controversy, the UK public were increasingly less positive towards scientific research and less trusting of scientists (Gaskell et al., 2003).

Scientists working in the field hoped that through communicating that genetic modification was based on 'sound science' they would win over public trust and support. However, a decline in trust towards scientists meant that this assurance was not enough to change public attitudes and understanding towards science. Research from this period remains relevant for the communication of risk, in that only providing expert information may not be enough to change public perception (Pidgeon et al., 2005).

A number of measures were taken in order to develop public understanding of science relevant to GM foods. Firstly a move towards openness and transparency was called for with the establishment of the Food Standards Agency in 2000. Secondly, the public were involved in policy discussions and the GM Nation debate launched in 2003 (Gaskell et al., 2003). The debates included open meetings, focus groups, online information, and public discussion events to explore attitudes towards genetically modified organisms and the roots of these beliefs.

Following the debate the government gave a tentative go-ahead to the growing and selling of GM crops in the UK but not to the application of GM science in animals. A recent international review of the public understanding of and attitudes towards GM crops indicated that since 2008 publics were marginally less concerned about ethical issues relating to GM science but there was no change in their intention to buy products relating to GM crops (Frewer et al., 2013).

An influential report, *Science and Society*, was issued by the UK House of Lords Select Committee on Science and Technology in 2000 in response to the problems in the public understanding of science field, which the report described as a 'crisis of confidence'. The report recognised that merely knowing more about science would not necessarily ensure that publics had positive attitudes towards science, and advised that a different approach would be needed to improve relationships between science and society. This report called for a more two-way conversation to emerge, for discussion to be facilitated between publics and scientists: a 'new mood for dialogue'.

Box 2. Public understanding of scientific controversies: Embryonic stem cell research in US

As the science of embryonic stem cell research developed at the turn of the 21st Century, questions were asked about government funding for this field and public unease grew about the applications and ethics of the studies. Attempts to alleviate growing tension in public attitudes to stem cell research again took the approach that "greater scientific understanding, it is assumed, will ensure that the public makes 'proper' judgements about science, that is assessments in line with those of scientists" (Nisbet, 2005, p. 90).

The controversy surrounding stem cell research in the US highlights the interplay between personal, emotional and religious beliefs, alongside knowledge and awareness about specific scientific issues (Nisbet, 2005). A survey of over 1000 adults over three years in the US indicated that public attitudes towards embryonic stem cell research were predominantly shaped by values including the degree to which a person was religious and their political ideology (Ho, Brossard, & Scheuffle, 2008). The same survey found that other factors contributed to attitudes towards stem cell research to a lesser extent, including attention to science news in the media.

The reactions of the US public around embryonic stem cell research illustrate the complexity of the formation of public attitudes towards science and understanding of scientific research. People's values, beliefs and behaviours are shaped by many influences, which are likely to differ between controversial topics.

The *Science and Society* report called for transparency in the way science was conducted and better communication around risk – a backlash from the scientific controversies of the previous decade. This paper marked the shift in the science communication field from the public understanding of science, to public engagement with science, a term which reflected the public's more active role in the interaction between science and society. This new approach and strategy for science communication led Miller (2001) to suggest that public understanding of science was 'at a crossroads'. Researchers felt that there needed to be a large culture shift in terms of the way science and society interacted, or else the situation could worsen.

The definition of public engagement has changed somewhat since it first became widely used following the UK House of Lords report (2000) and is now used broadly to describe any contact between the public and scientists (Dillon, 2011).

For example, public engagement might be defined as 'A multidirectional dialogue among people that allows all the participants to learn' (McCallie et al., 2009, p. 12).

Dialogue events were originally designed to inform policy and engage publics in decision-making, however following the call for two-way public engagement with science, many museums and science centres responded by developing discussion events and these dialogue activities became valued as a form of public engagement in their own right. Research into the impacts of dialogue events has indicated that they can be promising places for learning scientific content and can also influence attitudes, for example attitudes of publics and scientists towards biotechnology converged following a dialogue event and publics became more positive towards scientists (Zorn et al., 2010).

Later in the 2000s, with public engagement and dialogue still the aims of many science communication practitioners, researchers were calling for an even more active role of the public in science. Not only should non-scientists be active in the dialogue with scientists, but they should also be involved in scientific research, decision-making and policy. Wilsdon and Willis (2004) suggested that public engagement with science moved 'upstream'; that is that public consultation and input takes place before science policy and research is decided, and influences its direction and priorities.

An increased involvement in scientific research on the part of the public is reflected in the growth of citizen science. Popular inititally in the US but growing elsewhere more recently, citizen science involves non-scientists participating in activities which contribute to research alongside scientists. Examples of citizen science projects include the worldwide Galaxy Zoo, where citizen scientists are asked to classify galaxies, the Tree Health survey run by the Open Air Laboratories at the Natural History Museum London, UK, and the eBird survey of bird species run by the Cornell Lab of Ornithology and National Audubon Society, who started some of the first citizen science programmes.

Through participation in citizen science activities, non-scientists are actively participating in science – no longer restricted to only learning about it or discussing it as described above. There are a broad range of benefits of citizen science projects, perhaps explaining their popularity: scientists are able to collect more data, often over a wide geographical area; while volunteers have fun, can participate in real, authentic, scientific research, and often develop their scientific understanding and knowledge of the processes of research (Brossard, Lewenstein, & Bonney, 2005).

How much has Really Changed in the Public Understanding of Science?

The section above described the progression of the public understanding of science field, key turning points in the way science has been communicated and trends in terms of public knowledge about science and attitudes towards it. From this discussion it would seem that there has been much change in public relationships with science and the ways in which people engage with science. However, has anything really changed in public understanding of science?

In 2006, six years after the influential *Science and Society* report and its call for dialogue between scientists and publics, Brian Wynne (2006) reviewed the state of public understanding of science. Wynne described the situation as "hitting the notes, but missing the music" (p. 211), and said that, although intentions may be well informed and towards the direction of facilitating open dialogue between scientists and publics, the reality may be that power imbalances stay the same and communication follows the deficit model despite efforts otherwise.

It is not all bad news however. There has been a predominant shift in public attitudes towards science and scientists away from hostility and ignorance in terms of science, towards more positive, trusting and collaborative perceptions. Surveys of the UK population indicate that trust in scientists has increased in recent years, although trust in private industry and government scientists remains lower than for those working in other institutions (Ipsos MORI, 2014). Research with US adults also indicated that attitudes towards scientists were more positive in 2001 compared to 1983 (Losh, 2010). Whereas public trust in and understanding of science had reached a crisis point at the turn of the century, now it seems that science is facing a reaction of positive indifference from the public.

CURRENT PUBLIC UNDERSTANDING OF CLIMATE SCIENCE

How the Public's Understanding of Science Impacts their Attitudes Towards Controversial Scientific Issues

The impact of the public's lack of understanding of science can be seen clearly when exploring the topic of climate change. Confusions about scientific theories and terminology, combined with pockets of prior knowledge and belief systems, have led to a range of misconceptions and misunderstandings about this subject.

The public regard science as factual and scientists as being in search of the truth (Dillon & Hobson, 2013). Therefore, they are confused when there is a lack of consensus amongst scientists and when scientific statements change over time because they perceive these as lacking the certainty that they expect and want from science. This can result in the public simply dismissing certain scientific claims on the grounds that: scientists are not convinced, so why should they be (Shuckburgh, Robinson, & Pidgeon, 2012)?

It seems that scientists' lack of certainty in one aspect of climate science undermines the public's belief in climate change as a whole (CRED, 2009). For example, a lack of agreement over the exact nature and extent of climate change impacts causes the public to doubt whether the greenhouse effect is, in fact, occurring. The public assume that, if scientists have established that the greenhouse effect exists, they should be able to predict the consequences of it accurately. Words, such as 'may' and 'could', imply there is a lack of knowledge amongst scientists, rather than an absence of certainty (Shuckburgh et al., 2012).

This is compounded by a misunderstanding of scientific terminology. For example, when scientists refer to the greenhouse effect as a 'theory', non-scientists may interpret this term to mean one possible idea that scientists have posited, whereas scientists may in fact mean an accepted description of how greenhouse gases affect the planet based on the results of scientific experimentation and measurement.

In addition, some of the terminology scientists use to describe climate science concepts has been misinterpreted by the public. Interviews conducted with adult visitors to the Science Museum, London, revealed that the use of the term 'greenhouse effect' itself is misleading because it causes people to imagine that greenhouse gases congregate in a layer that acts like a roof over the Earth (Dillon & Hobson, 2013).

This is even more problematic when the public tries to incorporate the greenhouse effect into their existing mental model or conceptual understanding of climate change (CRED, 2009). According to Piaget, humans tend to reach a state of 'equilibrium' where they can make sense of the world around them by 'assimilating' new information into their existing mental models because it complies with what they already know (Piaget, 1952). When the new information does not comply with their prior knowledge, humans have to alter their mental models in order to 'accommodate' the different knowledge or else reject the new information. In the case of climate science, this has resulted in some members of the public misaligning the greenhouse effect with their prior knowledge about the hole in the ozone layer (Dillon & Hobson, 2013). This has led them to assume that the greenhouse gas 'layer' and the ozone layer are the same therefore a hole in the ozone layer is the same as the hole in the greenhouse gas layer or even that the hole in the ozone layer was a cause of climate change (CRED, 2009).

When mental models are combined with belief systems, this process of assimilation can result in 'confirmation bias' where the public misinterpret findings to comply with their existing incorrect mental models, rather than alter their conceptual schemas (CRED, 2009). Indeed, Shuckburgh, Robinson and Pidgeon (2012) found that, in their focus groups exploring the public's attitudes to climate science, "those with more firmly held views about climate change would tend to look for information that supported their own viewpoint" (p. 20). In relation to the ozone layer and the greenhouse effect, confirmation bias can lead climate change believers to think that the 'hole' allows more heat to reach the Earth causing more warming but climate change deniers to think that the 'hole' is enabling more hot air to escape and use it to counter the idea that climate change is occurring (CRED, 2009).

HOW CAN WE FIND OUT WHAT THE PUBLIC'S UNDERSTANDING OF A SCIENTIFIC TOPIC IS?

The public's understanding of climate science demonstrates that, before embarking on a science communication project, it is important to conduct research to establish the nature of the public's knowledge, understanding and beliefs in relation to the topic. This type of research was conducted successfully during the development of the *atmosphere…exploring climate science* gallery at the Science Museum, London which opened in 2010 (Dillon & Hobson, 2013). It involved:

- Desk based research
- In-depth interviews
- Focus groups

Desk based research. The Audience Research and Advocacy team at the Science Museum reviewed nationwide surveys of public opinion (see Table 1), academic literature and evaluations of climate science related exhibitions (Dillon & Hobson, 2013). This provided a useful starting point for understanding broad trends amongst a population, putting more localised findings into context and identifying areas requiring further research by the team.

Organizations in the UK, Europe and the US have conducted a range of nationwide surveys of the public in relation to both science as a whole, as well as various sub topics, such as climate change:

Table 1. Nationwide surveys on public understanding of climate change

Region	*Conducted by*	*Reports*	*Description*
UK	Ipsos MORI and the British Science Association for the Economic and Social Research Council and Department for Business, Innovation and Skills (2011 and 2014).	Public Attitudes to Science	A survey of the UK population's attitudes towards both science and scientists in general as well as three key topics, such as Robots or Big Data, which are different for each survey.
Europe	European Commission	Eurobarometer	Standard Reports: Twice yearly surveys of the public in the member states to track opinions on areas, such as health, defence, culture etc. Special Reports: More in depth explorations of themes, such as climate change (Special Barometer, 409).
USA	National Opinion Research Center at the University of Chicago	General Social Survey	The public's attitudes towards science and technology were surveyed by the National Center for Science and Engineering Statistics. Since 2010, these questions have been incorporated into the USA's General Social Survey, conducted biennially.

In-depth interviews. Thirty adult Science Museum visitors were interviewed about their awareness and interpretation of a range of words and theories relating to climate change, such as 'fossil fuels' and the 'carbon cycle' (Dillon & Hobson, 2013). Their answers enabled the researchers to compile a glossary of terms for those developing the gallery to refer to when determining the content of the exhibits and writing text. This included the scientific definition, the public's definition and a traffic light system to demonstrate which terms were well understood and could be used freely without the need for explanation (green lights), those which had some areas of misunderstanding or gaps in knowledge (amber) and those which were unheard of or vastly misunderstood so should always be explained when used (red).

Focus groups. Interviews were supplemented by focus groups with specific audiences. These allowed for more in-depth discussions around areas of understanding and confusion to occur to help explain the findings and trends identified through the other forms of front end research.

The front-end research for the gallery culminated in the team developing a diagram of a typical Science Museum visitor's mental model of climate change (Dillon & Hobson, 2013). This was used by the gallery team to decide which areas of climate science to focus on in the gallery in order to fill gaps in visitors' knowledge and address their misconceptions. Summative evaluation of the gallery showed that this approach was successful as the gallery deepened visitors' understanding of the subject; "in comparison to pre-visit interviews, visitors can talk more in post-visit evaluation about how, where and on what scale impacts might occur, the role of greenhouse gases and the complexity of the Earth's system" (Clipson & Hobson, 2011, p. 3).

SUMMARY

This chapter has reviewed the history and trends in the public understanding of science field, looking back at key turning points and progress made in the approaches taken to communicate science to and with the public. Public understanding of a particular area of science was then focused on, with a discussion of different methods of researching public attitudes and knowledge about climate change. The chapter illustrates the complexity of public attitudes towards science and the variety of approaches to bridging science and publics.

To conclude, we pose questions looking ahead to what might come for the field of public understanding of science. What will new formats and opportunities offer the relationship between science and the public, such as social media, digital technologies and crowd-sourcing software? What new challenges will these bring for the ways science and public audiences interact? Finally, what might be the next key areas of scientific controversy and how might they influence the field of public understanding of science?

FURTHER READING

Bauer, M. W., Allum, N., & Miller, S. (2007). What can we learn from 25 years of PUS survey research? Liberating and expanding the agenda. *Public Understanding of Science, 16*(1), 79–95.

Dillon, J. (2012). Science, the environment and education beyond the classroom. In B. J. Fraser, K. Tobin, & C. J. McRobbie (Eds.), *Second international handbook of science education* (Vol. 24, pp. 1081–1095). Netherlands: Springer.

Rowe, G., & Frewer, L. J. (2005). A typology of public engagement mechanisms. *Science, Technology & Human Values, 30*(2), 251–290.

Silvertown, J. (2009). A new dawn for citizen science. *Trends in Ecology & Evolution, 24*(9), 467–471.

REFERENCES

Bodmer, W. (1985). *The public understanding of science*. London: Royal Society.

Brossard, D., Lewenstein, B., & Bonney, R. (2005). Scientific knowledge and attitude change: the impact of a citizen science project. *International Journal of Science Education, 27*(9), 1099–1121.

Center for Research on Environmental Decisions. (2009). *The psychology of climate change communication: A guide for scientists, journalists, educators, political aides, and the interested public*. New York, NY: CRED.

Clipson, H., & Hobson, M. (2011). *Atmosphere ... exploring climate science: Gallery Summative Evaluation Report*. Science Museum (unpublished report)

Dillon, J. (2011). Science communication: A UK perspective. *International Journal of Science Education, Part B: Communication and Public Engagement, 1*(1), 5–8.

Dillon, J., & Hobson, M. (2013). Communicating global climate change: Issues and dilemmas. In S. Gilbert & J. Stocklmayer (Ed.), *Communication and engagement with science and technology: Issues and dilemmas. A reader in science communication* (pp. 215–228). New York, NY & London: Routledge.

European Commission. (2014). *Special Eurobarometer 409: Climate change*. Retrieved from http://ec.europa.eu/public_opinion/archives/ebs/ebs_409_en.pdf

Frewer, L. J., van der Lans, I. A., Fischer, A. R. H., Reinders, M. J., Menozzi, D., Zhang, X., van den Berg, I., & Zimmermann, K. L. (2013). Public perceptions of agri-food applications of genetic modification: A systematic review and meta-analysis. *Trends in Food Science & Technology, 30*(2), 142–152.

Gaskell, G., Allum, N., Bauer, M. W., Jackson, J., Howard, S., & Lindsey, N. (2003). *Ambivalent GM nation? Public attitudes to biotechnology in the UK, 1991–2002*. Life Sciences in European Society Report: London School of Economics and Political Science.

Ho, S., Brossard, D., & Scheufele, D. (2008). Effects of value predispositions, mass media use, and knowledge on public attitudes toward embryonic stem cell research. *International Journal of Public Opinion Research, 20*(2), 171–192.

Ipsos MORI. (2014). *Public attitudes to science survey 2014: Main report*. London, UK: Department for Business Innovation and Skills. Retrieved November 1, 2014, from https://www.ipsos-mori.com/researchpublications/researcharchive/3357/Public-Attitudes-to-Science-2014.aspx

Losh, S. C. (2010). Stereotypes about scientists over time among US adults: 1983 and 2001. *Public Understanding of Science, 19*(3), 372–382.

McCallie, E., Bell, L., Lohwater, T., Falk, J. H., Lehr, J., Lewenstein, B., Needham, C, & Wiehe, B. (2009). *Many experts, many audiences: Public engagement with science and informal science education*. Washington, DC: Center for Advancement of Informal Science Education.

Miller, S. (2001). Public understanding of science at the crossroads. *Public Understanding of Science*, (10), 115–120.

Nisbet, M. (2005). The competition for worldviews: Values, information, and public support for stem cell research. *International Journal of Public Opinion Research, 17*(1), 90–112.

Piaget, J. (1952). *The origins of intelligence in children*. New York, NY: International Universities Press.

Pidgeon, N., Poortinga, W., Rowe, G., Jones, T., Walls, J., & Riordan, T. (2005). Using surveys in public participation processes for risk decision making: The case of the 2003 British GM Nation? Public Debate. *Risk Analysis, 25*(2), 467–479.

Shuckburgh, E., Robison, R., & Pidgeon, N. (2012). *Climate science, the public and the news media.* Swindon, UK: Living with Environmental Change.

UK House of Lords Select Committee on Science and Technology. (2000). *Science and society*. London, UK: House of Lords Science and Technology Committee Publications.

Wilsdon, J., & Willis, R. (2004). *See-through science: Why public engagement needs to move upstream.* London, UK: Demos.

Wynne, B. (1992). Misunderstood misunderstanding: Social identities and public uptake of science. *Public Understanding of Science, 1*, 281–304.

Wynne, B. (2006). Public engagement as means of restoring trust in science? Hitting the notes, but missing the music. *Community Genetics, 9*(3), 211–220.

Zorn, T. E., Roper, J., Weaver, C. K., & Rigby, C. (2010). Influence in science dialogue: Individual attitude changes as a result of dialogue between laypersons and scientists. *Public Understanding of Science, 21*(7), 848–864.

SECTION VII

INCLUSIVE SCIENCE EDUCATION

JULIE A. BIANCHINI

33. EQUITY IN SCIENCE EDUCATION

This chapter provides a brief introduction to the research literature on issues of equity in science education. The construct of *equity* encompasses both providing all students adequate opportunities to learn science and expecting all students to meet high academic standards (National Research Council [NRC], 2012). It is rooted in the conviction – which is supported by psychological and anthropological research on learning – that all students can and should engage in the sense-making practices of science. I begin this chapter by examining reasons for advancing the equity agenda in science education – reasons why it is important to address the profound differences among diverse groups of students in their opportunities to learn and to achieve in science. I then move from reasons for attending to equity concerns to the dimensions of science education that promote inequities. These inequities include: (1) the marginalization of diverse student groups in the teaching and learning of science; (2) the failure to implement curriculum materials, instructional strategies, and assessments that build from the interests and experiences of all students; and (3) the uneven distribution of material, human, and social resources, including access to schools, highly qualified teachers, and collective decision-making. I end by pointing to additional texts readers can examine to gain deeper insight into equity challenges in science education and effective ways to address them.

WHY ADVANCE THE EQUITY AGENDA IN SCIENCE EDUCATION?

Few would argue that inequities in science education exist. Students who come from backgrounds and possess *funds of knowledge* (Moll, Amanti, Neff, & Gonzalez, 1992) that are not institutionally privileged have been traditionally marginalized from participation and achievement in science (NGSS Lead States, 2013). Researchers and policymakers have offered several different arguments for why attending to students from underserved groups in science education is needed. Two such arguments are examined here.

Some science educators and policymakers call for attending to equity issues so as to produce a larger scientifically and technologically literate citizenry (National Academy of Sciences, National Academy of Engineering, & Institute of Medicine [NAS, NAE, & IOM], 2011; Organization for Economic Co-operation and Development [OECD], 2014). They note that science achievement gaps persist across diverse groups in a single nation, such as the United States

K. S. Taber & B. Akpan (Eds.), Science Education, 455–464.

(National Center for Education Statistics, 2012), and across diverse nations around the globe, such as between China and Peru (OECD, 2014). The *STEM (Science, Technology, Engineering and Mathematics) pipeline* also remains problematic, leaking students from elementary school, through university, to careers (NAS, NAE, & IOM, 2011). As such, traditionally underserved students continue to be disadvantaged in acquiring high-prestige careers in STEM fields, participating in an increasingly science and technology savvy world, and understanding and attempting to address pressing science-technology-society issues. Widespread science and technology literacy is considered "critical to [both] the economic well-being of [a] nation and the personal well-being of its citizens in the 21st century" (Lee & Buxton, 2010, p. 9).

Other science educators concerned with equity underscore the need to re-envision what counts as effective science education so that all students can participate in and learn to use science in ways that they themselves find meaningful (Calabrese Barton, Tan, & Rivet, 2008; Carlone, Haun-Frank, & Webb, 2011). For example, Carlone et al. (2011) argued that the equity problem in science education must be addressed so that all students, including those traditionally underserved, can affiliate with science – so that all can view themselves as science people and actively engage in science learning. More specifically, *classroom culture*, the everyday classroom practices that imply particular meanings of science, and *normative identity*, what teachers and peers recognize as being scientific, must be broadened to include the experiences, worldviews, funds of knowledge, and interests of students from diverse backgrounds – to transform the science they are asked to learn, expand the ways they can productively engage in classroom work, and strengthen the expectations for all to competently participate. Along similar lines, Calabrese Barton et al. (2008) underscored the importance of creating *hybrid learning spaces* to facilitate all students' learning of science: Hybrid learning spaces are new, third spaces where underserved students merge the priorities and practices of their school science space with their home space to broaden what counts as meaningful participation and learning. Rivera Maulucci (2012) added that teachers who adopt a *social justice* framework can teach science as a tool to promote equity and empowerment for all – as a social activity underserved students can use to achieve personal and/or community transformation.

DIMENSIONS OF SCIENCE EDUCATION IN NEED OF REFORM

Ensuring equity in science education – providing all students adequate opportunities to learn science and supporting all students in meeting high academic goals – is a difficult charge. There are a dizzying number of factors tied to schooling that are in need of reform. Carlone et al. (2011) listed the following sources of inequities: teachers' limited knowledge, skills, and beliefs; policies that constrain teachers' agency and meaningful instructional time; historically enduring meanings of schooling; students' resistance to more demanding roles; the science curriculum's relevance; and misunderstandings about the nature of science. Lee and Buxton (2010) argued that addressing inequitable learning opportunities for underserved students should be organized around three themes: valuing the knowledge and experiences all students bring from their homes and communities, articulating students' funds of knowledge with disciplinary knowledge, and offering sufficient educational resources to support student learning. These three themes identified by Lee and Buxton are discussed in detail below.

The Marginalization of Diverse Student Groups

A growing number of studies in science education have focused on the marginalization of diverse student groups in the processes of teaching and learning science. Diverse student groups include girls, students from non-dominant racial and ethnic groups, economically disadvantaged students, students with disabilities, students with limited proficiency in the dominant language, students in alternative education programs, and students who are gifted and talented (NGSS Lead States, 2013). Investigating the intersection of student diversity and science education is complicated. A given student belongs to more than one group and each group includes diverse individuals; students cannot be lumped into insular categories and students within a given category cannot be treated as all the same (Nieto, 1999). Further, every underrepresented group experiences both unique and common issues. More specifically, despite differences in perceptions and contexts, common issues include marginalization, empowerment, curriculum relevancy, and the commitment of educators (Tal, 2012). Because the corpus of research on diverse student groups' experiences in science education is large, I summarize findings from one such group below: girls.

Since the late 1960s, science education researchers have spent concerted effort investigating gender issues: how and why girls and women continue to be underrepresented and underserved in a number of science and engineering disciplines (Brotman & Moore, 2008). These researchers emphasize that *gender* is distinct from *sex*: Gender refers to the cultural construction of what it means to be female or male, while sex refers to the biological features that make one female or male. (There is little existing research in science education that includes discussion of students who are *LGBTQ:* lesbian, gay, bisexual, transgender, and/or queer.) Early research on gender in science education tended to view girls and women through a deficit lens. In the 1990s, however, researchers moved from trying to "fix" girls to attempting to transform the science they were expected to learn. In the 1990s, researchers also began to move from investigating gender in isolation to looking at the complex interaction of gender with race, ethnicity, socioeconomic status, sexual orientation, language, religion, and/or sociocultural identity. Scantlebury (2012) cautioned that research focused solely on gender could be limited: Such studies sometimes found greater variation within girls or boys than between them and/or produced misleading results. Calabrese Barton et al. (2008) underscored the importance of conducting gender research that moved "beyond girls as a homogeneous population and beyond achievement as the only marker of success" (p. 72).

Brotman and Moore (2008) organized the approximately 40 years of gender research in science education along four dimensions: (1) *equity and access*, which documents differences between boys and girls in attitudes, achievement, and participation, as well as in their classroom experiences; (2) *curriculum and instruction*, which attempts to transform science teaching and learning to include girls' interests and experiences; (3) *the nature and culture of science itself*, which works to reconstruct how science is portrayed and viewed in school and in society; and (4) *student identity*, which investigates the ways both girls' views of themselves and others' views of girls in relation to science shape their engagement with and learning of science. For example, from their review of equity and access research, Brotman and Moore found that, irrespective of age or context, girls preferred the biological sciences and boys preferred the physical sciences. As a second example, from their review of research on curriculum and instruction, they found that teachers were largely uninformed about gender-inclusive approaches to teaching and did not necessarily see the need to revise their instruction along gender friendly lines.

Curriculum, Instruction, and Assessment

To help eliminate gaps in science learning outcomes for girls and for other diverse groups of students, science educators have called for classroom practices to be made more inviting and inclusive – for the teaching and learning of science to explicitly build on the funds of knowledge diverse students bring to the science classroom. Many of the current recommendations for transforming science curriculum, instruction, and assessment are squarely rooted in a view of science as a cultural activity. There are at least four ways science educators interpret the construct of *science as culture*: (1) as a multicultural endeavor that has historically appropriated the intellectual and physical tools of diverse cultures (Harding, 1998); (2) as a distinct culture, with its own language, norms, values, knowledge, and practices (Aikenhead & Jegede, 1999); (3) as composed of numerous disciplines and sub disciplines, each with overlapping and unique cultural practices (Stanley & Brickhouse, 2001); and (4) as separate from the culture of school science, the science that is traditionally taught in school classrooms (Brown, Collins, & Duguid, 1989). Some researchers have focused on the ways the culture of science and/or school science is separate from, foreign to, or discontinuous with students' home cultures. Other researchers have viewed the social, cognitive, and linguistic knowledge students bring to the classroom as resources to use to develop their understanding of science concepts and practices. (For discussion of both approaches, see Lee & Buxton, 2010.) In the remainder of this subsection, I emphasize the latter view – a view of science and students' cultures as mutually reinforcing and synergistic.

One recommendation routinely advanced by advocates of equitable science education is to broaden the definition of science taught in schools. Including *indigenous knowledge systems*, knowledge of how the world works grounded in a particular place, as part of science makes clear that science sense-making practices cut across time and cultures (Snively & Corsiglia, 2001). For example, Mpofu, Otulaja, and Mushayikwa (2014) encouraged the integration of traditional African plant healing (TPH) practices – traditional healers' use of medicinal plants supported by spiritualism to manage health – into modern science classrooms and provided a framework to do so. Including engineering practices and concepts as part of science also serves to broaden its definition (NGSS Lead States, 2013). Engineering can be used to foreground the historical contributions of other cultures, such as the Chinese's invention of the mechanical clock and of segmental arch structures, to offer opportunities for innovation and creativity, and to make science relevant to students' lives and future.

A second recommendation for addressing inequities in curriculum, instruction, and assessment is to implement culturally responsive approaches to the teaching and learning of science. As one example, Lee and colleagues (Lee & Buxton, 2010) developed a framework of *instructional congruence* to help teachers connect expectations for classroom interaction, the demands of science disciplines, and students' linguistic and cultural experiences. When using this framework, science teachers provide explicit instruction about the rules and norms for classroom behavior, discourse, and academic achievement. They also provide explicit instruction about science content, scaffolding students' participation in central science practices (e.g., questioning and argumentation) and students' movement from teacher-directed to student-initiated inquiry. As a second example, teachers are encouraged to implement place-based science education: to view students' homes and communities as central to students' learning, and thus, to connect science to students' *sense of place* (Chinn, 2006). By focusing on place, teachers and students can move beyond the confines of their classroom and school walls so as to make science more relevant and meaningful, as well as to facilitate the use of science as a tool for community transformation.

A third recommendation advanced by advocates of equity in science education is to pay greater attention to the linguistic demands of science – to the critical role language plays in science learning. The *academic language* of science differs from conversational language in multiple ways, from using discipline-specific vocabulary terms, to reading and writing complex texts, to engaging in evidence-based reasoning (Schleppegrell, 2004). Four of the eight science and engineering practices described in the recent *A Framework for K-12 Science Education* (NRC, 2012) and in the *Next Generation Science Standards* (NGSS Lead States, 2013) are explicitly identified as language-intensive (Quinn, Lee, & Valdés, 2012): developing and using models; constructing explanations (for science) and designing solutions (for engineering); engaging in argument from evidence; and obtaining, evaluating, and communicating information. For example, in the practice of arguing from evidence, students learn that "what counts [in science] as evidence is data and observations. Hence argumentation in science is not a purely verbal exercise. It is an exercise in the coordination of language and experience and thus another rich language learning opportunity" (pp. 4–5).

For many students, particularly for those learning science in a second or even third language, trying to understand science concepts and practices through the academic language of science is a challenging task. Science teachers must "purposefully enact opportunities for the development of language and literacy in and through teaching the core curricular content, understandings, and activities that teachers are responsible for (and, hopefully, excited about) teaching in the first place" (Bunch, 2013, p. 298). Brown (2006) found that underserved students experienced conflicts when attempting to engage in the discursive practices of science; students' own cultural and discursive identities required them to become both bicultural and bilingual in learning science. Brown and Ryoo (2008) demonstrated how students are better able to learn new science concepts by first using everyday language and then later transitioning to scientific terms. Further, Rosebery, Ogonowski, DiSchino and Warren (2010) made visible the ways students can productively use their home languages, such as Haitian Creole, to make sense of science concepts taught in English. They too underscored the centrality of talk in the learning of science.

Material, Human, and Social Resources

To eliminate gaps in science learning outcomes for diverse student groups, the transformation of classroom instruction is not enough. The uneven distribution of educational resources must also be addressed (NGSS Lead States, 2013; NRC, 2012). The academic success of underserved students depends heavily on the quality of the educational resources they are afforded. Yet, it is these very students who are the least likely to have adequate access. School resources to support student learning of science can be clustered into three categories: material, human, and social.

More specifically, material resources include time available for teaching, the professional development of teachers, instructional supports, curricular materials, equipment, supplies, and expenditures for school personnel (NGSS Lead States, 2013). As one example, in 2012, one in 10 children in developing regions of the globe – or 58 million children – did not even have access to education. Further, more than one in four children who began first grade dropped out before completing the last grade of their primary school (United Nations, 2014). As a second example, in the United States, time for science instruction has been virtually eliminated in low performing elementary schools. Because of policy and testing demands, these schools have pushed science aside in an attempt to develop basic literacy and numeracy (NGSS Lead States, 2013).

Human resources include teachers' knowledge and principals' leadership (NGSS Lead States, 2013). In Finland, for example, students' high science scores on international exams like the Programme for International Student Assessment, or PISA, are linked to highly educated and highly qualified teachers (Lavonen & Laaksonen, 2009). The country has made the education and professional development of its teachers a priority. Education authorities, policymakers, and parents also trust teachers and their professionalism: Teachers are considered experts in curriculum development, teaching, and assessment. In contrast, in the United States, there appears wide variability in teacher quality (National Science Board, 2014). Schools populated by large numbers of underserved students – low-performing and urban schools – require the most effective teachers to help narrow achievement gaps. However, these schools are the least likely to have highly qualified teachers. Instead, they often have teachers who are novices, have two or fewer years of teaching experience, and/or are teaching outside their field of expertise.

Social resources concern the relationships among individuals in a group or organization, including trust, collaboration, common values, shared responsibility, and collective decision-making (NGSS Lead States, 2013). For example, in Argentina, Furman (2012) found that teachers who taught students living in poverty shared low expectations for their own teaching (they had high levels of absenteeism) and for their students' learning (they viewed most as incapable). However, as a result of an intensive professional development program to support the teaching of inquiry-based science, teachers in these underserved schools began to build productive social resources: They learned to see themselves as a collective rather than as isolated individuals, to value the teaching of science as a way to transform children's lives, and to take responsibility for helping all their students achieve.

CHAPTER SUMMARY

- The construct of equity includes (a) providing all students adequate opportunities to learn science and (b) expecting all students to achieve academic excellence.
- Those student groups traditionally marginalized in science education include girls, students from non-dominant racial and ethnic groups, economically disadvantaged students, students with disabilities, students with limited proficiency in the dominant language, students in alternative education programs, and students who are gifted and talented.
- Science educators and policymakers offer different reasons for attending to equity issues in science education. Some argue for the production of a larger scientifically and technologically literate citizenry. Others emphasize the importance of students learning science in ways they themselves find meaningful.
- To achieve equity, science education must be transformed. Current inequities that must be addressed can be understood to fall into three categories: (a) the marginalization of diverse student groups in the teaching and learning of science; (b) the failure to implement curriculum, instruction, and assessment that build from all students' funds of knowledge; and (c) the uneven distribution of material, human, and social resources in classrooms and schools.

FURTHER READING

Readers are encouraged to further explore equity issues in science education. Overviews of this field of study, although grounded in the United States, are Chapter 11 of *A Framework for K-12 Science Education* (NRC, 2012) and Appendix D (including seven case studies of diverse student groups) in the *Next Generation Science Standards* (NGSS Lead States, 2013). Clear recommendations for equitable teaching strategies can be found in *Teaching Science to English Language Learners* (Rosebery & Warren, 2008) and the *Talk Science Primer* (Michaels & O'Connor, 2012). Descriptions of recent research, practices, and policies in this area can be examined in *Diversity and Equity in Science Education* (Lee & Buxton, 2010) and *Moving the Equity Agenda Forward* (Bianchini, Akerson, Calabrese Barton, Lee, & Rodriguez, 2012).

REFERENCES

Aikenhead, G. S., & Jegede, O. J. (1999). Cross-cultural science education: A cognitive explanation of a cultural phenomenon. *Journal of Research in Science Teaching, 36*(3), 269–288. doi:10.1002/(SICI)1098-2736(199903)36:3<269::AID-TEA3>3.0.CO;2-T

Bianchini, J. A., Akerson, V. L., Calabrese Barton, A., Lee, O., & Rodriguez, A. J. (Eds.). (2012). *Moving the equity agenda forward: Equity research, practice, and policy in science education.* Dordrecht, The Netherlands: Springer.

Brotman, J. S., & Moore. F. M. (2008). Girls and science: A review of four themes in the science education literature. *Journal of Research in Science Teaching, 45*(9), 971–1002. doi:10.1002/tea.20241

Brown, B. A. (2006). "It isn't no slang that can be said about this stuff": Language, identity, and appropriating science discourse. *Journal of Research in Science Teaching, 43*(1), 96–126. doi:10.1002/tea.20096

Brown, B. A., & Ryoo, K. (2008). Teaching science as a language: A "content-first" approach to science teaching. *Journal of Research in Science Teaching, 45*(5), 529–553. doi:10.1002/tea.20255

Brown, J. S., Collins, A., & Duguid, P. (1989). Situated cognition and the culture of learning. *Educational Researcher, 18*(1), 32–42. Retrieved from http://www.jstor.org/stable/1176008

Bunch, G. C. (2013). Pedagogical language knowledge: Preparing mainstream teachers for English learners in the new standards era. *Review of Research in Education, 37*(1), 298–341. doi:10.3102/0091732X12461772

Calabrese Barton, A., Tan, E., & Rivet, A. (2008). Creating hybrid spaces for engaging school science among urban middle school girls. *American Educational Research Journal, 45*(1), 68–103. doi:10.3102/0002831207308641

Carlone, H. B., Haun-Frank, J., & Webb, A. (2011). Assessing equity beyond knowledge- and skills-based outcomes: A comparative ethnography of two fourth-grade reform-based science classrooms. *Journal of Research in Science Teaching, 48*(5), 459–485. doi:10.1002/tea.20413

Chinn, P. W. U. (2006). Preparing science teachers for culturally diverse students: Developing cultural literacy through cultural immersion, cultural translators and communities of practice. *Cultural Studies of Science Education, 1*(2), 367–402. doi:10.1007/s11422-006-9014-0

Furman, M. (2012). International response for part V: Equity and diversity in science education and academia: A South American perspective. In J. A. Bianchini, V. L. Akerson, A. Calabrese Barton, O. Lee, & A. J. Rodriguez (Eds.), *Moving the equity agenda forward: Equity research, practice, and policy in science education* (pp. 351–354). Dordrecht, The Netherlands: Springer.

Harding, S. (1998). *Is science multicultural? Postcolonialisms, feminisms, and epistemologies.* Bloomington, IN: Indiana University.

Lavonen, J., & Laaksonen, S. (2009). Context of teaching and learning school science in Finland: Reflections on PISA 2006 results. *Journal of Research in Science Teaching, 46*(8), 922–944. doi:10.1002/tea.20339

Lee, O., & Buxton, C. A. (2010). *Diversity and equity in science education: Research, policy, and practice*. New York, NY: Teachers College.

Michaels, S., & O'Connor, C. (2012). *Talk science primer*. Cambridge, MA: TERC.

Moll, L. C., Amanti, C., Neff, D., & Gonzalez, N. (1992). Funds of knowledge for teaching: Using a qualitative approach to connect homes and classrooms. *Theory into Practice, 31*(2), 132–141. Retrieved from http://www.jstor.org/stable/1476399

Mpofu, V., Otulaja, F. S., & Mushayikwa, E. (2014). Towards culturally relevant classroom science: A theoretical framework focusing on traditional plant healing. *Cultural Studies of Science Education, 9*(1), 221–242. doi:10.1007/s11422-013-9508-5

National Academy of Sciences, National Academy of Engineering, & Institute of Medicine. (2011). *Expanding underrepresented minority participation: America's science and technology talent at the crossroads*. Washington, DC: The National Academies Press.

National Center for Education Statistics. (2012). *The nation's report card: Science 2011*. Retrieved from http://www.nationsreportcard.gov/science_2011/science_2011_report/

National Research Council. (2012). *A framework for K-12 science education: Practices, crosscutting concepts, and core ideas*. Committee on a Conceptual Framework for New K-12 Science Education Standards. Board on Science Education, Division of Behavioral and Social Sciences and Education. Washington, DC: The National Academies Press.

National Science Board. (2014). *Science and engineering indicators 2014* (NSB 14-01). Retrieved from http://www.nsf.gov/statistics/seind14/

Nieto, S. (1999). *The light in their eyes: Creating multicultural learning communities*. New York, NY: Teachers College Press.

NGSS Lead States. (2013). *Next generation science standards: For states, by states. Volume 1: The standards*. Washington, DC: The National Academies Press.

Organization for Economic Co-operation and Development. (2014). *PISA 2012 results: What students know and can do*. Retrieved from http://www.oecd.org/pisa/keyfindings/pisa-2012-results-volume-i.htm

Quinn, H., Lee, O., & Valdés, G. (2012, January). *Language demands and opportunities in relation to next generation science standards for english language learners: What teachers need to know*. Paper presented at the Understanding Language Conference, Stanford University, Stanford, CA.

Rivera Maulucci, M. S. (2012). Social justice research in science education: Methodologies, positioning, and implications for future research. In B. J. Fraser, K. G. Tobin, & C. J. McRobbie (Eds.), *Second international handbook of science education* (pp. 583–594). Dordrecht, The Netherlands: Springer.

Rosebery, A. S., Ogonowski, M., DiSchino, M., Warren, B. (2010). "The coat traps all your body heat": Heterogeneity as fundamental to learning. *Journal of the Learning Sciences, 19*(3), 322–357. Retrived from http://dx.doi.org/10.1080/10508406.2010.491752

Rosebery, A. S., & Warren, B. (Eds.). (2008). *Teaching science to English language learners: Building on students' strengths*. Arlington, VA: National Science Teachers Association.

Scantlebury, K. (2012). Still part of the conversation: Gender issues in science education. In B. J. Fraser, K. G. Tobin, & C. J. McRobbie (Eds.), *Second international handbook of science education* (pp. 499–510). Dordrecht, The Netherlands: Springer.

Schleppegrell, M. (2004). *The language of schooling: A functional linguistics perspective*. Mahwah, NJ: Erlbaum.

Snively, G., & Corsiglia, J. (2001). Discovering indigenous science: Implications for science education. *Science Education, 85*(1), 6–34. doi:10.1002/1098-237X(200101)85:1<6::AID-SCE3>3.0.CO;2-R

Stanley, W. B., & Brickhouse, N. W. (2001). Teaching sciences: The multicultural question revisited. *Science Education, 85*(1), 35–49. doi:10.1002/1098-237X(200101)85:1<35::AID-SCE4>3.0.CO;2-6

Tal, T. (2012). International response for part III: Reflections on context, place-based education and science for all. In J. A. Bianchini, V. L. Akerson, A. Calabrese Barton, O. Lee, & A. J. Rodriguez (Eds.), *Moving the equity agenda forward: Equity research, practice, and policy in science education* (pp. 211–218). Dordrecht, The Netherlands: Springer.

United Nations. (2014). *The millennium development goals report*. Retrieved from http://www.un.org/millenniumgoals/reports.shtml

DAVID BLADES AND ONOWA MCIVOR

34. SCIENCE EDUCATION AND INDIGENOUS LEARNERS

INTRODUCTION

Across the flat, prairie landscape in Canada in an area called, *Wanaskawin* (said Wa-na-skay-win; a Cree word meaning, "being at peace with one's self"[1]) are round boulders of various sizes. From a geological perspective, these rocks ("erratics") are believed to be deposits that formed as the vast glaciers that once covered most of North America retreated at the end of the last global ice age (Rutter, 2013). But to the Northern Plains people who have lived and gathered together in this area of the world for millennia, these boulders are considered to be "rock-people" that have spirits (Regier, 2015). In the taxonomic systems of the original inhabits of Wanaskawin, everything, including rocks, are alive.

Such differences in perspectives towards the natural world are common when comparing the knowledge-systems, or epistemologies, of Western science to traditional Indigenous epistemologies. By "Indigenous" this chapter refers to the First Peoples to live in an area of the world, those who are indigenous to these regions. These nations were and continue to be varied in political organization, resource use, lifestyle and culture. Amidst this diversity exist traditional knowledge, beliefs and orientations to the world that in some cases conflict with the epistemologies of Western science presented in school-based science education. By "Western" science we refer specifically to the form and approach of science developed in Europe and America that assumes that scientific knowledge is proven and reliable knowledge because it is objectively derived from experience (Chalmers, 1982); it is this form of science that dominates school science education world-wide.

Due to colonization, in order to be 'successful' in school, learners who have an Indigenous inheritance must demonstrate an understanding of the world through Western science that can be quite different than the traditional knowledge of their ancestral culture. This chapter outlines the importance of introducing *all* students to Indigenous ways of knowing and understanding the world; we argue that such inclusion in science education not only leads to a more socially-just, inclusive, decolonizing pedagogy but also helps students develop a more authentic and expansive understanding of the nature of science itself. We will share some key

K. S. Taber & B. Akpan (Eds.), Science Education, 465–478.

principles teachers can use to include Indigenous views in science education, some considerations when teaching Indigenous learners, and some resources to help teachers include the voices of all students, especially those of Indigenous learners; in this way teachers can model how all are welcome participants in the on-going search to understand the world we share.

MAKING SPACE FOR INDIGENOUS STUDENT'S VOICES

According to the Council of Councils[2] (Perkasa, 2015), 6% of the world's population has an Indigenous inheritance, which means that in any given school, it is likely that some of the students are Indigenous. Teachers in countries that encourage adoption of Indigenous perspectives in science education, such as Australia, New Zealand, and Canada or those looking to infuse more Indigenous perspectives in science education might be inclined to view Indigenous students in their classrooms as a possible source of ideas about Indigenous ways of knowing. This temptation must be avoided for two reasons. First, there is no one "Indigenous world view" that the student could share; Indigenous worldviews vary in history, traditions, practices and philosophies as much as any nations of the world. There are a few very general similarities Indigenous peoples share in their relationship with nature, which we discuss later in this chapter, but for the most part Indigenous societies are remarkably different and therefore a single worldview representing all Indigenous peoples would be superficial to the point of stereotyping. The second reason to avoid calling on an Indigenous learners for insights to Indigenous worldviews is due to the fact that individual students may have a wide variety of experiences of their ancestral histories and views. Some students, while Indigenous, have been raised with little to no contact with their ancestral homelands or knowledge of their traditions; while others have lived their entire lives on their traditional territories and learned aspects of their traditions, and others still, may have been raised in families that are opposed to teaching any aspects of the student's Indigenous background. Indigenous student's experiences are as varied as any students in a classroom; just as it would be unfair to ask a student with an English surname to share their insights of British thinking it is also unfair to put an Indigenous student in a parallel situation. Teachers should thus remember that *any Indigenous student cannot and should not be expected to represent all Indigenous students or share scientific and ecological knowledge of a particular Indigenous Nation.*

This does not mean, however, that Indigenous students may not have important views to share. Given that being silenced was one aspect of colonization, teachers may need to take a decolonizing approach in their classroom that *especially* provides an inviting space for Indigenous students to share the knowledge they may hold. One approach that may be helpful is the distinction between the "saying" and the "said" made by the philosopher Emmanuel Levinas. Levinas (1998) calls the "said" as that which has a hold over what someone is saying—in essence, the said is what we *know* and *understand* of what a person means or what they represent and this "said," which is formed at a cultural level, defines our thinking and approach to the Other. Now, with Indigenous students and, really, *all* students, teachers face a difficult challenge: Trying to ignore the said, what they think they know, about this student and to listen attentively to what the student is *saying*—not only in words, but how they act, interact, and live in the world. In other words, being open to the Indigenous student and not thinking of them, particularly, *as* Indigenous in order to allow space for the student to share their ideas and background, and to what the student wishes to *say* including (but not limited to) the student's thoughts and ideas from their Indigenous background. This 'listening' to the 'saying' of Indigenous learners in turn opens possibilities for bringing into science education ancient knowledge coming through the individual voices of the original peoples of that land, thereby introducing to science education new concepts and insights.

To understand more fully what Levinas means by 'the saying' and 'said', try this exercise: Choose someone you see on a regular basis. This could be fellow teacher, a professor at your university, a friend, a mentor teacher, a colleague—or someone with whom you have a close relationship, such as a family member or a partner. Now, when you next encounter this person, try to "bracket" everything you know about them, which Levinas calls the "said." In your mind, put your assumptions and anticipations aside and then open yourself to really listening to what this person is saying—try to do this without any judgment or prior interpretive framework based on your knowledge of this person. We call this listening, "being open to surprise" based on what the person is saying. Try this for a week and see if this changes your thinking about this person. In the same way, we advise teachers to approach all of their students, and in particular those with an Indigenous inheritance, in openness to what these students are saying and depending less on prior expectations and assumed knowledge about these students.

EXPANDING SCIENCE LITERACY TO INCLUDE INDIGENOUS VIEWS

In their study of science education and children of the Menominee First Nation (of the Wisconsin region, USA) Douglas Medin and Megan Bang (2014) found that Menominee children were as successful as their non-Indigenous peers in science, but in subsequent grades achievement in science fell in disproportion to their peers. This pattern seems characteristic of many Indigenous children around the world (Battiste & Barman, 1995; Krocker, 2004). A "deficit" approach to this apparent achievement trend in school science is a kind of "said" from the previous point in the assumption that there is something inherently wrong with Indigenous students, that these students need better work ethics, study skills, support from parents, etc. This approach maintains that Indigenous students are lacking in the skills or intelligence necessary to become scientifically literate. We examine this is more detail in the next point, but here we wish to remind teachers that Indigenous students are *not* lacking in science literacy, but that their particular literacy about the world may be different. For example, a student with an Indigenous inheritance may have been taught ancestral knowledge about which plants in a forest are suitable for preparing a healing tea, how to remove bark from a tree and not damage the tree, etc. but the student may not know the classification scheme for certain plants (e.g., if they are angiosperms). So, the student in this example is literate about the world, just in different ways. This does not, of course, preclude learning classification schemes to be even *more* literate and the reverse is true for students who do not have an Indigenous background by increasing *their* literacy to include the traditional ways plants were used. Anishinaabe scholar Michael Wassegijig Price explains: "Combining Indigenous plant knowledge with science and technology expands our breadth of understanding of plants and ecology, an understanding unlike that of our Indigenous ancestors" (2011, p. 12).

Including Indigenous perspectives in science education also provides an opportunity to discuss with all students how commonly-used patterns taught in science often originate from particular cultural perspectives. For example, when you look up into the night sky, what patterns do you see? If those patterns include constellations such as the Big Dipper or Orion, you are seeing connections that were first codified by the Greek astronomer Ptolemy using the common constellation patterns used by Arab sea merchants in the Mediterranean. However, the system he codified is not the only way to "join the dots" in the sky and all over the world people Indigenous to their regions used very different arrangements to form their own unique constellations. Teachers might try researching, for example, how the Inuit peoples of Northern Canada saw the stars in the sky of the North Polar Region, or the constellation patterns used for navigation by the peoples of Polynesia (see recommended resources at the end of this chapter). The Inuit peoples did not see the constellation "Orion" in the northern sky; instead, the "belt" of Orion was seen as an arrow and the top two stars of Orion's "arms" were seen as part of a completely different constellation related to stories of hunting and family (see *the Arctic Sky* under "Recommended Resources"). Teaching alternate constellations reveals to students in science education that how we "read" the world very much depends on how we are taught to read the world. A fun and effective cross-curricular exercise is to give students a night sky with magnitude 4.0 or brighter stars and have them in groups develop and name their own constellations and also ask students to write a story about how the constellation became known by this name.

Patterns in science extend beyond the night sky. During David's early years as a professor he taught in the Western Canadian province of Saskatchewan. In this region of the world, the Indigenous peoples used a method of classification of nature that is radically different from the systems used in science education. As mentioned in the introduction, traditional Indigenous classification systems consider all objects to have a spirit, therefore *everything* is alive. As well, plants and animals in this system are categorized by their function and use to the community; for example, "poison-biting beings" may include spiders and snakes. Embedded in this system is profound respect for every part of nature, but especially the older members, such as trees and rocks, which are considered ancient and therefore have special forms of wisdom they can communicate. While this classification is clearly different from taxonomies taught in traditional, Western science classrooms, if one lived out on the land in this area of the world the traditional schemes could be far more useful that knowing the phylum and class of, for example, a particular animal. Instead, traditional classification schemes, which themselves are very elaborate in some Indigenous Nations, are really useful for knowing what plants one can eat, which are useful for making rope, which animals are dangerous and which are useful for eating. Speaking of the Anishnaabe First Nation, Prince (2011) conveys that,

> Indigenous names of plants are descriptive, metaphoric and intertwined with the intricacies of the landscape: they indicate relationships to animals and birds; they describe how the Anishnaabe utilized each plant according to its physical characteristics. (p. 5)

The key point in this section is that Indigenous students may bring some of their worldview and traditional knowledge about nature to their science classes; these students *are* literate about nature, their knowledge is just *different* from what is usually taught in science classes. Yet this difference is never apparent to students as their science education increasingly focuses on specific content knowledge and memorization, especially after Grade 4, that bears little to no resemblance to the ancestral knowledge of Indigenous peoples. In her study of why some Indigenous students turn away from science, Nikki Krocker observed in her interviews with Indigenous students that they found school science "continually presented as facts students must memorize and regurgitate" (p. 104), that is, "disconnected from the lives of students and irrelevant to their community and cultural background" (p. 104). The increasing focus on Western scientific knowledge from Grade 4 onwards in most school science curricula crowds out any other knowledge about the world, including Indigenous knowledge. Krocker and many others argue that including Indigenous ways of reading the world would help to create a science curriculum that is more familiar and more useful to Indigenous students and that this would in turn increase the retention and success of Indigenous students with science education (Berkowitz, 2001; Cajete, 1999; Henderson, 1996; Krocker, 2004). We agree, but add that the exclusion of other forms of knowledge from Grade 4 onwards is not a focus on scientific knowledge but a *delimiting* of what passes as science to the discoveries of "Western" scientists. Inviting alternate views of constellations or classification, we argue, *expands* scientific knowledge to include ways of understanding the world from the perspectives of Indigenous peoples. In other words, science education becomes, we believe, *more* scientific if this education does not exclude forms of understanding that fall outside the present curriculum focus on European/American scientific discovery. Indigenous scientists Cheung (2008) and Price (2011) agree that combining the best of Western science and Indigenous knowledge of place is a superior combination from just one or the other and that such a science curriculum would better enable students to tackle the challenges of this millennia.

Another key consideration for teachers is how the pedagogy of science changes after Grade 4. In the primary school years of Kindergarten to Grade 3 science education tends to be more hands-on, inquiry-based learning, such as discovering the properties of magnets or planting beans to determine how they grow towards light. Intermediate elementary school science shifts its pedagogy towards content memorization that increases with each grade as students learn classification schemes used in Western science, weather patterns, the names of types of rocks, etc. The shift is from what scientists practice, their form of inquiry, towards what Western science has discovered. While we recognize that sharing and memorizing knowledge is a part of science, we also support curriculum guidelines for science that emphasize the *processes* of a place-based science. Place-based science is defined as science that is rooted in the territory one is studying, connecting both to the geographic qualities such as rivers, lakes, mountains, ocean, desert, sky as well as the teachings and ancient wisdom that exists about that place from the original inhabitants of that territory. A more inquiry and place-based approach to science at higher grades would move from an increasingly transmissive, written approach that discourages oral ways of demonstrating understanding to a more holistic approach. This emphasis on written over oral methods of learning and teaching affects all students but particularly Indigenous students whose ancestral cultures tend to value oral approaches more highly. We argue, then, it's not Indigenous students who need to change or adapt to science education but the pedagogy of science education itself that needs to become more hands-on, inquiry based at all levels (elementary through to post-secondary) for the benefit of all students.

BEYOND BORDER CROSSING

Western science, as mentioned in the previous section, was born in the development of inductive, experimental approaches to understanding nature first by Islamic scholars then picked up in the work of Occam, Bacon and other Europeans. The spectacular discoveries of science as an experimental approach were linked to economic prosperity early in the European mind, especially with discoveries in shipbuilding, navigation and materials development enabled movement of European capitalist interests to all regions of the world (Johnson, 1991). In this way, science became a exported ideology of conquering nature as per Biblical injunction as a means to a better life, where "better" meant economic prosperity through the development of nature. Those closest to the epistemology of progress through Western science and technological invention understood, valued and appreciated Western science. Thereby, the children of these cultures, or those at least familiar with this particular view of science, were at an advantage in science education to those children from cultures with different approaches to understanding the world. Indigenous students, in particular, find that their ancestral cultures do not share the same values of the dominant societies and, in particular for this chapter, experience different approaches culturally towards nature.

For example, most Indigenous cultures, as a broad generalization, approach Nature with an attitude of *respect*. There is not a fundamental split between humans and nature that informs Western science; instead, Indigenous cultures tend to see themselves as *part* of Nature, or creation. In addition, Indigenous peoples do not see themselves as above nature, that is rocks, trees, animals, insects, waterways, but rather beneath them—recognizing that as humans we are useless and helpless without these things—a vast difference to the hierarchy of humans as the top creatures in the natural world according to Western philosophies that assume that humans have a superior thought processes. Another fundamental difference is the assumption among Indigenous cultures that Nature is infused with spirits that play an active, historical role in relationships (MacIvor, 1995; Ermine, 1995). These two features, the assumption of being part of Nature that is also spirit-filled, marks what Arun Agrawal (2004) calls, "striking differences" (p. 2) between Indigenous cultural approaches to the world than the approaches of Western science.

Many science educators argue that one way to encourage the success of Indigenous students in science is for teachers to become "brokers" of the two cultures by guiding Indigenous students on how to cross over from the borders of their culture to the borders of science, much like learning to live in a country different from one's own. For example, Glen Aikenhead, the leader in making this argument, claims that "learning Western science for most Aboriginal students is a cross-cultural event" (2001, p. 340) as students move from their, "everyday cultures associated with home to the culture of Western science" (p. 340). It is the job of teachers, argues Aikenhead, to enable students to be exposed to and understand the world of Western science while at the same time retaining space and respect for their Indigenous worldviews; in this way teachers can serve as cultural brokers for Indigenous students (Aikenhead, 2006).

We are troubled by this pedagogical advice for several reasons. First, as we point out earlier, some Indigenous students may come from homes where their everyday culture is essentially the same as the dominate culture, so for these students there really are not any borders, at least in terms of science education, to cross. But a deeper concern is also evident in the call for teachers to be brokers of such border crossing. Essentially, border crossing is a pedagogy that does not challenge or critique the crossing itself, which in turn makes this 'uncritical brokering' a form of colonization. It is as if teachers are saying, "Here's science, let me show you how to understand science *because you are Indigenous*." We acknowledge that those advocating for border crossing respect and admire Indigenous cultures and have the best interests of Indigenous students at heart, i.e., helping these students be successful in their negotiation of the dominate culture, but essentially border crossing is a pedagogy of accommodation to the Western world that leaves unchallenged the culture of science.

We advocate instead for *cultural encounters* and that these encounters with Indigenous perspectives are important *regardless whether there are Indigenous students in the science classroom or not.* To explain our idea, we invite the reader to consider the well-known classroom demonstration of the magnetic field of a bar magnet where a standard bar magnet is covered with a large sheet of paper and some iron filings (or small paper clips) are sprinkled over the paper. The resulting pattern indicates the magnetic field around the magnet; see Figure 1.

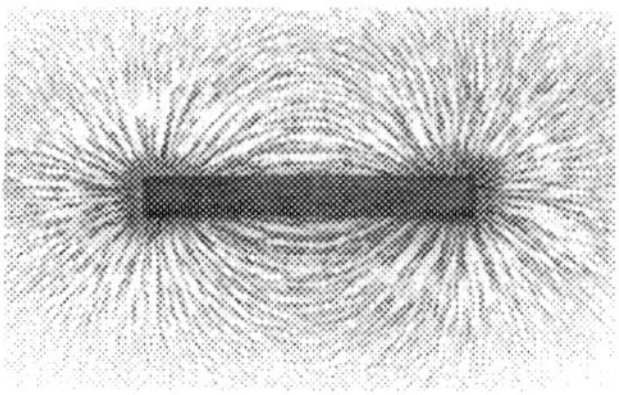

Figure 1. Bar magent with iron filings

The Earth, according to Modern Science, is also a giant magnet and has similar magnetic fields, which is why a compass works to point out the north and south poles (although the magnetic poles are about 20 degrees away from the geographic poles). Now, according to science, when particles from the Sun collide with the Earth's atmosphere, these particles are attracted to the north and south magnetic poles of our planet, and the interaction of these particles with the atmosphere in those locations causes the Northern Lights (aurora borealis) and the Southern Lights (aurora australis). However, the Indigenous Nations in Northern Canada, for example, teach a completely different interpretation of this phenomenon. Many Nations teach that the lights are the appearance of departed ancestors, sometimes dancing, sometimes playing a ball game in the sky.

The Indigenous interpretation of the Northern or Southern Lights begs a question: Which interpretation is true? On one hand, the explanation by Western science can be tested and examined, so it can be found to be true within the parameters of experimentation set out by science. But is the Indigenous explanation false? Is there any way to determine if this explanation is true or false? If not, then the lights we see could be *simply* the interactions of particles in the sky or, and this is where we reach the limits of scientific inquiry, these lights *could also* be ancestors dancing—Western science is limited in this way in that it is void of interplay with the spiritual world, which is equally real and valued by Indigenous peoples as part of the natural world.

Encountering traditional Indigenous views of the Northern lights thus reveals the limits of science to particular interpretations that can be tested. The belief that *only* views that can be tested through science can establish what is true or not is called, "scientism." In effect, but likely not the intention, school science promotes scientism by excluding any other interpretations of phenomenon. This is an issue precisely because there are questions beyond the limits of scientific examination, such as questions concerning the existence of a Creator. These questions are larger than science, or "metaphysical" in nature. So, one benefit of introducing students to alternate views of phenomenon is to teach students that Western science is helpful, of course, but also limited; to introduce to students to the idea that Western scientific approaches are not the only way to understand the world we share.

There is much to be learned by other cultural ways of approaching nature. As Michael Michie notes, "Indigenous science reminds me that there are other ways of looking at the world and that knowledge is valued in different ways. Indigenous science gives me another perspective in the world" (2002, pp. 36–37). For example, dreamcatchers are made by some Indigenous nations in North America and the idea of a dreamcatcher has spread beyond Indigenous cultures as well. One form of making a dreamcatcher involves using shoots of a red willow tree. However, before cutting branches to make the dreamcatcher, the person must offer thanks to the tree for the sacrifice of the branch and leave a gift (in prairie cultures this is often tobacco). This practice is in response to the belief that the tree has a spirit that requires respect and therefore acts of gratitude. Indigenous peoples relationship with nature was an understanding to never take too much from one place or one plant, so as to ensure sustainability and not to overly disrupt the interactive and synergetic relationships within every ecosystem.

There is no possible way to carry out an experiment on whether the tree is animated by a spirit or not, but consider: What would our modern world be like if each time humans came to use a tree, or mineral, we gave thanks for its life, or considered how our use may disrupt the ecosystem in which it lives? Is there not, embedded in this ancient practice, a lesson of value on how to live with care and thoughtfulness on Earth? Bringing Indigenous views into science education also brings alternates to exploitation of natural resources and a form of environmentalism that could assist to avoid some of the catastrophic disasters that at present seem inevitably ahead. We further advocate that science education would better serve humanity if respect for nature and being part of nature were a fundamental part of all science curricula.

Finally, encounters with different cultural perspectives can serve to bring students to a more authentic and holistic view of science. The origins of Western science lie in openness to questions, a sense that there lies behind every explanation more to learn. One of the most important philosophers of science of the 20th century, Karl Popper, argues that science never arrives at a true explanation of how nature works, but works to develop more and more useful and accurate explanations (Popper, 1963). He argues against premature adoption of any explanation as a complete and final explanation. The history of science supports Popper's caution; for example, the declaration by Lord Kelvin near the end of the 19th century that there was nothing new to discover in physics proved to be entirely premature with the discovery of quantum mechanics in the century that followed. Being open to questioning and understanding lies at the heart of science. Indigenous perspectives can compliment Western science and assist with forming new questions to explore or in some cases reinforce discoveries in science with human experience. Instead of delegating Indigenous views as "mythological" or "superstitious" scientists and science educators could respectfully honour this wisdom shared by a "[p]eople that have a long tenure within a particular region [who] have gained much knowledge about the ecology of place" (Price, 2011, p. 11). Actually *authentically including* Indigenous perspectives in science education opens up explanations of nature and thus reveals new possibilities for inquiry. This is not "border crossing" or accommodation of Indigenous learners to the "said" of science but widening the circle of human understanding of nature by *including* Indigenous views.

LEARNING TO INCORPORATE INDIGENOUS WORLDVIEWS IN SCIENCE EDUCATION

Increasingly teachers are being asked to include Indigenous knowledge in their science curriculum. How might teachers, the majority without an Indigenous inheritance, authentically bring Indigenous views into their science classrooms? In this section, we offer some practical advice for those who wish to expand the concept of science with their students.

a. *Indigenous knowledge is place-based*

While some Indigenous scientific knowledge transcends place (such as the discovery of the foundation for the drug Aspirin from the bark of a tree having a numbing effect leading to a now widely used commercial remedy), much knowledge remains place-based. When science teachers realize and acknowledge that the knowledge they may be seeking for inclusion in their curriculum and pedagogy lives on the land and is connected to the territories in which they live, it may prompt them as teachers to (re)discover that place. A good start would be would be learning which plant and animals species are native to that place and which were introduced, and sharing this information with students in meaningful ways.

b. *Indigenous knowledge is held in community*

There are Indigenous scientists and authors who publish their knowledge about Indigenous science. Gregory Cajete, a Tewa scholar, and Michael Wassegijig Price are excellent examples of this practice, having found ways to share the knowledge of their people and territory appropriately and authentically. However, the majority of Indigenous 'scientific' knowledge and worldviews 'lives' in community and on the land. This knowledge is specific to the places they are from and are generally held by the eldest generations of a community. Growing relationships with the local people, becoming a trusted ally worthy of having access to this knowledge, and further still—involving these respected Elders and knowledge keepers in your classroom and school community—is a privilege worth earning.

c. *Build trust*

Genuine relationships built on time spent in respectful presence and trust is key. In order to gain access to local, place-based knowledge teachers will need to build authentic relationships with the local people of the territory. Getting to know who the knowledge keepers are and how to respectfully and appropriate request help (see next section) and build up one's own knowledge is foundational to meaningful inclusion of Indigenous worldviews in science education.

d. *Learn local protocols*

Be sure to observe local protocols as they are being conducted. Respectfully inquire as to how you can come to know the Elders and knowledge keepers in a community and be so fortunate as to learn from them and perhaps even have them share their knowledge in your school or classroom. Humility and patience is key here. One must understand that Indigenous people's cultures are fundamentally different at the core and even ways of asking for and gaining access to knowledge can be different from Western ways of knowing and being. For instance, some Indigenous cultures believe certain plant knowledge can only be passed on through ceremony with those who are well suited and prepared to receive it to avoid exploitation and possible harm. Sammel (2005) explains:

> It is also essential to learn how to incorporate Elders and their knowledge into the educational system. Their experiences and knowledge are based on an oral tradition. Although the knowledge of some Elders has been included in books, Elder knowledge cannot be solely found in books. The Elders are the keepers of knowledge. It is their job to protect that knowledge and relay that knowledge in appropriate situations. It is knowledge to be shared, if and when it is appropriate. But one must ask to be taught in ways that are respectful and appropriate to the traditions from which that Elder comes. (p. 6)

In summary, we advise science educators to tread lightly on the land beneath your feet, and handle the relationships with the original people of that land with care. Do not let fear or 'not-knowing' stop you; that is not a good excuse. Certainly you have entered uncharted territory in other parts of your life at different times. Be respectful and realize you are on a journey of understanding: You can do it, one small personal action at a time.

SUMMARY

1. Avoid calling on Indigenous learners as sole representatives of their own or other Indigenous cultures.
2. Bracket the "said"; what you think you know about Indigenous peoples, to make space for what you might be able to learn.
3. In order to more effectively engage learners past Grade 4, aim for a more interactive, hands-on, inquiry-based, place-based approach to teaching students science.
4. Rather than a deficit model of teaching Indigenous learners, remember that Indigenous learners and their family/community members may have vast place-based knowledge and alternative ways of seeing the world; this contribution increases the science literacy of all students.

5. Teaching Indigenous learners is more than border crossing; genuine contributions from both Indigenous and Western scientific worldviews provide greater understandings of the world for all students.

NOTES

[1] Or, "living in harmony" depending on pronunciation.
[2] A foreign policy initiative by G20 countries for research and exchange of information about global trends.

RECOMMENDED RESOURCES

Books

Bridging cultures: Indigenous and scientific ways of knowing nature (2011) by
Glen Aikenhead & Herman Michell; Don Mills, Ontario, Canada: Pearson Education; ISBN-13: 9780132105576

Integrating Aboriginal perspectives into the curriculum: A Resource for curriculum developers, teachers and administrators (2003). Manitoba Education and Youth; ISBN: 0771124716.

Keepers of the Earth: Native American Stories and Environmental Activities for Children (1997) by Michael J. Caduto and Joseph Bruchac; Fulcrum Publishing; ISBN-10: 1555910270.

Keepers of the Night: Native Stories and Nocturnal Activities for Children (1994) by Michael Caduto and Joseph Bruchac; Fifth House Books; ISBN-10: 1895618398.

Lighting the seventh fire (1994) by David Peat; Carol Publishing; ISBN-10: 1559722495.

The Arctic Sky: Inuit Astronomy, Star Lore, and Legend (1998) by John Macdonald; Royal Ontario Museum/Nunavut Research Institute; ISBN-10: 0888544278.

The New Patterns in the Sky: Myths and Legends of the Stars (1988) by Julius D. W. Staal; McDonald and Woodward Publishing Company; ISBN-10: 0939923041.

They Dance in the Sky: Native American Star Myths (2007) by Ray A. Williamson and Jean Guard Monroe; HMH Books for Young Readers; ISBN-10: 0618809120.

Journal articles; see references as well as the following:

Blades, D. (1997). Towards a postmodern science education curriculum-discourse: Repetition of a dream catcher. *Journal of Educational Thought, 31*(1), 31–44.

Brandt, C. B., & Kosko, K. (2009). The power of the earth is a circle. In W. M. Roth & K. Tobin (Eds.), *The world of science education: Handbook of research in North America* (pp. 389–407). Amsterdam: Sense Publishers.

Chinn, P. (2008). Connecting traditional ecological knowledge and Western science. In A. J. Rodriguez (Ed.), *The multiple faces of agency: Innovative strategies for effecting change in urban school contexts* (pp. 1–27). Amsterdam: Sense Publishers.

Oberg, A., Blades, D., & Thom, J. S. (2007). Untying a dreamcatcher: Coming to understand possibilities for teaching students of Aboriginal inheritance. *Educational Studies, 42*(2), 111–139.

Snively, G., & Corsiglia, J. (2001). Discovering Indigenous science: Implications for science education. *Science Education, 84,* 6–34.

Websites

Aboriginal Curriculum Integration Project: Science Lessons Index
http://abed.sd79.bc.ca/acip/indexfiles/science_lessons_index.html
Aboriginal Nations Education Division – Resources (Science)
https://aned.sd61.bc.ca/resources.aspx
Living Knowledge: Indigenous Knowledge in Science Education
http://livingknowledge.anu.edu.au/html/educators/
The Windspeaker Classroom
http://www.ammsa.com/content/classroom-edition
15 Strategies for Teachers of Aboriginal Students
http://www.ictinc.ca/blog/15-strategies-for-teachers-of-aboriginal-students

REFERENCES

Agrawal, A. (2004). Indigenous and scientific knowledge: Some critical comments. *IK Monitor, 3*(3), 1–9.

Aikenhead, G. (2001). Integrating Western and aboriginal sciences: Cross-cultural science teaching. *Research in Science Education, 31,* 337–355.

Aikenhead, G. (2006). *Science education for everyday life.* New York, NY: Althouse.

Battiste, M., & Barman, J. (1995). *First nations education in Canada: The circle unfolds.* Vancouver, Canada: UBC Press.

Berkowitz, P. (2001). Western science meets Mi'kmaq knowledge. *University Affairs* (University of Victoria), 16–20.

Cajete, G. A. (1999). *Igniting the spark: An Indigenous science education model.* Asheville, NC: Kivaki Press.

Chalmers, A. F. (1982). *What is this thing called science?* Queensland, Australia: University of Queensland Press.

Cheung, M. (2008). The reductionist-holistic worldview dilemma. *MAI Journal: A New Zealand Journal of Indigenous Scholarship, 3(5), 1–7.*

Ermine, W. (1995). Aboriginal epistemology. In M. Battiste & J. Barman (Eds.), *First Nations education in Canada: The circle unfolds* (pp. 101–112). Vancouver: UBC Press.

Henderson, M. (1996). A Mohawk vision of education. *Green Teacher, 49,* 15–18.

Johnson, P. (1991). *The birth of the Modern.* New York, NY: Harper-Collins.

Krocker, N. (2004). *Discussions on science curriculum: Stories told from northern places.* (Unpublished Masters Thesis). Department of Curriculum and Instruction, University of Victoria, Victoria, Canada.

Levinas, E. (1998). *Otherwise than being* (A. Lingis, Trans.). Pittsburgh, PA: Duquesne University Press.

MacIvor, M. (1995). Redefining science education for Aboriginal students. In M. Battiste & J. Barman (Eds.), *First nations education in Canada: The circle unfolds* (pp. 73–98.) Vancouver: UBC Press.

Medin, D. L., & Bang, M. (2014). *Who's asking?: Native science, western science, and science education.* Cambridge, MA: MIT Press.

Michie, M. (2002). Why indigenous science should be included in the school science curriculum. *Australian Science Teachers' Journal, 48*(2), 36–40.

Perkasa, V. (2015). *The world conference on indigenous peoples: A view from Indonesia.* Council of Councils. Retrieved from http://www.cfr.org/councilofcouncils/global_memos/p33476

Popper, K. (1963). *Conjectures and refutations: The growth of scientific knowledge.* London: Routledge.

Prince, M. W. (2011). Indigenous taxonomy: Ethnobotany and sacred names. In P. Boyer (Ed.), *Ancient wisdom, modern science: The integration of Native knowledge in math and science at tribally controlled colleges and universities* (pp. 1–13). Pablo, MT: Salish Kootenai College Press.

Regier, J. (2015). *Integrating first nations and metis content and perspective: Grade 4 Earth and space sciences; rocks, minerals, and erosion*. Duck Lake, Saskatchewan, Canada: Prairie School Division No. 246.

Rutter, N. W. (2013). Glaciation. In *The Canadian Encyclopedia* (Anthony Wilson-Smith, Pub.). Toronto: Historica Canada. Retrieved from http://www.thecanadianencyclopedia.ca/en/article/glaciation/

Sammel, A. (2005). *Aboriginal perspectives into the teaching and learning of science education: Beginning the conversations in Southern Saskatchewan*. Regina, Saskatchewan, Canada: SIDRU Publications.

MANABU SUMIDA

35. SCIENCE EDUCATION FOR GIFTED LEARNERS

INTRODUCTION

The rapid expansion and qualitative changes in the field of science from the late 20th century to the 21st century have had a great impact on the social structure, cutting down routine jobs and physical labour while increasing jobs dealing with abstract and creative challenges. The expansion and diversification of science in the 21st century is likely to throw up a variety of issues concerning the conventional style of science education and classes. For example, given that over one million scientific papers are published annually, how do we define the basis and basics of science education? What are the talents and skills that need to be developed in science education intended for all children in compulsory education? While scientific research has become increasingly global, there is a dearth of scientific specialists. What kind of science education can develop people who can put forth new ideas and pass on the benefits of innovative science and technology? In addition, from what perspectives can we improve the quality of conventional science education that is taught to students at large? Research on science education for the gifted is a key research area in education that seeks to eliminate the gap between children's everyday life, school science, and real science.

No matter how naturally gifted a child (or an adult for that matter) is, some kind of educational support is necessary for the inherent talent to bloom. This opportunity to receive educational support must be provided to all regardless of gender, race, place of residence, or household income. Gifted education is a kind of special education as well as an equal education. Although the number of countries where 'giftedness' is included as a part of the national law is limited, including Europe (e.g. Australia, Switzerland, Germany, Spain, Hungary, Poland, Romania, and Slovenia), almost all countries have some type of legislative regulations and guidelines related to providing for gifted education from early childhood (Mönks & Pflüger, 2015). These can be in a variety of formats, such as early entrance into kindergarten, grade skipping, cooperation with companies or non-profit organisations, extra-curricular activities, personal mentoring, participation in school internal/external competitions, psychological counselling, and summer camps.

Partially because the IQ test has been historically used to certify gifted children, the certification is often based on indicators, such as IQ based on intelligence test or creativity, that demonstrate general talent rather than aptitude/advanced ability in any specific area. Simple indicators such as the achievement of the student

K. S. Taber & B. Akpan (Eds.), Science Education, 479–491.

are mainly used to measure the results of gifted education. However, educational research focusing on talent in specific areas is becoming popular; science education, in particular, is drawing attention in recent years to gifted education as an attractive research area. Moreover, the idea of developing learning support to enable all children to enhance their personal characteristics and abilities by focusing on social factors in the educational settings and carefully studying special skills and teaching styles of excellent teachers—rather than merely regarding the learning characteristics of gifted children as natural or innate—is finally catching on.

In addition to taking a global look and introducing facts about a multitude of gifted children's talents, such as special abilities related to science, learning styles, interests, and concerns, this chapter will describe methods to understand the characteristics of children's talent in science, science education programmes to promote such talents, and measures for science learning to accommodate individual needs. The fruits of science education, such as the different international science and technology competitions and Olympiads that are rolled out globally, are closely related to society and everyday life, and they include many opportunities for acquiring skills through observation and experiment and participation in cooperative activities that help improve the quality of science education normally provided in a classroom. This chapter will also direct attention to social factors that help in grooming talent—for example, the support provided by a science teacher in particular—and as well discuss the educational challenges required to bring out children's talent in science in the 21st century.

CHARACTERISTICS OF GIFTED STUDENTS IN SCIENCE

Whereas it is possible to identify children worldwide who show a strong interest in natural phenomena from an early age and demonstrate an outstanding ability to think creatively and in abstract terms, there is no single universally accepted definition of the terms 'gifted', 'talented', or 'giftedness'. The traditional method of identifying a gifted child in many countries has been an IQ test. An IQ test score (specifically, a score of 130 or higher) is still used as a partial basis for identification. In some cases, a certain upper percentage of scores (e.g. the top 1% or 10%) is still used as the standard, based on the notion that a gifted child is someone who performs better than other children of the same age.

However, a great variety of methods beyond the traditional IQ test are currently being promoted as ways of identifying gifted children. More importantly, even a child identified as gifted is not regarded as 'perfect' or able to demonstrate excellence in every area. In reality, such children may exhibit imbalances in their socio-emotional development, experience difficulties in their interpersonal relationships, or be underachievers in fields not of interest to them. Cross and Frazier (2010) insisted that the socio-emotional needs of gifted students even in state-supported residential high schools for gifted students or specialized residential 'Science, Technology,

Engineering, and Mathematics (STEM)' schools are important. In order to support these characteristics of gifted children, educational programmes exist to cater to these differences.

In the U.S.A., for example, the proportion of children currently identified as gifted is on average between 6% and 10% nationwide (National Association for Gifted Children, 2015). This is approximately the same percentage of children that require special education, and is therefore not to be ignored in ordinary school education activities. School education of the gifted in the U.S.A. was originally raised in the 1972 Marland Report to Congress. This was the first time that giftedness was defined, and schools were formally encouraged to define it broadly; in addition to academic and intellectual talent, the definition included leadership ability, visual and performing arts, creative or productive thinking, and psychomotor ability.

In identifying gifted children, it may be helpful to include broader cognitive and behavioural characteristics, such as the possession of a large vocabulary, good expressive abilities, mental agility, a sense of humour, and good concentration and attention span. However, it is normal for people to be stronger in some areas and weaker in others. The domain-specific, dynamic nature of science, which encompasses a wide range of different fields of study, can accommodate children's varied areas of interest, making it an ideal subject in which they can exhibit and educators can identify their giftedness.

There are widely-known and reputable screening tools, such as the Scale for Identifying Gifted Students (SIGS) and the Screening Assessment for Gifted Elementary and Middle School Students, K-8-Second Edition (SAGES), which include a limited number of items related to science. Some proposed behavioural characteristics of children gifted in science include: 'is imaginative', 'uses numbers often when expressing ideas', 'displays curiosity by asking relevant questions', and 'goes beyond obvious answers'. Sumida (2010) developed a behaviour checklist for use among Japanese primary school children in science classrooms in the non-Western context. The 60 items include items like 'reports clearly the result of an observation and experiment' and 'tries to do things in his/her own way, not according to the instructions given'. On the basis of such analysis, three styles of giftedness in science may be identified, namely, (1) a spontaneous style, (2) an expert style, and (3) a solid style. Sumida and Ohashi (2015) further noted that 'participation' in local science events and informal science projects was a useful indicator for broadly identifying children gifted in science from the bottom up.

Returning to the above-mentioned issue of imbalance, among the scientists who have made their mark on history, quite a few are known to have had not only outstanding talent and brilliance but also some kind of learning difficulty. Similarly, as earlier pointed out, even a child identified as gifted based on multiple criteria may also have special educational needs. Such unique children, who possess both gifts and challenges, are referred to as 'twice-exceptional' (2E) or 'dual-exceptional' children. The domain-specific, dynamic nature of science can once again accommodate such children, with their varied areas of interest. Sumida (2012) used his original gifted behavioural checklist in the field of science and implemented science lessons for 2E primary school children, insisting that inclusive science lessons could enable both 2E students and regular students to study together for their mutual benefit.

SCIENCE CURRICULUM AND INSTRUCTION FOR GIFTED LEARNERS

Gifted and talented students are sometimes maladjusted to regular classrooms with standard curriculum and teaching methods, and sometimes show underachievement in some subjects or learning difficulties in the classroom. In general, it is recommended that the curriculum needs to be more in-depth, abstract, and complex for the gifted. Moreover, a curriculum for high-ability and gifted learners demands not only higher level thinking and inquiry-based materials, but also materials that demonstrate adequate depth and complexity while providing for individual rates of learning. Johnson, Boyce, and VanTassel-Baska (1995) analysed commercial science curricula such as Great Explorations in Math and Science (GEMS) from three phases, (1) general curricular features, (2) exemplary science features, and (3) tailoring for special populations. They found that existing basal science textbooks fail to meet the needs of students, particularly for high-ability learners. The William & Mary Center for Gifted Education promotes 'advanced content', 'high-level process and product', and 'overarching concepts' as the three core dimensions to develop Problem-Based Learning (PBL) Science Units and showed its positive effects of implementation not only on students, but also on teachers (VanTassel-Baska, Ries, Poland, & Avery, 1998).

Acceleration and Enrichment

There are two main forms of gifted education programmes. They are acceleration and enrichment. Acceleration aims to provide students an opportunity to study the content in the upper grade curriculum and earn credits towards a university degree. Skipping grades, special grouping in a specific subject, Advanced Placement (AP), and dual enrolment are examples of acceleration. Table 1 summarises the change of the number of high schools providing AP and students taking AP exams between 1955 and 2014 (College Board, 2015). For about 50 years, from 1955 to 2000, the number of high schools providing AP increased from 104 to 13,680, an approximately 130 fold increase. In 2014, 21,594 schools provided AP and 4,478,936 students took AP exams. AP is recognized in the admissions process by more than 4,000 universities worldwide, and outside the U.S., more than 600 universities in more than 65 countries recognize qualifying AP Exam scores. Early entrance and skipping grades are also recommended for gifted children, but there is some opposition to this kind of acceleration because it is not easy (nor recommended) to reverse the skipped grade after the early entrance.

Table 1. Annual advanced placement programme participation 1956–2015

Year	*Schools*	*Students*	*Exams*	*Colleges*
1955–56	104	1,229	2,199	130
1965–66	2,518	38,178	50,104	1,076
1975–76	3,937	75,651	98,898	1,580
1985–86	7,201	231,378	319,224	2,125
1995–96	11,712	537,428	843,423	2,895
2005–06	16,000	1,339,282	2,312,611	3,638
2014–15	21,594	2,483,452	4,478,936	4,154

Enrichment, another form of a gifted education programme, provides children an opportunity to study interdisciplinary and/or extended content. This includes personal learning, project learning, centre approaches, Saturday/Summer/Winter programmes, and contests. Figure 1 summarises the key programmes of acceleration and enrichment.

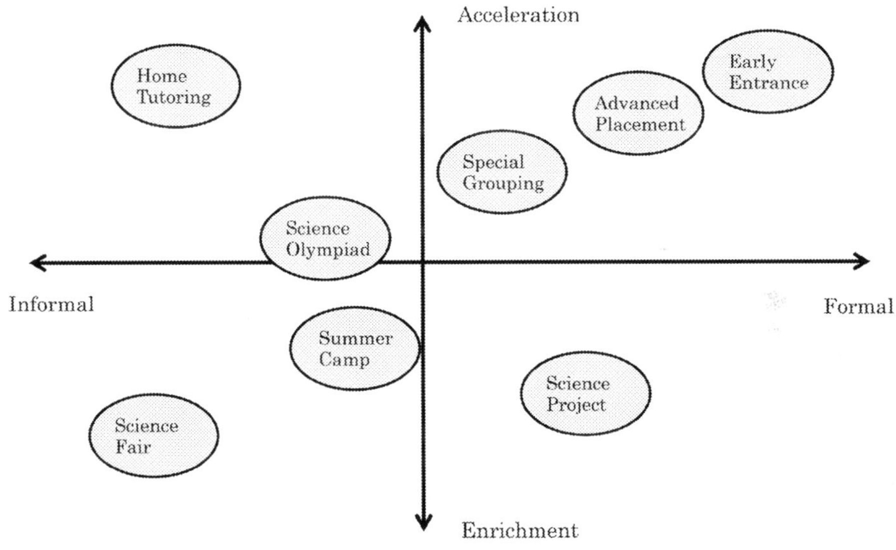

Figure 1. Forms and contexts of science education for gifted learners

Enrichment requires special mentoring in an extra area so it is not easy to implement in a formal education setting. However, enrichment is practical with the cooperation of scientists and scientific associations to provide students with science lectures, to provide students the opportunity to conduct science projects, and the chance to give a presentation at a science conference. For example, through an enrichment programme of university, Japanese high school students won the number one spot in a Science and Engineering Challenge and have had articles published in international academic journals. In Japan, the Ministry of Education, Culture, Sports, Science and Technology encourages student challenges in science Olympics in cooperation with science associations and universities. The total number of participants in the Japan Science Olympiad, Physics Challenge, Chemistry Grand Prix, Japan Earth Science Olympiad, and Japan Biology Olympiad increased from 1,524 to 9,774, a 6.4 fold between 2004 and 2014 (Sumida, 2017).

Parallel Curriculum & Differentiation

There are several schools that provide special classes for gifted students wherein lessons are taught at a faster pace compared to regular classes. However, if the enacted curriculum for the gifted children is similar to the curriculum for regular students, gifted students would not be able to fully realize their inherent potentials and, consequently, unable to optimize on their talents most of the time. Therefore, it is necessary for the students to have curriculum and instruction that promote an understanding of broad-based interdisciplinary concepts, foster the development of higher level of thinking skills, guide the students toward expertise, and nurture students' self-understanding, self-direction, and interpersonal skills.

The Parallel Curriculum Model (PCM) can be tailored to address the demanding needs of gifted education through redesigning science curriculum. It is composed of four interrelated facets (core curriculum, curriculum of connections, curriculum of practice, and curriculum of identity) that are projected to nurture and raise the children's level of abilities (Tomlinson et al., 2002). The core curriculum involves the teaching of core concepts, principles and skills of a discipline. It is designed to help the students comprehend essential knowledge that eventually leads to children's expertise of the discipline. The curriculum of connection is designed to help the students connect their learned knowledge to new content, content areas and disciplines. The curriculum of practice is designed to enable the students to function with increasing skills and competencies at par with the professionals. The curriculum of identity is designed to aid the students to explore and participate in a discipline that is related to their own interests, goals and strengths, both in the present and future. It also allows gifted children to use their prior knowledge that will direct them toward significant and appropriate solutions.

Faustino, Sumida, Fajardo, and Pawilen (2010) applied the three facets of the Parallel Curriculum Model (PCM) in designing the science curriculum to develop the problem-solving skills of Grade III gifted children in the Philippines. They analysed and summarized the possibility of science curriculum in a cross-subject manner. For example, the topic of sense organs can successfully integrate the building of cognitive skills such as memory, creativity and intuition as well as contexts/concepts pertaining to other curricular subjects such as Language, Arts, Mathematics, Home economics and Culture. In addition, curriculum practice can be reflected through building connections with medical practitioners, health workers and social workers. Figure 2 summarizes the example of the science curriculum on sense organs using PCM.

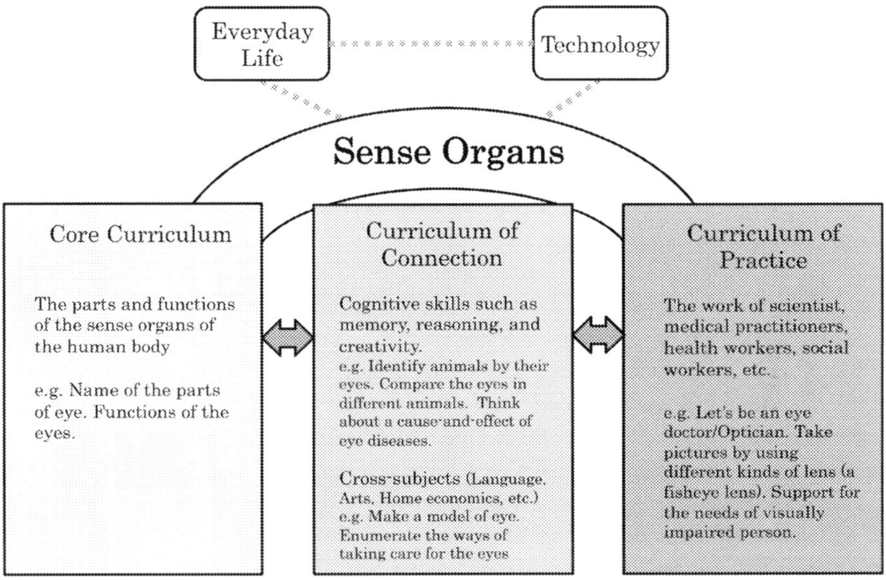

Figure 2. An example of primary science curriculum using PCM

Children in the present school system are recognised with varying levels of academic readiness, learning styles and socio-cultural background. Differentiation is a very useful instructional strategy to design science lessons for gifted children in regular classrooms. Differentiation is defined as working to address the abilities, interests, and needs of individuals through the means of providing students with a variety of learning activities and distinctly appropriate instructional strategies to understand content, to process ideas and to develop products.

Sondergeld and Schults (2008) designed and implemented a third-grade differentiated unit on simple mechanics. The unit, taught over the course of 3 weeks, addressed four forms of differentiation:

1. Content—The use of different materials based on student ability.
2. Process—The use of the hands-on approach to learning or reading based on students' performance.
3. Product—Students' choice in the end product as evidence of learning.
4. Environment—Quiet independent study areas and small group work areas.

When compared with traditional instruction, they reported that 'more fun', 'choices', 'learned more/better understanding', 'work at own speed', and 'experiment and create' as positive responses to differentiated instruction while 'needed to learn to work in small groups', 'teacher had to do a lot of preparation work' and 'distracting' as negative responses from students. In addition, they affirmed that teachers gained satisfaction in observing students' demonstration of learner autonomy in differentiated instruction.

LONGITUDINAL ASSESSMENT ON SCIENCE EDUCATION FOR THE GIFTED

In gifted education programmes, children identified with high ability are screened first so it is not easy to assess the 'further' development of their high ability in a short-term by using the criteria commonly in regular classrooms. Some years of implementation of a new curriculum for the gifted may be insufficient for an innovation to impact the total school culture. In this section, some important findings of a longitudinal study on science education for the gifted are described.

A leading longitudinal study on education for the gifted is Terman's study. Terman (1954) selected children based on an intelligence test (in the top 1%) and followed their careers for 30 years. He extracted the gifted men into those scientists and non-scientists (e.g. lawyers, social scientists, non-college group) as well as the division of the scientist subgroups (e.g. physical science research, engineers, medical-biological specialists). The culmination of fame such as entry into 'Who's Who', the 'American Men of Science' and membership in the National Academy of Sciences as well as evidences of intelligent workings such as university enrolment, doctorate achievements, publications and patents were used as indicators for the comparison. The results revealed that the scientist group had more numbers in the American Men of Science, university enrolment, doctorate achievements, publications and patents. It is noteworthy that the two gifted groups were closely matched in their high IQ's, but otherwise they demonstrated differences with respect to scientific promise. This result would present the need to develop tests or ratings for special abilities and interests specific to scientific talent.

Tirri (2000) followed Olympians in Finland, who participated in academic 'Olympics' competitions as students in mathematics (N=73), physics (N=50), and chemistry (N=35) with their ages ranging between under 21 to 50. The results indicated that 43% completed their doctorate degrees, and more than half of the Olympians choose the professions in science and engineering as researchers (39.3 %), engineer (10.1%), or computer specialist (10.1%). They published articles, books and research papers aside from holding patents. Campbell and Feng (2011) studied the science Olympians aged from 16 to 53 and revealed that a conducive home environment provided by parents is very effective to develop successful adult Olympians.

In addition, Wai, Lubinski, Benbow, and Stiegler (2010) assessed participation in various educational opportunities such as science fair/math competitions, research apprenticeships, academics club, inventions and projects, and accelerated classes among 1,467 individuals who had been identified as gifted in math at the age of 13. They found that those who had been involved in more of these educational opportunities (a higher "STEM dose") had, at age 33, a higher rate of notable accomplishments in STEM, such as earning a doctorate degree, writing publications, obtaining patents, or securing tenure. Lubinski and Benbow (2006) conducted a 35-year longitudinal study on mathematically precocious youth and focused on the critical personal antecedents for developing outstanding scientific careers. They screened students by an achievement of their schools and the Scholastic Assessment Test (SAT) score when they were 12 or 13 years old, and found that 21.7 % of them who were in tenure-track positions in the top 50 U.S. universities were already full professors by their mid-30s. Moreover, an excessive number of talent search participants earned especially high incomes (e.g. US$500,000+). Moreover, they reported the talent search participants who had earned patents were well beyond base-rate expectations.

PROFESSIONAL CHARACTERISTICS OF SCIENCE TEACHERS FOR GIFTED CHILDREN

The role of the teacher is critical in science lessons, as is commonly recognised among researchers and practitioners in gifted education. Hanson and Feldhusen (1994) compared teachers trained in gifted education (43 primary and 11 secondary school teachers) to teachers without such training (11 primary and 17 secondary school teachers) and found a number of significant differences. They showed that the teachers trained in gifted education scored significantly higher than the untrained teachers, and the primary school teachers scored significantly higher than the secondary school teachers, in terms of teaching skills, including subject matter coverage, clarity of teaching, motivational techniques, pace of instruction, student-directed activities, variety of student experiences, teacher-student interaction, higher level thinking, creativity, teacher planning, and learning aids.

The National Association for Gifted Children published the 'Teacher Preparation Standards in Gifted and Talented Education' (NAGC, 2013). These seven new standards cover (1) Learner Development and Individual Learning Differences, (2) Learning Environments, (3) Curricular Content Knowledge, (4) Assessment, (5) Instructional Planning and Strategies, (6) Professional Learning and Ethical Practice, and (7) Collaboration, and include detailed guidelines for each standard, particularly aimed at new teachers in gifted education. This standard also includes an appendix with explanations of fundamental technical terms of gifted education (e.g. acceleration). Research in this field led the William and Mary Centre for Gifted Education to develop the Classroom Observation Scales (Teacher Observation) (VanTassel-Baska, Quek, & Feng, 2007). This observation checklist includes 25 check items relating to a teacher's behaviour, classified in terms of (1) General Teaching Behaviours – Curriculum Planning and Delivery, (2) Differentiated

Teaching Behaviours – Accommodations for Individual Differences, Problem Solving, Critical Thinking Strategies, Creative Thinking Strategies, and Research Strategies.

Although the characteristics and competencies of teachers of gifted learners have been investigated in a range of different contexts, research on educational support provided to the gifted by science teachers remains limited. A case study by Sumida, Shirahata, and Kato (2010) identified teachers who had over the preceding eight years instructed numerous winners of prominent prizes at a regional science fair in Japan and modelled their mentoring as shown in Figure 3.

	Stage 1	Stage 2	Stage 3
Students	Inspiration, imagination, and rich sensitivity	Fastidiousness and clarification of the purpose of research	Logical thinking, creativity, and good composition in reporting
Teachers	Understanding of learners' characteristics and individual differences	Planning, continuity, methodology, skills, and research environment	Assessment, organization and communication
Teachers	Communication with students' family		
Family Support			

Figure 3. A model of students' talent and teachers' mentoring in science fair (data from Sumida, Shirahata & Kato, 2010)

Stage 1 in Figure 3 reflects the idea stage of learners' scientific research, underscoring the importance of inspiration, imagination, and rich sensitivity among gifted children. At this stage, science teachers need to support gifted students' independence, understand their specific scientific interests, and provide appropriate resources with a thorough understanding of learners' characteristics and individual differences. Stage 2 is the accomplishment phase of the research, in which gifted students' fastidiousness is to be noted. It is also important for learners to clarify the purpose of their research at this stage. Science teachers might support gifted students by opening science laboratories for the research, consulting on research methods and equipment, or sharing the required chemicals for the research. Such educational support might focus on planning, continuity, methodology, skills, and research environments. Finally, Stage 3 is the presentation stage of the research, characterised by assessment and communication. Logical thinking, creativity, and good composition in reporting findings are required. Science teachers need to evaluate gifted students' efforts and the process of their research, as well as their achievement, in appropriate ways, and enhance their sense of self-fulfilment.

Many science teachers mention the importance of family support in students' scientific research, and it is crucial to establish good relationships with students and family throughout the above stages (Sumida, Shirahata, & Kato, 2010). In this regard, Cho and Cambel (2011) investigated family related factors among students with a talent for science in Grades 4 to 12 and Korean Science Olympians, revealing that family support played a role in both achievement and the nurturing of talent in mathematics and science, even among high school children.

CONCLUDING REMARKS

Rather than addressing the issue of gifted education in a narrow or shortsighted manner, we should examine it within a broad context from both synchronic and diachronic perspectives. Science educators have the potential to innovatively respond to the diverse needs of young students and to add breadth and depth to their ideas and skills through dynamic exploratory activities and cooperative projects. Scientific study is an appropriate context for children to use and develop their gifts in a practical setting and for teachers to discover and support these skills. Expectations are high that gifted science education will be a field of study in which researchers can propose new theoretical foundations and practical directions for developing individuality and ability in children.

In fact, science education research is already playing a leading role in Science Olympiad events, science fairs, and international studies on education. Education policies aimed at facilitating the development and operation of special science high schools are also becoming more widespread globally. An academic environment is rapidly emerging in which students are able to attend lectures by scientists visiting their schools, use the labs at universities and companies in order to conduct their own research or receive guidance from university professors, present their work to academic societies, and file patents. Regardless of where they study, they are increasingly able to use information and communications technology, construct complex models, and analyse big data. At the same time, the more attention we pay to individual differences in the interests, skills, and abilities of students, the greater the demand becomes for high-quality education and increases in staffing and funding. Factors including the school, teachers, and family of individual students impact this process. It is therefore possible that various types of inequality and difference in educational opportunity will emerge.

While the study of gifted education is one type of special education research, it is also a field that has the potential to contribute to equality in education. For example, the study and implementation of science education for 2E children, who have both learning disabilities and giftedness, has the potential to engender new types of inclusive education. Furthermore, using the lens of gifted education to re-examine how differences in gender, race, region, and family income impact science education can enable the discovery of new evidence and the proposal of new solutions, as well as the uncovering of latent human resources and the building of networks and resource pools. In other words, the knowledge gained through gifted education, which is offered to children separated out for special educational opportunities, can provide us with important perspectives for improving science education for children on the non-gifted track, and even more broadly, science education for all children.

In the 20th century, gifted education played an important role in improving the quality of education and providing more diverse options. The enhancement and individualisation of science education was seen as an effective foundation for education for gifted students with high levels of ability, as was the provision of special educational opportunities for gifted children in a specific setting. By contrast, in the 21st century world of science, researchers are expected to work cooperatively in heterogeneous big groups whose members may differ in age, area of expertise, and background (Sumida & Ohashi, 2015). Moving forward, a key challenge for those involved in gifted science education will be determining how to strike a balance between providing students with individualised education and nurturing their ability to cooperate with students who are different from them.

FURTHER READING

Johnsen, S. K., & Kendrick, J. (Eds.). (2005). *Science education for gifted students.* Waco, TX: Prufrock Press, Inc.

Sumida, M., & Taber, S. K. (Eds.). (2017). *Policy and practice in science education for the gifted: Approaches from diverse national contexts.* Routledge.

Taber, K. S. (Ed.). (2007). *Science education for gifted learners.* London: Routledge.

Taber, S. K., & Sumida, M. (Eds.). (2016). *International perspectives on science education for the gifted: Key issues and challenges.* Oxon: Routledge.

Taber, S. K., Sumida, M., & McClure, L. (Eds.). (2017). *Teaching gifted learners in STEM Subjects: Developing talent in science, technology, engineering and mathematics.* Routledge.

REFERENCES

Campbell, J. R., & Feng, A. X. (2011). Comparing adult productivity of American mathematics, chemistry, and physics Olympians with Terman's longitudinal study. *Roeper Review, 33*, 18–25.

Cho, S., & Cambell, J. R. (2011). Differential influences of family processes for scientifically talented individuals' academic achievement along developmental stages. *Roeper Review, 33*, 33–45.

College Board. (2015). *Annual AP program participation 1956–2015.* Retrieved from https://secure-media.collegeboard.org/digitalServices/pdf/research/2015/2015-Annual-Participation.pdf

Cross, T. L., & Frazier, A. D. (2010). Guiding the psychosocial development of gifted students attending specialized residential STEM schools. *Roeper Review, 32*, 32–41.

Hansen, J. B., & Feldhusen, J. F. (1994). Comparison of trained and untrained teachers of gifted students. *Gifted Child Quarterly, 38*(3), 115–121.

Faustino, J. B., Sumida, M., Fajardo, A. C., & Pawilen, G. T. (2010). Parallel curriculum for the development of problem-solving skills of gifted children in grade III science. *Bulletin of the Center for Education and Educational Research the Faculty of Education Ehime University, 28*, 51–65.

Johnson, D. T., Boyce, L. N., & VanTassel-Baska, J. (1995). Science curriculum review: Evaluating materials for high-ability learners. *Gifted Child Quarterly, 39*(1), 36–44.

Lubinski, D., & Benbow, C. P. (2006). Study of mathematically precocious youth after 35 years: Uncovering antecedents for the development of math-science expertise. *Perspectives on Psychological Science, 1*, 316–345.

Mönks, F. J., & Pflüger, R. (2015). *Gifted education in 21 European countries: Inventory and perspective*. Retrieved from https://www.bmbf.de/pub/gifted_education_21_eu_countries.pdf

National Association for Gifted Children. (2013). *2013 NAGC-CEC teacher preparation standards in gifted education*. Washington, DC: National Association for Gifted Children.

National Association for Gifted Children. (2015). *Gifted education in the U.S.* Retrieved from http://www.nagc.org/resources-publications/resources/gifted-education-us

Sondergeld, T. A., & Schultz, R. A. (2008). Science, standards, and differentiation. *Gifted Child Today, 31*(1), 34–40.

Sumida, M. (2010). Identifying twice-exceptional children and three gifted styles in the Japanese primary science classroom. *International Journal of Science Education, 32*, 2097–2111.

Sumida, M. (2012). Meeting the needs of twice-exceptional children in the science classroom. In S. Wichian (Ed.), *Learning disabilities* (pp. 149–174). Rijeka, Croatia: InTech.

Sumida, M. (2017). Gifted science education in the context of Japanese standardization. In M. Sumida & K. S. Taber (Eds.), *Policy and practice in science education for the gifted: Approaches from diverse national contexts*. Routledge.

Sumida, M., & Ohashi, A. (2015). Chemistry education for gifted learners. In J. Garcia-Martinez & E. Serrano-Torregrosa (Eds.), *Chemistry education: Best practices, opportunities and trends* (pp. 469–487). Weinheim: Wiley-VCH.

Sumida, M., Shirahata, A., & Kato, T. (2010). *Development of science teacher training program for the gifted. 1: Model of thinking and behaviours of science teachers of science contest winner students* (pp. 355–356). Proceedings of the 34th Annual Conference of Japan Society for Science Education, Japan.

Terman, L. M. (1954). Scientists and non-scientists in a group of 800 gifted men. *Psychological Monograph: General and Applied, 68*(7), 1–43.

Tirri, K. (2000, August). *Actualizing mathematical giftedness in adulthood.* Paper presented at the ECHA Conference, Debrecen, Hungary.

Tomlinson, C. A., Kaplan, S. N., Renzulli, J. S., Purcell, J., Leppien, J., & Burns, D. (2002). *The parallel curriculum: A design to develop high potential and challenge high-ability learners*. Thousand Oaks, CA: Corwin Press.

VanTassel-Baska, J., Bass, G., Ries, R., Poland, D., & Avery, L. S. (1998). A national study of science curriculum effectiveness with high ability students. *Gifted Child Quarterly, 42*(4), 200–211.

VanTassel-Baska, J., Quek, C., & Feng, A. X. (2007). The development and use of a structured teacher observation scale to assess differentiated best practice. *Roeper Review, 29*, 84–92.

Wai, J., Lubinski, D., Benbow, C. P., & Steiger, J. H. (2010). Accomplishment in science, technology, engineering, and mathematics (STEM) and its relation to STEM educational dose: A 25-year longitudinal study. *Journal of Educational Psychology, 104*(4), 860–871.

BEN AKPAN

36. SCIENCE EDUCATION FOR SUSTAINABLE DEVELOPMENT

INTRODUCTION

The focus of this chapter is on how to use science education as a tool for facilitating sustainable development. The chapter opens with a review of the status of the global environment within the framework of the imminent dangers posed by climate change. An argument is made for the use of education for sustainable development in attaining sustainable development goals. Science education is shown to fit in very well with the tenets of education for sustainable development. The use of topic study, an active learning approach, in teaching and learning of climate change by science educators is advocated. The chapter ends with recommended actions for individuals towards a sustainable environment.

THE STATE OF THE WORLD'S ENVIRONMENT

A Canadian academic and environmental activist, Dr. David Suzuki, has spent several years advocating the sustainable use of Earth's resources. A Foundation he has co-founded, the *David Suzuki Foundation* sets as priorities sustainability, climate change and clean energy as well as oceans, sustainable fishing and protection from oil spills. Dr. Suzuki's environmental programme has led him to produce television documentaries and several books. According to Wikipedia, one of his books *The Sacred Balance* explores society's impact on Earth where he discusses issues of the planet's balance, toxic pollution, and global warming; and also narrates how dependent humanity is on natural resources such as vegetation, water, soil, and sunlight. According to him:

> Human use of fossil fuels is altering the chemistry of the atmosphere; oceans are polluted and depleted of fish; 80 percent of Earth's forests are heavily impacted or gone, yet their destruction continues. An estimated 50,000 species are driven to extinction each year. We dump millions of tonnes of chemicals most untested for their biological effects, and many highly toxic, into air, water and soil. We have created an ecological holocaust. Our very health and survival are at stake, yet we act as if we have plenty of time to respond. (Suzuki, UD)

K. S. Taber & B. Akpan (Eds.), Science Education, 493–504.

Elsewhere, he also says:

> We're in a giant car heading towards a brick wall and everyone's arguing over where they're going to sit. (Suzuki, UD)

Wikipedia also reports that Dr. Suzuki has been worried about his own carbon footprint. As he travels constantly to spread his message, Dr. Suzuki has unwittingly exceeded his carbon limit and so has now chosen to avoid vacations overseas, cluster his speaking engagements together, and where feasible to use video conferencing.

Dr. Suzuki is raising a very legitimate concern. He is speaking the minds of millions of people. Issues surrounding dwindling forests, unsustainable consumption patterns, species extinction, carbon emission, and biodiversity loss have been on top of global agenda for quite some time. What has been lacking has been a joint, concerted and sincere efforts to mitigate these problems. Much of these problems result in climate change, a concept to which I next turn.

THE SCIENCE OF A PHENOMENON

What essentially is the science behind climate change? This term is often used interchangeably with global warming. According to the WWF Global (ND), life on Earth depends on the continuous flow of heat from the Sun. This warms the surface of the Earth and the Earth in turn sends back some heat energy in the form of infrared radiation into the atmosphere. Some gases such as carbon dioxide, nitrous oxide, hydrocarbons (such as methane), ozone, and water vapour which occur naturally in the atmosphere and acting as a blanket, absorb some of this heat thus ensuring that the Earth's temperature is maintained at an average of 14°C. If these gases were not there, the Earth's temperature could probably plummet to -18°C at which many forms of life would not exist. The natural warming effect caused by these gases is commonly referred to as greenhouse effect. Enhanced (or anthropogenic) greenhouse effect may occur due to increased atmospheric carbon dioxide caused by human activities such as the burning of fossil fuel, cement production, and destruction of tropical forests. When this occurs, the result is global warming or climate change. Climate change can lead to drought, heatwaves, rising sea levels, storms, and floods which are capable of causing untold damage to infrastructure, agriculture, and tourism. These ultimately lead to increase in living costs and poor living and health standards.

The issue of climate change has therefore received global attention at the highest level. World leaders, scientists, environmentalists, and other stakeholders have over the years sought to work in concert to mitigate the impact of human-generated greenhouse gases on the natural environment. The main approach has been how to get various countries to cut down on carbon emission. Several declarations have been made in the past but the latest was the *Paris Climate Change Conference* which took place in 2015 where world leaders in consort with other stakeholders agreed to strengthen the global response to the threat of climate change, in the context of sustainable development and efforts to eradicate poverty through several measures such as:

- holding the increase in the global average temperature to well below 2 °C above pre-industrial levels and to pursue efforts to limit the temperature increase to 1.5 °C above pre-industrial levels so as to significantly reduce the risks and impacts of climate change;
- increasing the ability to adapt to the adverse impacts of climate change and foster climate resilience and low greenhouse gas emissions development, in a way that would not threaten food production; and
- providing finances in conformity with a pathway towards low greenhouse gas emissions and climate resilient development (United Nations, 2015).

It does appear that the key to tackling climate change is sustainable development. In the following section, this concept is examined in detail.

SUSTAINABLE DEVELOPMENT

Sustainable development is development that meets the needs of the present without compromising the ability of future generations to meet their own needs (United Nations, 1987). Grace (2010) contends that in looking at sustainable development, three areas should be of concern. These are economic development, social development, and environmental protection. Interestingly, the United Nations (UN) has shown a lot of interest in this subject matter by issuing a set of goals towards the attainment of sustainable development (United Nations, 2016). In what follows, I provide a summary of the position of the UN on sustainable development:

- *Poverty*: According to the UN, 837 million people live in extreme poverty as of 2014. Hunger, malnutrition, illiteracy, discriminatory practices are all indicators of poverty. The UN is pushing for inclusive economic policies in the various countries to assure sustainable living conditions by 2030.
- *Hunger and Food Security*: End hunger, achieve food security and improved nutrition and promote agriculture. This is the message from the UN in view of the degradation of soils, oceans, forests, freshwater, and biodiversity as a result of human actions that promote climate change. It urges nations to work towards profoundly changing food consumption patterns and agricultural systems by 2030.
- *Health*: There is need to ensure good health for people of all ages as a prelude to sustainable development by making concerted efforts to eradicate a wide range of diseases such as tuberculosis and polio. It is envisaged that by 2030, there will be an end to preventable deaths of newborns and infants.
- *Education*: An overarching need for qualitative and lifelong education as a foundation for improvement of people's lives and assurance of sustainable development should be anchored on equitable access to education for all by 2030.

Globally, a stunning 103 million people lack access to basic literacy, of which about 60 per cent are women. There is need to reverse this trend according to the UN.

- *Gender equality and women's empowerment*: Although the UN admits gender equality is not a fundamental human right, it does see it as a necessary foundation for a peaceful, prosperous, and sustainable world. To that extent, the UN calls for the provision of equal access to education, employment, health care, and representation in economic and political matters to women and girls. This is more so because in Sub-Saharan Africa, Western Asia, and Oceania, girls do still have difficulties gaining access to primary and secondary education.
- *Water and Sanitation*: Even though there is sufficient freshwater to serve humanity, bad economic policies and lack of infrastructure have resulted in millions of people (mostly children) dying every year from water-borne diseases and poor sanitation as well as scarcity of water. As of 2014, 1.8 billion people used water that was contaminated with faeces while 2.4 billion people lacked access to basic sanitation services.
- *Energy*: Since energy is central to every area of human endeavour, it is essential that every person has access to it. Sadly enough, 3 billion people use coal, animal waste, charcoal, and wood for domestic chores. Indeed, energy accounts for about 60 per cent of total global greenhouse emissions and occupies the top unenviable position as lead contributor to climate change. The UN is of the view that the reduction in carbon emissions is a strategic initiative towards mitigating the impact of greenhouse effect and thus protecting the Earth's climate.
- *Economic growth*: With half of the population of the world living on about US$2 per day, sustainable economic growth demands that various countries facilitate conditions that create quality jobs and stimulate the economy without posing dangers to the environment.
- *Infrastructure and industrialisation*: Transport, irrigation, information and communications are critical areas requiring investments to ensure rapid and sustainable development.
- *Inequality*: A significant majority of households in developing countries are experiencing more disparate and unequally distributed incomes than the levels in the 1990s. A major plank of attaining a positive turn-around on this issue is to place a high premium on the needs of disadvantaged and marginalised groups.
- *Cities*: Cities should be made more inclusive, safe, resilient and sustainable as it is projected that by 2030, nearly 60 percent of people in the world will reside in urban areas. This is critical as currently the cities in the world constitute 3 percent of Earth's land, but account for over 60 percent of carbon emissions.
- *Sustainable consumption and production*: The UN has the slogan *doing more and better with less* as the key to sustainable consumption and production given the gloomy picture that if global population rises to 9.6 billion by 2050, then humanity

will require the equivalent of three planets to provide the natural resources needed to sustain people at present consumption levels.

- *Climate change*: Changing weather patterns and rise in sea levels are indicators of climate change caused by greenhouse gas emissions from activities of humans. Unfortunately, the impact of climate change knows no national boundaries. That is why nations are working together and have adopted a global framework (United Nations, 2015) on global change.
- *Oceans*: As oceans cover three quarters of Earth's surface; contain 97 percent of Earth's water; absorb about 30 percent of carbon dioxide produced by humans thus mitigating the effect of global warming; they, along with seas and marine resources, should be conserved and used sustainably.
- *Biodiversity, forests, and desertification*: Noting that forests are home to more than 80 percent of all terrestrial species of animals, plants, and insects; 12 million hectares of land that could grow 20 million tons of grain are lost to drought and desertification; and that of the 8,300 animal breeds known, 8 percent are extinct while 22 percent are at risk of extinction, the UN is pushing for sustainably-managed forests, a halt to biodiversity loss, and measures aimed at combating desertification and land degradation.
- *Peace and justice*: According to the UN, sustainable development at national and international levels is promoted by the rule of law.
- *Partnerships*: Inclusive partnerships between government, private sector, and civil society which are built upon mutually beneficial visions and values are key to sustainable development.

EDUCATION FOR SUSTAINABLE DEVELOPMENT

There is no question as to whether there is need for sustainable development. However, one of the issues that has arisen has been how to create awareness about the need for sustainable development. A major approach has been through Education for Sustainable Development (ESD). According to UNESCO (2014, p. 19), 'ESD empowers everyone to make informed decisions for environmental integrity, economic viability, and a just society for present and future generations, while respecting cultural diversity.' Indeed, to underpin the importance of ESD, 2005–2014 was declared as the *UN Decade for Sustainable Development*. Over that decade, UNESCO took major steps to facilitate ESD, especially by:

- Working towards ensuring that people had access to basic education;
- Getting various countries to reform education policies so as to incorporate sustainable development issues;
- Mounting enlightenment campaigns to assure global awareness of the issues involved in ESD; and
- Facilitating capacity building to raise the required human capital to support ESD.

In the light of these goals, UNESCO implemented a multiplicity of strategies to assure the success of ESD. These included (but not limited to) advocacy programmes, consultations, networking, monitoring, and evaluation (UNESCO, 2005). It should be noted that prior to DESD, there were several efforts by the UN in pursuance of sustainable development. Some of these were the 1992 *UN Conference on Environment and Development* (UNCED) also known as the *Rio Earth Summit*; the 2000 *UN Millennium Summit* where world leaders declared their commitment to the attainment of *Millennium Development Goals* (MDGs) by 2015; and of course, the 57th Session of the UN General Assembly where 2005–2014 was adopted as *United Nations Decade for Education for Sustainable Development*. According to Wikipedia, the MDGs are the eight international development goals that were established following the *Millennium Summit* referred to above. The 189 UN member states at the time (currently there are 193) and at least 23 international organisations, committed to help achieve the following MDGs by 2015:

- To eradicate extreme poverty and hunger.
- To achieve universal primary education.
- To promote gender equality and empower women.
- To reduce child mortality.
- To improve maternal health.
- To combat HIV/AIDS, malaria, and other diseases.
- To ensure environmental sustainability.
- To develop a global partnership for development.

Each MDG goal had specific targets and dates for attainment of those targets. Wikipedia notes that critics of the MDGs complained of lack of analysis and justification of the objectives, especially the difficulty of precise measurements for some of the stated goals as well as uneven progress. Even though developed nations' aid for attaining the targets rose during the earlier years of implementation, much of such assistance went for debt relief with a significant part of the remainder used for natural disaster relief and military assistance, instead of the required further development. What has been illustrated here is the nexus between MDGs and ESD.

Beyond 2005, several activities took place in furtherance of DESD. These included the 2007 *4th International Conference on Environmental Education Towards a Sustainable Future* ; the 2009 1st DSED Global Monitoring and Evaluation Report; the 2012 *2nd DESD Global Monitoring and Evaluation Report*; and the 2013 37th Session of UNESCO General Assembly which endorsed the *Global Action Plan* (GAP) on ESD as a follow-up programme on ESD (UNESCO, 2014).

THE ROLE OF SCIENCE EDUCATION

Science Education has a critical role to play as the world seeks concerted efforts on promoting sustainable development through ESD. If according to UNESCO (2014, p. 19) 'ESD requires participatory teaching and learning methods like critical thinking, imagining future scenarios and making decisions in a collaborative way in order to empower learners to take action for sustainable development', the corollary is that science education is well placed to facilitate ESD. From the above position of UNESCO, it is clear that the educational strategies that support ESD are very much in use in science education. Some of these are:

- Ability to ask critical questions so as to clear any doubts;
- Creative thinking as may be applied in individuals imagining more positive and sustainable futures;
- Ability to apply subject matter learnt to new situations as is usually accomplished through transfer of learning; and
- Ability to adjust to changes in society in general and the environment in particular as may be required in adopting new habits towards the environment in the face of human-created problems especially the generation of greenhouse gases.

The challenge for science education has been on identifying an approach that is suitable for implementing ESD. One such approach is 'topic study' which I discuss in the next section.

TOPIC STUDY

Topic study is an approach that is useful in teaching interdisciplinary subjects such as global warming. Essentially, a topic study requires teachers and learners to imagine a context or somewhere where something can occur. Everybody joins in describing it, drawing it, and writing about it as much as they can. They all learn about 'models' of the world they have in their heads. The group shares ideas, questioning each other's ideas in turn. The result is the generation or modification of ideas. That way, the topic gets more dynamic as ideas on the topic are changed and improved as stories are told. The teacher's role is to assure that learners work out what best to do as well as getting them to enjoy the excitement and fun associated with topic study. By so doing, the learners have a sense of ownership of the learning process as it is their 'topic'. The teacher is seen principally as a facilitator.

Topic story was created at Jordanhill in Glasgow, Scotland. It is based on the active learning model of education which engages children in problem-solving activities in order to promote skills of analysis, synthesis, and evaluation. In implementing topic study, the facilitator takes learners through a set of written texts to which the learners respond to either as a group or sub-groups depending on the circumstances of the learning environment. The following set of texts is adapted with permission from Okebukola and Akpan (1997, pp. 22–23) and the activities are suitable for students who are 15 years old and above:

Text 1: A dying planet – The Earth was dying. The forests were gone, the sea polluted and fish driven to extinction. Now, the crops were failing because of soil loss, changing weather patterns, new and unexpected diseases. Global warming was unstoppable. Millions if not billions faced a slow death by malnutrition, starvation, or genocide, as social order broke down. Those with power and money looked for an escape. Could they use their wealth to survive? One solution was to seek new lands but there were none left … on Earth. They paid scientists to search the skies for a new planet; one where they could, given time, recreate Earth. The search was successful. There was a suitable planet. The problem was that it was so distant that it would take 250 years to reach. Or, as scientists could be very exact about these things, 248 years, 34 days, and 6 hours! The problem was to design and build a spaceship which could make the journey. Teams of scientists were selected to advise and help.

Text 2: Space convoy – The huge spaceship was designed as modules which could be joined in space held together by passageways like the spokes and rim of a large wheel. Each module was similar but adapted for its particular function. Spaceship … {place name of desired country} had modules for energy and power, recycling and recovery, a species bank, accommodation, education, health, passengers in hibernation, workshops, entertainment, and a command centre. At the centre of the wheel was the life support system. Plumbing and other services were visible on the outside of the modules. So each module was easily recognized. Inside, the walls and rooms were arranged to suit the function of the module. Each module was therefore allocated to a separate design team, so that all would be ready by the launch date. If the deadline was missed by one module, the whole spaceship would never leave Earth orbit.

Text 3: The crew and passenger – Many more people wanted to go than there were places for. Committees were set up to determine the criteria for selection. There were two categories: crew and passengers. The crew would be there because their skills were essential for the success of the voyage. They would also need to be replaced when they became old and so no longer able to function in that capacity. The passengers included the rich and powerful. They decided that they needed others to make life on the new planet a success. Some passengers would come aboard 'alive' some in deep sleep at low temperatures, and some frozen in the deep freezer in the hope that they could be revived. The organisers asked groups to draft proposals for the selection criteria. A consensus would be established.

Text 4: Life's necessities – Life on board the ship would not be much fun. Luxuries would not be allowed; only the necessities of life. Although this rule was established early in the planning, it became clear that different teams and people had different ideas. A plan had to be found to develop a list which all could agree to. They asked for suggestions, seeking one page summaries from each team. If no better suggestion came up, then each team would create its list and present it to the others, who would accept, modify, or reject items. The finished combined list would still need to be shortened.

Text 5: Sourcing necessities – Even simple 'everyday' items are the result of many activities, indeed a chain or 'tree' of processes. The organisers knew that the spaceship would be too far away for help, so they had to make sure they forgot nothing that could endanger the voyage. They approached consultants. Each item selected as an 'essential' was considered. Its parts were identified, and sourced. But where in turn did these parts come from? How had they been made? Where had the factory, the raw materials, and the energy come from? Experts said that everything could be traced back to minerals and elements and energy sources, usually the Sun. For example, to explain the process, they selected the ball point pen. It was made of three kinds of plastic, a brass writing head, a tungsten ball, and filled with ink. How had these parts been made, with what techniques, what resources, and what energy? They put their heads together and thought about the matter.

Text 6: Energy sources – The importance of the Sun was a matter of concern. It would only be a few months before the spaceship was too far from the Sun to be able to harvest its energy. All fuels were considered. Whatever fuel system was to be used, they would have to take care not to be wasteful. Scientists told them that energy did not disappear, it was never destroyed; but just converted into other forms and spread out. 'Using energy' was a process which took stores of energy and scattered them. The organisers asked if the scientists could think of ways to recover the metals and plastics and other materials. It was difficult for the scientists to explain this. Why can't the scientists recycle energy?

Text 7: Sustainable life support system – Once the energy source had been decided upon, the teams of scientists were asked to submit designs for the core of the spaceship – the sustainable life support system. The message from the ecologists was this: simple systems can easily swing out of balance and need very close monitoring and control; yet complex systems seemed to be able to create stability from their very complexity. The choice, therefore, was to create a computer-monitored and controlled simple system, with all the risks; or to establish a complex ecosystem transplanted from Earth. Was there any other way?

Text 8: Space pirates – There had always been rumours that there were other convoys in space. These had been poorly designed and their life support systems failed to meet the needs of all who boarded. Fighting had broken out, and (so the rumour went) a ruthless group had seized control. They had stopped trying to make their spaceship work, and instead had adapted them for attack. They prowled through space, seeking defenceless spaceships to plunder, so that they could survive themselves. Unknown to everyone was a new danger. After the convoy had left, the rich who had been rejected, the drug lords, and other criminals of the world, had designed attack ships. These were on their way.

Text 9: Memories of Earth – In the school, the children who had been studying Earth history were asked to write an imaginary poem describing what living on Earth might have been like. The structure they used was a sense poem. The teacher gave them headings:

- see
- smell
- touch
- fear

- hear
- feel
- do
- wish

When the class or groups had come up with good expressive phrases such as:

I see the remnants of the once alluring forests
I smell the stinking odour of polluted air
I touch the skeleton of extinct species
I fear the total collapse of the ecosystem

the teacher rubbed out the first two words on each line. Now, they had a poem.

Text 10: Messages to Earth – After many years of journeying, the crew decided that they should send messages to Earth. They wanted to say that they had learned to live with limited resources, never wasting materials or energy, and making sure that they recycled everything that could be recycled. They thought it was proper for those on Earth to be told. Now, they realised that Earth itself was a spaceship travelling through empty space with no hope of help from outside. The 'crew' of Earth had to make their spaceship work.

Text 11: A message from space – Earth had not done well in the years since the spaceship left. The message was too late. There was no one who could do anything, for it was a question of individual survival. There was no hope, no one to tell, all was lost. The stinking hulk of the once beautiful Earth would forever drift in space. Nature itself continued and new species filled the niches – so life continued, but not humans. It need not have been so. The message from space could have been learned in time. What are the implications globally, nationally, locally, and individually of the message from space? Select a part of another group's message and write a response to it under the four headings:

- Individual Implications
- Local Implications
- National Implications
- Global Implications

INDIVIDUAL ACTION TOWARDS SUSTAINABLE ENVIRONMENT

The issue of climate change is one that requires everyone's attention and action. The following are recommended as actions that should be taken by individuals in support of efforts against climate change:

- Patronise low carbon diets. This minimizes the emissions released from the production, packaging, processing, transport, preparation, and waste of food. Eat less industrial meat, dairy, and food produced industrially. Go for locally grown food items.

- Protect trees and plant new ones in the yard, along the roads, and public parks as a means of promoting the absorption of carbon dioxide from the atmosphere.
- Use low energy household appliances, electronics, and office equipment. Look out for *ENERGY STAR* label on products. This is an international standard for energy-efficient consumer products. Devices carrying the label use about 25 percent less energy than conventional products.
- Use vehicles which get high gas mileage to reduce emissions of carbon dioxide, or avoid using vehicles where possible.
- Promote the movement away from fossil fuels by using renewable energy to meet personal power needs. Use of solar photovoltaics or wind to generate energy is a step in this direction.
- Reduce personal energy use by: a). turning down temperature on refrigerator, washing machine, and water heater; and b). using public transportation, walking, or biking when possible.
- Keep garbage out of landfills by composting scraps from kitchen and trimmings from the garden.
- Recycle paper, plastic, metal, and glass.
- Ask your province to impose carbon taxes. These taxes make polluting activities more expensive and green solutions more affordable. Such action promotes energy-efficient businesses and enables households to save money.
- Reduce the frequency of flights that you take as air travel promotes carbon emissions. Use of buses, trains, or videoconferencing to stay in touch with people are some options.
- Stay informed about current developments on climate change.

SUMMARY

In this chapter, we have discussed the following:

- The state of the world's environment especially with regards to the dangers posed by human-generated greenhouse gases
- The scientific explanation for climate change (or global warming)
- The views of the United Nations on sustainable development and the need for education for sustainable development
- How to use topic study for the implementation of education for sustainable development in science education using global warming or climate change as an example.
- Actions to be taken by individuals to protect the environment.

USEFUL WEBSITES

UNESCO: http://en.unesco.org/sdgs
United Nations: http://sustainabledevelopment.un.org
World Wide Fund for Nature: http://worldwildlife.org

REFERENCES

Dillon, J., & Huang, J. (2010). Education for sustainable development: Opportunity or threat? *School Science Review, 92*(338), 39–44.

Grace, M. (2010). Education for sustainable development. *School Science Review, 92*(338), 27–30.

Hogg, M. (2010). Using scientific enquiry to make sense of global challenges. *School Science Review, 92*(338), 45–49.

Okebukola, P., & Akpan, B. B. (1997, December). Spaceship Nigeria: A topic study for global warming, greenhouse effect, and ozone layer depletion. *Science Education International, 8*(4).

Suzuki, D. (u.d.). Retrieved March 12, 2016, from http://www.climate-change-guide.com/quotes-on-climate-change.html

UNESCO. (2013). *Education for sustainable development: A sound investment to accelerate African development.* Retrieved January 19, 2016, from http://archive.ias.unu.edu/resource_centre/TICADV-ESD-flyer-2p.pdf

UNESCO. (2014). *Shaping the future we want: UN decade for sustainable development (2005 – 2014), Final Report.* Retrieved January 18, 2016, from http://unesdoc.unesco.org/images/0023/002301/230171e.pdf

UNESCO. (2015). *Global action on education for sustainable development.* Retrieved January 18, 2016, http://www.unesco.org/new/en/unesco-world-conference-on-esd-2014/esd-after-2014/global-action-programme/

United Nations. (1987). *Report of the world commission on environment and development: Our common future.* Retrieved January 19, 2016, from http://www.un-documents.net/our-common-future.pdf

United Nations. (2015). *Framework convention on climate change.* Retrieved January 18, 2016, from http://unfccc.int/resource/docs/2015/cop21/eng/l09.pdf

United Nations. (2016). *Sustainable development goals: 17 goals to transform our world.*

WWFGlobal. (u.d.). Retrieved January 18, 2016, from http://wwf.panda.org/about_our_earth/aboutcc/how_cc_works/

SECTION VIII

SCIENCE TEACHER EDUCATION

MARISSA ROLLNICK AND ELIZABETH MAVHUNGA

37. PEDAGOGICAL CONTENT KNOWLEDGE

INTRODUCTION

In this chapter we explore an important dimension of teacher knowledge, called pedagogical content knowledge (PCK). PCK is specialised knowledge about the content to be taught possessed only by teachers. This knowledge is often hidden in that teachers do not realise that they have it or that it is important. So we use different ways of trying to expose this knowledge, make you aware of it and equip you to apply it to other science topics, not explored in this chapter.

WHAT IS PCK?

In 1986, Dr Lee Shulman delivered an address to the American Educational Research Association. He was concerned about how teachers were trained and felt there was too much emphasis on teaching methods at the expense of content. He asked, "Where did the content go?" Most importantly he pointed out the difference between a teacher and a content specialist. This difference, he said was due to specialised knowledge that teachers possessed to transform content knowledge into teachable form. He thus invented the term, 'pedagogical content knowledge' (PCK) which he described in these words,

> Within the category of PCK I include, for the most regularly taught topics in one's subject area, the most useful forms of those representations of those ideas, the most powerful analogies, illustrations, examples, explanations and demonstrations – in a word, the ways of representing and formulating the subject that makes it comprehensible to others. PCK also includes an understanding of what makes the learning of specific topics easy or difficult: the conceptions and preconceptions that students of different ages and backgrounds bring with them to the learning of those most frequently taught topics and lessons. If those preconceptions are misconceptions, as they so often are, teachers need knowledge of the strategies most likely to be fruitful in reorganising the understanding of learners, because those learners are unlikely to appear before them as blank slates. (Shulman, 1986, pp. 9–10)

Since then, there has been much published on PCK but the most difficult challenge has been to define it precisely and give examples so that new teachers can understand

K. S. Taber & B. Akpan (Eds.), Science Education, 507–522.

it and use it in their teaching. We have learnt that this is much easier when one defines PCK as specific to every topic taught and to every class you teach. It is not possible to prepare new teachers for every group of students they may teach, nor for every topic. However it is possible to give specific examples of PCK in specific topics and thus provide tools for developing PCK.

PCK is understood to be multidimensional, which means there are various levels at which one can explore the construct. For our purpose, there are two types of PCK which are of interest to this discussion – personal PCK and canonical PCK (Gess-Newsome, 2015). Personal PCK varies from teacher to teacher and from context to context. It also refers to the knowledge teachers have and how they put it into practice. Gess-Newsome refers to personal PCK as

> the knowledge of, reasoning behind, and planning for teaching a particular topic in a particular way for a particular purpose to particular students for enhanced student outcomes. (p. 36)

Canonical PCK is PCK that depends on the topic being taught and is part of the body of knowledge established by the science education profession as good practice. Thus it means that we can determine the quality of canonical PCK and share common findings about its nature. Thus the discussion henceforth refers predominantly to canonical PCK unless otherwise mentioned.

Looking at Shulman's definition above we can easily see several components that make up PCK, especially if we look at PCK topic by topic. PCK in a topic is considered another dimension of PCK which is different from the broader discipline perspective (Veal & MaKinster, 1999). In this chapter we will be using a topic-specific approach to understand PCK using examples from different science topics in the curriculum. Four components that clearly emerge from Shulman's definition are:

- *Representations*: including analogies, illustrations, examples, explanations and demonstrations
- *What makes the learning of specific topics easy or difficult*
- *Learner prior knowledge*: the conceptions and preconceptions that students bring with them. These could be preconceptions or misconceptions
- *Knowledge of teaching strategies most likely to be fruitful*

We also include a fifth one that, after Geddis (1993), we call *Curricular Saliency*. Curricular saliency refers to the ability to identify the big ideas that hold a topic together. Big ideas are statements formulated to express the main concepts of the topic. Curricular saliency includes how those big ideas relate to other ideas in the topic and how to sequence them in teaching. It also includes an understanding of what topics are taught before and after the topic and why it is important to teach the topic. Finally curricular saliency includes an understanding of what not to teach at a particular stage as well as what to teach.

Table 1. sets out the components of PCK that are important in teaching any science topic. (The reader may wish to consider a topic they know well, and consider suitable entries for each component.)

Table 1. Identifying components of PCK in a topic

Component	*Examples from any topic*
Learner prior knowledge	
Curricular Saliency	
(i) Big ideas	
(ii) Sequence of big ideas	
(iii) Prior concepts needed	
What is difficult to understand	
Representations	
Teaching strategies	

CAPTURING AND PORTRAYING PCK

In the introduction we said that PCK is hidden knowledge, so we need to find ways to make it visible. Two well-known ways of doing this are using Content Representations (CoRes) and cases. Shulman called for the use of cases to bring teachers' knowledge to light. Below is a case of Mr Banda teaching chemical equilibrium to 17 year olds. We first provide an overview with examples of the nature of Mr Banda's classroom practice and further demonstrate how the quality of his PCK in this particular topic could be captured for display.

Mr Banda spends a whole lesson on the nature of chemical equilibrium, trying to establish two important concepts – that of open and closed systems and that of dynamic equilibrium. He wishes the class to have a clear conceptual understanding of the big idea that dynamic chemical equilibrium takes place in a closed system where the rate of the forward reaction is equal to the reverse reaction. Both the forward and reverse reaction continue to happen but externally there is no apparent change in the concentrations of the various substances. He begins the lesson by establishing understanding of open and closed systems through the use of examples. He then establishes with the class that chemical equilibrium takes place only in a closed system. He then seeks to establish understanding of the term dynamic equilibrium. The case below focuses on the second part, establishing understanding of dynamic equilibrium.

Establishing understanding of dynamic chemical equilibrium: Mr Banda begins the lesson by reminding the students that in the previous lesson they discussed open and closed systems and began to talk about dynamic chemical equilibrium. He asks the class (B- Mr Banda S – Student answers):

B: Can I get someone who would like to explain to us, it could be in your own words – what you think dynamic equilibrium is. Yes, sir.

S1: I think dynamic equilibrium is when two things are equal but it might be… (hesitates) How can I say it? Can somebody help me?

S2: My definition of dynamic equilibrium is where the reactants and the products are moving in a constant concentration

B: So it's when the concentration of the reactants and the products are the same?

S2: No, they are constant. They are not the same, because if they are the same it means they are equal. So, they are not the same.

B: I get what you are saying. You are saying that the concentrations are not necessarily equal, but they are constant. OK. You guys are giving good answers.

At this point Mr Banda turns to some drawings that he put on the board before the lesson, illustrated in Figure 1 he says:

B: If you look behind me on the board… All right? Have you ever played a seesaw?

S: Yes

B: What could happen if we have the same weight and we are on the seesaw? It's gonna be balanced. Right? And we have come to understand that equilibrium is actually a state of balance. But today I actually want us to look at the difference between two types of equilibrium. Dynamic and static, or dynamic and physical equilibrium.

B: I would request that you give me your own examples of static equilibrium. All right? We have given the first one. The first one is a seesaw, whereby two people of the same mass are just sitting on the seesaw and they are in a state of balance. So the seesaw is not moving. It's just balanced. OK? Where else have you ever observed static equilibrium? Yes?

S: Tug of war

B: Tug of war. OK. So you have 2 people, so you have people on the left and the right and they are just pulling a rope.

B: If the other side is pulled, right, then we do not refer to that as static equilibrium. But where will we refer to it as static?

S: When you have 3 people with the same mass this side, and three people with the same mass this side, and the one is pulling the other like this and the balance. Everyone is standing in the same position pulling straight, …

B: So there is no movement and the people on the left are pulling with the same force as the people on the right.

After exploring some further analogies including that of an escalator he moves to chemical equilibrium:

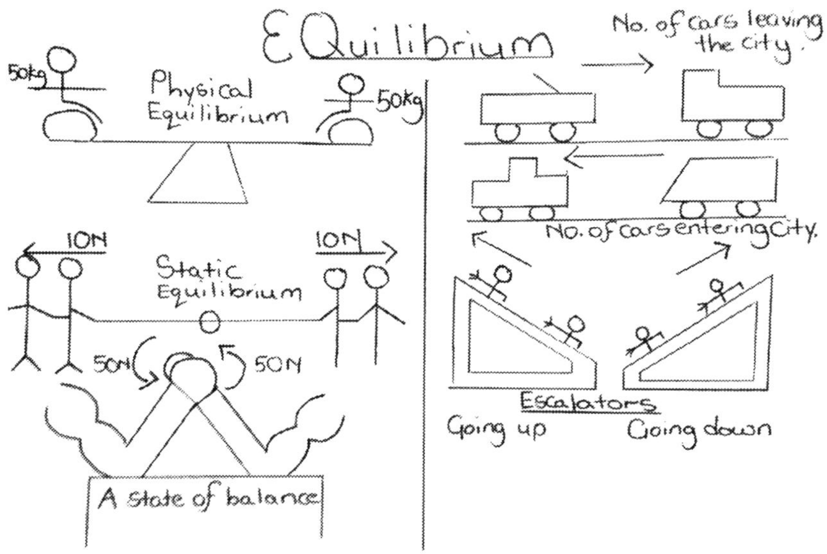

Figure 1. Mr Banda's analogies of chemical equilibrium

B: But now we want to narrow our study to dynamic equilibrium of chemicals. We have gotten an understanding that when the concentration of the reactants and products are constant, then the system is in dynamic equilibrium. I'm going to give you an example again of nitrogen gas, reacting with hydrogen gas, to give two molecules of ammonia. If we were to represent this, using sub-microscopic representations, you would have to denote keys for nitrogen, hydrogen and ammonia. All right?

On the board Mr Banda draws a box with 2 molecules of nitrogen and two molecules of hydrogen and asks the class if what he has drawn is correct. The class tells him that the equation needs to be balanced. He then works with them to balance the equation and finally writes the following representation on the board:

He then moves to what is present in the mixture at equilibrium:

B: This is the result. OK? This is what we would expect from a normal reaction. But now we are speaking of equilibrium. A reaction has happened and we discover that we are able to detect some nitrogen and we are able to detect some hydrogen, and obviously the product which is ammonia. So now the whole idea changes – we no longer have nitrogen reacting with hydrogen to produce ammonia, but we have something different. We have nitrogen gas again reacting with 3 molecules of hydrogen gas

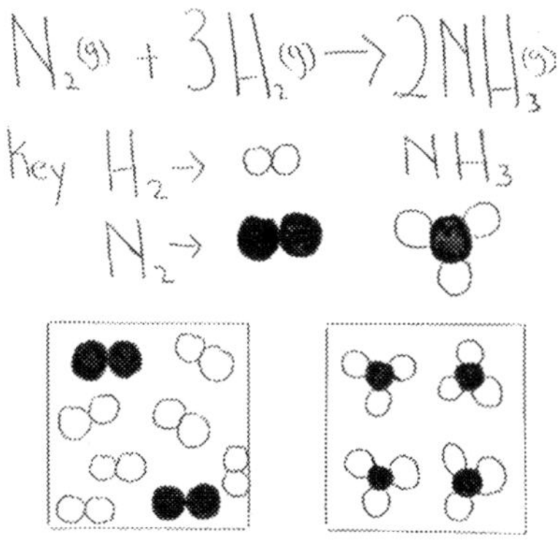

Figure 2. Mr Banda's representation of ammonia formation

to form 2 molecules of ammonia gas. But now the arrow is like this. The arrow – what is the arrow showing us now?

S: reverse

B: They are showing us that the reaction is reversible. And there are some contents within the system of nitrogen, hydrogen and ammonia. So, yes, we do accept that nitrogen and hydrogen have reacted, but they are still in the system. OK? So what we would expect within the system is molecules of nitrogen, molecules of hydrogen, and molecules of ammonia. OK? … You have hydrogen, you are able to detect that there is hydrogen, you are able to detect that there is nitrogen, you are able to detect that there is ammonia. Is this reaction complete?

S: It's not complete

B: OK. Why is it not complete?

S: 2 molecules of ammonia… and 6 hydrogen atoms – it's complete

B: OK. That's one view. Who has another view?

S: I see … the molecule that was supposed to be formed is already formed. It's just we still have the reactant inside

After some more interchanges he continues:

B: So, if a complete reaction, tells us that nitrogen reacts with hydrogen and they are completely used up to form ammonia – so in this system you will not find any nitrogen and you will not find any hydrogen. But now we react these substances together and we find the nitrogen, hydrogen and the ammonia within one system and we are saying it's complete?

B: It's not complete.

Table 1 could be used to identify components of PCK evidence in this lesson segment. (Readers may wish to see which elements of PCK they could identify in the lesson extract, and consider how they might have made alternative choices.)

CONTENT REPRESENTATIONS (CORES)

As indicated earlier, we can capture the quality of PCK demonstrated by Mr Banda by constructing a CoRe. Often CoRes are used to plan teaching but in this case we are using it as a backward looking tool, to analyse teaching. To construct a CoRe one needs first to identify the big ideas for a topic (part of curricular saliency) and then unpack them using a series of prompts as shown in the table below. Identifying big ideas is not easy and we will be dealing with this in more detail below. For the lessons on the topic of chemical equilibrium described above there are four big ideas:

- What is Chemical Equilibrium?
- Factors that affect Equilibrium
- Equilibrium Constant
- Application of Equilibrium Principles

The CoRe illustrated in Table 2 is based on one big idea as an illustration. You will also see that the prompts require other components of PCK to be made explicit. The table shows the CoRe for Mr. Banda's teaching as described in the case above. (The reader may find it useful to relate these entries back to the lesson extract presented above.)

Table 2. CoRe for chemical equilibrium

Big idea	*What is Chemical Equilibrium? equilibrium, open and closed systems, A reverse reactions, Dynamic equilibrium*
1. What do you intend the students to learn about this idea?	• Define Equilibrium • Open and closed systems (the difference) • A Reverse reaction • Dynamic equilibrium.
2. Why do you think it is important for students to know this?	• Learners need to understand the story of chemical equilibrium, where chemical equilibrium occurs and the difference between dynamic and static equilibrium which helps in preventing a lot of misconceptions as to how learners see chemical equilibrium. It is also important because it informs their understanding on what comes next because knowledge builds on other concepts.
3. What else do you know about this idea that you do not intend students to know yet?	• The definition of chemical equilibrium using the rate idea
4. What are the difficulties associated with teaching this idea?	• It is hard to teach students that the reverse reaction is the same as the backward reaction if they cannot picture it

(Continued)

Table 2. (Continued)

Big idea	*What is Chemical Equilibrium? equilibrium, open and closed systems, A reverse reactions, Dynamic equilibrium*
5. What knowledge can you share about student's thinking that influences your teaching of this idea?	• Students will have difficulties with the language associated with this big idea, e.g. equilibrium position • Students confuse dynamic and static equilibrium. They think that when a reaction has reached equilibrium there will be no further change because they compare it with the static equilibrium of the see saw. • Students find it difficult to understand that the chemical reactions do not carry on to completion
6. Are there any other factors that would influence your teaching of this idea?	• Lack of secure knowledge of earlier topics, e.g. students unable to state that balanced chemical equations represent the rearrangement of atoms. Students have difficulty in recognizing and describing instances of physical and chemical change
7. What teaching procedures would you employ?	• I would do a demonstration where I boil water in a see through pot with a loose lid where students are able to see boiling and the steam in a closed system • On teaching chemical reversibility students would have to do a practical between NO_2 and N_2O_4 gases. • On closed and open systems I would first clarify what they are and then show students different examples and ask them to identify whether the system is open or closed. • There would be symbolic and sub-microscopic representations so that I would deal with any confusion and provide an in depth understanding of the topic.
8. What strategies could you use to ascertain student's conceptions/ misconceptions of this idea?	• I would check their knowledge by asking the following questions: • What is meant by chemical equilibrium? • What is meant by a reversible reaction? • What is the difference between dynamic and static equilibrium? • What is meant by a closed system?

The value of the CoRe lies in the display of the most important considerations that a teacher has made about teaching a big idea. It is however not meant to measure the quality of PCK but display the reasoning behind key aspects of the topic. From Mr Banda's CoRe another teacher would be able to access important knowledge of teaching a big idea in a topic.

THE FIVE COMPONENTS OF PCK

We now look in more detail at the 5 components of PCK within a topic using different topics as examples. It must however, be borne in mind that ultimately PCK requires the interaction of all the five components all focusing onto a particular topic. In the discussions below we look at the individual components for the purpose of establishing a better understanding of their nature.

Learner Prior Knowledge

Learner prior knowledge refers to the ideas (correct and incorrect) that learners bring to the teaching of the topic. This kind of knowledge also includes an understanding of the learner context, including their language background, what they like and what they do not like and their interests.

Research has been done into many of the common topics taught at school (e.g. Canal, 1999; Dekkers & Thijs, 1998; Sanger & Greeenbowe, 1997) so it is possible to find out about the well-known misconceptions that learners throughout the world have in various topics. However as a teacher you may find particular misconceptions or preconceptions that your learners have that have not been documented. One well-known topic where learners have misconceptions is in kinematics (see for example, Lemmer, 2013; McDermott, Rosenquist, & van Zee, 1987).

The graph in Figure 3 represents the position as a function of time for a moving object.

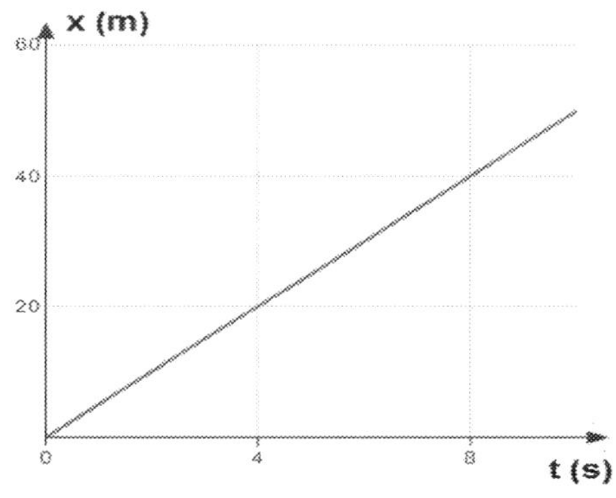

Figure 3. Position vs time graph

Which of the following is true?

a. The object increases its velocity
b. The object decreases its velocity
c. The object's velocity stays unchanged
d. The object stays at rest
e. More information is required

Given such a question, it is common for some students to select response A. There could be a number of reasons behind this (perhaps the reader may wish to consider possible reasons why a student might select this option).

It seems likely a student selecting option A has an inappropriate understanding of graphs in relation to kinematics. She has either failed to notice that the graph represents position vs time as opposed to velocity vs time or mistakenly thinks that the fact that the line moves in an upward direction indicates an increase no matter what the graph represents.

As a teacher you need to provide more than just the correct answer. You first need to acknowledge the misconception, then confront it and explain why it is wrong. In your explanation you will need to emphasize the actual concept that makes it incorrect and the accurate one. You may also use representations that illustrate the specific issue being explained. Finally you provide the correct answer. So an appropriate response in this case could go along these lines:

> The graph shows position vs time. This means that the position of the object is changing at a constant rate with respect to time. Notice that it is the position that is changing, not the velocity. A straight line graph shows a constant increase. If the position of an object increases at a constant rate with respect to time, then it is moving with a constant velocity. If the velocity were increasing then the graph would appear as a curve not a straight line

Curricular Saliency

As indicated already, curricular saliency involves the identification of the big ideas for teaching a topic. This is a very important step in beginning to teach a topic. Big ideas are more than topic headings. Identifying big ideas involves seeing the most important concepts that hold the topic together. Without a comprehension of the big ideas, the learner will not understand the topic.

The following are big ideas that have been identified for the human circulatory system (Loughran, Berry, & Mulhall, 2006)

a. It is useful to explain the circulatory system using the model of a continuous closed system
b. The circulatory system functions to service the needs of the individual cells
c. Body systems are very dependent on each other for their proper functioning
d. Blood is a complex substance
e. The heart is a pump that maintains the movement of blood around the body
f. Different types of blood vessels perform different functions
g. Membrane permeability enables diffusion for supply and removal of materials to/from cells

The big ideas are arranged in their optimal teaching sequence. Below we further unpack big idea c.

Sub concepts: body parts and systems are interdependent. Damage to one system/ part will affect, to some extent, all others. Exchange occurs between the circulatory systems, e.g. the respiratory (oxygen/carbon dioxide); digestive (products of digestion) and renal systems (metabolic waste).

Importance of the big idea: Even though the blood circulatory system is a "closed circuit" it requires exchange with other systems for "life" to be maintained. This emphasises a need to maintain all body systems.

What you would not teach at this stage: Details of the other systems it depends on. Prior concepts that are needed for teaching the identified big idea: Understanding of the circulatory system as a continuous closed system.

Big ideas can be identified in different science topics. Here we offer suggestions for the big ideas in two key science topics: genetics and chemical bonding. The reader may wish to consider what they would pull out as curricular saliency aspects of one or other (or both) of these topics before reading on.

Big Ideas for Teaching Genetics

a. All organisms have genetic information that is hierarchically organized and that is replicated during cell division
b. The genetic information specifies protein structure
c. Proteins have a central role in the functioning of all living organisms and they are the connection between genotype and phenotype

Big Ideas for Teaching Chemical Bonding

a. A chemical bond is an electrostatic interaction between a positive and a negative charge.
b. Energy is required to break bonds and energy is released when bonds are formed.
c. Chemical bonds form as the resulting entity has greater stability.

You would notice that the successful analysis of a topic in terms of both the misconceptions and curricular saliency components of PCK require the teacher to have a sound understanding of the content knowledge of the topic. Misconceptions do not present themselves out loud as "I am a misconception" but need to be recognized, similar to the understanding needed to distinguish main concepts to be expressed as big ideas from subordinate concepts in a topic.

Representations

Representations and analogies are vital in quality teaching of science topics and each topic has effective analogies. Two examples are provided below one from chemical equilibrium and one from electric circuits.

When teaching the particulate nature of matter one of the most important ideas to represent is the sub-microscopic representation of matter. Figure 4 shows one such useful representation:

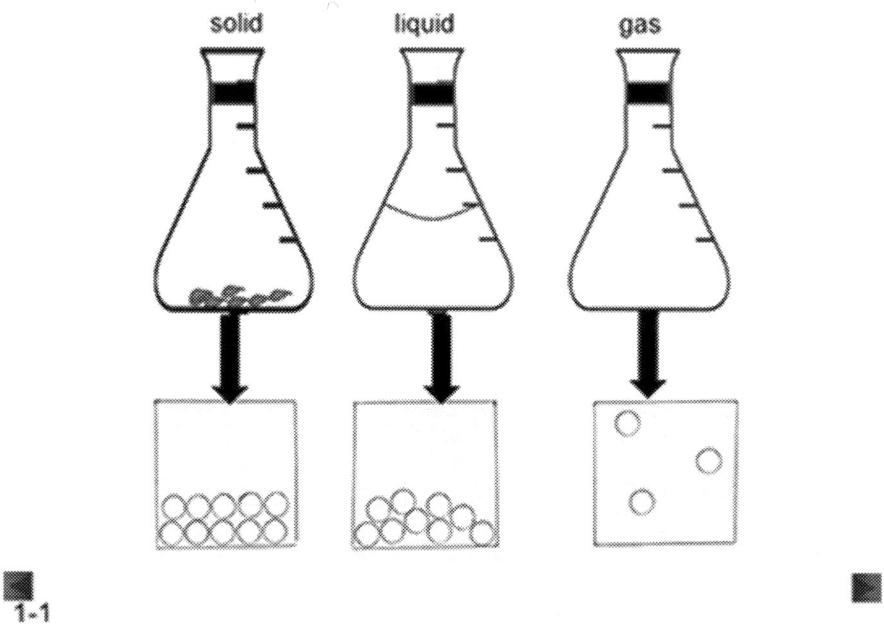

Figure 4. Representation for teaching the particulate nature of matter

A common analogy used in the teaching of electric circuits is a comparison to water flowing in pipes as illustrated by the comparison to a fish tank in Figure 5.

You will notice that the value of a representation lies in the identification of the particular aspect intended to represent, as representations are limited in nature. Thus, it is important when using representations to be conscious of the content aspect to be highlighted. There will often be 'negative' aspects of any representation or analogy (apects that do not reflect what is being represented) and it is important for the teacher to be aware of these and be clear about which aspects of the representation or analogy learners are expected to take across to the target concept. (Readers may wish to consider which features they would identify as positive or neative if teaching with these examples.) Whenever teachers meet examples of unfamiliar representations (including analogies, metaphors, similes) which they might use in teaching they should look to identify which aspects students are meant to map to the target, and which aspects may be salient for learners, but are not relevant to the target science concept.

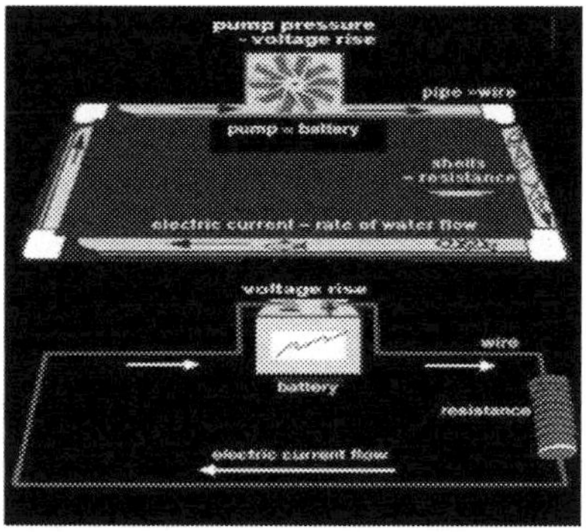

Figure 5. Representation of electric circuits using a water analogy http://www.windows2universe.org/physical_science/physics/electricity/circuit_analogy_water_pipes.html&edu=high (Permission from windows to the Universe, National Earth Science Teachers' association, USA)

What is Difficult to Teach?

Identifying learner difficulties in specific topics is a key aspect of PCK and it is linked to knowing learner prior knowledge. Two examples are provided below from different topics:

For Big Idea C in the circulatory system above three teaching difficulties have been identified – firstly that smaller systems link together to form bigger systems; secondly the barriers are permeable and therefore appropriately carried bits can penetrate (i.e. diffusion) and thirdly students may not know much about other body systems.

In electrochemistry difficulties arise in deciding on the positive and negative terminals in electrochemical and electrolytic cells. Students also have difficulty understanding that the cell remains neutral throughout the process. Thus what could be difficult to learn may not necessarily be a misconception but an area of potential conflict with another concept. For example, in understanding the dynamic nature of chemical equilibrium the issue of simultaneous forward and reverse reactions is difficult to understand as often the introduction of unidirectional chemical reactions could have been taught as one of the prior concepts needed. There are similar areas of learning difficulty in most if not all science topics (and readers might wish to reflect on topics they know well to consider the nature of common learning difficulties they have come across).

Conceptual Teaching Strategies

This last component is the most important of all as it brings together all the other components. Mr. Banda's lesson above gives some idea of an overall teaching strategy but one further example is given below:

Calculate the average relative atomic mass of a sample of oxygen gas that contains the following isotopes of oxygen atoms (see Table 3):

Table 3. Isotopes of Oxygen

Element	*Isotopes*	*Isotopic mass*	*Percentage abundance*
Oxygen	^{16}O	16	99,790
	^{17}O	17	0,037
	^{18}O	18	0,173

A learner's response:

O-16: 16
O-17: 17
O-18: 18
16+17+18=51
51 ÷ 3= 17amu

Following the learner's response, how will you teach a lesson on calculating the average relative atomic mass?

One teacher provided an excellent response as shown in Figure 6:

The most important thing to consider is that the percentage abundance will influence the average atomic mass. In this example the vast majority of the isotopes in this sample are ^{16}O isotopes, so the average relative atomic mass will be a decimal value closer to 16 than 17 and 18. To explain this I would give the example of calculating the class average after a test. I think that learners almost instinctively know how to do this calculation. When doing this calculation one has to consider how many learners got a particular mark - if 98 out of 100 learners got 16, 1 learner got 17 and 1 learner got 18, then the calculation would be $\frac{(98x16)+(1x17)+(1x18)}{100}$ = 16,03. Similarly when calculating relative atomic mass the calculation is $\frac{(99{,}790x16)+(0{,}037x17)+(0{,}173x18)}{100}$ =16,00383

Figure 6. Teacher's response to task

Notice the teacher takes students' prior knowledge into consideration and reasons conceptually rather than merely going through calculations which can be referred to as an algorithmic approach.

In all cases the strategy is mainly influenced by the considerations made about content knowledge rather than pedagogical knowledge. Thus these are conceptual teaching strategies. It is also in the conceptual teaching strategies that we see the interplay of the components of PCK in a topic as they support each other to build a concrete effort to enhance the understanding of the learners.

DISCUSSION

As mentioned earlier the purpose of this chapter is to illustrate the knowledge related to PCK in a specific topic and ways of capturing this kind of knowledge for display in order to understand the nature of thoughts that bring it to the fore. The important point to notice is that the knowledge of the five constituent components of PCK provides a framework that could be used for teaching a range of different topics. This PCK framework leads to indispensable PCK that would be common across contexts. The manner in which each teacher uses the PCK framework would be by considering the particular students and learning context thus bringing in the element of personal approach. While we consider the value in using the PCK framework in teaching, it does not replace the professional function of the teacher to make decisions on the job and continually reflecting and modifying own practices.

FURTHER READING

All the references below are useful resources for PCK and learner prior knowledge in science topics. The list below contains other books and websites

Berry, A., Friedrichsen, P., & Loughran, J. (Eds.). (2015). *Re-examining pedagogical content knowledge*. Oxford: Routledge

Bishop, K., & Denley, P. (2007). *Learning science teaching: Developing a professional knowledge base*. Maidenhead: Open University Press.

PCK Summit Dissemination Site http://pcksummit.bscs.org/ contains a number of resources, papers and videos connected to the PCK summit held in Colorado in 2012.

Rachinger, B. Diagnostic/Remedial Tests in Introductory Physics http://www.compadre.org/portal/items/detail.cfm?ID=8350

REFERENCES

Canal, P. (1999). Photosynthesis and 'inverse respiration' in plants: An inevitable misconception? *International Journal of Science Education, 21*(4), 363–372.

Dekkers, P., & Thijs, G. (1998). Making productive use of students' initial conceptions in developing the concept of force. *Science Education, 82*(1), 31–51.

Geddis, A. N. (1993). Transforming subject-matter knowledge: the role of pedagogical content knowledge in learning to reflect on teaching. *International Journal of Science Education, 15*(6), 673–683.

Gess-Newsome, J. (2015). Teacher professional knowledge bases including PCK: Results of the thinking from the PCK Summit. In A. Berry, P. Friedrichsen, & J. Loughran (Eds.), *Re-examining pedagogical content knowledge* (pp. 28–42). Oxford: Routledge.

Lemmer, M. (2013). Nature, cause and effect of students' intuitive conceptions regarding changes in velocity. *International Journal of Science Education, 35*(2), 239–261.

Loughran, J., Berry, A., & Mulhall, P. (2006). *Understanding and developing science teachers Pedagogical content knowledge*. Rotterdam, The Netherlands: Sense Publishers.

McDermott, L. C., Rosenquist, M. L., & van Zee, E. H. (1987). Student difficulties in connecting graphs and physics: Examples from kinematics. *American Journal of Physics, 55*(6), 503–513.
Sanger, M., & Greeenbowe, T. (1997). Common student misconceptions in electrochemistry: Galvanic, electrolytic and concentration cells. *Journal of Research in Science Teaching, 34*(4), 377–398.
Shulman, L. S. (1986). Those who understand: Knowledge growth in teaching. *Educational Researcher, 15*(2), 4–14.
Veal, W., & MaKinster, J. (1999). Pedagogical content knowledge taxonomies. *Electronic Journal of Science Education, 3*(4).

JOANNE K. OLSON

38. TEACHER PREPARATION FOR SCIENCE EDUCATION

INTRODUCTION

The preparation of science teachers has received increasing attention as the field of science education has grown and developed over the last several decades. A temptation naturally exists to begin a discussion of science teacher preparation by articulating what practices "work" to "best" prepare science teachers to teach effectively. However, of central importance in any discussion of science teacher preparation is a realization that the details of how to prepare science teachers are fundamentally connected to a clear vision of the purposes of schooling and our goals for science teacher education. If the goal is to prepare a teacher to seamlessly fit into a school system that emphasizes recall of vocabulary words, a very different teacher education program is warranted than if the goal is to prepare a science teacher who has high levels of autonomy and is expected to teach children to think critically and design and conduct laboratory investigations. Given the wide variance in school systems across the globe, any general statement about whether a particular program or practice is "effective" must consider the context—effective for what? This does not imply that "anything goes" in teacher preparation. In fact, much has been learned about what practices and program structures are well suited for particular ends. This chapter provides research findings about the preparation of science teachers in the context of commonly advocated goals for science education. An examination of science teacher education programs built upon these research findings reveals a variety of structures, content, and program emphases. This variety may wrongly lead to a perception that the field is in disarray. However, a close look at the conceptual orientations that underlie programs reveals three distinct modes of thought that are used to frame intellectually coherent programs and explains why they differ from one another, but still utilize the research base regarding science teacher preparation. This chapter concludes with a description of these conceptual orientations present in science teacher education.

The reader is advised to carefully consider the context of science teacher education in which one is engaged, and to deeply consider broader discussions about the purposes of schooling, the purpose of learning about science as part of compulsory schooling, and modes of thought that are present. Several references

K. S. Taber & B. Akpan (Eds.), Science Education, 523–537.

in the Further Reading section at the end of this chapter are provided to serve as a starting point for looking into these issues. As the old saying goes, if we don't know where we are going, then any route will do. Without a well-considered purpose for education, a clear vision of the role of science education, and an awareness of various conceptual orientations that frame teacher education, then any teacher preparation route will do, with potentially unfortunate consequences.

SETTING CONTEXT: PURPOSES FOR SCHOOLING AND SCIENCE EDUCATION

Because public schools are supported by the public, their purposes are impacted by the public—through local school boards, elected or appointed public officials, and parents. Thus, the purposes of schools, science teaching, and by extension, science teacher education, exist in a delicate balance between serving public needs and wants, and responding to more transcendent visions for these endeavors. In my country (the USA), the public once viewed schools as places where children would learn to become engaged and loyal citizens; this view was called the "Melting Pot"—a vision of schools that assimilated children from diverse backgrounds into a unified citizenry. (This vision is still reflected on our money, through the phrase *E pluribus unum*—Out of many, one.) More recently, the public has shifted its views from a focus on the public role of schooling toward personal benefits, viewing schools through a lens of economic utility—schools exist to help children become gainfully employed after graduation. Neil Postman criticized economic utility as not compelling to children, shallow, and impractical. Consistent with the tensions between public desires and more noble purposes of schooling, schools in my country also face reform documents that advocate ends for schools that go beyond gainful employment. In science, desired experiences reflect several (sometimes conflicting) purposes of schooling; reform documents call for students to understand the nature of science, engage in scientific inquiry, examine the influence of science on society and the engineered products and processes of our time, understand cross-cutting concepts in science, and develop science practices, in addition to learning science content. Clearly this is more than job preparation, yet my country still struggles to agree upon a coherent vision for schooling that is compelling to students, advances humanity, serves the public, and yet is not limited to what the public views as the next quick fix for societal ills. Postman and others, such as DeNicola, provide compelling possibilities for schooling that are worth reading (see Further Reading).

Despite tensions between public policy and more transcendent ends related to the purposes of schooling, substantial consensus exists among the science teacher education community and among public school teachers about desired outcomes of science in compulsory public schooling, even across countries and cultures. These outcomes include:

Table 1. Goals for students in science education (modified from Clough & Olson, 2016)

- Demonstrate robust understanding of fundamental science ideas.
- Exhibit an accurate understanding of the nature of science, technology and engineering.
- Effectively identify and solve problems.
- Be creative and curious.
- Use critical thinking skills.
- Effectively use communication and cooperative skills.
- Participate in working towards solutions to local, national, and global problems.
- Set laudable goals, make decisions, and accurately self-evaluate.
- Access, retrieve, and use credible scientific knowledge in socio-scientific decision-making.
- Convey self-confidence and a positive self-image.
- Express how a robust science education can promote personal and societal well-being.

The general consensus regarding these goals is possible because such goals can be obtained within multiple visions for schooling. For instance, those who view schooling through a lens of critical race theory will value communication and cooperative skills because they are useful for particular groups to attain a more equitable society. Those who have a vision of schooling as a way to unite humanity to survive on a planet with limited resources will view communication and cooperative skills as valuable for different reasons. Thus, goals for students exist to serve greater ends, and while consensus generally exists for the goals, the broader ends may vary. Many science teacher education programs and research on science teacher preparation are based on the assumption that the goals in Table 1 are desired. Thus, the science teacher education research reported in this chapter reflects these goals, but the reader is cautioned to consider the specific context in which such findings are to be applied. Purposes of schooling may differ widely across cultures, and some variation in goals may also be seen.

SCIENCE TEACHER PREPARATION PROGRAM STRUCTURES

Structuring a science teacher education program requires consideration of what teachers need to know in order to promote desired education ends. Knowledge domains widely considered important in teaching include: content knowledge, pedagogical knowledge, pedagogical content knowledge, curricular knowledge, knowledge of learners and learning, and knowledge of schools and schooling (e.g., Shulman, 1986). While most domains are fairly self-explanatory, pedagogical content knowledge (PCK) refers to subject matter knowledge specific for teaching, including ideal analogies, examples, explanations, activities, assessments, the best representations of specific subject matter for students, as well as what makes that subject matter particularly easy or challenging for students to understand. Perspectives vary about whether PCK should be the focus of teacher preparation programs, or whether PCK is largely developed through experience. Merely possessing knowledge in these domains is insufficient, because prospective teachers must also learn to enact teaching practices that are effective in promoting desired goals. Therefore, science teacher education must address both knowledge and practice.

The domains of knowledge and practice are reflected in the content of science teacher preparation programs. These tend to include a requirement for general education, science content courses, and professional education. General education includes broad coursework outside of science and education courses, such as mathematics, history, literature, etc. and is often completed as part of an undergraduate degree. Science content courses may focus on one science field, or may be required across multiple science fields, such as chemistry, geology, biology, and physics. Professional education includes methods of teaching science and other education coursework that addresses knowledge of how people learn, foundations/history of education, educational technology, and strategies for teaching diverse learners (this can include linguistic, academic, socioeconomic, and/or racial/cultural diversity). Most all teacher preparation programs require extensive field experiences, which can include classroom observations, practice teaching in limited settings (often called *practicum* experiences – e.g., small group of students or single lessons), and more extensive *student teaching*, which involves substantial teaching responsibilities. In some cases, this can involve assuming increasing responsibility of the classroom over the course of a 12–16 week term or full school year.

Despite the overarching commonality of requirements among science teacher preparation programs, their structures vary widely. The preparation of primary teachers (who are licensed to teach students usually up to age 11) has included apprenticeship models and "normal schools" (typically a two year program designed to provide subject matter knowledge, knowledge of pedagogy and the history of education, and field experiences in a school), 4–5 year undergraduate degrees, and graduate-level programs ranging from a few months to 2 years or more. The preparation of secondary science teachers (who teach students ages 11–18) includes undergraduate programs that are either stand-alone majors, taken concurrently with a science major, or completed after obtaining a science degree. Governments typically have authority over requirements for earning a teaching license, and often dictate minimum requirements for teacher preparation programs, including the number of university courses in particular science areas, what courses are to be taken in professional education, and a minimum number of clock hours that are required for practice teaching in a school.

Primary teachers tend to teach science in addition to other subjects, and if so, they are usually prepared as generalists, although in some countries primary teachers who teach science must specialize in science. Secondary teachers in many countries are licensed in a specific science area, such as physics or biology, and must complete additional coursework and preparation in order to teach other science subjects. In other locations, governments have established general science teaching licenses that permit secondary teachers to teach multiple science subjects. In places where the demand for science teachers exceeds supply, governments have filled these positions in a number of ways, many of them detrimental to ensuring high quality teachers (Olson et al., 2015). In the United States, for example, all states have university-based teacher education programs with specific requirements, but also have "alternative" routes for individuals to earn a license without completing coursework that is required of university-based preparation programs. Over 130 different routes have been created. Thus, we see many entities running teacher preparation programs, including school districts and for-profit agencies. Some include as little as five weeks of pedagogical preparation prior to teaching full time. Some governments/licensing agencies allow science teachers to teach science courses for which they are not licensed, or simply possess an undergraduate degree and have work experience in a science industry—bypassing professional education coursework. Other licensure pathways are fully online, with the exception of mandated field placements. In Canada, primary teachers can be assigned to secondary science classrooms when science teacher shortages occur. Public concerns about schools and schooling often extend to criticisms of science teacher preparation, but rarely do such concerns take into account the variety of ways that teachers are allowed to earn a teaching license. This variability is most pronounced across countries, but can also occur within countries, particularly where teaching licenses are granted by the province/state, rather than the federal government.

Research makes clear that science teachers' understanding of science content is crucial for helping students learn science. Teachers cannot effectively teach what they do not know, and numerous studies have shown that the teachers' knowledge of the subject matter they teach is positively associated with more sophisticated teaching practices, and student learning of those concepts (e.g., Floden & Meniketti, 2005). In addition, teachers who understand the specific science concepts they teach and know what misconceptions students are likely to have about those concepts (i.e., have both content knowledge *and* PCK) show even greater gains in student learning than those with content knowledge alone (Sadler et al., 2013). However, research has not established definitive claims regarding the number of science courses that are needed for optimal impact on student learning. Clearly, a minimum number of courses is necessary, but at some point, additional coursework becomes inconsequential. When considering the number of courses needed, we must also consider what topics are addressed within these courses, as well as the quality of instruction. Sadler and Sonnert (2016) recommend that science coursework for prospective and inservice teachers should target the specific topics that will be taught to their students. They note that teachers' subject matter knowledge in a particular area (e.g., physical science) may be rather high, but if teachers have "holes" in their understanding of a particular concept, students' learning is detrimentally impacted. To address science content requirements, many programs require an undergraduate degree in the science area to be taught (e.g., biology), but others will allow as few as two introductory-level courses in the discipline, particularly when the teacher has completed a high number of science courses in another area. Concerns have been expressed about allowing teachers to teach science with low numbers of courses completed in the discipline they are teaching. In addition, the nature of science is also widely included in standards documents for public schools, yet few teacher preparation programs require coursework in the nature of science.

With regard to the professional education component of science teacher education, the structure of the program impacts the quality of science teachers' practices. The number of science methods classes, duration of the program, field placement experiences, and socialization experiences within the program are positively associated with science teachers' practices (Roehrig & Luft, 2006). Studies have shown that secondary science teachers' beliefs and classroom practices are more sophisticated after having completed three or more sequential science methods courses over a four semester period rather than a single science methods course (Tillotson & Young, 2013; Herman, Clough, & Olson, 2013; Bergman, 2007; Krajcik & Penick, 1989). Some improvements have been reported with as few as two science methods courses (Clough & Numedahl, 2002), with more lasting changes on teachers' practices occurring with 3–4 science methods courses (Herman et al., 2013).

Programs that utilize a cohort model have been found to be superior to programs without this structure; cohort-based programs require prospective teachers to complete program coursework together and include extensive requirements and opportunities to collaborate. Graduates of cohort-based programs are more likely to continue professional collaborations with each other after the program (Tillotson & Young, 2013), and these collaborations help graduates maintain research-based practices, even when faced with severe constraints at the school (Ihrig, 2014). Those involved in support networks composed of fellow program graduates are more likely to teach the nature of science at high levels, and have better general science teaching practices (Clough & Olson, 2012; Herman et al., 2013; Ihrig, 2014).

When prospective science teachers are placed in school settings where the cooperating teacher uses practices congruent with what is taught in the science teacher education program, the prospective teachers are more likely to use those practices after graduation (Tillotson & Young, 2013). Programs with extensive field experiences and three sequential science methods courses result in graduates with more student-centered views (Salish, 1997) than their counterparts in programs with fewer methods courses and field experiences.

Student teaching is usually placed at the end of a teacher education program, but successes have been documented in programs where student teaching is in the middle of the program, and students take additional coursework after the experience (Herman et al., 2013). What is likely far more critical to the success of student teaching than where it is placed within a program is the quality of the cooperating teacher and supervisor. Arguments have been made that student teaching can negate the effects of the teacher education program and inculcate student teachers into the status quo of the school system. However, researchers have found that if the student teaching placement and supervision practices support the goals of the teacher preparation program, graduates can implement practices during their first years of teaching that overcome a preponderance of lower-quality practices among their colleagues during graduates' first year of teaching (Ihrig, 2014). Ihrig (2014) found that teacher education programs can have a positive impact on teachers' practices even when they are in school settings that are unsupportive of, and even hostile toward, research-based practices of teaching science as and through inquiry, but only when those science teachers were involved in support networks of colleagues who shared the same values as their teacher education program, a finding supported by Herman et al. (2013).

SCIENCE TEACHER PREPARATION CONTEXTS AND CHALLENGES

Studies on current and prospective science teachers prepared as generalists at the primary level indicate that science is the subject they often feel least prepared to teach, that they generally prefer non-science subjects, and tend to use teaching strategies associated with other subjects rather than strategies more compatible for science teaching (e.g., Banilower et al., 2013; Appleton, 2006). In addition, primary teachers tend to have misconceptions about the nature of science. However, issues regarding the teaching of science are not limited to primary teachers. At the secondary level, in many countries, science teachers may be required to teach out of their specific science field, and opportunities for professional development may be limited. In small communities, secondary science teachers may be asked to teach multiple science subjects and may be the only science teacher, resulting in heavy workloads and professional isolation. Like their primary colleagues, they are also likely to have misconceptions about the nature of science.

Science teachers at all levels tend to enter teacher education programs with misconceptions about science, teaching in general, and science teaching in particular. For example, many prospective teachers wrongly think that teaching difficult concepts can only be accomplished through telling, and that activities, if used at all, are to keep students motivated to pay attention in class, or to verify that was what presented in a lecture is correct. Other prospective teachers believe that activities themselves teach students, and that the main role of the teacher is to keep students on task with an activity so that learning will naturally occur. These perceptions, along with their misconceptions about science concepts and the nature of science, create a substantial challenge for science teacher education. Programs and instruction must be structured in a way that promotes conceptual change in order for prospective teachers to teach science in a way that helps students deeply learn desired science concepts and meet the goals in Table 1.

SCIENCE TEACHER PREPARATION PROGRAM CONTENT

Assembling coursework and field placement experiences is only a part of effective science teacher preparation. Sufficient time is required to help prospective teachers become aware of their own conceptions, undergo conceptual change, develop the knowledge required to be successful in the classroom, and in many cases, be prepared to improve the status quo in the schools where they will work. Thus, the quality and nature of the experiences in coursework and field placements is of utmost importance.

Who is responsible for the preparation of science teachers is a crucial issue to consider. The instructors of courses and supervisors of student teachers may be professors of science education, graduate students, current or retired science teachers, current or retired school administrators, and even professors of disciplines outside of science education (e.g., math education or social studies education faculty members, or scientists). Each may have vastly different understanding of what is required to teach prospective science teachers. Practicing or retired teachers may or may not have been effective, and can often simply share what worked for them, which may or may not transfer to other contexts. University science education faculty members may have a deep theoretical knowledge base, but some may have had little or no public school teaching experience, or taught poorly, or spend little or no time in schools and are disconnected from the realities of classrooms, teachers, and children. Given that science education is an applied field, instructors and supervisors should have a sophisticated understanding of how to prepare science teachers, and successfully bridge theory and practice. Some graduate programs provide extensive preparation of graduate students to teach and supervise science teacher candidates, including seminars, apprenticeship opportunities with faculty, supervised teaching experiences in science methods courses or supervised experience in student teacher supervision, and other opportunities.

Given what is known about coursework and field experiences for science teacher preparation, what accounts for the vast array of teacher preparation program structures and emphases? To address this question, these differences must be placed in a broader context that includes the purpose and conceptual orientation of any particular science teacher education program. McKeon (1994; 2016) identified four broad intellectual traditions, and science teacher education programs have been influenced by at least three of these. These intellectual traditions, or modes of thought, reflect differing ways of organizing knowledge. When applied to teacher education, they result in differing program values, emphases, structure, and content (Owen, 2003). For those who study science teacher education, they also provide order and meaning to a diverse landscape of teacher preparation practices. By recognizing the mode of thought of a particular program, practices begin to make sense; we can understand differences across programs and better understand our own values and those of others.

Construction

Construction is a mode of thought that begins with small units and progressively builds a system from those pieces (hence, the term *construction*). Construction is based on the intellectual tradition of Euclid, Democritus, Newton, Locke, Hobbes, Skinner, and Thorndike. When applied to teacher education programs, the assumption is made that learning to teach is done by examining what a teacher needs to know, followed by breaking that knowledge into its component parts, and organizing the parts from simple to complex in a carefully arranged sequence. For example, science content knowledge could be learned as a prerequisite to a science methods course; classroom management could be learned in a separate course; classroom observation hours may precede small group practicum teaching; lesson planning could be taught as a series of discrete steps (e.g., develop a measurable objective, follow a template of lesson stages, etc.). Science teacher preparation programs that have separate courses to teach multiple aspects of teaching (e.g., science content, foundations, science methods, assessment, classroom management, technology) followed by student teaching where the prospective teacher is expected to integrate these components into practice is one example of the construction mode of thought. Methods courses that follow from this mode of thought will be heavily focused on beginning with simpler topics and building on them as the course and/or program progresses.

An analysis of science methods textbooks designed for use in the U.S. found that 81% of randomly selected texts were aligned with the construction mode of thought (Wilcox & Olson, 2014). These books divided tasks of teaching into discrete acts, such as writing a behavioral objective, following prescribed steps to write a lesson plan, introducing strategies to assess student work, etc. One science methods textbook recommended that prospective teachers study a lesson plan designed to teach children the skill of tying a shoe, and used this science content-free example of skills instruction as a basis for learning lesson planning in science. Textbooks for methods courses are likely to follow a construction mode of thought due to the linear nature of texts and the need for chapters and sections. These textbook structures favor construction in much the same way as popular presentation software biases presenters toward headings and bulleted lists. Many of us are familiar with construction due to our own experiences as learners (research indicates that many of us experienced science instruction as learners in a construction-oriented environment), and in several countries, schools and curricula are structured using this mode of thought. Teachers prepared through a construction-oriented teacher education program may fit seamlessly into such school systems. However, other modes of thought do exist, and can also be used to structure strong science teacher education programs.

Resolution

In this mode of thought, emphasis is placed on defining and resolving problems. Resolution is based on the intellectual tradition of Aristotle, Aquinas, Rousseau, and Dewey. Applied to science teaching, the task of the teacher is to use informed judgment to diagnose problems in the classroom and to use professional knowledge (from research as well as desired educational ends and goals) to make prudent decisions about how to proceed; thus, teaching is viewed as decision-making. Problems include all aspects of teaching: What content is appropriate to teach? How do I engage all students? How do I diagnose student thinking? What do I do with their misconceptions? How do I teach particularly difficult concepts? How do I find appropriate resources? The list could go on and on.

Some programs that are congruent with this mode of thought have multiple methods courses coupled with field placements (often occurring early in the program), during which prospective science teachers are expected to engage in extensive analysis of their teaching practice, and work through critical incidents and issues that arise during their teaching practice during coursework. Courses often employ the use of video cases so that classrooms can be analyzed, problems articulated, and decisions made about how to proceed. Rather than teaching a set of "best practices," these programs emphasize the context and how to make a decision in that context—perhaps there are several practices that may "best" solve a particular problem. As contexts change, the problems change, and the range of solutions may change. The result is not radical relativism, but a deliberate study of multiple tools and the jobs for which they are best suited. For instance, a learning cycle lesson plan structure may be taught (in addition to several other models) in the context of teaching science concepts. If the goal is the learning of a skill, such as how to use a microscope, a different model may be more appropriate, such as direct instruction. An assessment tool sometimes used by such programs is the "rationale paper" or Research-Based Framework paper—a document that articulates goals for students and the decisions teachers must make to promote those goals, supported by research on particular strategies as well as research on how children learn. Oral defenses may be used as an assessment tool where prospective teachers are presented with critical incidents, exposing their decision-making to the instructor who provides further questions to both diagnose and teach the prospective teacher.

Resolution in science teacher education was present in only one science methods textbook examined by Wilcox and Olson (2014), and not surprisingly, that text was the only one that had a question in the title: "Introducing Students to Scientific Inquiry: How do we know what we know?" (Etheredge & Rudnitsky, 2003). The text paid little attention to the "skills" of teaching, unless they were embedded within the context of scientific inquiry. The majority of the textbook described full units implemented by teachers, including the teachers' and authors' reflections and questions about what could have been done differently and what could have made the unit better (note the emphasis on decision making). The text is framed as a set of nested problems: What is inquiry? Where does scientific knowledge come from? How does one teach scientific inquiry to children? Despite the paucity of resolution-oriented textbooks, this mode of thought can be seen in the research literature, particularly Krajcik and Penick (1989), Tillotson and Young (2013) and Clough,

Berg, and Olson (2009). Clough et al. (2009) published a graphic organizer that structures their courses and the entire science teacher education program, and not surprisingly called it a "decision-making framework."

Discrimination

Discrimination (also called "perspectival") is a mode of thought that emphasizes the perspective of the knower. The position is based on the intellectual tradition of Kant, Schrodinger, Descartes, and Freud, and asserts that what is known depends upon the knower; each person has a unique view of the world, which must be distinguished from others' views of the world. Carl Rogers exemplifies this mode of thought in education, as well as many who place a central focus on social justice issues. Science methods courses based on this mode of thought will raise issues such as: Whose science is being learned? How is scientific knowledge developed? What worldviews do students bring to the classroom? How do we teach science to students while taking into account their cultural background and ways of knowing? Who are the students and how do we connect with them? Who am I, and how does my background influence my teaching and the way I perceive my students?

Of the science methods textbooks examined by Wilcox and Olson (2014), 14% of the selected texts represented the discrimination mode of thought. One such textbook was titled "Science Stories: Teachers and Children as Science Learners" (Koch, 1999), and another had a chapter titled, "Your Personal Context" with a section titled, "Who are You, and How Do You Feel about Science?" (Fraser-Abder, 2011). Other expressions commonly (but not exclusively!) used by science teacher education programs with a discrimination mode of thought include an emphasis on personal relevance and teaching in a way that is "culturally responsive" or "equitable."

Assimilation

Assimilation is a fourth mode of thought that receives little attention in the literature and was not found in science methods textbooks (Wilcox & Olson, 2014). However, the assimilation mode of thought may exist in science teacher education programs, and it may serve as an organizing framework for teacher educators who find that the previous three modes of thought do not reflect how they organize knowledge. Assimilation (called "holistic" by Owen, 2003) is exemplified by Plato and Einstein. Reality is viewed as a whole. This may be an idealized whole that may not exist in our personal experience, but it is the ideal that explains individual events, and the ideal that defines what we are working toward. Using this mode of thought, the focus is on the whole, not on its parts. Any perceived differences among the parts illustrate a lack of understanding; when the parts are properly understood, they are assimilated into the whole. Perhaps the best example of assimilation-based teacher education occurred in the early days of public schooling when prospective teachers served as an apprentice with an experienced teacher. Learning to teach was not divided into separate courses or skills; the novice learned to teach in the classroom with all of its complexity.

MODES OF THOUGHT IN SCIENCE TEACHER EDUCATION

Science teacher education programs can be structured very well within any of these modes of thought. Each mode of thought views the *means* of teacher education differently, and each is more likely to align with particular purposes of schooling (the *ends*). The mode of thought used to structure a teacher education program is largely based on the values and expertise of the teacher educators, and the context of the program (including the values of the communities in which graduates will teach). A mode of thought identifies what becomes an organizing focus for the program, and does not completely exclude other values—those values may appear in the program, but in a different context. For example, a science teacher education program based on construction is not inherently racist because it is not organized around discriminating between perspectives of the knower. Racial issues may be strongly addressed in such a program, but they will be accomplished in a way consistent with breaking knowledge into its component parts and gradually developing sophisticated understanding over time. The issue for each mode of thought is what is placed in the foreground as a central way to organize the program and course experiences. Different modes of thought will organize a variety of experiences differently in a teacher education program. For example, should classroom management be integrated within a science methods course, or taught as a stand-alone course? Should student teaching occur at the end of the program? Should a practicum experience be provided prior to student teaching, and what should it look like? If prospective teachers analyze themselves on video, what are they looking for (e.g., particular instructional decisions, issues of race or gender, the efficiency of a transition between activities)? Should a course on the foundations of education focus primarily on social foundations, philosophical foundations, historical foundations, or all three equally?

Importantly, instructors within a single teacher education program can have different modes of thought. In such cases, the teacher education program can appear to be a disjointed set of courses that may operate at cross-purposes—what Posner (2003) calls "garbage can eclecticism." This was evidenced by an elementary education major in one of my studies who lamented that her math methods professor taught her to teach completely differently than how she had learned to teach in a literacy methods course, and she was concerned that she had to be a different teacher each time a new subject was being taught. Since that didn't make sense to her, she questioned whether her professors knew what they were doing. Teacher education programs that do not help their students navigate a pluralistic program can unwittingly create graduates who reject some, or all, of the program. Such prospective teachers may view the teacher education program as a marketplace where consumers can pick and choose isolated strategies that they like, and avoid those they dislike. Solutions to this challenge include teaching prospective teachers the multiple modes of thought operating in their program and how to navigate across these traditions, or creating a program that reflects a single mode of thought. Programs with intellectual coherence can be very powerful (e.g., Salish, 1997; Tillotson & Young, 2013; Herman et al., 2013), but this requires instructors who understand that intellectual tradition as well as possess a genuine commitment to a program built upon that mode of thought. Unfortunately, in too many cases, groups with a particular mode of thought can try to bully their colleagues into creating a science teacher education program with a single focus, despite several faculty members working from a different mode of thought. This creates hostility rather than understanding. As Dewey noted, "It is easier to see the conditions in their separateness, to insist upon one at the expense of the other, to make antagonists of them, than to discover a reality to which each belongs" (1902, p. 4).

FINAL THOUGHTS

Science teacher preparation is a complex landscape that is influenced by the political and societal conditions in which the program is situated and regulated. That said, much consensus exists regarding specific goals for public school students in science classes. Promoting these goals requires teaching practices that have been shown to be associated with these outcomes; unfortunately, such teaching practices are not at all intuitive or favored by many prospective teachers (Clift & Brady, 2005). Therefore, science teacher education programs need to focus on conceptual change, a process that requires sufficient time, mental engagement, conceptual development, and experiences in schools to promote lasting change. Programs that have been successful in this endeavor have a structure that requires strong science content knowledge, 2–4 science-specific methods courses, have extended field-based teaching experiences, use a cohort model, and engage prospective teachers in learning about students, how they learn, and the misconceptions they bring to the classroom. Several successful programs also include coursework in the nature of science. How such programs arrange course sequences and experiences within courses will be influenced by one or more mode of thought. When multiple modes of thought are present in a program, teacher educators are advised to make these differences apparent to students and make efforts to understand and appreciate the intellectual traditions of others.

FURTHER READING

Purposes of Schooling

DeNicola, D. R. (2012). *Learning to flourish: A philosophical exploration of liberal education*. New York, NY: Continuum International.
Eisner, E. (2002). *The educational imagination* (3rd ed.). New York, NY: Macmillan.
Postman, N. (1995). *The end of education*. New York, NY: Vintage Books.

Goals for Science Education

DeBoer, G. E. (1991). *A history of ideas in science education: Implications for practice.* New York, NY: Teachers College Press.

Goodlad, J. I. (1983). A summary of a study of schooling: Some findings and hypotheses. *Phi Delta Kappan, 64*, 52–57.

McComas, W. F., & Olson, J. K. (1998). The nature of science in international standards documents. In W. F. McComas (Ed.), *The nature of science in science education: Rationales and strategies.* Dordrecht: Kluwer Academic Publishers.

Teacher Education

Appleton, K. (2006). *Elementary science teacher education: International perspectives on contemporary issues and practice*. Mahwah, NJ: Lawrence Erlbaum and the Association for Science Teacher Education.

Carlsen, W. S. (1999). Domains of teacher knowledge. In J. Gess-Newsome & N. G. Lederman (Eds.), *Examining pedagogical content knowledge: The construct and its implications for science education* (pp. 133–144). Boston, MA: Kluwer.

Cochran-Smith, M., & Zeichner, K. M. (2005). *Studying teacher education: The report of the AERA panel on research and teacher education.* Mahwah, NJ: Lawrence Erlbaum and the American Educational Research Association.

Gabel, D. L. (1994). *Handbook of research on science teaching and learning.* New York, NY: Macmillan and the National Science Teachers Association.

Modes of Thought

McKeon, R. (1994). *On knowing: The natural sciences* (D. B. Owen & Z. K. McKeon, Eds.). Chicago, IL: The University of Chicago Press.

McKeon, R. (2016). *On knowing: The social sciences* (D. B. Owen & J. K. Olson, Eds.). Chicago, IL: The University of Chicago Press.

Rogers, C., & Freyberg, H. J. (1993). *Freedom to learn* (3rd ed.). New York, NY: Merrill.

Rousseau, J. J. (1764/1979). *Emile: Or, on education* (A. Bloom, Trans.). New York, NY: Basic Books.

Skinner, B. F. (1948). *Walden two.* New York, NY: Macmillan.

REFERENCES

Banilower, E., Smith, P. S., Weiss, I., Malzahn, K., Campbell, K., & Weis, A. (2013). *Report of the 2012 national survey of science and mathematics education.* Chapel Hill, NC: Horizon Research, Inc.

Bergman, D. J. (2007). *The effects of two secondary science teacher education program structures on teachers' habits of mind and action* (Unpublished doctoral dissertation). Iowa State University, Ames, IA.

Clift, R. T., & Brady, P. (2005). Research on methods courses and field experiences. In M. Cochran-Smith & K. M. Zeichner (Eds.), *Studying teacher education: The report of the AERA panel on research and teacher education* (pp. 309–424). Mahwah, NJ: Lawrence Erlbaum and the American Educational Research Association.

Clough, M. P., Berg, C. A., & Olson, J. K. (2009). Promoting effective science teacher education and science teaching: A framework for teacher decision-making. *International Journal of Science and Mathematics Education, 7*(4), 821–847.

Clough, M. P., & Numedahl, P. (2002, January). *Conceptions of learning and teaching expressed by students at two stages in their preservice program.* Paper presented at the annual meeting of the Association for the Education of Teachers in Science, Charlotte, NC.

Clough, M. P., & Olson, J. K. (2016). Connecting science and engineering practices: A cautionary perspective. In L. Annetta & J. Minogue (Eds.), *Connecting science and engineering education practices in meaningful ways—building bridges. Contemporary trends and issues in science education series*. Dordrecht, The Netherlands: Springer.

Dewey, J. (1902). *The child and the curriculum*. Chicago, IL: The University of Chicago Press.

Etheredge, S., & Rudnitsky, A. (2003). *Introducing students to scientific inquiry: How do we know what we know?* Boston, MA: Pearson Education, Inc.

Floden, R., & Meniketti, M. (2005). Research on the effects of coursework in the arts and sciences and in the foundations of education. In M. Cochran-Smith & K. M. Zeichner (Eds.). *Studying teacher education: The report of the AERA panel on research and teacher education* (pp. 309–424). Mahwah, NJ: Lawrence Erlbaum and the American Educational Research Association.

Fraser-Abder, P. (2011). *Teaching budding scientists: Fostering scientific inquiry with diverse learners in grades 3–5*. Boston, MA: Pearson Education.

Herman, B. C., Clough, M. P., & Olson, J. K. (2013). Teachers' NOS implementation practices two to five years after having completed an intensive science education program. *Science Education, 97*(2), 271–309.

Ihrig, L. (2014). *The effects of socialization on beginning science teachers' pedagogical decision making and science instruction* (Unpublished doctoral dissertation). Iowa State University, Ames, IA.

Koch, J. (1999). *Science stories: Teachers and children as science learners*. Boston, MA: Houghton Mifflin Company.

Krajick, J., & Penick, J. (1989). Evaluation of a model science teacher education program. *Journal of Research in Science Teaching, 26*, 795–810.

McKeon, R. P. (2016). *On knowing: The social sciences* (D. B. Owen & J. K. Olson, Eds.). Chicago, IL: University of Chicago Press.

Olson, J. K., Tippett, C. M., Milford, T., Ohana, C., & Clough, M. P. (2015). Science teacher education in a North American context. *Journal of Science Teacher Education, 26*(1), 7–28.

Owen, D. B. (2003, November). *The foundations course: Depth versus Breadth?* Paper presented at the annual meeting of the Midwest Philosophy of Education Sciences, St. Louis, MO.

Posner, G. J. (2003). *Analyzing the curriculum* (3rd ed.). New York, NY: McGraw-Hill.

Roehrig, G. H., & Luft, J. A. (2006). Does one size fit all?: The induction experience of beginning science teachers from different teacher preparation programs. *Journal of Research in Science Teaching, 43*(9), 963–985.

Sadler, P. M., Sonnert, G., Coyle, G., Cook-Smith, N., & Miller, J. L. (2013). The influence of teachers' knowledge on student learning in middle school physical science classrooms. *American Educational Research Journal, 50*(5), 1020–1049.

Sadler, P. M., & Sonnert, G. (2016). Understanding misconceptions: Teaching and learning physical science. *American Educator, 6*(2016), 26–32.

Salish I Research Project. (1997). *Secondary science and mathematics teacher preparation programs: Influences on new teachers and their students* (Final report). Washington, DC: U.S. Department of Education and Office of Educational Research and Improvement.

Shulman, L. S. (1986). Those who understand: Knowledge growth in teaching. *Educational Researcher, 15*(2), 4–14.

Tillotson, J. W., & Young, M. J. (2013). The IMPPACT project: A model for studying how preservice program experiences influences science teachers' beliefs and practices. *International Journal of Education in Mathematics, Science and Technology, 1*(3), 148–161.

Wilcox, J. L., & Olson, J. K. (2014, January 15–18). *An analysis of ontological and epistemological underpinnings of science methods textbooks and syllabi*. Paper presented at the annual meeting of the Association for Science Teacher Education. San Antonio, TX.

YASHWANTRAO RAMMA, AJEEVSING BHOLOA AND
BRINDA OOGARAH-PRATAP

39. RESEARCH PERSPECTIVES AND SKILLS FOR SCIENCE EDUCATION

INTRODUCTION

In this chapter we examine the extent to which research in science education can improve teaching and learning of science, in particular physics and mathematics in schools. The advancement of scientific knowledge, conceptual understanding and skills are important elements for the development of critical minds in learners. There is ample evidence that the traditional teacher-centered approach to teaching and learning science still prevails despite the wealth of data from research in science education about the benefits of adopting a learner-centered approach (Ramma, Samy, & Gopee, 2015). We further contend that teachers have a major role to play in situating the existing schemas of learners with the intention to facilitating learners' knowledge-acquisition and knowledge-construction processes.

Teaching (good teaching) is considered to be an art, while learning relates to a process (Wright, 2001) which encapsulates the intrinsic element referred to as reflection or as Schön (1987) puts it – reflection-in-action. To be able to reflect purposefully and to drive their thinking towards a specific goal, teachers have to possess a strong knowledge base (content knowledge – CK), in addition to pedagogical content knowledge (PCK) and curricular knowledge (Shulman, 1986). Furthermore, by being engaged in critical thinking and reflection, teachers develop connections between content knowledge and pedagogical content knowledge in a bid to establish concepts as an interconnected web of processes. Teaching and learning science as discrete packets of information is doomed to become meaningless as many learners concentrate more on learning 'recipes for examination success' without making the effort to develop understanding of the underlying concepts.

There is a strong claim that science teachers and science education researchers have the obligation to bridge the research-practice gap (Bulterman-Bos, 2008) to ensure that a synergy is created between the researcher's (teacher-researcher) findings and the teacher's classroom practice. For learners to develop conceptual understanding, the teacher-researcher has to bring meaning to an abstract science concept by using the appropriate curricular and pedagogical content knowledge and contextualizing it into a concrete, consistent and logical flow of ideas (Bayrakar, 2009). Teachers are central to students' development of scientific understandings, competencies and dispositions; however, several studies have demonstrated that

K. S. Taber & B. Akpan (Eds.), Science Education, 539–549.

science students and even graduates hold various misconceptions and face a number of learning difficulties (Ramma, Bholoa, Watts, & Ramasawmy, 2014; Schneps & Sadler, 1988) despite years of instructions or experience. We recognize that the role of the teacher is primordially important in guiding learners in the construction of purposeful knowledge structures. Looking across a wide body of research on students' difficulties and on how to address them, we posit that teachers, as agent of change, will become sources of inspiration for learners to develop conceptual understanding. Prensky (2005) elaborately argues that one of the noticeable causes of students' disinterest in their studies is the result of monotonous lessons which are teacher-centered. It becomes therefore imperative for the development of science education programmes to include conceptual change orientation, firmly grounded on student learning.

To enable teachers within and across disciplines to develop appropriate insight into the various dimensions that similar concepts taught in different subject areas are consistent, collaboration among the teachers becomes the necessary predicament. For instance, the *modulus of elasticity* (or Young Modulus) learnt in physics may slightly differ in terminology when the same concept is learnt in a mathematics lesson. In physics, the modulus of elasticity bears unit N/m^2 whereas in mathematics, the unit Newton (N) is employed by some examination bodies. Collaboration among the physics and mathematics teachers may help to identify the semiotic in variation in the two subject areas and to create a shared understanding among the teachers of that situation, which otherwise may result in confusion. It is of utmost importance for educators of different subject areas to introduce concepts located in various disciplines in such a way that students are also able to identify notational consistency. If not addressed, it may become breeding ground for the development of misconceptions in teachers and which are eventually transmitted to students (Akgun, 2009).

TEACHER EDUCATION & RESEARCH

Teacher variable is an important element which significantly affects students' learning outcomes and contextually relevant teacher education is therefore crucial. Teacher education lies at the heart of teacher development. Moreover, a growing interest in professionalism of teachers lies in their engagement in undertaking reflection-based research (Leeman & Wardekker, 2015). Taking a reflective stance enables the identification of teachers' misconceptions, course design, and pedagogical approach amongst others. Some teachers are so immersed in the use of the traditional approach to teaching that reflection is often ignored. However, research findings provide evidence that with reflection, in addition to having improved students learning outcomes, relationships between teacher and students are more positive and constructive (Wubbels & Korthagen, 1990). Furthermore, McIntyre (2005) acknowledges the existence of a research-practice gap in the work of teachers and to address the problem he proposes that findings from research should provide clear indication to teachers of how they might be able to improve their practice.

In this chapter, a new teacher education model for science educators (Figure 1) is discussed, which incorporates the elements of research findings and collaborative learning. The collaboration under discussion here is among science (physics) and mathematics educators during a professional development course. Figure 1 depicts the model of teacher professional development (TPD) in teacher education. It encompasses lesson delivery through reflection, incorporation of research findings into lesson conceptualization and, finally, improvement or review of CK and PCK.

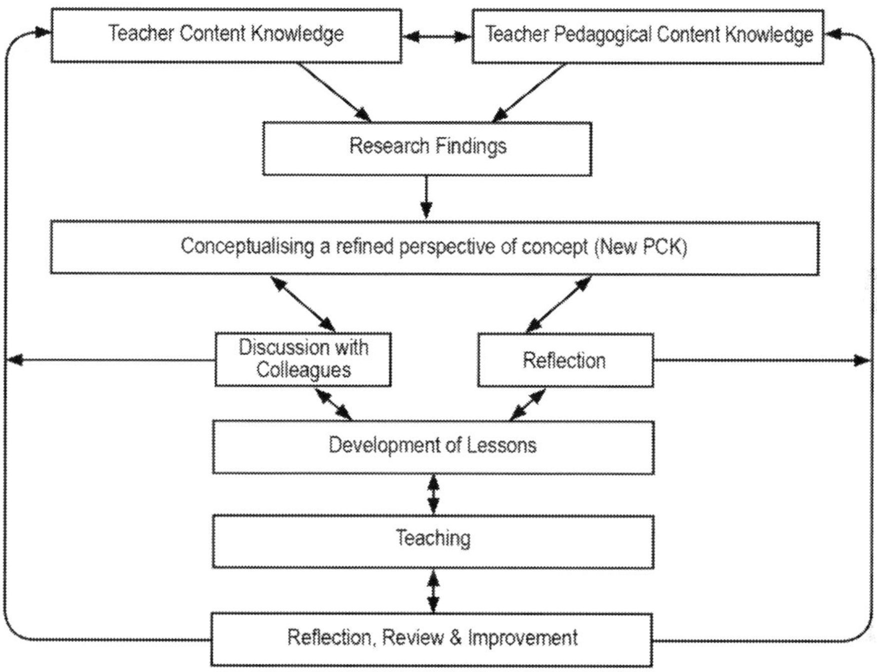

Figure 1. Teacher Professional Development (TPD) Model

To elaborate on the proposed TPD model, in this section, integration of the concept of simple harmonic motion (SHM) in physics and mathematics is discussed within a research-based perspective. The scenario described in the subsequent section depicts the various facets of notational consistency of mathematics and physics concepts. Teachers' CK, PCK and research findings on the content and on pedagogy (which includes curriculum requirement and instructional experiences) form the basis of the TPD model. During the professional development course, teachers had the opportunity to employ the model in a continuous reflective manner in order to refine or reflect on their previously held CK and PCK. Such a reflection took place during intensive discussion and collaboration in face-to-face sessions. Thereafter, teachers could proceed with the development of their physics or mathematics lessons grounded in research. Once the lessons had been implemented, further reflection by teachers was expected as evidence of engagement in a research culture (within subject area and across).

A CASE STUDY OF THE TPD MODEL

Because students acquire skills in mathematics through repeated practice (Stingler & Hiebert, 1999), their ability to transfer the mathematical knowledge to physics becomes quite problematic. Table 1 provides essential insights into the process of an inquiry lesson on SHM taught to mathematics and physics teachers at the same sitting. The teachers were registered in a Post Graduate Certificate in Education (PGCE) programme (part time) in a teacher training institution and the lesson was implemented for the module Subject Didactics – Physics and Subject Didactics – Mathematics.

The first two authors from the departments of Science Education and Mathematics Education respectively have made a comprehensive selection of the difficulties educators face in a number of physics and mathematics concepts. They investigated the topics of Newton's second and third laws, projectile motion, Hooke's law and SHM in which mathematics is prominent and is a pre-requisite for the study of physics.

For the purpose of this chapter, the case of simple harmonic motion is discussed. Usually, SHM is introduced by educators using the traditional approach starting with its definition, that is, acceleration is proportional to the displacement of the particle undergoing SHM (and directed towards a fixed point) and the trigonometric function is the resulting equation. Such an approach is also adopted in many physics textbooks. However, the link between the mathematical notation and the physics notation is rarely made. For instance, in mathematics, SHM is introduced as $a = -n^2x$, whereas in physics the equation $a = -\omega^2x$ is considered. To ensure that students do not develop alternative conceptions, a physics educator has to engage the students right at the start of the lesson to identify consistency in notation of a mathematics concept when such a concept is introduced in the physics lessons. The physics and mathematics trainee teachers were provided with reading materials related to misconceptions in physics and mathematics, inter-relationship between mathematics and physics and students' understanding of SHM.

The physics and mathematics trainee teachers, working collaboratively, were required to make a presentation of how they would have conceptually analysed SHM and then proceed with the teaching while making prominent the connections between the physics and the related mathematics parts. Table 1 illustrates a summary of the physics and mathematics trainees' development of SHM during their presentation. Soon after the presentation each group was required to reflect on its performance based on feedback received from peers and the researchers.

Table 1. Insight from the inquiry SHM lesson

Mathematics	*Physics*	*Missing link*
Students learn about the trigonometric function as early as from Form 3 (Grade 9, age 14).	Students (Grade 12, age 16) are introduced with the trigonometric function for the first time during the learning of SHM.	In the very first instance, no link was made in the physics lesson from the acquired mathematical knowledge of $y = asin\theta$ to physics knowledge of $y = asin\omega t$. During the presentation reference was not made about the interplay between the dependent and independent variables (see Chapter 16).
Educators use a table of values and plot the corresponding graph without any notion of its application. Basically, emphasis is on the functional form of the equation $y = asin\theta$ and is not progressive.	Educators start the lesson with the definition of SHM and relate the displacement of a particle undergoing SHM with the trigonometric function without even relating it to mathematics. The equation $y = asin\omega t$ is introduced as being progressive and the path traced by a particle performing SHM.	Furthermore, both physics and mathematics trainee teachers' difficulties arise in the interpretation of θ and ωt. The relationship between θ, measured in radians and ωt, also measured in radians but with the time dimension contributed to confusion. The trainees could not relate the two quantities as being identical (mathematics) and dimensionally consistent (physics). Elements of reflection was missing and they were resorting to memorised concepts, such as number of cycles per second and frequency, angular velocity rather than using the new information received during the discussion session and coming up with a conceptually appropriate understanding and interpretation of θ and ωt.
		Teachers are required to have recognised the equivalence of θ and ωt in context and about their consistency in their dimensions. Emphasis has to be directed towards representing ω as cycles per unit time (mathematics) and angular velocity (physics) to make the desired link.

(Continued)

Table 1. (Continued)

Mathematics	*Physics*	*Missing link*
Teachers introduce the angle while referring to a particle moving in a circular trajectory describing the angle at any instant. 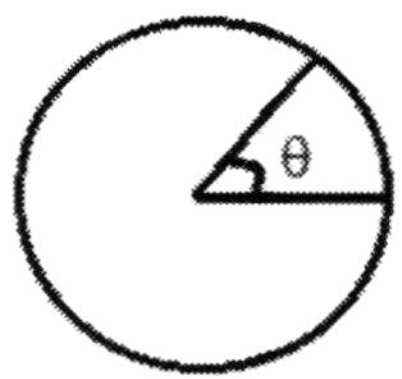	Motion is considered in either x direction or y direction, but not in both is presented in the physics lesson.	No link was made from circular motion introduced in mathematics to circular motion in physics. Circular motion can be represented as motion in one direction (linear motion) if the projection of the particle is considered on a screen. But what happens when the screen is stretched with the inclusion of the time dimension was missing. 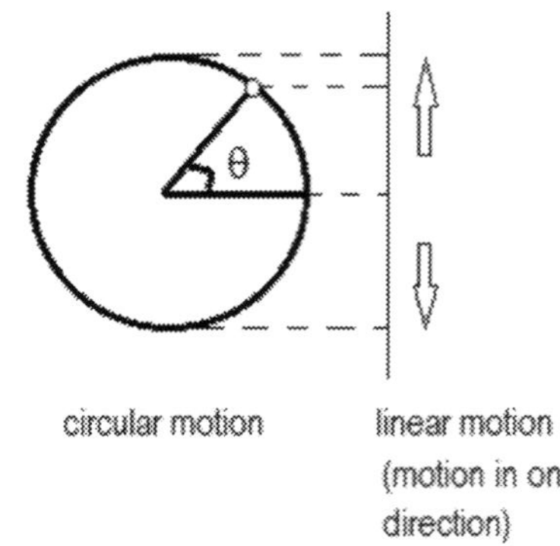

Circular motion Sine curve	Linear motion Sinusoidal relationship	The relationship among circular motion, linear motion and time-dependent motion is illustrated in the diagram. 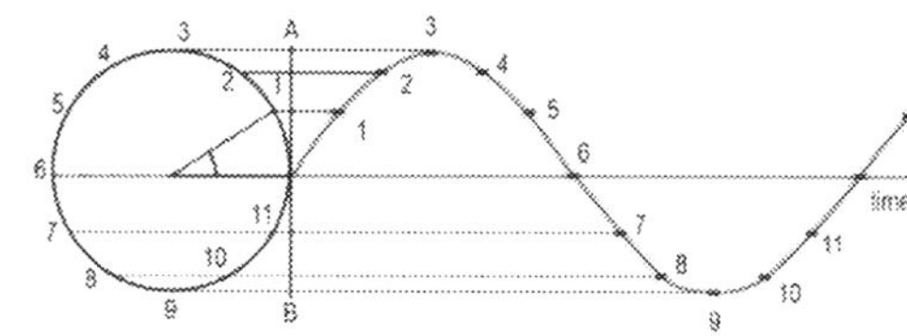 As the particle moves in a circular path, its linear progression along AB, when the time dimension is introduced, is described by a sine or cosine relationship. Additionally, a hands-on activity which illustrates the formation of a sine curve when a pen or pencil is moved sideways in a vertical position on a piece of paper and the latter is pulled slowly in a horizontal position makes the ultimate link from known to unknown. 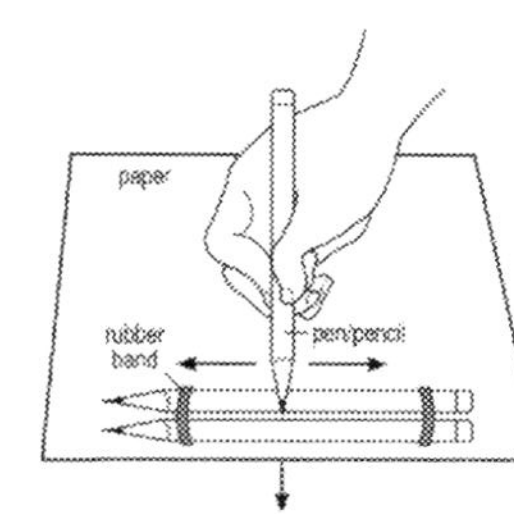

DISCUSSION

The proposed model of teacher professional development (Figure 1) focuses on the content knowledge (CK) and pedagogical content knowledge (PCK) of both groups of physics and mathematics trainee teachers and their research skills developed during their professional development courses to conceptualise a refined perspective of their CK and PCK. With the acquired experience of using the model, the teachers were expected, through constructive discussions and reflections at their school level, to engage more deeply in the preparation of their lessons by designing inquiry activities to supersede their students' inclination and need for rote-learning. During the process of engagement with the TPD model, the physics and mathematics teachers recognised commonalities and differences between mathematics and physics concepts, and synthesised the knowledge gained to model the construction of meaningful knowledge during the teaching. For example, in the SHM case, the commonality of the graphical representation and features of sine curve could be highlighted while discussions about the differences in the generation of the curve in the two contexts (that is, mathematics and physics) were used as opportunities for students to enhance their conceptual understanding of trigonometric functions. In the case of a mathematics lesson, the trigonometric curves are usually generated in terms of angular consideration while in a physics class, for the case of SHM, the curves are produced when the variable time is considered as part of the motion of a particle.

The implementation of the TPD model, while setting the basis for a shift from the conventional approaches to teaching within and across subject areas, would not stand into a sustained relationship between research and teaching if teachers do not regularly engage into reflection after their lesson implementation. In our case, despite being provided with research-based articles it was found that elements of research findings were still remote in the trainees' discussion, reflection and inference. In addition, the content of the research papers appeared to be abstract for the trainees such that they could not internalize the relevant information.

There has been calls from various quarters to innovate on the current teaching practices in science as teaching is no longer seen as a process of transmission of knowledge – rather one that facilitates the construction of knowledge by learners. The recent BERA report (2014) highlights the fact that improvement in student's achievement is associated with the quality of teaching and teachers remain a predominant figure to create conditions for students' engagement. Moreover, an emerging perspective from the report is that there is a need to create classroom environments that build on research-informed practice.

The case for integrating research into teaching is further warranted since teaching is seen to be subjective – dependent on a teacher's beliefs, disposition and comfort towards the subject matter and the topics to be taught. Teachers often take intuitive decisions as they are unable to find a common ground to map out their CK and PCK with their professional development setting. For instance, in the case of SHM, a formula based approach is commonly favoured for convenience at the expense of inquiry activities. Infusing a research-based approach, on the other hand, would enable teachers to analyse their decisions based on non-intuitive recommendations and make reflections to assist them in reviewing and improving their practices (see Figure 1). Marzano (2003) suggests that there is a significant improvement in students understanding through inquiry approaches based on research rather than on approaches which favour drill and practice and rote memorization.

There is also the challenge to create the right conditions for research-based teaching to be accepted among practitioners. Teachers can be unmotivated or indifferent or possess negative attitudes to linking research to their teaching. One approach to deal with these situations is that research-related collaboration between teachers should be promoted within discipline and across discipline as well. The model proposed in this chapter strongly favours such collaborative endeavour taking the disciplines of mathematics and physics as a case study.

While some researchers (Pocklington & Tupper, 2002) argue that research can be a detraction from the quality of teaching, others (Lee, 2004) claim that the quality of teaching would necessarily be enhanced when elements of research are included. Lee further asserts that, although using research findings for enhancing is welcoming, it is more befitting if teachers can use their own research findings rather than others in their practice. In addition, Goodwin (2005) argues about the heterogeneity of teaching and research whereby a variety of approaches can be made to enable the teaching-research links to be made. Griffiths (2004) considers three distinct types of research intervention in teaching, namely, research-led, research-oriented and research-based. The research-based teaching emphasizes inquiry-based activities which, in turn, have been shown to enhance the cognitive development of the learners. While the generic arguments are in favour in establishing a solid basis for connecting research and teaching, it is the type of the links that is varied.

In this chapter, we have proposed a research-based link (see Figure 1) that centers around the collaboration between the physics and mathematics teachers to design inquiry activities. The reflective cyclical process of the research-teaching link proposed is an attempt, not only to support teachers in bringing about change in their practice, but also to ensure a smooth and positive experience for their students. Along with the changes in current practice through reflection, review and improvement are essential for integrating and maintaining the proposed model in the long term.

CONCLUSION

The goal of professional development is to firstly enhance teachers' CK and PCK and transfer the added knowledge gained from both training and research inputs for improving students' achievements. Comparatively, a difference is made when teachers get the opportunity to analyse and reflect over their weaknesses and challenges to teaching science that demand good mastery of mathematical concepts. Such an opportunity was given to a group of physics and mathematics teachers to engage in collaborative research activities during their professional development course. The research model (TPD) is founded on continuous collaboration and reflection among the teachers in an attempt to review and reflect existing beliefs and disposition towards teaching. At the outset, the TPD model enables the notification of marked differences in approaching the sinusoidal relationship for SHM; else these differences would have gone unnoticed. The proposed model attempts to address a number of factors prevalent in the teaching and learning of science.

FURTHER READINGS

Aydin, S., Demirdogen, B., Tarkin, A., Kutucu, S., Ekiz, B., Akin, F. N., Tuysuz, M., & Uzuntiryaki, E. (2013). Providing a set of research-based practices to support preservice teachers' long term professional development as learners of science teaching. *Science Education, 97*(6), 903–935.

Fraser, B. J., Tobin, K. G., & McRobbie, C. J. (Eds.). (2012). *Second international handbook of science education*. Dordrecht, The Netherlands: Springer.

Lederman, N. G., & Abell, S. K. (Eds.). (2014). *Handbook of research on science education* (Vol. II). New York, NY: Routledge.

REFERENCES

Akgun, A. (2009). The relation between science student teachers' misconceptions about solution, dissolution, diffusion and their attitudes toward science with their achievement. *Education and Science, 34*(154), 26–36.

Bayrakar, S. (2009). Misconceptions of Turkish pre-service teachers about force and motion. *International Journal of Science and Mathematics Education, 7*, 273–291.

BERA. (2014). *The role of research in teacher education: Reviewing th evidence interim report of the BEAR-RSA inquiry.*

Bulterman-Bos, J. A. (2008). Will a clinical approach make education research more relevant to practice? *Educational Researcher, 37*(7), 412–420.

Goodwin, A. (2005). Linking educational research and the teaching of science. *School Science Review, 86*(316), 119–123.

Griffiths, R. (2004). Knowledge production and the research-teaching nexus: The case of the built environment disciplines. *Studies in Higher Education, 29*(6), 709–726.

Lee, R. (2004). Research and teaching: Making – or breaking – the links. *Planet, 12*, 9–10.

Leeman, Y., & Wardekker, W. (2015). Teacher research and the aims of education. *Teachers and Teaching, 20*(1), 45–58.

Marzano, R. (2003). *What works in shools. Translating research into action.* Alexandria: VA: ASCD.

McIntyre, D. (2005). Bridging the gap between research and practice. *Cambridge Journal of Education, 35*, 357–382.

Pocklington, T., & Tupper, A. (2002). *No places to learn: Why universities aren't working.* Vancouver: University of British Columbia Press.

Prensky, M. (2005). Engage me or enrage me: What today's learners demand. *Educause Review, 40*(5), 60–64.

Ramma, Y., Bholoa, A., Watts, M., & Ramasawmy, J. (2014). Pre-and in-service physics teachers' content and pedagogical content knowledge: Implications and challenges. *American International Journal of Research in Science, Technology, Engineering & Mathematics, 1*(8), 30–34.

Ramma, Y., Samy, M., & Gopee, A. (2015). Creativity and innovation in science and technology: Bridging the gap between secondary and tertiary levels of education. *International Journal of Educational Management, 29*(1), 2–17.

Schneps, M., & Sadler, P. (Directors). (1988). *A private universe* [Motion Picture]. Retrieved August 27, 2015, from http://www.learner.org/resources/series26.html

Schön, D. (1987). *Educating the reflective practitioner: Towards a new design for teaching and learning in the professions.* New York, NY: Basic Books.

Shulman, L. S. (1986). Those who understand: Knowledge growth in teaching. *Educational Researcher, 15*(2), 4–14.

Stigler, J. W., & Hiebert, J. (1999). *The teaching gap.* New York, NY: Free Press.

Wright, D. (2001). Creativity and learning: Creative work and the construction of learning. *Reflective Practice, 2*(3), 261–273.

Wubbels, T. H., & Korthagen, F. A. (1990). The effects of a pre-service teacher education program for the preparation of reflective teachers. *Journal of Education for Teaching, 16*(1), 29–43.

COLIN SMITH, BODIL SVENDSEN AND PETER GRAY

40. FURTHER PROFESSIONAL DEVELOPMENT OF TEACHERS

INTRODUCTION

This Chapter is about teacher professional development (TPD) for science teachers, a term which covers a wide range of in-service activities, from the acquisition of functional information to deep philosophical reflection about teaching and learning. We see TPD as essential if science teachers are to react to changing circumstances. It is a process of lifelong learning, which should also stimulate their pupils to become lifelong learners, to engage with science and perhaps to take up science-based careers.

First, we provide an overview of two teacher professional development models for lifelong learning: '*teachers as technicians*', delivering set curricula using established methods, and '*teachers as pedagogical problem solvers*', supporting their students in dealing with local learning issues, using Loughran's (2010) distinction between professional development and professional learning.

We then argue that teachers are 'professionals in situ', making pedagogical decisions in unique contexts. Decision-making can be enhanced by analytical tools, such as Bloom's taxonomy, or the SOLO (Structured Observation of Learning Outcomes) Taxonomy (Biggs & Collis, 1982) to support student learning, experiment with practice and develop teachers' repertoires of actions. These models are prescriptive, however, as they present learning as a hierarchy of processes or outcomes. Instead, we argue for conceptual tools that allow teachers to think reflectively, by posing questions such as: *What am I doing now and why? What might I do instead and why?* We suggest that such tools embed learning in teachers' practice.

Finally, we provide a 'roadmap' showing what an ideal science TPD system would look like, bearing in mind that TPD around the world is far from ideal (OECD, 2013) and how teachers can work with researchers, school management, students and other stakeholders to achieve it. This provides material for debate and reflection for prospective teachers and TPD providers.

WHAT IS TPD?

Teacher professional development (TPD) evolves through thinking about teaching and teachers. Historically, teacher education took place in specialised colleges and was regarded as a one-time activity. Over the last thirty or so years, there has been

K. S. Taber & B. Akpan (Eds.), Science Education, 551–560.

an increasing recognition of the need to make teaching more professional, and to promote lifelong learning within that profession (EC, 2013). This has led to the rise of teacher professional development (TPD). TPD should not be regarded as synonymous with continuing professional development (CPD) or in-service education and training (INSET), which is often associated with the delivery of factual content, such as health and safety information. As we argue below, the term 'teacher professional learning' (TPL) may be more appropriate, but we use the term TPD because it is internationally recognized.

We are not, however, suggesting that content-based training is irrelevant. In science education, teachers frequently need to know how to use innovative items of equipment, or new experimental techniques. This area of training is often handled via 'resource centres' and their associated online resources. In situations where schools cannot provide the full range of science education facilities, the use of regional science resource centres is an effective way of delivering high-quality training, with the added benefit that teachers are able to meet and discuss activities with colleagues.

TPD is, then, about teachers and their conceptual systems rather than about teachers and factual information. Since metaphors form a major part of conceptual systems or "metaphorical schemas" (Lakoff & Johnston, 1999), one way of categorising teachers is to use metaphors. The range of possible metaphors is extensive and cannot be fully explored here. To simplify matters, we discuss two of the most important metaphors and their related metaphorical schemas, which determine the consequences of the metaphor for practice. One example would be the "time as a quantity" metaphorical schema, in which time can be wasted, spent or saved. The metaphors here are *teachers as technicians* and *teachers as pedagogical problem solvers.*

These two metaphors form the ends of a continuum, over which a debate continues to flourish. For many years, teacher educators have promoted reflection and reflective practice in teaching, partly as a reaction against behaviourist approaches during the so-called 'cognitive revolution'. Conversely, policymakers have tended to adopt increasingly instrumental approaches to teaching and learning, in which outcomes have been measured by quantitative indicators.

WHY ENGAGE IN TPD?

A recent European Commission report (EC, 2013) identifies three components, which need to be present in education systems in order to encourage teachers to engage in TPD. These are stimulation, assessment and provision.

Stimulation refers to a form of self-understanding in which teachers recognise that TPD is relevant to their own needs and can have positive outcomes for them and their students. This often requires a catalyst in the form of an external intervention, since teachers rarely have time to initiate the necessary activities without some kind of support. This can come from school leaders, from external agencies or projects, or from government policies, although this is rare.

The question of assessment relates either to qualification structures, where certain levels of TPD may be required to maintain certification, or to self- and peer assessment, formal or otherwise. However, the ultimate aim, in our view, is to make science teaching a research-based profession in which all teachers routinely draw on the results of research and carry out research, in a wide sense, into their own practices. Assessment therefore becomes self-assessment, which in turn becomes practitioner research.

Regarding the provision of TPD, in the, *teachers as technicians* metaphor, teaching is seen as the delivery of set curricula using approved methods and TPD is seen as upgrading teachers' skills related to those methods. This metaphorical schema presents TPD as doing things *to* or *on* teachers and teaching as doing things *to* or *on* students. Students are largely passive in this model. This is particularly unfortunate in science education, where there are extensive opportunities for activity and practical work, even where laboratory facilities are absent. It is important to note that 'activity' and 'hands-on' science are not synonymous. Inquiry-based activity can often involve simply reframing the presentation of a topic so that progress is driven by students' questions.

The metaphor of *teachers as pedagogical problem solvers* is about doing things *together with* teachers, and teaching as doing things *together with* students. Under this schema, TPD is about supporting *teachers* in inquiring into how to support their *pupils* in learning e.g. through inquiry methods in science. We recognize that these two 'umbrella metaphors' cannot cover the full scope of teaching, but they serve our purposes here.

Both these schemas are compatible with Elmore's (2002, p. 8) conception of TPD:

> Professional development, in the consensus view, should be designed to develop the capacity of teachers to work collectively on problems of practice, within their own schools and with practitioners in other settings, as much as to support the knowledge and skill develop-ment of individual educators.

There is nothing in either model against the sharing of practices that have been found to be successful in other contexts. The difference is that in the *technician* schema, these practices are handed down as 'recipes' while in the problem-solving scenario they are critically discussed as possible solutions. Both are also compatible with TPD as a lifelong process from initial teacher education to retirement (Villegas-Reimers, 2003).

The *pedagogical problem solver* metaphor acknowledges the need for TPD to move beyond simple achievement of subject knowledge and skills (Hewson, 2007; Vescio, Ross, & Adams, 2008). This should be based on a shift from acquiring knowledge to researching teaching practice, from single training sessions to learning that takes place over time, and from individual to collaborative learning (Borko, 2004; Watson & Manning, 2008). Traditional transmissive approaches fail to meet the professional needs of science teachers in particular (Ostermeier, Prenzel, & Duit, 2010) and in any case are incompatible with modern views of science teaching as inquiry.

Furthermore, traditional forms of TPD often treat teachers as isolated practitioners, thus limiting their exposure to the work of colleagues (Elmore, 2002). They also underplay the complexity of schooling (Ball & Frozani, 2007) and, consequently, need to be more sensitive to the local contexts in which teachers work and learn. Teacher learning can be described as a complex system, involving systems within systems (Opfer & Pedder, 2011). Also, there is a shift in some countries towards collaborative inquiry approaches, often within schools, in which teachers work in partnerships and support and learn from each other in professional learning communities, in which teachers work together with a focus on improved learning and teaching (Harris & Jones, 2010). This has been very successful, for example in the SINUS programme in Germany.[1] In science education, this indicates that inquiry should be a priority at all levels within the system, including pupils, teachers and school leaders.

Based on the above, we are suggesting that TPD can become professional learning through the active solving by teachers of pedagogical problems that they identify as relevant to the need of their students. This has been tested in a number of EU funded projects designed to support the widespread adoption of inquiry based teaching and learning in science subjects. Here, we discuss two interventions arising from the S-TEAM (Science-Teacher Education Advanced Methods) project (S-TEAM, 2013). These were PISCES (Promoting Inquiry-based Science in Scotland) (Smith et al., 2013) and SUN (*Skolebasert utvikling i realfag*)[2] (Svendsen & Marion, 2013). Both of these were collaborative, multi-session activities with initial input from researchers but mainly sustained by teachers' active involvement in applying new ideas in the classroom.

In these activities, instead of applying top-down prescriptions, the teachers focused on sharing knowledge and skills about how to meet the learning needs of students. Teachers needed conceptual tools to help them with this process and to think about such questions as *What am I doing now and why? What might I do instead and why?* This form of TPD treats them as *professionals in situ* (Smith et al., 2013), who are best placed to judge the needs of their own students and to seek ways to meet those needs. This involves providing teachers with a language and matrix of concepts to describe, discuss and explain what they are doing, what they might do instead, and why this is worth trying. Our suggestion is that this is more powerful than other forms of TPD in that it embeds a wider repertoire of knowledge and action in teachers' practices. This repertoire is the basis of practitioner theories about what will support particular student groups, topics and educational aims (Smith et al., 2013).

Sometimes, as in PISCES, participating teachers from different schools may work with individual student groups to solve their pedagogical problems. The practices developed may be used to solve similar problems in the future. Support and collaboration, however, occurs through meeting over a period of time with each other and supporting researchers. Or, as in SUN, they may come from the same school or from other schools connected in science networks and collaborate (again with non-prescriptive support from academic researchers) in reflecting upon what they already do, what they might do and how to support each other in making changes.

Either way, teachers work together with a focus on learning and teaching in their own contexts, and on generating new professional knowledge. They may form what Bolam et al. (2005) describe as "professional learning communities". These are communities where teachers continuously seek, share and act upon learning. Supportive school leadership is necessary for these professional learning communities to be effective (Robinson, Lloyd, & Rowe, 2008), and this might involve school leaders themselves forming such learning communities.

So, to sum up, it is possible for TPD (a) to move from top-down prescription of practice, (b) to support teachers in solving pedagogical problems identified locally and to providing conceptual tools that help them in relation to those problems and (c) to move from a language of development to a language of learning and mutual support or co-operation. Let us expand our metaphors into two models of how TPD might be implemented.

TWO MODELS OF TPD

Loughran (2010) provides a description of the differences between teacher development and professional learning that we can use to further illustrate the evolution of TPD. We have adapted Loughran's model as an alternative model of how teachers might develop theories of practice.

Model 1 (*Practitioner Theory Developing Through Professional Development*) is a deficit perspective in which teachers lack skill to deal with changes in policy, need to be shown 'best practice' by 'experts', and, by implication, are incapable of developing practitioner theory for themselves. In contrast, in Model 2 (*Practitioner Theory Developing Through Professional Learning*) professional learning occurs through reflection, discussion and experiments in practice. Teachers, as professionals in situ, focus on developing practices appropriate to their aims and contexts.

It might be that model 1 is somewhat a caricature but, even so, when one of the authors presented it to teachers in Scotland and Latvia, they described it as being all too familiar. Model 2 differs from model 1 in that it begins with teachers' existing knowledge and beliefs, and the learning that follows is personally shaped by these beliefs. This is in line with conceptions of autonomy that emphasise that what one does emanates from the self, is self-authored, relates to one's own interests and involves choice in actions (Su & Reeve, 2011). Although the models described here are generic, in the sense that all teachers can move to a professional learning model, they are of particular interest in science teaching. This is because Model 2, in particular, is aligned with what is regarded as the 'nature of science', in that it

Table 1. Two models for developing Practitioner Theory

Model 1- Practitioner Theory through Professional Development	*Model 2- Practitioner Theory through Professional Learning*
Professional development arises during changes such as new curricular initiatives, involving the assumption that teachers need to be 'up-skilled'	Assumes that the teachers are committed to these changes and that they can implement them through collaborative reflection and inquiry.
Experienced by teachers as having something done to them	Teachers *experience growing self-confidence and reflective abilities leading to a feeling of greater autonomy* since they bring expert judgement to changes in contexts and practices.
Delivery of TPD involves the deliverer as an active changer of teachers' attitudes, beliefs, skills and knowledge. Models of 'best practice' are handed down.	TPL involves working *with* teachers, to develop attitudes, beliefs, skills and knowledge congruent with shared objectives. Both facilitator and teacher may experience changes in attitudes, beliefs, skills and knowledge. *Provision of external resources such as conceptual models empowers teachers to reflect upon their own contexts and what they want to achieve.* i.e. a shift from 'best practice' to 'appropriate practice for my/our goals.'
Teachers are trained in changes mandated from above that are presented in the form of a Practitioner Theory stated in terms of technical requirements.	Teachers learn to make changes for themselves and develop their own Practitioner Theories aimed at achieving goals valued in their own contexts, and share them, so learning from each other.
A top-down approach with experts (depending on country) from Universities or government agencies setting out the changes required.	*A partnership approach* between 'actors' with different forms of expertise working towards a shared objective and with teachers working out the changes required in their own contexts and for their own students.
Changes sometimes little more than relabelling existing practices.	Change more likely to be qualitatively genuine and innovative.
Functions in a similar way at all levels in education from the central bureaucracy to school level.	Requires a mindset in which the school supports this form of learning by its teachers and each level thinks about how to support the one below in achieving this end.
Practitioner theory is not owned by the teachers, and freedom for them to deal with complexity in its local forms is limited.	Teachers own their practitioner theories, are constantly developing them to deal with variations in the complexity they experience and share them professionally.
A top-down approach that treats teachers as technicians implementing handed down techniques/practices.	*A bottom-up approach that treats teachers as professionals in situ solving pedagogical problems in their own unique contexts.*

is tentative, dependent on observation, theoretical reflection and analysis, and constantly open to new thinking. Model 2 is, therefore, science in action!

Practitioner theory is conceived here as science teachers' theories of what will work for particular purposes, with particular groups of students and why. Rich practitioner theories are based upon:

- A growing repertoire of practices associated with educational purposes;
- A good understanding of educational aims and how conflicts can be resolved (For example, the conflict between the aim of using more inquiry in science and the aim of getting through a crowded syllabus);
- Conceptual tools that adequately analyse what teachers are doing, and the learning needs of their students;
- Readiness to acquire more conceptual tools for different purposes;
- Collaboration with other science teachers, in reflecting on educational aims and current practice, in using and developing conceptual tools, and in sharing of practice and its results.

We suggest that a model in which teachers' practitioner theory is developed through professional learning is congruent with the metaphor of teachers as pedagogical problem solvers and is also more supportive of the development (by the teachers themselves) of practitioner theories of this type. As they develop such theories, teachers empower themselves to make more decisions in their unique contexts. The role of teacher educators in model 2 is to bring conceptual models, together with their own experiences, to the collaborations. Two points can be made.

First, it will be noticed that participation in collaborative TPD has been described as voluntary. This is because the teachers in our projects commented that this was desirable. They believed that the experience would have been diminished had it been compulsory – they would not have embedded the changes into their practice, or approached the TPD with the right learning intentions. This could present a dilemma, if we believe that some form of TPD should be part of science teachers' lifelong learning. It does imply that Model 2 TPD has to gain enough momentum for teachers to participate voluntarily. Our suggestion is that, if model 2 is applied widely enough, it will gain this momentum as teachers see the benefits of participation. This implies the need for political support in providing time and space for professional development and nurturing professional learning communities.

Secondly, although teacher educators and educational researchers have a role in bringing conceptual tools to TPD, the sort of tools we have in mind are open to adaptation and further development by the teachers themselves to fit their local contexts. The next section looks at a roadmap for teacher professional development and how it might be used. In terms of science teaching, this roadmap may meet resistance from those teachers with a background in natural science, who are not accustomed to reflecting on their own practice and who may see inquiry as a distraction. Our experience is that teachers frequently move from initial reluctance to considerable enthusiasm, providing that the ethos of inquiry is shared across all those involved, and that there is time within the curriculum to purse both TPD and the resulting changes in practice.

A ROADMAP: FROM TEACHER PROFESSIONAL DEVELOPMENT TO TEACHER PROFESSIONAL LEARNING

1. All levels in an educational system should move from top-down delivery of TPD to bottom-up teacher learning, replacing development (doing things to teachers) with mutual learning and support.
2. All levels should therefore support teachers as they inquire into: *What am I doing now and why? What might I do instead and why?* This creates the conditions for voluntary participation in TPD and establishes teachers as professionals in situ, solving local pedagogical problems.
3. Inter- and intra-school professional learning communities should be formed, in which participating teacher educators and researchers provide teachers with tools to think about *What am I doing now and why? What might I do instead and why?* These tools should not be too prescriptive but should provide relevant sub-questions for teachers to think about. They should be open to amendment by the teachers to fit local contexts.
4. The whole process of professional learning should be part of lifelong learning for participants, helping them to build ever-richer practitioner theories of what will work, and why, for particular educational aims and outcomes. With rich practitioner theories, teachers empower themselves to make more pedagogical decisions relevant to their own contexts and to the benefits of their students.

CONCLUSIONS

Professional development, or professional learning, for science teachers has not been a policy priority for governments across the world. This is unfortunate, because the vast majority of governments prioritise science education, along with mathematics, as a key factor for economic growth and social progress. As we have shown, the principles of effective professional learning are relatively simple. The main objective for those concerned with professional learning in science education should be to make time and space for teachers to learn, and to empower them to research their own practice, and the practice of others, in order to make improvements. Providing content-based training can also be important, but the availability of online content and resources has made it easier for teachers to acquire content and learn how to perform technical operations.

The pedagogical aspects of science teaching cannot just be 'delivered', and require not just collaboration, but also a sense of responsibility for one's own learning and for improving practice generally. In turn, the introduction of more inquiry-based methods means that responsibility for learning has to be shared with students, and that student learning should be assessed in new ways, outside the confines of traditional assessment systems. We hope this chapter will help readers to achieve real results!

NOTES

[1] See "International report on Implementation Strategies" at: http://www.mascil-project.eu/resources/reports-and-deliverables

[2] In English: School-based Professional Development in Science http://www.ntnu.no/skolelab/sun-prosjektet

FURTHER READING

Hattie, J. (2009). *Visible learning: A summary of over 800 meta-analyses relating to achievement.* London: Routledge.

Essential reading for anyone in education, provides a strong evidence base for changing practice.

Hewson, P. W. (2007). *Teacher professional development in science.* In S. K. Abell & N. G. Lederman (Eds.), *Handbook of research on science education* (pp. 1179–1203). Mahwah, NJ: Lawrence Erlbaum.

A standard text, useful for all science teachers.

Hoveid, M. H., & Gray, P. (Eds.). (2013). *Inquiry in science education and science teacher education: Research on teaching and learning through inquiry based approaches in science (teacher) education.* Trondheim: Akademika Forlag.

Provides some alternative thinking about teacher professional development, in line with this chapter.

McNally, J., & Blake, A. (2010). *Improving the professional learning of teachers.* London: Routledge.

A useful source for professional learning in general, based on an empirical study of new teachers in Scotland.

WEBSITES AND ONLINE RESOURCES

Education Scotland: government agency providing a wide range of teaching and learning resources: http://www.educationscotland.gov.uk/professionallearning/

Association for Science Education (ASE) (United Kingdom): huge range of online resources for science teachers and their development: http://www.ase.org.uk/home/

National Science Teachers' Association (USA): similar to ASE, but even larger range of resources. http://www.nsta.org/

Pedagoo: run by teachers, illustrating the learning community approach: http://www.pedagoo.org/

European Science Education Research Association: main conference for science educators internationally, with extensive web resources: http://www.esera.org/

REFERENCES

Ball, D., & Forzani, F. M. (2007). What makes education research 'educational?' *Educational Researcher, 36*(9), 529–540.

Bolam, R., McMahon, A., Stoll, L., Thomas, S., & Wallace, M. (2005). *Creating and substaining effective professional learning communities* (Research Report 637). Bristol: University of Bristol.

Biggs, J., & Collis. K. (1982). *Evaluating the quality of learning: The SOLO taxonomy.* New York, NY: Academic Press.

Borko, H. (2004). Professional development and teacher learning: Mapping the terrain. *Educational Researcher, 33*(8), 3–15.

EC (European Commission). (2013). *Supporting teacher educators for better learning outcomes.* Brussels: European Commission, Directorate-General for Education and Training.

Elmore, R. (2002). *Bridging the gap between standards and achievement: The imperative for professional development in education.* Washington, DC: Albert Shanker Institute.

Harris, A., & Jones, M. (2010). Professional learning communities and system improvement. *Improving Schools, 13*(2), 172–181.

Lakoff, C., & Johnston, M. (1999). *Philosophy in the flesh: The embodied mind and its challenge to Western thought.* New York, NY: Basic Books.

Loughran, J. (2010). *What expert teachers do: Enhancing professional knowledge for classroom practice.* London: Allen and Unwin.

OECD (Organisation for Economic Cooperation and Development). (2013). *TALIS 2013 results: An international perspective on teaching and learning.* Paris: OECD.

Opfer, V. D., & Pedder, D. (2011). Conceptualizing teacher professional learning. *Review of Educational Research, 81*(3), 376–407.

Ostermeier, C., Prenzel, M., & Duit, R. (2010). Improving science and mathematics instruction: The SINUS Project as an example for reform as teacher professional development. *International Journal of Science Education, 32*(3), 303–327.

Robinson, V., Lloyd, C. A., & Rowe, K. J. (2008). The impact of leadership on students outcomes: An analysis of the differential effects of leadership types. *Educational Administration Quarterly, 44*, 635.

Smith, C., Blake, A., Kelly, F., Gray, P., & McKie, M. (2013). Adding pedagogical process knowledge to pedagogical content knowledge: Teachers professional learning and theories of practice in science education. *Educational Research e-Journal, 2*(2), 132–159.

S-TEAM. (2013). *Firing up science education.* Retrieved May, 2013, from http://www.s-teamproject.eu

Su, Y. L., & Reeve, J. (2011). A meta-analysis of the effectiveness of intervention programs designed to support autonomy. *Educational Psychology Review, 23*(1) 159–188.

Svendsen, B., & Marion, P. V. (2013). *School-based teacher professional development in science: Participants' reports on perceived impact.* Paper presented at ESERA Conference 2013, Nicosia, Cyprus. (Science Education Research for Evidence-based Teaching and Coherence in Learning.)

Vescio, V., Ross, D., & Adams, A. (2008). A review of research on the impact of professional learning communities on teaching practice and student learning. *Teaching and Teacher Education, 24,* 80–91.

Villegas-Reimers, E. (2003). *Teacher professional development: An international review of the literature.* Paris: UNESCO, International Institute for Educational Planning.

Watson, R., & Manning, A. (2008). Factors influencing the transformation of new teaching approaches from a programme of professional development to the classroom. *International Journal of Science Education, 30*(5), 689–709.

BEN AKPAN

41. NATIONAL AND INTERNATIONAL LINKAGES FOR SCIENCE EDUCATION THROUGH SCIENCE TEACHER ASSOCIATIONS

INTRODUCTION

In July of 1993, UNESCO (The United Nations Educational, Scientific, and Cultural Organisation) held an international forum of Project 2000+ in Paris, France in furtherance of its efforts to promote science and technology education for all by 2000 and beyond. Of particular significance was the *Declaration* by the participants at the forum urging non-governmental organisations such as *science teacher associations* to: (i) enter into partnership with and make their knowledge and experience available to United Nations and other inter-governmental bodies as well as establish innovative programmes in a common effort to achieve the goal of scientific literacy and technological literacy for all; and (ii) participate in national, regional and international programmes for the enhancement of scientific and technological literacy for the improvement of the quality of life in all societies and for the achievement of sustainable development. More recently, UNESCO (Fensham, 2008) has drawn attention to the important role of science teacher associations, where its members not only have the insights and experience, but also the interest in helping science teacher colleagues.

This chapter highlights the various contributions possible from science teacher associations (STAs) in the development and delivery of high quality science and technology education in a world that is increasingly driven by the outputs of science and technology. We will consider the meaning, examples, functions, and structural model of STAs. The chapter ends with a strong recommendation for teachers to join STAs.

What are STAs?

Generally, science teacher associations are non-political, non-religious and not-for-profit professional organisations with the goal of improving teacher effectiveness in various science subjects. Some STAs are national in outlook while others are international. The National Science Teachers Association in the USA, Association for Science Education in UK, Australian Science Teachers Association, New Zealand Association of Science Educators, and the Science Teachers Association

K. S. Taber & B. Akpan (Eds.), Science Education, 561–575.

of Nigeria are examples of national STAs. The Commonwealth Association for Science, Technology and Mathematics Educators; International Organisation for Science and Technology Education, and the International Council of Associations for Science Education are examples of STAs at the international level. In what follows we briefly describe these organisations in turn.

National Science Teachers Association (USA). The National Science Teachers Association (NSTA) is based in Arlington, Virginia, USA. It was founded in 1944. With a membership of 55,000 spread over several countries, NSTA is the largest science teacher association in the world. NSTA's guiding principles are geared towards modelling excellence; championing science literacy; valuing scientific excellence; embracing diversity, equity, and respect; enhancing teaching and learning through research; collaborating with partners, and exemplifying a dynamic professional organization. The Association has recently established a new 5-year strategic plan, *NSTA Strategic Goals 2015* with the goal of raising the status of science education and science teaching as a profession by advocating high quality science education; enhancing the professional learning of science educators by providing a suite of tools, resources, and opportunities that support long-term growth within a collaborative learning environment; using the Next Generation Science Standards (see chapter 14) to revitalize science education to boost student achievement and science literacy; nurturing scientific curiosity among children in the earliest grades; enriching the NSTA membership experience; and updating infrastructure, as well as providing support to staff. Many organisations such as the National Association for Research in Science Teaching and the Association for Multicultural Science Education are affiliated to the NSTA.

Association for Science Education (UK). The history of the Association for Science Education (ASE) can be found in Layton (1984). It dates back to January 1901 when the Association of Public School Science Masters was formed. The ASE was established much later in 1963 by the merger of the Science Master's Association and the Association of Women Science Teachers. In its current Charter of Incorporation, ASE seeks to promote education by improving the teaching of science, providing an authoritative medium through which opinions of teachers of science may be expressed on educational matters; and by affording means of communication, among all persons and bodies of persons concerned with the teaching of science in particular and with education in general. Headquartered in College Lane, Hatfield, the ASE is a leading science teacher association in the world.

Australian Science Teachers Association. The Australian Science Teachers Association (ASTA) was founded in 1951. Based in Canberra, ASTA is a federation of science teacher associations from Australian States and territories with the following objects:

i) provide for professional stimulation of science teachers by developing professional standards, providing opportunities for professional development and facilitating networking nationally and internationally;
ii) provide information and advice to science teachers regarding industry links, equity in science education, resources, teaching methods and curriculum;
iii) provide science teachers with resources;
iv) provide information and advice to those who influence science education, including developers of curricula and resources in science, teacher educators and supervisors, policy makers whose decisions relate to science education and those engaged in research into science education;
v) recognize significant contributions to science education and to provide information and advice to the general community regarding the aims of science teachers and social issues of a scientific nature; and
vi) promote the value of science education to the community.

New Zealand Association of Science Educators. The New Zealand Association of Science Educators co-ordinates and supports many organisations. It has the following objects:

i) To promote the development of science education throughout New Zealand.
ii) To facilitate liaison and cooperation between regional science teachers' Associations.
iii) To assist regional science teachers' Associations in their efforts to sustain and expand their activities.
iv) To disseminate information, articles and other material related to science education through newsletters, journals and other means.
v) To represent the interests and concerns of people involved in science education and to enhance their skills and interest.
vi) To develop links with international science education Associations.
vii) Such other objects as are deemed by the Council to be ancillary or related to the objects already stated.

Science Teachers Association of Nigeria. The Science Teachers Association of Nigeria (STAN) was established in 1957 with the following aims:

i) To promote co-operation among science teachers with a view to raising the standard of science education in the country.
ii) To provide a forum for discussion by science teachers on matters of common interest.
iii) To help science teachers keep in touch with developments in science and its application to industry and commerce
iv) To popularize science
v) To co-operate with and affiliate to other societies and bodies with related interests

vi) To do or perform such other functions incidental to or necessary for the realization of these objectives.

Commonwealth Association of Science, Technology, and Mathematics Educators (CASTME). CASTME began as CASME (the technology was not originally part of the name) in 1974 through the effort of Maurice Goldsmith who facilitated the institution of the then Guinness Awards scheme for science and mathematics teachers. Mr. Goldsmith had developed interest in popularizing science and mathematics in the British Commonwealth. The Guinness Awards was for teachers who entered winning essays that described innovative ways of teaching science in a social context. Chisman (2004) reports that in April 1974 Mr. Goldsmith and several colleagues involved in the Guinness Awards Scheme took part in a conference of the Association of Science Teachers of Jamaica during which there was a decision to establish CASME. Mr. Goldsmith was the founding President and a full meeting of CASTME was held a year later in London. The Guinness Awards became CASME awards. These were presented annually. Also, the Magazine, *The Science Teacher* earlier published privately by Mr. Goldsmith was renamed CASME journal although its editorship and publication were still carried out by him. According to Chisman, from that time onwards, CASME, and later CASTME – has become very active and later developed branches in some regions – Africa, Asia, and Europe.

Currently, CASTME links science, technology, engineering, and mathematics educators across the commonwealth. CASTME works in partnership to do research, support awards and scholarships, and run projects in commonwealth countries. It also works to advance the social relevance of the teaching of science, technology and mathematics through networking of educators in these subjects. The CASTME Journal currently emphasizes this link to social relevance. An electronic newsletter, first published in 2010, updates members on developments in science, technology, and mathematics education. CASTME has also developed a number of capacity-building scholarships with the University of Westminster in the UK for suitably qualified applicants to pursue Master's Degree programmes.

International Organisation for Science and Technology Education. The International Organisation for Science and Technology Organisation (IOSTE) was established to advance the cause of education in science and technology as a vital part of the general education of the peoples of all countries and to provide scholarly exchange and discussion in the field of science and technology education. It advocates the peaceful and ethical use of science and technology in the service of humankind. IOSTE's origins can be traced to a symposium on world trends in science education convened in August 1979 in Halifax, Nova Scotia, Canada. At the third symposium held in Brisbane, Australia in 1984, the informal circuit of 'world trends' was transformed into a formal organization with members from over 60 countries. Today, IOSTE has members from about 80 countries.

International Council of Associations for Science Education (ICASE). In the 1950s and 1960s there was a general movement in various parts of the world to reform science education (see Chapter 14). There were at the time some challenges in the sector and core teachers were required in each country to help generate and implement ideas. It was thought that the ideal body for such intervention was science teacher associations (STAs) with membership spanning the entire educational spectrum. STAs could represent the authentic voice of the science teaching profession. Fortunately, they were already in existence in some countries. International co-operation was required to strengthen existing associations. UNESCO was anxious to act in this direction. So it was that at the UNESCO Regional Workshop on the Teaching of Integrated Science held in the Philippines in 1970, UNESCO was requested to facilitate exchange of information among STAs in Asia. As a first step, UNESCO collaborated with the Science Teachers Association of Singapore and the Singapore National Academy of Sciences to organize a meeting of leaders of STAs with Singapore as venue in 1972. The main outcome of that meeting was the call for the establishment of an international federation of STAs, providing a forum for groups of people with similar ideas and goals and with similar problems to share their concerns and hopefully find solutions. Events moved very swiftly afterwards and ICASE was formally inaugurated in April 1973 at the University of Maryland in the USA with eleven members while Professor David Lockard, Director of the Science Teaching Centre at that university was elected as the first President of ICASE and Mr. Dennis Chisman as Secretary and Treasurer. Within ten years the membership grew to 44 member associations. The objectives of ICASE have since then been to:

- Extend and enhance the work of its member organisations.
- Provide and support activities and opportunities to enhance formal and non-formal science and technology education worldwide.
- Establish and maintain an international communication network.
- Encourage and support the establishment and development of professional science and technology organization, especially where none currently exists in a country.

The Governing body of ICASE is its General Assembly, consisting of one delegate from each member association together with any members of the Executive Committee who are not delegates. The Governing Body delegates the operation of ICASE to an elected Executive Committee, comprising of a President, President-Elect, Past- President, Secretary, and Treasurer and up to eight members elected on a geographical basis. The Executive Committee has the power to appoint chairmen of standing committees to further specific ICASE activities.

At its General Assembly in Kuching, Malaysia in 2013, ICASE approved a strategic plan for the future. This includes re-examining the goal of ICASE in order that by its 50th Anniversary in 2023, the vision of the Association to provide the foundation and leadership in *Delivering Excellence in Science Education Worldwide* could be realized. Thus moving forward, ICASE envisions its role as helping to develop and sustain science teacher associations so that all science teaching is enhanced through collaboration, innovative methodologies and connections throughout the globe. In this direction, the ICASE mission is to deliver and co-ordinate, enact and disseminate research and resources that enhance the impact and growth of science education and science teacher associations throughout every continent.

FUNCTIONS OF SCIENCE TEACHER ASSOCIATIONS

Science Teacher Associations (STAs) have been known to play significant roles in science, technology and mathematics education (STME). Silber (in King 1991:47) sees STAs as performing the following functions:

- Communications — through Journals, conferences, publications.
- Representation — to teachers and government, liaison with other groups and participation in international activities.
- Services — continuing education, employment, low cost equipment and out-of-school activities.
- Leadership — curriculum development, teacher benefit, guidance on new developments in science education.

In fact, STAs such as NSTA and ASE have tremendous influence on STM education not only in their countries but also in foreign nations. For instance, Holbrook and Chisman (1988) report that in the U.K. the ASE was responsible for producing *Science in Society* and *Science in a Social Context* courses.

In *Africa*, Bajah and Yoloye (1981:27) while evaluating *Science Education Programme for Africa* had this to say:

> *A powerful kind of organization that stimulated development in science education in practically all the countries was the association of science teachers. Every one of the countries studied had such an association in one form or another. There were variations in the magnitude of contributions made by these organisations from country to country. Gambia, Liberia and Lesotho report negligible contribution. At the other end the Ghanaian and Nigerian associations have exerted tremendous influence on the training of science teachers, the curriculum, and educational policies. STAN in Nigeria produced its own books in integrated science for the first two years of secondary schools.*

Curriculum development. The Association for Science Education, for instance, has been heavily involved in curriculum development efforts in the UK. Members of the ASE played influential roles in the Nuffield projects of the 1960s. By the 1970s the ASE had published the Study Series which provided teachers the required updates on science curriculum issues of the day. *Science and Technology in Society*

(SATIS), *Education through Science,* and *APU Science Reports* are among many other curriculum documents/books published by the ASE. More recently, ASE was pivotal in producing the influential report *Beyond 2000: Science education for the future.*

In the USA, the National Science Teachers Association (NSTA) has been heavily involved in curriculum development over the years, not least being the star role it played in the various curriculum projects of the 1960s and 1970s. When Project 2061 was initiated in 1986 by the American Association for the Advancement of Science, NSTA collaborated effectively. The NSTA also initiated the Scope, Sequence, and Coordination Project at the instance of its then Executive Director Bill Aldridge (see the chapter on science curriculum development initiatives). NSTA's role was also very significant in the build up to, initiation, and implementation of the National Science Standards which at its core were based on the principles that science is for all students, learning science is an active process, school science reflects the intellectual and cultural traditions that characterize the practice of contemporary science; and that improvement in science education is part of a systematic reform in education. More recently, the NSTA has collaborated with the National Research Council and Achieve in initiating the Next Generation Science Standards (NGSS). With the adoption of the NGSS in many States in the USA, NSTA has been devoting its resources in ensuring effective implementation. It is leading the way in showing teachers how NGSS are different from the national science education standards and how to use science to connect to the common core, for example, by using science to support literacy in English Language Arts. NSTA is therefore introducing a different way of thinking to the science classroom by assisting teachers plan an NGSS curriculum, design units and lessons, select teaching/learning materials, and conduct assessments.

In Australia, the Australian Science Teachers Association (ASTA) is equally involved in curriculum development and implementation. Currently, in partnership with Education Services Australia (ESA), ASTA has designed several units of work in support of teachers to enable them implement the science curriculum in that country. The units are developed by experienced teachers using online resources.

In Africa, STAN is probably a good example of how professional bodies have intervened in curriculum development. Among other curriculum development efforts, STAN initiated the Integrated Science Project in Nigeria (see Chapter 14). The work of STAN in curriculum development in Nigeria has been such that it has been regarded as a curriculum development agency. In the words of Ivowi (1993:353):

> *In appraising the performance of the curriculum development agencies* (in Nigeria), *five such bodies, namely, the Nigerian Educational Research and Development Council, West African Examinations Council, National Teachers Institute, National Commission for Colleges of Education, and the Science*

Teachers Association of Nigeria have been singled out. STAN, a professional association that has contributed much to curriculum development in Nigeria is here regarded as a curriculum development agency. It is a very typical and foremost example of such professional association in Nigeria.

Production of textbooks. Both the NSTA and ASE have published a wide range of books cutting across the various science subjects and educational levels. Other STAs such as ICASE, ASTA, and STAN have also published textbooks and/or resource books.

Organization of in-service training for teachers. All STAs listed so far organize in-service training for members regularly. These in-service training programmes take the form of annual conferences/meetings, workshops, seminars, etc. Typical programmes of annual conferences include international roundtable exchange, commissioned lectures, pre-bookable training courses, workshops, panels, demonstrations, special reports, briefings, commercial workshops, and exhibition. The ASE holds its conference in the first week of January each year, NSTA in March or April, and STAN in August. ICASE runs a world conference on science and technology education every three years and this holds in October or November. The 2016 ICASE World conference was held in Antalya, Turkey. The ICASE conference brings all the STAs together to exchange ideas. It is an event every science teacher who is willing and financially capable should endeavour to attend.

Popularization of science. Most STAs popularize science through competitions. For teachers the NSTA currently runs the Shell Science Lab Challenge which encourages teachers who have found innovative ways to deliver quality lab experiences with limited school and laboratory resources. For students, NSTA offers a variety of fun and friendly competitions to engage them in the pleasures of science beyond the curriculum. These include the DuPont Challenge Science Essay Competition, Cyber mission, and Toshiba/NSTA ExploraVision.

The ASE has designed a programme, *Science Across the World*, which brings an international dimension to science education in schools and colleges. Students are able to get a global perspective on scientific issues related to their personal lives, their impacts on the environment and the varying cultural impacts of science on people in different countries. In order to participate, teachers choose a topic, make a contact with other teachers to be partners, decide on the method to use for engaging ideas and information, work on the topic with their group of students, and share the findings with others and also encourage students to discuss what they learn from the exercise.

In Nigeria, STAN popularizes science through STAN quiz and STAN projects competitions at science fairs. These provide opportunities for individuals and groups to display the various science projects which they have undertaken. A project may set out to make a discovery, develop new ways of demonstrating important principles or attempt to demonstrate practical applications of known principle (Figure 1).

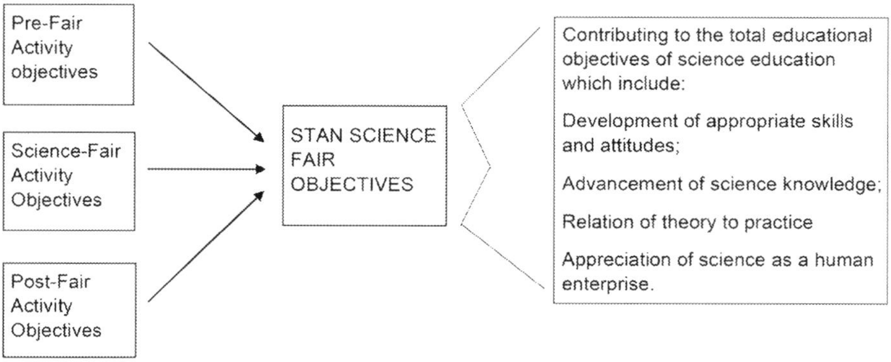

Figure 1. Objectives of STAN Annual Science Fair.
Source: Obioha (1983:82)

Publication of periodicals. STAs have also embarked upon the publication of periodicals such as Journals, bulletins (newsletter) and proceedings of conferences and workshops. These publications feature articles, research reports, innovations, science notes, reviews, approaches to science teaching, trends in science teaching worldwide and updates on members.

The ASE publishes several journals. These include: *School Science Review*, *Post – 16 Science Issues*, *Primary Science*, *Education in Science*, and *Science Teacher Education*. The NSTA also publishes a wide range of journals. These include: *Science & Children*, *Science Scope*, *The Science Teacher*, *Journal of College Science Teaching*, *Quantum*, and NSTA *Reports*. On its part ASTA publishes *Teaching Science* as its flagship journal. In New Zealand, NZASE publishes the *New Zealand Science Teacher* while STAN publishes the *Journal of Science Teachers Association of Nigeria*, a *Bulletin* that carries teaching notes as well as a *Newsletter*. The STAN journal is available online for free (www.stanconference.com/jstan). ICASE is the publisher of *Science Education International* which is available online as an open access journal (www. icaseonline.net/seiweb).

Research work. All the STAs are involved in some research efforts although to different extents. The NSTA has the series *on 'What Research says to the Science Teacher'* among other research-oriented publications to its credit. The ASE is also heavily involved in research and has a very viable research committee.

STRUCTURAL MODEL OF STAS

The following guidelines are adapted from ICASE (1998).

a. *The Basic Philosophy of Science Teachers Association (STAS)*: STAs should think through the basic purposes of teaching science in their own particular context. STAs should play a decisive role in all aspects of science education improvement and in developing the kind of teachers that are needed to fulfill such purpose. An association with a varied membership represents an authentic voice of the science teaching profession. Having established its basic philosophy, the STA can then set about fulfilling its purpose.
b. *The Purpose of STAs*: For the purpose of improving science education, there have been and are many on-going programmes for science teaching and for the popularization of science. To implement the ideas which have been generated in these programmes, there is a need for a nucleus of enthusiastic people in each country who can be involved both in the process of generating new ideas and also in the much lengthier implementation process. The ideal body for such a task is a Science Teachers Association (STA).
c. *Membership of STAs*: STAs could include the following type of members:

- practicing teachers of various science subjects at all levels – (i.e. from pre-school to tertiary)
- decision makers within schools – (e.g. head teachers and principals)
- decision makers for the school – (e.g. Ministry of Education Personnel)
- staff of college/university science and education departments
- institutions/industries interested in promoting good science education
- libraries

Such a body is in a position to draw on the ideas of all its members and to consider science teaching at all levels of its educational system.

d. *Type of Objectives*: Associations are strongly advised not to pursue objectives of the following kind:

- regulation of relations between employer and employee
- imposing of restrictive conditions on the conduct of any trade or profession
- provision of pecuniary/financial benefits for members.

e. *Structure and Management of STAs*: All associations require a management organisation and structure. Generally, organisations are managed by a central council or executive committee elected by association members. In associations of large areas or countries, a regional structure with regional representatives may also be needed. The management council might be elected at a central or national meeting by regional representatives. Honorary members can be included in the council to act as advisors or to give influence to the association by virtue of their position in society.

Committees, sub-committees and working parties can be established to work in specific areas of responsibility. These bodies can be established as needed, and exist for different periods. Some standing committees may be permanent to deal with on-going matters, but with rotation of committee members. Other ad hoc committees or working parties can be established for a limited term to deal with particular projects or specific issues.

The management council/executive committee provides, along with sub-committees and working parties, the mechanism through which the organization functions. The number of sub-committees and/or working parties required in each association depends on the tasks to be undertaken.

All committees, sub-committees and working groups should maintain a close relationship with the council/executive committee, so that coherence and good communications are maintained. This is often done by having a member of the council/executive working with the committees or working parties.

Flexibility should also be built into the management organization and structure to be able to respond to new situations and initiatives. *Involvement* of as wide a variety of members as possible is recommended.

Figures 2 and 3 show the suggested organizational structures for small and large Associations respectively.

A SUGGESTED MANAGEMENT STRUCTURE OF A SMALL ASSOCIATION

(A few hundred members or less)
COUNCIL/EXECUTIVE COMMITTEE
Chairperson
Vice-Chairperson
Secretary/Minutes Secretary
Treasurer
Editor
Membership Secretary
Course/Activities Organiser
Special Representatives from Sub-Committees
COMMITTEES, SUB-COMMITTEES, WORKING GROUPS
Curriculum Development
Finance
Awards
Publications
Special projects/activities/areas

Figure 2. Structure of a small association (adapted from ICASE, 1998)

A SUGGESTED STRUCTURE OF A LARGE ORGANISATION
(A few thousand members)
Honorary Officers
Chairperson, Chairperson-Elect, Immediate Past Chairperson
Secretary, Treasurer, Editor
Annual Meeting Secretary, Schools Liaison Officer
Sub-Committee Representative
SUB-COMMITTEES AND WORKING GROUPS
Executive/Administrative Committee
Headquarters Administration
Co-ordination of Council Officers
Executive and Finance Committees
Journal
Annual Meeting
Regional Committees
Committee for each geographical region
Local Activity Planning
Education/Co-ordination (E/C) Committee
Co-ordinates the work of sub-committees, concerning for example
Apparatus
Examinations
Further Education
Girls and Science Education
Health and Safety
Technology
Representation on other bodies
Education Conference with (E/R)
Publications with (E/R)
Education/Research (E/R) Committee
Research and development promotion working parties
Awareness of Science
Language
Policy review
Education Conference with (E/C)
Publications with (E/C)

Figure 3. Structure of a large Association (adapted from ICASE: 1998)

Trouble Shooting

The following are some of the constraints likely to be faced by STAs:

- *Lack of Funds* – virtually all the STAs complain of lack of funds.
- *No Office or Headquarters, Dependence on Voluntary Activity* – Both the NSTA and ASE have well established Headquarters offices. However, except for STAN, there is no STA in Africa with a permanent paid Secretariat. Other STAs are run by the elected officers only on part-time basis. Indeed, as an Association grows, more demands are made on the few council and committee members. Until

the association grows sufficiently to afford premises and staff, little significant progress can be made in solving this problem. This according to ICASE (1998) is the critical breakthrough point. In Africa, it would appear from this that only STAN in Nigeria has reached this point.

- Shortage of secretariat and clerical services.
- Insufficient members actively involved.
- Publication difficulties.
- Official recognition.

Following are suggested solutions to the constraints

a. *Lack of Funds*: STAs should:

- Monitor organisations that may be prepared to contribute funds in general or towards specific activities.
- Promote the authorship of books – some associations have successfully used active members to write books, and the association can collect royalties after payment to the writers of suitable honoraria.
- Obtain sponsorship for such ventures as science fairs, courses, and journal publications
- Promote sales of journal to institutions and libraries
- Take commissions on curriculum development
- Rent exhibition space at annual conferences to publishers and manufacturers
- Sell advertising space in the association's publication
- Negotiate with Ministry of Education or Local Authorities for the payment of fees and travelling cost of association members who attend updating courses.

b. *No office or headquarters, dependence on voluntary (often isolated) activity from individual working in their own homes*: Sometimes a teacher centre or training institution is willing to supply premises. Some Ministries of Education, Local Education Authorities or University Departments may be prepared to consider the loan of such a facility. The Association would need to present a well prepared case for its use and be responsible for the organization and supervision of such a facility.

c. *Shortage of Secretarial and Clerical Services*: The availability of premises immediately enables more members to become actively involved and increases the scope of the association. Clerical activity can be centralized and used more efficiently.

d. *Insufficient Members Effectively Involved Leading to too many Demands on the Few Committee Members*: The sharing of tasks and responsibilities is most important in a growing association although this can occasionally create inefficiencies and delays. This is an especially acute problem for associations in a large or scattered community. The establishment of a communication hub or a formalized communication network pattern may help. This may be achieved, for instance, through the establishment of zonal branches.

e. *Publication Difficulties*: Publishers tend to give low priority to small print runs. Try to arrange regular publication to coincide with slack times in the printing trade. Cost of publication often inhibits greater activity in this field. See lack of funds above.
f. *Official Recognition*: Most science teacher associations receive official recognition from national to local education authorities as professional organisations. Until this recognition is obtained it is difficult to maintain the desired co-operation. However, perseverance and adherence to declared professional aims, along with the avoidance of undesirable aims will increase the likelihood of an association gaining recognition.

Why you Should Join STAs

We strongly recommend that a science teacher or educator joins one or more science teacher associations because they provide the following benefits to members:

i) STAs will enable members develop professionally through in-service training programmes, annual conferences, workshops, seminars, symposia, curriculum guides, and general guidance on content/pedagogy.
ii) Members have renewed confidence and competence in teaching as they engage in exchange of information and skills through interaction with colleagues.
iii) Some STAs offer free journals and other periodicals to members; others provide the publications on discounted rates.
iv) Members are constantly updated on developments in science and technology education especially as they relate to content and pedagogy
v) There is an opportunity for international meetings which further broadens one's horizon on science and technology education issues within the framework of a globalizing world.
vi) Members also have opportunity to share their ideas with others
vii) Members contribute to consultations and debates that make it possible for policy makers to listen to the science teaching profession.
viii) Some STAs provide free access to online resources for members
ix) Members are able to join specific interest groups such as subject panels – physics, chemistry, Biology, mathematics, etc.
x) Some STAs ensure they are represented in important national committees set up by government thus providing an opportunity for such representatives to feature at that level.
xi) Members may be given awards as a form of recognition for some achievement thus boosting their morale.
xii) Members get to form a personal network of friends who share common vision and goals.

SUMMARY AND CONCLUSION

In this chapter, we have discussed the following:

- meaning of STAs;
- a description of some STAs at national and international levels;
- functions of STAs;
- structural model of STAs; and
- the need for science teachers to join STAs.

We need to reflect on the various issues raised and set agenda for ourselves. We consider it important to advocate support from policy makers for existing STAs and assistance in the establishment of STAs in countries where none exist.

USEFUL WEBSITES

National Science Teachers Association: www.nsta.org
Association for Science Education: www.ase.org.uk
Australia Science Teachers Association: www.asta.edu.au
New Zealand Association of Science Educators: www.nzaze.org.nz
International Council of Associations for Science Education: www.icaseonline.net
National Association for Research in Science Learning: www.narst.org
Science Teachers Association of Nigeria: www.stanonline.org

REFERENCES

Akpan, B. B. (2010). Innovations in science and technology education through science teacher associations. *Science Education International*, *21*(2), 67–79.

Bajah, S. T., & Yoloye, E. A. (1981). *A report of twenty years of science education in Africa*. Ibadan, Nigeria: *Science Education Programme for Africa*.

Chisman, D. (2004). CASTME: The first 30years: Memories and nostalgia. *CASTME Journal*, *24*(2).

Fensham, P. J. (2008). *Science education policy making: Eleven emerging issues*. Paris: UNESCO.

ICASE. (1998). *A guidebook for science teacher associations*. Limassol, Cyprus: Executive Secretary, ICASE.

Ivowi, U.M.O. (Ed.). (1993). *Curriculum development in Nigeria*. Ibadan: Sam Bookman.

King, W. R. (1991, April 15–19). *Ensuring quality science and technology education: The role of agencies outside the school* (pp. 44–53). Proceedings of the CASTME/COL North American Regional Seminar on Quality in Science, Technology, and Mathematics Education, University of British Columbia, Vancouver, Canada.

Layton, D. (1984). *Interpreters of science: A history of the association for science education*. London: John Murray.

Obioha, N. E. (1983). Science projects and fairs in science education. *Journal of the Science Teachers Association of Nigeria*, *22*(1), 79–84.

CONTRIBUTORS

Ian Abrahams is Head of the School of Education and Professor of Science Education at the University of Lincoln. Having previously been a teacher of physics and head of department he undertook a PhD in Education and then, in 2007, moved into Higher Education working at the IoE (now UCL Institute of Education), the University of York and then the University Leeds. Ian's research interests relate to practical work, misconceptions, Continuing Professional Development (CPD) – in particular the effectiveness of such programmes – and the assessment of practical skills. Ian has directed the evaluation of various large-scale national interventions, including the 'Getting Practical' programme (2009–11), a Nuffield funded project to improve the effectiveness of practical work (2012), and the Wellcome Trust funded pilot CPD programme for primary science specialists (2012–14). Ian's research has led him into collaborations with colleagues in China, Germany, Ireland, Israel, Singapore and Turkey.

Agustín Adúriz-Bravo has a degree in physics teaching from the Universidad de Buenos Aires (Argentina) and a Ph.D. in didactics of science (science education) from the Universitat Autònoma de Barcelona (Spain). He is currently Senior Researcher of the CONICET (National Council of Research in Argentina) at the Research Institute CeFIEC of the Universidad de Buenos Aires and the Institute IEC of the Universidad Nacional de Tierra del Fuego. He is head of Research Group GEHyD, which focuses on the various contributions of the philosophy and history of science to science education. His research interests include: science teacher education in the philosophy of science, nature of science, epistemics of scientific argumentation, construction and validation of historical narratives, and scientific models and modelling.

Sudhakar Agarkar (born on 1 July 1951) has M.Sc. (Chemistry) and Ph.D. (Science Education) to his credit. He was a faculty member of the Homi Bhabha Centre for Science Education, a national centre of the Tata Institute of Fundamental Research for science and mathematics education from 1976 to 2013. Working with the slum dwellers in Mumbai and tribal children in Western Ghat of India he has designed Remedial Instructional Strategy to teach science to socially deprived students. He went further to develop Open Educational Resources for children, their teachers and parents to facilitate learning of school science among first generation learners. Dr. Agarkar is presently working as Professor and Dean at VPM's Academy of International Education and Research, Thane (Maharashtra State) in India. In collaboration with various professional organisations of teachers and teacher educators he is striving to build bridges among science educators from different countries.

Ben Akpan is the Executive Director of the Science Teachers Association of Nigeria (STAN). He served as President of the International Council of Associations for Science Education (ICASE) for 2011–2013 and currently serves on the Executive Committee of ICASE as the Chair of World Conferences Standing Committee. Ben's areas of interest include chemistry, science education, environmental education, and support for science teacher associations. He is the editor of Science Education: A Global Perspective published by Springer.

Christopher Andersen (Ph.D., Columbia University) is an Associate Professor in the Patton College of Education at Ohio University (USA). Previously, he has worked as a K–12 classroom teacher in Venezuela and New York City, as a consultant to companies that produce educational television, instructional materials, and games, and as a faculty member and administrator at Ohio State University. As a cognitive developmental psychologist, his research interests are centered around the role of metacognition in children's learning of science and other school subjects, and its role in the development of reasoning strategies in particular.

Mehmet Aydeniz is an Associate Professor of Science Education at the University of Tennessee, Knoxville (UTK). After receiving his doctoral degree at Florida State University he started to work at UTK. Dr. Aydeniz's research focuses on pre-service and in-service science teachers' adoption of reform-based instructional practices and assessment of student learning. Most recently, he has started a research line that focuses on K-12 students' computational thinking skills and pre-service science teachers' pedagogical knowledge for teaching engineering. He is currently investigating the impact of a RET program on STEM teachers' pedagogical knowledge of engineering. He is the Editor of the Journal of Research in STEM Education.

Ajeevsing Bholoa is a Senior Lecturer in Mathematics Education at the Mauritius Institute of Education, which is a teacher training institute. He is currently the Programme Coordinator for pre-service B.Ed honours and is also involved in curriculum development at the primary level. His research interests are related to the integration of technology as a pedagogical tool in teaching and learning of mathematics and the identification of content knowledge and pedagogical content knowledge of teachers.

Julie A. Bianchini is a professor of science education at the University of California, Santa Barbara (UCSB). Her research investigates prospective, beginning, and experienced teachers' efforts to learn to teach science to all students in equitable and reform-minded ways. She serves as Faculty Director of UCSB's CalTeach Initiative, a University of California System-wide program designed to encourage science, mathematics, and engineering undergraduates to consider K-12 science and mathematics teaching as a career. She is the principal investigator on two teacher education research and development grants funded by the National Science Foundation: CalTeach Physical Sciences and Engineering (CTPSE) and STEM Teachers for English Language Learners: Excellence and Retention (STELLER). She is also an editorial board member for the Journal of Science Teacher Education.

Paul Black is Professor Emeritus of Science Education at King's College London. After 25 years spent on teaching and research in physics at the University of Birmingham, he moved to a chair in science education in London. He has made many contributions in both curriculum development in the Nuffield Curriculum Projects at primary and secondary levels, and to research into learning and assessment. He has served on advisory groups of the USA National Research Council and as visiting professor at Stanford University. His work on formative assessment with Dylan Wiliam and the King's assessment group has had widespread impact.

David Blades is Professor of Science Education and Curriculum Theory in the Faculty of Education at the University of Victoria. His scholarship examines the theory and philosophy of science education, including a recent exploration of aesthetics in science education. David was born in Victoria and in his spare time, which is never enough, loves to explore the islands of the Salish Sea.

Richard Brock studied physics at Durham University, before teaching in South Africa, Japan and a special needs school. After completing a PGCE, he worked for eight years teaching physics at a secondary school in Essex. During this time, he worked as an associate lecturer on an MEd programme. His MPhil research at the University of Cambridge into the role of insight and intuition led to an invitation to present his work at the Cambridge University TEDx conference 2015. Richard has acted as a research assistant on the Learning About Science And Religion (LASAR) project and contributed to work on papers exploring students' understanding of the boundaries between science and religion. He is currently working on a PhD at the University of Cambridge focusing on the process of making sense in physics education.

Joanne Broggy is a Lecturer in Science Education in the University of Limerick, Ireland. She is also a Project Officer for the Teaching and Learning of Science in EPI·STEM National Centre for STEM Education. Her research interests include Inquiry Based Learning, Knowledge Representation and Scientific Literacy.

Gultekin Cakmakci is an Associate Professor of Science Education at Hacettepe University, Turkey. After working as a Chemical Engineer in the industry for about a year, he decided to become a researcher and changed his career paths. He earned his master's degree in Science Education from the University of York, UK and PhD in Science Education from the University of Leeds, UK. He earned a Postdoctoral Research Fellowship to carry out his research at Stanford University in 2011–2012. His research interests focus on STEM communication, nature of STEM and engineering education. He is a member of the scientific committee of the "Public Communication of Science and Technology (PCST)" and member of the Editorial Advisory Board of "Studies in Science Education".

Michael P. Clough is a professor of science education at Iowa State University where he teaches The Nature of Science and Science Education, Secondary Science Methods I, Secondary Science Methods II, and Restructuring Science Activities. He is the recipient of several awards for his teaching (at both the university and secondary school level), scholarship and service. His scholarship is directed at the nature of science and its implications for robust science learning, effective science teaching, and science teacher education; and the synthesis, criticism, and clarification of extant knowledge and research in science education. He is a past president of the International History, Philosophy and Science Teaching (IHPST) organisation.

Beatriz Crujeiras is a Science Education teaching fellow at the University of Santiago de Compostela in Spain. She is engaged in teacher education, teaching Chemistry and Physics Education. Her doctoral dissertation focused on high-school students' engagement in scientific practices in the chemistry laboratory. Her research interests focus on students and teacher learning, including inquiry-based teaching and learning, scientific and epistemic practices as well as scientific competencies. She is involved in professional development about scientific competencies for high school science teachers.

James de Winter works at the Faculty of Education, University of Cambridge leading the Secondary Physics PGCE initial teacher education course as well as teaching on the Primary PGCE course. He currently also leads the country-wide teacher fellow and educational research programs for the Ogden Trust. Previously, for many years, he worked as a secondary school physics teacher and then for the Science Learning Network developing, writing and leading continuing professional development (CPD) courses regionally and nationally with a focus on supporting specialist and non-specialist teachers of physics. James has been published widely including academic papers, textbooks, practical guides and classroom resources for primary and secondary aged students as well as in-service and pre-service teachers.

Shannon L. Dubois is an Assistant Professor of Science Education, in the Department of Education, at Valparaiso University, Valparaiso, Indiana, USA. As a former high school biology teacher, her work is influenced by a desire to impact the lives of science teachers and students. Her research focuses on the continuum of teacher learning, including science teacher education, induction, and professional development. She received the Janice Hooper Hold Outstanding Dissertation in

Research on Teacher Induction Award in 2015. Her work has been published in The Journal of Science Teacher Education, Studies in Science Education, The Science Teacher, and Science Educator. She recently co-edited a book, Newly Hired Teachers of Science: A Better beginning, which was published by Sense Publishers.

Ingo Eilks, Prof. Dr. FRSC, studied chemistry, mathematics, philosophy and education at the University of Oldenburg. He is a full trained grammar school teacher, having a PhD and the Habilitation in chemistry education. Since 2004 he is professor for chemistry education at the Institute of Science Education (IDN) at the University of Bremen. His research interests encompass various areas. One area is research-based curriculum development, founded on the application of Participatory Action Research to science education. Foci of corresponding projects deal with understanding the particulate nature of matter, cooperative learning, ICT in science education, promoting scientific and scientific media literacy, education for sustainable development, or the socio-critical and problem-oriented approach to science teaching. A second major domain of research focuses science teachers' cognition and learning. Different studies are carried out about science teachers' beliefs, pedagogical content knowledge, and continuous professional development. A third domain focuses innovations in higher chemistry education.

Sibel Erduran is Chair of STEM Education & Director of EPI·STEM National Centre for STEM Education in the University of Limerick, Ireland. She also holds a Distinguished Chair Professor position at National Taiwan Normal University, Taiwan. She is an Editor for International Journal of Science Education and Section Editor for Science Education. Her research interests focus on the applications in science education of interdisciplinary perspectives on science, particularly the epistemic practices of science.

Merce Garcia-Mila (Ph.D., Columbia University; BSc in Chemistry and BA in Education, University of Barcelona) is a professor of developmental and educational psychology in the School of Psychology at the University of Barcelona (Spain). She started as a science teacher in high school in Barcelona, and she now teaches courses in the psychology of education and developmental psychology. Her research background is on scientific reasoning, use of external representations and argumentation, all addressed to science learning. She has also been a visiting scholar at Columbia University and Stanford University. She was the research coordinator of the Research Institute of Childhood and Urban World (CIIMU, Barcelona, Spain) for six years.

Peter Gray is European Projects adviser for the Department of Teacher Education (PLU) at the Norwegian University of Science and Technology (NTNU) in Trondheim, Norway. He completed his Ph.D at the University of Stirling, where he also worked on the Early Professional Learning project, funded by the Teaching and Learning Research Programme (TLRP) as well as several EU-funded projects. He started working with NTNU in 2007 and was proposal writer and subsequently project manager for S-TEAM. Since 2012 he has worked on a significant number of European collaborative projects and proposals. Currently his research interests include responsible governance in research, the theory-practice gap in teacher education and the theory and practice of learning spaces.

Emily Harris originally a secondary science teacher, has worked for the last nine years with informal learning and heritage education. She spent two years as Education Officer at the Houses of Parliament and five years at the Horniman Museum and Gardens where she set up new learning programmes aimed at connecting the museum's indoor collections with its outdoor spaces. More recently, as a consultant, Emily has delivered education and interpretation projects for several large heritage organisations including the Royal Botanic Gardens Kew. At Cambridge University Emily has been employed as Research Assistant on an EU funded STEM project exploring 'Tinkering' as a pedagogical approach for learning in STEM. She previously worked for Cambridge University on the EU funded PATHWAY project, which focused on inquiry-based learning in STEM. Emily's professional and research interests are in science education, informal learning, museum and heritage education, outdoor learning and family learning.

Marie Hobson is a Learning Researcher at the Natural History Museum, London. She conducts and commissions evaluation and research to inform exhibition and digital development at the museum, having previously worked in the audience research team at the Science Museum, London. Marie is also chair of the Research and Evaluation group of the Ecsite network of European museums and science centres.

Avi Hofstein was the head of Department of Science Teaching and Head of the Chemistry Group in this in the Weizmann Institute of in Israel. His research is mainly focused on issues related to issues related to teaching and learning high school chemistry focusing on laboratory work, curriculum development (e.g. making learning chemistry more relevant) and implementation, attitudinal, and interest research studies. He was also intensively involved in issues related to chemistry teachers' professional development. In 2016 he was awarded the ACS award for his research on teaching and learning chemistry in the laboratories. In the last 4 years he edited two books (in collaboration with Professor Ingo Eilks) published by Sense on pedagogical issues related to the teaching and learning of chemistry.

Jack Holbrook is a visiting professor at University of Tartu, Estonia, with considerable experience in science education worldwide. His PhD is in Chemistry, but he also has a school science teaching background, extensive experience in pre- and in-service teacher education; running in-service training workshops in a range of countries in Asia, Latin America, the Middle East and Africa, as well as in Europe. He has acted as an education consultant in a range of Asian and African countries. His research covers curriculum development in science subjects, teacher education and assessment and evaluation, as well supervising PhD students in use-inspired research, scientific literacy, inquiry learning and changing teacher beliefs. He has written and adapted numerous classroom teaching materials and published books for teachers, as well as science textbooks for use in school. He has been and continues to work as curriculum and assessment consultant for the Ministry of Education in Bangladesh (2002-) and from 2013 has been contracted by the World Bank for handling the science curriculum component of a project in Kuwait. He is the past president of ICASE (International Council of Associations for Science Education) and currently a member of executive committee responsible for international projects and besides having lead the ICASE involvement in the European Commission FP7 projects – PROFILES and ENGINEER, has played a role as editor of Science Education International as well as the ICASE newsletter.

María Pilar Jiménez-Aleixandre is Science Education professor at the University of Santiago de Compostela, in Spain. Before that she was biology teacher for a decade. Her research interests include argumentation, epistemic and scientific practices and how to scaffold them in science classrooms. Methodologically she is interested in classroom longitudinal studies, in order to examine learning processes. She has authored or co-authored about 70 papers in refereed journals, in particular about argumentation, and about 50 books or book chapters, such as the edited volume with Sibel Erduran Argumentation in Science Education: Perspectives from classroom-based research published by Springer in 2008.

Julie A. Luft is an Athletic Association Professor of Science and Mathematics Education, in the College of Education, at the University of Georgia, Athens, Georgia, USA. In the field of science education, she has held many national positions, published and presented her work nationally and internationally, and she has received awards for her research and teaching. She is a former middle and high school science teacher, which influences her research and teaching interests. Her current research interests are in the area of science teacher education with a focus on newly hired teachers of science and the process of professional development. Within these areas, she is interested in how teachers progress professionally and how they build their subject matter knowledge.

Joseph Makokha is a Systems Analyst/Technology Specialist at Stanford University Libraries (SUL) at Stanford University in the United States. Joseph has a Masters from Stanford University School of Education's Learning Design and Technology program. He also has credentials in Math, an MA in Urban Education (Holy Names University), plus a BS in Electrical & Electronic Engineering from the University of Nairobi. His professional experience includes teaching mathematics in schools, creating maths and robotics curriculum, and setting up computer labs and training in Kenya.

Rachel Mamlok-Naaman earned her BSc in chemistry from the Hebrew University in Jerusalem (1966), her MSc and PhD (Suma com laude) in Science Education from Bar Ilan (1992 and 1998 respectively). After conducting her postdoctoral research at the University in Michigan (2000), she joined the Weizmann Institute of Science at the Department of Science Teaching, where she is the coordinator of the Chemistry Group, and the head of the National Center for Chemistry Teachers at the Weizmann Institute, and the coordinator of the chemistry group at the Department of Science Teaching. Dr. Mamlok-Naaman is engaged in development, implementation, and evaluation of new curricular materials, and research on students' perceptions of chemistry concepts. She is the coordinator of a chemistry teachers' programs in the framework of the Rothschild-Weizmann MSc program for science teachers, and of projects in Israel in the framework of the European Union (the FP7 programs).

Elizabeth Mavhunga is a senior lecturer in chemistry education at Wits University, South Africa. Dr Mavhunga completed a BSc Hons. in chemistry. After completing her honours she spent two years in industry and subsequently moved into chemical education. She has been involved in a variety of education development activities, including the development of innovative materials for supporting science teachers. She completed her PhD at the end of 2012. On graduation she was appointed to the staff at the Marang Centre for Maths and Science Education as a senior lecturer. She currently leads a project linking content courses in the teacher preparation curriculum to methodology. One of her publications in the African Journal of Research in Mathematics, Science and Technology Education has been identified as heading the list of most downloaded articles in 2014.

Alan J. McCormack is Professor of Science Education (Emeritus), San Diego State University, San Diego, CA, USA. He holds the degrees of BA (State University of New York), MA (Harvard University), and Ed.D (University of Northern Colorado). He is a Past President of the National Science Teachers Assn. (NSTA); of the National Association of Biology Teachers (NABT) and the Council for Elementary Science-International (CESI). His research interests: include Visual/Spatial Thinking; Development of Creative Thinking through Science Activities; and Uses of Discrepant Events in Science Programs.

Onowa McIvor is maskiko-nihiyaw from Norway House Cree First Nation in Northern Manitoba, Canada and also Scottish-Canadian. Onowa is an Assistant Professor and the Director of Indigenous Education in the Faculty of Education at the University of Victoria. Onowa's scholarship focuses on Indigenous language and cultural revitalization and pre-service teacher education for both Indigenous and non-Indigenous students.

Shiho Miyake is an Associate Professor at the Department of Biosphere Science, School of Human Sciences, Kobe College, Japan. Her teaching areas include environmental education, social aspects of science education, and teacher training for primary and secondary school science. After a doctorate in science education at Kobe University, her researches focused on both an individual and a community based learning among wide range of people in museums and social educational facilities. Her recent interest is a function of science education to develop sustainable society. She has been a member of a board of the Japan Society for Science Education since 2015 and a secretariat of an editorial board of the Society of Japan Science Teaching since 2012.

Louise Newby, formerly Head of Science in a secondary school, is a Teaching Associate at the University of Cambridge and teaches on the PGCE course in initial teacher education. Her research interests include teacher education, classroom environment, and science education, encompassing biology education, and inquiry-based learning. Louise worked on the EU funded PATHWAY project, which focused on inquiry-based learning in STEM.

Joanne K. Olson is a professor of science education at Iowa State University, and earned her Ph.D. from the University of Southern California. Her research interests include science teacher education practices that promote advanced decision-making, and the role of the nature of science/nature of technology in science education. Dr. Olson is the former president of the Association for Science Teacher Education, and has received numerous awards for her research, teaching, and program development. She co-developed the Master of Arts in Teaching program at Iowa State University, and continues to co-direct that program. She began her career in Los Angeles, teaching middle school physical science and earth science, and a multiage primary class (grades 3–5). When not working with students or analyzing data, she can usually be found playing the mandolin.

Brinda Oogarah-Pratap is an Associate Professor in Health and Nutrition Education at the Mauritius Institute of Education. She is currently the Programme Coordinator for the B.Ed (Hons) Primary and is involved in curriculum development for both primary and secondary levels. Her research interests include innovative practices in teacher education, including integration of online technologies. She is currently a member of a core interdisciplinary research team for a study on the content knowledge and pedagogical content knowledge of pre- and in-service trainee teachers.

John O'Reilly works as a lecturer in Initial Teacher Education in the department of Education and Professional Studies at the University of Limerick, Ireland. His research interests are focused on Inquiry Based Science Education (in particular the affordances of technology, power relations and pedagogical content knowledge) and Negotiated, Integrated Curriculum (in particular student voice, empowerment, and ownership of inquiry oriented leaning).

Yashwantrao Ramma is Head of the Research Unit at the Mauritius Institute of Education. As a physicist, he has worked on several research projects related to technology integration and misconceptions of teachers and students in physics. Currently, he is leading research projects on exploring teachers' content knowledge and pedagogical content knowledge across various subject areas and also on students' transitions from secondary to university and teacher training.

Miia Rannikmäe is the head of the Centre for Science Education, University of Tartu, Estonia and has considerable experience in science education in Estonia, Europe and worldwide (Fulbright fellow – University of Iowa, USA). She has a strong school teaching background, extensive experience in pre- and in-service teacher education and has strong links with science teacher associations worldwide. She has been a member of a European Commission high-level group associated with the publication of the 2004 report on 'Europe needs more Scientists'. She is/has been running a number of EC funded projects and is the grant holder for an Estonian research grant determining the change of levels of students` scientific literacy at gymnasium level. Recently completed grants are on modelling science education for relevance, exploring teachers' understanding of the Nature of Science, and Stakeholders views on the goals of science education and its impact on students' career choices in science. Her PhD students are involved in research areas such as scientific literacy, relevance, creativity/reasoning, inquiry teaching/learning and the nature of science – stakeholder (including industrialists) interactions with the science education community. She has supervised 3 international postdoctoral students and is a member of international research organisations – ESERA; NARST, IOSTE and a member of the editorial board for several science education related journals (e.g vice editor for the Journal of Baltic Science Education). She was the organiser of a 2010 ICASE & UNESCO World Science Education Conference on innovations in science and technology education and the organiser of the Eurovariety conference in 2015. She is the past secretary for ICASE (International Council of Associations for Science Education) and is currently a member of the executive committee, chairing a standing committee for University liaison.

Andrea Revel Chion holds a first degree in biology teaching (Buenos Aires, Argentina) and a Ph.D. in didactics of science (science education) from the Universidad Nacional de Catamarca (Argentina). She is currently Lecturer at the Research Institute CeFIEC of the Universidad de Buenos Aires and a member of Research Group GEHyD, which focuses on the various contributions of the philosophy and history of science to science education. Her research interests include: conceptions and models of health and disease for health education, language in the science classes, fostering academic writing, and construction and validation of historical narratives.

Fran Riga is a teaching and research associate in the Faculty of Education at the University of Cambridge, where she has worked on a number of research projects in areas such as: science education for the gifted, assessment, inquiry-based science education, dialogic approaches in secondary education, and computer-based adaptive learning. She comes from a background of teaching science and mathematics in secondary schools, and in her PhD research she investigated secondary students' conceptual development in astronomy topics. Areas of particular interest to her are: conceptual development and associated thinking processes, inquiry-based science education, science for the gifted, and dialogic approaches to teaching and learning.

Marissa Rollnick holds has taught at school level in South Africa and at teachers' college and university level in Swaziland before returning to South Africa in 1990. Since then she has worked at Wits University and has been Chair of Science Education since 2005 in the Marang Centre for Maths and Science Education, Wits School of Education, University of the Witwatersrand, Johannesburg South Africa. She is engaged in research into subject matter for teaching, or pedagogical content knowledge.

Anwar Bhai Rumjaun is an Associate Professor in the Science Education Department at the Mauritius Institute of Education in Mauritius. He is engaged in teaching and programme development, school curriculum development and research in education. He is the coordinator for a B.Ed programme run jointly with the University of Mauritius. He also co-supervises students registered for a Master and Doctoral programme run in collaboration with University of KwazuluNatal (South Africa) and University of Brighton (UK). His research interests are in Biology Education, EE and ESD.

Amy Seakins is a Research Associate at King's College London. Her work focuses on museum education and public engagement with science, and her doctoral research explored the impacts of visitors meeting scientists within a museum. Amy works on the Enterprising Science project which explores the concept of science capital and the role of museums in engaging young people with science.

Colin Smith is a former biology teacher and researcher. He worked in Scottish schools for over thirty years, but also took time out to complete his Ph.D under the supervision of Noel Entwhistle at the University of Edinburgh, where he currently teaches educational psychology to adult learners. He was a teacher-researcher on the

Early Professional Learning project at the University of Stirling, which studied the induction of teachers in Scotland (2004–2007), and was also involved in the Science-Teacher Education Advanced Methods (S-TEAM) project (2009–2012), where he was responsible for innovative forms of professional development involving learning communities. His research interests include teacher learning, categories of profesional knowledge and the didactics of biology

Manabu Sumida is Professor of Science Education at the Ehime University in Japan. He holds a BA in Chemistry from Kyushu University and Ph.D. in Science Education from Hirohima Univesity. He was a visiting researcher at Univesity of Georgia in 1998 and visiting scholar at the University of Cambridge in 2012. He has been a Director of Kids Academy Science (a special science programme for gifted young children) for 7 years. He is a committee member of the Japan Student Science Award. He was a chief editor of the first special issue of science education for the gifted in the Japan Society for Science Education journal in 2012. He is currently Regional Representative for Asia of the International Council of Association for Science Education and Country Delegate of the Asia-Pacific Federation of the World Council for Gifted and Talented Children.

Bodil Svendsen is a teacher educator at the Department of Teacher Education (PLU) at NTNU. She recently completed her Ph.D, which was a study of professional learning in the context of inquiry-based science education in Norway. She has worked as a science teacher and is currently involved in science teacher professional development through the Norwegian SUN project. Her research interests are in science education, science resource centres and collaborative teacher development, as well as marine biology. She works with activity theory and has used the concept of mediating artefacts in the study of science teacher development, the artefact in this casr being the well-known 5E model (Engage, Explore, Explain, Elaborate and Evaluate), developed by Roger Bybee and colleagues.

Keith S. Taber is Professor of Science Education at the University of Cambridge, England. After reading chemistry, Keith qualified as a teacher of chemistry and physics, and taught in comprehensive secondary schools and further education. He undertook part time research degrees (MSc, PhD) looking at topics related to his professional work as a teacher, and was involved in mentoring new science teachers. He joined the Faculty of Education at Cambridge in 1999, initially working primarily in initial teacher education. He currently teaches educational research methods across a range of courses, and supervises research student projects, mostly in science education. His own research is largely concerned with student thinking and learning in science. He is the chair of the Science, Technology and Mathematics Education academic group. He is the editor of the journal Chemistry Education Research and Practice, and the book reviews editor of Studies in Science Education.

Gráinne Walshe is the Coordinator of the Science Learning Centre at the University of Limerick, Ireland. She also works as a postdoctoral researcher in EPI-STEM – National Centre for STEM Education in the University of Limerick. Her research interests are in developing curriculum models for integration of science and mathematics, investigating mathematical practices in science, teacher-based curriculum development, teacher subject identity, and physics education research for higher education.

Mark Winterbottom, originally a secondary science teacher, is a Senior Lecturer in science education at the University of Cambridge. He teaches on the PGCE course in initial teacher education, and on Masters and PhD programmes. His research interests include teacher education, classroom environment, and science education, encompassing biology education, and inquiry-based learning. Mark is the principal investigator on an EU funded STEM project exploring 'Tinkering' as a pedagogical approach for learning in STEM. He was also principal investigator on the EU funded PATHWAY project, which focused on inquiry-based learning in STEM. Mark has written a variety of school textbooks, and authored and contributed to a number of books for teachers and Masters students.

INDEX

Printed in the United States
By Bookmasters